P9-DNB-060

# ECONOMICS

# *The Science of Common Sense*

**ELBERT V. BOWDEN**

*Professor of Economics*

*State University of New York*

*College at Fredonia*

**H90**

*Published by*

**SOUTH-WESTERN PUBLISHING CO.**

CINCINNATI WEST CHICAGO, ILL. DALLAS PELHAM MANOR, N.Y.
PALO ALTO, CALIF. BRIGHTON, ENGLAND

ISBN: 0-538-08900-8

Library of Congress Catalog Card Number: 73-81097

1 2 3 4 5 6 7 8 K 1 0 9 8 7 6 5 4

Printed in the United States of America

# PREFACE

Well hooray. Here's another economics book. Big deal.

"Claims to be different?"
"Of course."
"Easy to understand?"
"Sure."
"Student oriented?"
"You bet."
"Interesting? Relevant? Innovative?"
"Highly! Highly! Highly!"
"A thoroughly differentiated product?"
"Definitely!"
"Just the right book has finally been written?"
"Right!"

How many times have we all read these claims? Too many. And how many times have we all been disappointed? Too many. So let me claim none of these things. Let me just tell you what this book is all about. Then you decide.

## The Philosophy Of This Book

A person who writes an innovative textbook must have some ideas about what's wrong, and some convictions about what's needed. I think the innovative-textbook-writer should know what he believes, and I think he should share his beliefs with his readers. So here goes.

This book is based on the following propositions:

- that basic economic concepts and principles are for *everybody*—not just for the scholarly few, but for everybody who wants to learn them;

- that the freshman-sophomore economics courses should be among the most interesting, relevant, rewarding, and *popular* courses on every campus—community colleges, major universities *everywhere;*
- that it is neither necessary nor desirable to "flash by" the beginning student everything he would need to know for the Ph.D. degree—that the first objective should be to get the student *turned on* to economics, and then to get the basic concepts and principles *thoroughly understood* and *permanently integrated into his way of thinking;*
- that the sophisticated analytical tools of the economist should not be permitted to deny students an understanding of the concepts and principles of economics—that *the concepts and principles should not be subordinated to the tools of analysis,* but vice versa;
- that scholarly vocabulary and prose, although delightful to the scholarly elite, is not the language of today's youth—that economics won't be understood by most students until it is explained in simple everyday language—that is, *in the vernacular of today's young people;*
- that *simple explanations* do not have to be *simple-minded* explanations—that economic principles can be explained with simple words, easy conversational prose and familiar examples *without any sacrifice of depth or rigor or precision;*
- that if there is any possible way for a student to miss the point or to misunderstand, some will be sure to do so—that *it's better to explain two or three times in two or three different ways* than to leave some students confused about an important concept or principle; and
- that economic principles cannot be understood by memorizing definitions—that the student first needs to see and feel each principle, each concept, as it works within the narrow horizon of his own life and world—that only after the concept or principle is understood will a precise "memorized definition" have any real meaning.

In short, this book is based on the proposition that *all college students can understand the fundamental concepts and principles of economics* if the explanations are made with familiar language and with examples that the student can see and feel and relate to—that the economics textbook writer should try to come as close as he can to achieving a "frictionless communication" of economic ideas. If he can make it *fun,* so much the better.

### The Specific Objectives Of This Book

Specifically, this book offers interesting, easily understandable explanations of the essential concepts, principles and issues of economics.

I have tried to design and write the book so that:

1. *no one* will find anything difficult to read;
2. *everyone* who reads it *really will* understand;
3. no one will *ever* be misled, or misunderstand;
4. everyone will find the explanations really *interesting*, and
5. all who study the book will gain a *permanent* understanding of and interest in economics.

That's a big order. I don't think any other economics textbook writer ever aimed for such a set of objectives. But I would rather fail while trying for that, than to succeed at something less.

### Designed For The Student's Success

This book is not designed to help the instructor to "weed out" half the class. It's designed to help every student to succeed. The purpose of this book is to pull students in, to turn them on, *to start them off right in economics*.

This book is addressed to the kinds of students *I know*—students who are very human, with all the problems, pressures, and frustrations of young people—students who have no intention of dedicating their lives to the study of economics, but if it makes sense and isn't too demanding, they will give it a reasonable try. I think that's fair enough. If a student will do that, using this book, I believe he will gain *permanent knowledge* and a *lifelong interest* in economics.

Many who use this book will take more courses in economics. Some will sense the urgent relevance and excitement of economics and will decide to make it their major field of study. They will be well prepared to do so. But the ones who don't take any more economics will have something of real and lasting value, too. Isn't that as it should be?

### Designed For All Students

This book is designed for *all students* who are taking their first course in economics. It contains everything needed for a well-balanced and interesting one-semester course. The instructor who wishes can drop some chapters and place more emphasis on other chapters, and on supplementary materials and current issues. The book is written so that some chapters can be deleted (or taken up in different order) without loss of continuity. The instructor's manual suggests several alternatives.

The instructor who uses this book as the basic text for his two-semester (or two- or three-quarter) course will have great flexibility in building his course. This book will provide the *solid theoretical core* of concepts, principles and issues and will leave the instructor plenty of time to introduce supplementary sources and to deal with current issues and problems. If the instructor of the two-semester course desires to advance his students as rapidly as possible in economic theory, then this book can be supplemented with an intermediate text.

With all the alternatives which exist (as suggested in the instructor's manual), I do not believe there is *any* beginning course in economics in which this book would not prove highly effective. It will serve excellently in a "terminal, general economic education" role. It will serve equally well as the basic building block to get the student launched on his journey to becoming an economist. I believe that the student who studies this book will gain *a true conceptual perspective of economics*—and I think that's an essential, yet usually unattained, objective.

**Thanks For The Help Of My Friends**

I suppose every author, in building a book, goes into debt to his friends. A few words of acknowledgement in the preface hardly seems adequate. Let's just call this a down payment.

**To Katsuhiro (Ken) Otsuka, Cartoonist.** This book wouldn't be the same without the tasteful and sensitive art work of Katsuhiro (Ken) Otsuka, our young Japanese cartoonist in Tokyo. He deserves not only my thanks but the thanks of every student who will read this book and find it more interesting and more fun because of Ken's excellent illustrations. Thank you, Ken, for the many outstanding illustrations. I'm proud to have your cartoons in my book.

**To My Colleagues And Students.** My sincerest thanks go to the several of my colleagues at the State University College at Fredonia, who have used the evolving manuscript in their freshman and sophomore courses and have helped me to make many improvements. To Marwan El Nasser, Ann N. P. Fisher, Warren Fisher, Kanji Haitani, Stan Hart, and Norm Starler, thank you. Thanks also to Larry Mansfield of the University of Miami who read the original manuscript and made several valuable suggestions. In addition, more than a thousand students have used the manuscript and have filled out questionnaires and offered comments and suggestions on every aspect of the book. Students, my sincerest thanks to all of you for your criticisms and your encouragement. Surely I couldn't have done it without you.

**To Mary Ann Burgess, Extraordinary Secretary.** Finally, I owe my greatest debt of gratitude to Mary Ann Burgess. Cheerfully, speedily, expertly she has taken this book through draft after draft after draft. My sincerest thanks to you, Mary Ann, for making a joy of what could have been the most tedious part of the job. I will never forget you for that.

## Dispelling The Dismalness

For many years I've been acutely aware that we must try to do something different for our beginning economics students. We must try to give them something that's really interesting *to them*—something they can relate to, really understand, really learn from. I'm not sure this book will be thoroughly successful in doing that. But I am sure that now there's an economics book which *really does try.*

I hope that those of my colleagues who may disagree with my approach will not be too harsh on me. And I hope that those of you who agree will join me in trying to bring a little blue sky and sunshine—a little warmth and friendliness—into the economics classroom. Perhaps, working together, we can dispel some of the inherent (and some of the inherited) dismalness of our "queen of the social sciences"—of our "science of common sense."

Now, a final word. Sydney J. Harris has said:

> Mankind will not be able to grasp essential truths until it is able to make the subtle distinction between the simplicity of the simple-minded and the simplicity of the sage.

Into which category does this book fall? I leave that to you to decide.

Elbert V. Bowden

# CONTENTS

# PROLOGUE: WHY ECONOMICS?

Why should anybody want to learn about economics? What good is it? Does it help to understand anything? If so, what? And why should *you* care about that, anyway? Let's take a minute to talk about that.

## Economics Is Everywhere

Where is economics? It's everywhere. It's at work all the time, in everything. Once you learn to recognize it, you can never escape it. It's just everywhere!

When you're doing things or making things, cooking breakfast or washing your car, going places, studying math, playing basketball, using up your time and energy and money and things to do one thing or another—all those things have economics mixed in. If economics is involved in everything, then how can you ever get away from economics? That's the point. You can't. You just can't.

Economics is mixed in with all the things you've been doing all your life. So how have you been able to get along all these years without knowing any economics? You haven't. The fact is, you know quite a lot of very basic economics already. Everybody does. Chances are you already know, right now, fifty or sixty or maybe eighty or ninety percent of the economics you will know when you get to the end of this book. But right now you don't call it "economics." You call it "common sense."

## Economics Can Sharpen Your Own Common Sense

So why should you study economics? Because it can expand and sharpen your common sense? Yes! You're going to be surprised how much it can do that. But that's not all. It can do much more than that. It can let you see more clearly and deeply many things that are going on in your life and world.

This book starts with sharpening your own common sense—talking about how you solve your own personal "economic problem." Then it expands the horizon more and more until the first thing you know you'll be working with and understanding the "economic problem" from the point of view of your society, your nation, your entire world.

**You're Always Facing An Economic Problem**

I suppose you already know that when you're trying to decide if you'll save your money or spend it, you're facing an "economic problem." Or whether to pay your tuition or go to Ft. Lauderdale or someplace, that's an "economic problem" too. But what about deciding whether to study math or play basketball? or cook breakfast or wash the car? an "economic problem" too? Sure. An economic problem is a problem of having to choose—like whether to have your cake or eat it.

When are you facing an economic problem? All the time! Every moment of every day of your life you're having to decide what to do with yourself—with your time and energy and thoughts—and what to do with your possessions—your money and your things. Right now you're reading this book. But there are a dozen other things you *might have* been doing right now. Right? Sure.

So many, many choices we all have! We're choosing all the time: to do this or that, to make this or that, to use up this or that, to save this or that. There's just no end to it! And now, already, just from reading this first couple of pages you can see your own personal "economic problem." It's your problem of choosing—of deciding what to do with all the different things available to you. That's *your* economic problem.

Does your family have an economic problem? Like who will wash the dishes or the car and who will get to use the car and who will get to spend how much of the money for what? Sure. What about your church? Your club? Any organization? Do they have an economic problem? Sure. And what about your city? Your local school board? The corner grocery? The burger drive-in? Your college? Do they have an economic problem too? A problem of deciding what to do with what they have? and of who will get to use how much of what? Of course they do.

**Every Society Is Always Making Choices**

How about your whole society? Does the society face an economic problem? A problem of choosing, of deciding who will do what, who will get to have how much of what, which things to use up and which to save, and all that? Of course it does!

It's pretty easy for you to see how you decided to spend your time right now, getting started in this book. It isn't so tough seeing how each person decides. He thinks about the different things he *might* do, and then he tries to make the best choice for himself. But what about the society? What about trying to see how "the society as a whole" solves its "economic problem"? How does *the society* make its choices? How does it decide what to do and what to make, and what to use up and what to save, and who will get to have how much of what?

Seeing how the society solves its economic problem is not so easy. Not unless you understand economics. But if you do, then it's easier. That's one of the things you're going to learn from this book.

### Society's Problems Are Economic Problems—And More Than That

Every society has always faced problems. As times change the problems change, but some problems always seem to be there. These days people are very concerned about the problem of too much population, and of the destruction of the environment. Are problems of this sort *economic* problems? Yes. They're economic problems, but not *just* economic problems. They're economic problems *and more than that.*

All real-world problems—all the problems that face the human race—are economic problems. Sure. But they're more than that. You'll never find a real-world problem that isn't both an economic problem and more than an economic problem. Pick any current problem you're concerned about: the crime rate, inadequate housing, drug abuse, unemployment, poverty—all economic problems? Yes. All economic problems *and more than that*. What about the problem of international tensions? wars? rising prices? school integration? urban blight? Are these *economic* problems? Of course. Not *just* economic problems, but economic problems to be sure. Economic problems and more than that.

### Common Sense, Without Some "Scientific Economics," Can Mislead You

In the final part of this book (Part Eight) you'll be reading about some real-world problems. By the time you get to Part Eight you will be able to see each of the problems in a new and different light. The nature of each problem will be more clear to you than ever before. That's one of the main things you're going to get out of all this. That's one of the main reasons why it's a good idea for everyone to study economics.

The study of economics is going to enable you to *really understand* so many things that are going to concern you, all your life—things which your common sense alone couldn't handle. Your common sense alone might even *mislead* you. There are so many things a person really can't understand without some "scientific knowledge" of economics.

If the study of economics is going to do all these good things for you, then it's worth some effort. Right? But suppose it would turn out to be fun, too! *Impossible*, you say? Not at all. Just you wait and see.

# PART 1

YOU CAN'T HAVE YOUR CAKE AND EAT IT, TOO!

Each person, business, and society
must choose between one thing and another,
must decide what to use up and what to save,
what to do and what to make, and what
to trade for what.

# 1 ECONOMICS IS THE STUDY OF CHOOSING

*The problem is: How can I get to where I want to get to, with no more than I have to work with?*

Economics is the study of choosing. Who chooses? Everyone. Each person, each family, each business, each educational or religious or political organization, each society, each nation. Everyone. The successful ones—the ones who really wind up getting as close as they can to their objectives—are the ones who are making the right choices.

**Economizing Is Making The Right Choices**

Economizing is the science and the art of making the right choices. Which choices are the right ones? Those which best serve the wishes of the chooser. Everyone has his own desires and objectives. If your choices are helping you as much as possible to move toward the fulfillment of your desires and objectives, then you are making the best choices for you.

Why choose? We *must* choose. The things we have to work with are limited, so when we use up something in one way, obviously we can't use it in some other way. So we must choose one way, or the other. If we're smart, or lucky (or maybe both), we will choose the way that is most important to us.

**Most Things Are Scarce**

Too bad so many things are scarce. If things weren't scarce, it wouldn't be necessary to choose and "economize." But people just seem to want *more* of most things. That means most things are "scarce." If cake wasn't scarce—if unlimited amounts were available—then you could have all the cake you wanted and eat all the cake you wanted, too. We would all be up to our knees in cake! Are you beginning to get the idea that scarcity is an important concept in economics? I hope you are, because it surely is. Let's talk about scarcity.

## ★ SCARCE THINGS ARE WANTED, AND LIMITED

To say that something is scarce is not to say that there is a *shortage* of it. Scarcity only means that the available amounts are not completely unlimited. For example, out on the desert, water is very scarce. You wouldn't use up the water in your canteen washing your face and hands before lunch! You *economize*. That is, you make the best choice about how to use your water. You limit its use (conserve it) so it will be used only for the most important purpose—moistening your parched throat. If your canteen runs dry, you will pay all the money you have just for enough water to make it back to civilization. Water is not nearly so scarce in the city as it is out on the desert. But even in the city, it's a little bit scarce. People in the city do pay something for water. They do "economize" the use of water a little bit. Why? Because water in the city is a *little bit* scarce.

The more scarce something is the higher will be its economic value (its exchange value, or price), and the more it will be economized. If something is not scarce at all then it will have no "economic value" (no price). It will be free. It will not be economized because there is no need. Enough is available to serve all the possible uses, so there is no need to choose the best use and conserve it for that purpose.

Suppose our desert wanderer stumbles into an oasis and finds a bubbling spring and a big lake. Suddenly water becomes a free good. He will drink it, wash his face and hands, water the donkey, splash around in the lake, then splash some water on the donkey just for fun. Why not? Nobody economizes or conserves or limits the use of free goods!

### Marginal Utility Means "Additional Satisfaction"

Suppose, while our happy wanderer is splashing around in the lake, someone comes along and offers to sell him a drink of water. Will he buy? Obviously not. He would not pay a penny for a hundred gallons. Does this mean that water is of no value to him? Again, obviously not. It only means that *more* water (more than he already has) is of no value to him. Additional water would bring him marginal utility (additional satisfaction) of zero.

Is it true that the best things in life are free? Well, for some people, maybe so. It all depends on what you like best, and on whether or not the amount of it is *limited*.

We're into an important concept in economics. It's this: the economic value of something to you (that is, the price you would be willing to pay for it) is never determined by the "total usefulness" of it. The value is determined by *how useful it would be for you to have more of it than you already have.*

## Free Goods Have Zero Marginal Utility

Suppose you have all of something you could possibly ever use. Would you like to have more? Of course not. Any additional amount would have zero value to you. The more of something you already have, the less important it is to you to get even more of it. What we're talking about is called the principle of diminishing marginal utility.

"Marginal utility" is the "extra satisfaction" you get from having a little more of something. Does marginal utility "diminish" as you get more and more? Sure. As more and more of something is available to you, the additional utility you could get from having even more, gets smaller and smaller. If *unlimited* amounts are available, then the marginal utility (the extra utility from having more) drops all the way to zero. You saw what happened to the marginal utility of water when our thirsty wanderer stumbled into the oasis. Right?

You can see that the price a person would be willing to pay for an additional something reflects the importance of that additional something to him. If he thinks he's going to gain a lot of additional satisfaction (marginal utility) from whatever he's thinking of buying, then he is willing to pay a high price. If he thinks he will gain very little additional satisfaction (marginal utility) from it, he will buy it only if the price is very low. If he expects the marginal utility to be zero, he will not pay anything for it.

What if nothing was scarce? Then everything would be free. There would be enough of everything for everyone to have all he could ever want, and some left over. We would have no reason to worry about economics. There would be no need to choose, or conserve, or economize. But that's just not the way it is. Almost everything is scarce. Scarcity is an inescapable fact of life. So, like it or not, we must choose and conserve and economize.

We are forced to live with economics. We have no choice about that! So we might just as well learn a little something about it—about "the *science* of common sense." That's what you're going to be doing throughout this chapter, and throughout this book. First we're going to talk a little more about free goods and scarce goods and marginal utility, just to be sure you really understand these basic concepts. Then we'll get into "the economic problem."

## Only Scarce Things Have Marginal Utility And Economic Value

Of all the things that people need, air tops the list. Yet, to the individual, air has no economic value. Why not? Because it is not scarce. No one would pay to get any more air than he already has. Would you? Of course not. You already have more than you can use! The additional satisfaction (marginal utility) you could get from an additional jugful of air would be zero. Therefore, the amount you would be willing to pay is zero.

Why is air not scarce? Because there is so much of it around. Suppose there's only a little bit of something around. Will it be scarce? and valuable? Not necessarily. There probably aren't a dozen No. 2 cans of Mississippi mud in the entire world. The same could be said of cans of mosquito-wings, dried-fleas, toadstools, apple-cores, broken glass, and rotten peaches. Very, very few. Right? But scarce? No.

Something is scarce only if people want more of it than they can get for free. If people want something enough to pay for it, or work for it, or trade for it, then it's scarce. To the individual, air is not scarce (and therefore is of no economic value) because no one wants any more of it than he already has. No. 2 cans of Mississippi mud are not scarce (and therefore are of no economic value) because nobody wants any of these things anyhow. Air has no marginal utility and therefore no economic value, for the same general reason that canned mud has no marginal utility and economic value. No one wants any more air and no one wants any more canned mud than he already has. So neither is scarce.

Both of these qualifications must be met if something is going to be called "scarce": (a) people must want it, and (b) the amount of it must be limited. The more it is limited and the more it is wanted, the more scarce it is. As something gets more scarce, it gets more valuable. If only a small amount of something is available and many people want very much to have it, then it is very scarce. Its marginal utility will be high and its price will be high. Its use will be carefully conserved, limited, economized.

## Economics Is About Conserving And Producing Scarce Things

Most things are scarce. People are willing to work and pay to get scarce things, and people economize the use of scarce things. Also, people work hard to produce more of the scarce things. Essentially, that's what economics is all about. *Economics is about conserving scarce things, and producing more of the scarce things. It's about choosing which scarce things to conserve and which to produce—which things to save, and which to use, and for what.*

## THE ECONOMIC PROBLEM IS THE PROBLEM OF CHOOSING

The economic problem is the problem of economizing—that is, of carefully choosing what to do with our scarce things. The problem faces each individual, each family, each business. It faces each society, each nation. More and more it is coming to face the entire world as a whole. How are these choices made? How do the individual and the business and the society decide what to do with their scarce things? And what difference does it all make, anyway? These are the questions of economics. You're going to be finding out about the answers all through this book. Some of the answers may surprise you.

The economic problem is simple enough. All we need to do is answer this question: How can we best use what we have to fulfill our desires? To reach our objectives? To achieve our goals? That's what it's all about.

### No One Can Escape From The Economic Problem

This economic problem—the problem of deciding what to do with the things we have—faces every individual in each of his roles in life. As a parent, an executive, a workman, a student, a vacationer, a spouse, a patron of the arts, a homeowner, a concerned citizen or whatever, each person is constantly facing this problem. All of us must decide what to do with what we have—our money, time, mental and physical energies, everything. If we make the best choices, we will make the most progress toward our objectives—toward whatever it is that we want most to achieve.

This same economic problem faces every organization—business, religious, social, educational, political, governmental. It faces every society—totalitarian or democratic, capitalistic or socialistic, primitive or advanced, civilized or barbaric, Christian or Buddhist, Anarchistic, Communistic, Utopian, or what have you. The economic problem is truly universal. No person, no organization, no society can escape it. Those who do the best possible job of solving this problem get the most they can of what they want. Those who don't, don't.

### What's Your Most Pressing Economic Problem?

For most students, the most pressing economic problem is trying to decide how to use their most valuable resources—their time and "mental energy." How about you? Everyone has lots of conflicting objectives. You want to go out and have fun, get plenty of rest, daydream about your new love, and earn enough money to stay in college. Last (hopefully, not least) you would like to get a certain amount of education. Someday you'd like to graduate. Lots of conflicting objectives, right?

It's pretty obvious that you aren't going to be able to satisfy all your desires. You can't spend all your time trying to satisfy any *one* of your desires unless you are willing to give up *all* of the others. Whenever you use some of your time (say, an hour) for one objective, you can't be using that same hour to do some other thing. See what a vital and serious matter this economic problem is? You've been struggling with it all your life. So has every other person, every organization, every society.

One more time, what's the economic problem? Simply this: You can't have your cake and eat it too. The resources you use up making, improving, repairing, protecting, enjoying, learning, or doing any *one* thing cannot be used for doing, making, improving, learning (etc.) any *other* competing thing. Each time you choose to do one thing or to go one way, you have to give up the chance to do the other thing or go the other way. Frustrating? You bet. But that's what economics is all about.

## Which Things To Give Up? And Which To Have?

People (and organizations, and societies) want to *do* more and they want to have more than they can do, or have. So what happens? We all have to learn to accept the fact that we can't do and have everything we'd like. We learn to forego our desires for some of the things we want. Which are we going to give up? Which are we going to have? That's the economic problem. It's the problem of making choices among the available alternatives.

The economic problem is everywhere. It's inescapable. It results from the simple physical fact that nobody and nothing can be in two different places, or be doing two different things, at the same time. You can't spend this morning studying for a sociology test and also studying economics. Your father may have discovered that he can't be selling insurance policies at the same time he's mowing the lawn. Your mother has discovered that she can't spend all her money on the children's clothes and still have money for school lunches. Your school has found that if it puts all its money into the athletic program there's nothing left for library books.

The United States is finding that we can't overcome poverty and pollution and urban slums and crime and all the other major problems of the society, and do everything else we might want to do, all at the same time. Something must be given up. But what? The space program? Military power? Foreign aid? Low-cost housing? Universal college education? You can see that these are tough questions. Which are we going to have? Which are we going to give up? And how are we going to decide?

There are all kinds of ways of defining the economic problem. Perhaps most of us, as individuals, think of it this way: "How am I going to make it? How can I achieve what I want to achieve, with no more than I have to work with?" That's our economic problem—yours, mine, everybody's.

### Choosing One Thing "Costs You" The Opportunity To Have Some Other Thing

The most frustrating thing about economics—about "solving the economic problem"—is the fact that each choice to do, to make, or to have one thing is automatically a choice *not* to do, *not* to make, or *not* to have some other thing.

This is always true. Sometimes we try to pretend that it isn't. We like to just think about what we're *getting*. (It's no fun to think about what we're giving up!) But, like it or not, each time we decide to do or to have one thing, sure enough, it's going to *cost* us the *opportunity* to do or to have some other thing.

Suppose you decide to go skiing this weekend instead of studying your English. Then suppose your English grade slips down from C to D. You might say that the ski trip "cost you" your C in English. (You lost your C because you went skiing.) But maybe that's not all. Suppose you had been planning to buy a new speaker for your stero, but you spent the money on the ski trip instead. The ski trip cost you your new speaker, too. Right? Maybe the trip also cost you your "true love," who found somebody else to do English (and other things) with. Seems that the "costs" of the ski trip are beginning to get out of hand! Maybe you're about to decide that you made a bad choice.

It isn't always easy to make the best choices. Everybody makes bad choices sometimes, but we all must keep on choosing just the same. So it is with each person, organization, society, and nation. When something is used one way, it can't be used some other way. The opportunity you choose to give up is truly the "cost" of whatever you choose to get. If you make a good choice, that means *what you're getting* is worth more to you than *what you're giving up*. It would be just great if all of us could do that, all the time!

### The Opportunity Cost Is The Opportunity Lost

Suppose a city urgently needs a new firehouse and a new school building, but it has only enough money to build one, or the other. If the city builds the firehouse, the "cost" is the opportunity to have the school building; if the city builds the school building, the "cost" is the opportunity to have the firehouse. Can you see why economists call this the opportunity cost concept?

The idea of "opportunity cost" is that in order to have whatever you decide to have, you must *give up the opportunity* to have the "something else" you also wanted. If you are craving a chocolate malted and also a maple-walnut sundae, and if you only have enough money for one or the other, what happens? You must choose. If you choose the malted, you give up the opportunity to have the sundae. The "opportunity cost" of the malted is the sundae. If you choose the sundae, the "opportunity cost" of the sundae is the malted.

During World War II, the American economy was producing about as much as it could with the labor, materials, machinery, and other things available. More of everything was badly needed. Soldiers were needed. But every time a steelworker became a soldier, it "cost" the country one steelworker. Every time steel was put into making a ship, it "cost" the nation the opportunity to make more tanks or guns or aircraft engines.

### No Country Can Produce Any More Than It Can Produce

It's obvious that an individual can't be in two different places, doing two different things at the same time. The same idea applies to the whole society. The society is limited in how much it can produce. If all the labor is busy and all the factories, mines, railroads, machines, and everything else are operating at "absolute maximum output," then that's all the economy is capable of producing. If that's all it can do, then that's just all it can do!

Suppose a country's economy is clicking along at top speed. Then what happens if the society wants to do more to try to discover a cure for cancer? Or to clean up Lake Erie? Or to send a man to Mars? Or to produce more power mowers and color TV sets? Or to build a new bridge across the Mississippi at Vicksburg? You know the answer. Something else has to go. Whatever goes, that's the *opportunity cost* the society must "pay" for what it gets.

## MANY THINGS IN ECONOMICS CAN BE SHOWN ON GRAPHS

Economists use a very simple graph to illustrate the opportunity cost concept. You know the idea. It's: "the more you get of one thing, the less you can have of some other thing." This relationship—"the more of one, the less of the other"—is easy to picture on a graph.

Graphs really can be a lot of help in economics. Often, two things will be "related to each other" in such a way that when *one* changes, the *other* also changes. Whenever this kind of a situation exists, a graph can be used to show the relationship. When it's shown on a graph, you can see exactly how much *each* one changes as the *other* one changes.

## One Graph Is Worth A Thousand Words—If You Understand It

It's been said that one picture is worth more than ten thousand words. I'm sure that one good graph in economics is worth at least a thousand words. Maybe more. But a picture is no good unless you take time to understand it. If you aren't in the habit of reading and drawing graphs, now is a good time for you to get started off right.

When you come across a graph for the first time, study it for about five minutes. (Time yourself.) Then close the book and try to draw the graph and explain it to yourself. Keep doing this until you really understand it and can draw and explain it without looking at the book. Pretty soon you'll find that you're completely familiar with graphs.

Now here comes a chance to practice all this good advice. The next page shows a graph illustrating the economic problem—that is, the "opportunity cost" concept. Study the graph, then practice drawing and explaining it to yourself until you have it down pat. You'll be surprised how clearly you can see the concept, once you learn to draw a graph illustrating it.

## A Graph Always Shows The Relationship Between Two Things

The "expenditure possibility" graph (Figure 1-1) is only one of the several kinds of graphs economists use. But don't let that worry you. Graphs are all pretty much alike. "When you've seen one, you've seen them all," sort of.

A graph always shows the relationship between two things. One thing is measured along the horizontal line (the "x axis"); the other thing is measured up the vertical line (the "y axis"). Any point in the graph shows a certain amount of "the x thing" and a certain amount of "the y thing."

## A "Curve" May Have Positive Slope Or Negative Slope

When a line (a "curve") is drawn in a graph and labeled, it tells you something. It tells you how much of "the y thing" goes with each amount of "the x thing," or vice versa. A curve in a graph is an "if, then" line. It tells you "*if* you have 'this amount' of x, *then* you have 'this amount' of y."

Sometimes the curve slopes downward as it moves to the right (as in Figure 1-1). This is called "negative slope." It means that as you get more of x, you have less of y. Or as you get more of y, you have less of x.

Some curves have "positive slope." As you get more x, you get more y. For example, if you make a graph showing "miles traveled" on the x axis and "gasoline used" on the y axis, the curve in the graph will have

**Fig. 1-1 THE EXPENDITURE POSSIBILITY CURVE**

You can't buy anymore than you can pay for, so you must choose.

This shows the concept of opportunity cost in spending money. You have five dollars weekly to spend for lunch and you must choose between the "counter-special lunch" and the "giantburger-shake combo."

If you only have $5 to spend for lunches this week, you can only spend $1 per day. This graph shows your choices.

The *expenditure possibility curve* is also called the *consumption possibility curve*, the *budget constraint line*, and the *constant outlay line*.

Which combination will you choose? You already know the answer to that—the one that best suits your tastes! You'll choose the combination that brings you the most satisfaction from your five dollars.

If you really go for the combos, maybe you will buy *all* combos. But this you know: each day when you order the lunch, it's costing you the combo; each day when you order the combo, it's costing you the lunch—*opportunity cost*, that is.

a positive slope. It will show that if "miles traveled" increases, then "gasoline used" increases also. Can you picture it?

Maybe it would be a good idea for you to draw some graphs showing the relationships between some things you know about. Could you put study-time on one axis and your exam grades on the other? Relate how much you eat to how much you weigh? The number of people you smile at to the number who smile at you? Hours of the week spent on study to hours spent on other things? Time in the sun and amount of suntan? (or maybe sunburn)? Sure. Take a few minutes and draw some graphs. If you learn to draw and read these "pictures" right now, you will gain a new and valuable tool to work with, and you will dispel all fear of graphs forever.

## PRODUCTION POSSIBILITY GRAPHS SHOW THE ECONOMIC PROBLEM

In Figure 1-1 you learned about the "expenditure possibility" or "consumption possibility" curve. You had a limited amount of lunch money ($5 a week) and had to choose what to eat for lunch each day. Now you're going to see another graph just like Figure 1-1, only it will show a "production possibility" curve.

A "production possibility" curve is exactly like an "expenditure possibility" curve, except that instead of a limited amount of *money to spend to buy something,* now you have a limited amount of *resources to use to produce something.* Now you are going to decide how much of each product to produce. The more you produce of one, the less you can produce of the other.

### Opportunity Cost, Again?

Production possibility curves illustrate the concept of "you can't have your cake and eat it too," only maybe it would be better to say it this way:

> If you use all your flour to make bread, you can't make any cakes. If you use all your flour to make cakes, you can't make any bread. The more bread you make, the less cake you can make.

Obviously? Of course. *Opportunity* cost again? Sure.

The graphs on the following pages explain and illustrate different kinds of production possibility curves. Each is explained right on the page on which it appears. If you did a good job of learning Figure 1-1, the following graphs will be a breeze. But take your time and learn them well.

Fig. 1-2 **THE STRAIGHT-LINE PRODUCTION POSSIBILITY CURVE**

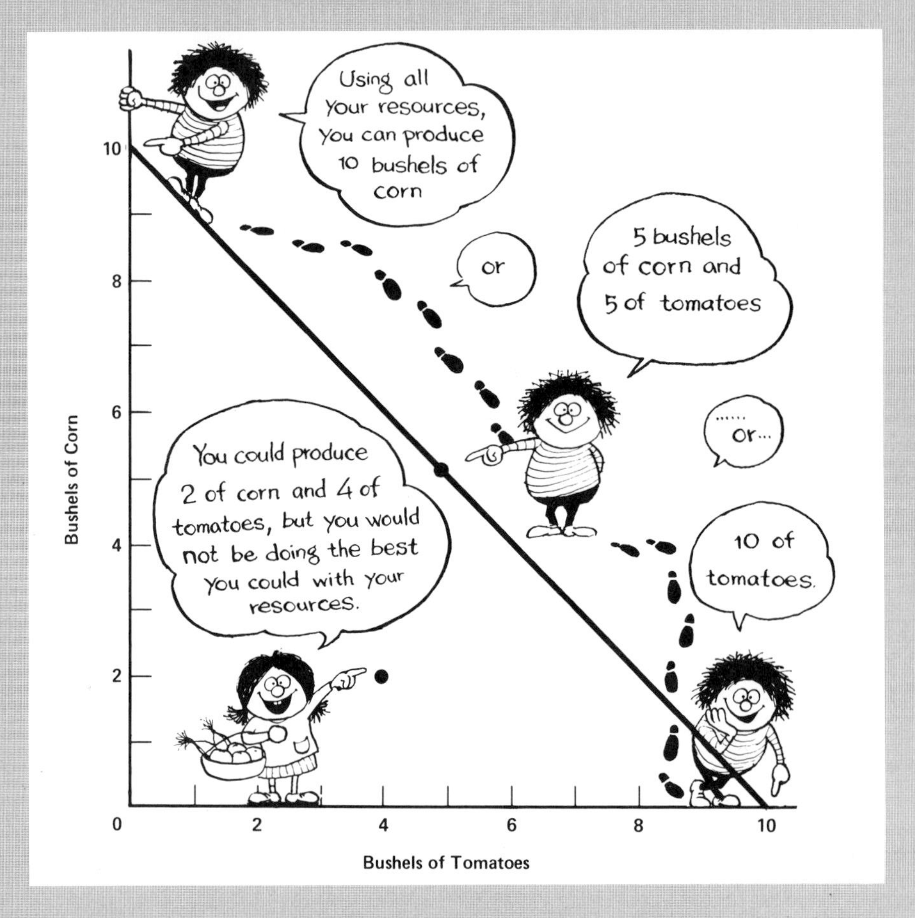


The "Transformation Ratio (Opportunity Cost Ratio) Between Corn And Tomatoes Is Constant."

You have a garden plot, some gardening tools, enough money to buy some seed and fertilizer, and enough free time and energy (labor), to grow a patch of corn and/or tomatoes.

This graph shows your "production possibilities" between corn and tomatoes. It shows that the trade-off ratio (the opportunity cost) is one bushel of corn for one bushel of tomatoes, and that the ratio does not change as you shift output from one to the other. Whenever the "curve" is a straight line, it shows that the ratio does not change. It might be one-for-one, or two-for-one, but whatever it is, if the curve is straight, the ratio will not change as you shift from one product to the other.

The production possibility curve is also called the *"transformation curve"* because it shows how, through shifting your efforts and resources from one to the other, you can (in effect) "transform" corn into tomatoes. In this graph the *transformation ratio* is one for one.

Fig. 1-3 **THE CONVEX—IN-SAGGING—PRODUCTION POSSIBILITY CURVE**

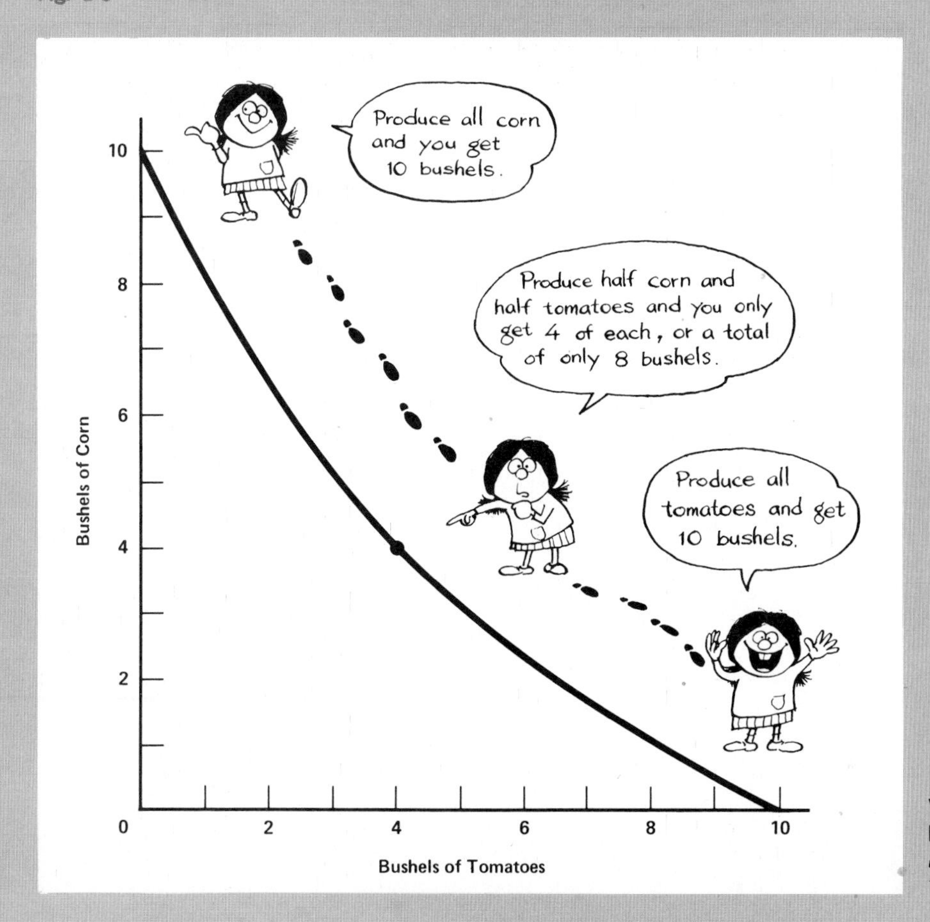


You Will Get The Most Product By Producing All Corn Or All Tomatoes But Not Both.

Notice that if you are producing all corn (10 bushels) and decide to produce half corn and half tomatoes, you must give up six bushels of corn (from 10 bushels, down to 4 bushels) in order to get 4 bushels of tomatoes. But then, if you decide to produce all tomatoes, you can get six more bushels just by giving up that last 4 bushels of corn.

This curve shows that you can produce more total product by specializing entirely in one product or the other. There could be several reasons for this.

Perhaps your seed, fertilizer, bug spray, etc. will cost you more if you buy in smaller quantities. Or maybe you need a big patch of one or the other for better pollination. Or maybe you work more efficiently in preparing the land, planting, weeding, and harvesting, if you specialize in one crop or the other. There are many reasons why specialization might increase your productiveness, or "productivity." More on this, later.

**Fig. 1-4 THE CONCAVE—OUT-BULGING—PRODUCTION POSSIBILITY CURVE**

You Will Get The Most Product By Growing Corn In The Cornfield And Tomatoes In The Tomato Patch.

Sometimes you can produce most by producing the best combination of products. Here's a case where half of your land and other resources are best for growing corn, and the other half are best for growing tomatoes. You get the most product by producing corn on the best corn land and producing tomatoes on the best tomato land. When your best corn-producing resources are specializing in corn, and your best tomato-producing resources are specializing in tomatoes, your total output is greatest.

Your maximum output combination is 7 bushels of each. Suppose you decide you want 9 bushels of corn. You must give up about three bushels of tomatoes to get the two extra bushels of corn. Then suppose you decide you want *all* corn. To get that 10th bushel of corn, you must give up about four bushels of tomatoes.

### The Slope Of The Curve Shows The "Transformation Ratio"

Now that you know about the different shapes of production possibility curves, let's talk about them a little bit. First, the straight-line "curve." The straight-line production possibility (transformation) curve shows that your productive resources and energies are all "equally substitutable" between "product x" and "product y." The "trade-off ratio" between the two is always the same, no matter what combination of x and y you choose. The ratio might be one x for one y, or two x for one y, or ten x for one y, but whatever the ratio is, it will always remain the same no matter what combination you choose. Why? Because the "curve" is a straight line, so its slope will be the same at all points on the line—at any combination of x and y you choose.

If the production possibility curve is convex—that is, if you are looking at it from the zero point and the curve is sagging down toward you—then it means you can have the greatest total output if you produce all "the x thing" *or* all "the y thing." But if the curve is bulging out (concave, when looked at from the zero point), that means you have some resources which are better at producing "the x thing" and other resources which are better at producing "the y thing." You can get the most output by using each kind of resource in its most productive way. If you want to get maximum output you must produce some of one thing and some of the other.

The concave (out-bulging) curve is the one which best describes the production possibility for the society as a whole. Whenever a nation contains a great variety of "productive inputs"—land and natural resources, factories and machines, skilled and unskilled people, etc.—some of these "inputs" will be best at producing some things, others will be best at producing others.

Suppose the economy is running along producing at maximum output. It's producing a normal combination of "consumer goods" and "industrial products." Now suppose that for some reason the society decides it wants more output of industrial products. The output of consumer products must be cut back. If the consumer goods output is cut back, inputs will be released. These "released inputs" can be used to produce the extra industrial output the society wants.

*How much* must the consumer goods output be cut back to release enough "inputs" to produce the extra industrial products? That depends on the "transformation ratio" between the two. Figure 1-5 shows that the more we cut back our consumer goods output (y) and expand our output of industrial products (x), the higher goes the opportunity cost of the extra industrial products. We must forfeit more and more of our consumer goods output flow to get each additional increase in our output flow of industrial products. Now if you'll study Figure 1-5, I believe all this will become very clear to you.

**Fig. 1-5 PRODUCTION POSSIBILITY CURVE FOR A NATION**

Even The Nation Can't Have Its Cake And Eat It Too.

If we want more consumer goods, we must produce fewer industrial products. But if we want to build more factories and machines and college libraries and highways and such things, we must cut back on consumer goods.

If the society uses its best consumer-goods inputs to make consumer goods and its best industrial-goods inputs to make industrial goods, it will get a maximum total output. The further it moves away from this combination of output, the smaller the total output gets. Each move from this maximum output position means society is giving up more of one than it's getting back of the other.

### Why The Marginal Rate Of Transformation Diminishes

Figure 1-5 shows that the nation can produce a lot more total output if it will produce its "usual mixture" of industrial products and consumer goods. As the combination moves from this "most appropriate output mix," what happens? As we transfer more and more of our labor, machines, factories, and natural resources out of the production of consumer goods and into the production of industrial products, pretty soon we get to the place where the inputs we are transferring out of consumer goods production are very highly specialized. This means they are especially designed to be very productive in making consumer goods and that they are not very good at doing anything else. For example, you can stop making canned peas and use the same farms and processing plants and people to make gear grinders. But you will lose a lot more canned peas than you will get back in extra gear grinders!

Economists call this concept the diminishing marginal rate of transformation. As you keep shifting your resources out of the production of canned peas and into the production of gear grinders, you keep getting back less and less extra gear grinders for each truckload of canned peas you give up. The number of extra gear grinders you get keeps on diminishing. This is the reason why the production-possibility curve for a large nation will always be concave (bulging away from the zero point). The outward bulge shows that the biggest output comes when those workers and machines and resources which are best at producing industrial products are specializing in the production of industrial products, and those which are best at producing consumer goods are specializing in consumer goods.

The total output is greatest when each of society's resources is being used to produce that product which it is *most efficient* at producing—growing corn on the best corn land and tomatoes on the best tomato land; using the dairy farms to produce dairy products; using the canning plants to make canned peas instead of gear grinders.

### Sometimes The Marginal Rate Of Transformation Increases

Some economic units (individuals, families, businesses, small regions, or even very small nations) are not very diversified. For these small units there may not be much variety in the kinds of resources available. It might be best for such an economic unit to produce only one thing and then trade to get the other things it wants. By specializing in this way, it can become more efficient in the production of its specialty. In this situation the production possibility curve would be convex (in-sagging toward the zero), showing the *smallest* output when a combination of goods is produced, as you saw in Figure 1-3.

The convex (in-sagging) production possibility curve faces most individuals. An individual can specialize in doing one thing and usually get a higher total output and income than if he tries to spend his work time doing a variety of things. There are exceptions, but generally a person who concentrates on his profession or skill is likely to be better at it than someone who is a jack-of-all-trades. A person's total output (and income) will probably be greatest if he specializes on something he is good at, and then works to improve his skills in his own special field.

### The Transformation Curve Highlights The Economic Problem

The production possibility (transformation) curve is nothing more than an illustration of the economic problem. It highlights the concept of opportunity cost. It shows very clearly that something must be given up in order to get more of something else. The curve can be used to illustrate the basic economic choice for an individual, a business, or a society. It shows how much of one we must decide to *not* have, in order that we *can* have the chosen amount of the other. The curve is an illustration of the fact that every choice involves the decision to "not get" or to "not do" or to "not have" something that we might like to get, do, or have.

Perhaps it's a bit sad to realize that all throughout our lives we (as individuals, and as a society) are going to have to be constantly deciding to "not get" and "not do" some things, so that we *can* get and do certain other things. But if we always give up the things which are *least* important to us and choose to get (and do, and have) the things which are *most* important to us, then we will have no cause for regrets.

The sadness comes when we fool ourselves into thinking we can have our cake and eat it too—when we move blindly ahead doing and having the things we want without considering the opportunity costs. Then when the opportunity cost comes along and smacks us in the face, we realize we made the wrong choices—we gave up the best opportunities and settled for second best, or third best—or maybe sometimes even worse than that!

Maybe all the kids have expensive new jackets, but there's no money left for school lunches. Suddenly their mother realizes that she has not economized. She has made a bad trade-off. She has "traded off" the opportunity for lunches in order to get the most expensive jackets. A little less spent on jackets and a little more spent on lunches would have made everybody happier. (Perhaps their mother needs a course in "consumer economics.")

Just as individuals can make bad choices, so can businesses and other organizations. Not everybody understands "the science and the art of economizing"—of using their available resources to maximize their objectives. Not everybody has learned to plan and manage efficiently.

Even the society as a whole can (and does) make unwise choices. No society has ever achieved perfection in solving its economic problem. It isn't likely that one ever will.

Some people now are questioning whether or not the United States really should have been putting so much of its resources into urban thruways, instead of building more efficient mass transit systems; some people question the use of resources in space exploration when the opportunity cost for that is some other kind of progress—maybe finding a cure for cancer. Many criticize the use of resources for military purposes, charging that the opportunity cost is the chance to solve some of our pressing domestic problems—maybe the problem of the ghetto areas of the big cities. Some say the American economy makes too many gadgets and that it should put more effort into protecting the environment.

The final part of this book (Part Eight) gets into several of these current problems. But for now we must dodge these vital issues and move on to the question of *how the "economizing choices" are made*, by individuals, by businesses, and by the society. That's what the next few chapters will be talking about. First, though, let's talk a little more about this new subject you're getting into—about economics.

## THE ROLE OF MATH IN ECONOMICS

You probably have heard that math is often used in economics. It is. But in this book, you have nothing to fear from math. The simple graphs you studied and learned a few pages back are as high-powered as any math you will see in this book. You will see some of these graphs again, and you will see a few other kinds of graphs. All will be very simple.

It is essential that you take the time to learn each graph when you come to it. If you have not yet learned the graphs in this chapter, you are now using your time unwisely. Economize your time. Stop reading this and go back and learn the graphs. If you have already learned the graphs and the concepts they illustrate, congratulations! Read on. You are ready to understand what comes next.

The graphs you have seen in this chapter illustrate the basic economic problem—the problem of "you can't have your cake and eat it too." Each of the curves shows the trade-off ratio between one thing and another. By looking at the graph you can see *exactly* how much of one thing you must give up to get an extra unit of the other. See how helpful graphs can be?

### Equations Can Deal With Many Variables At Once

Sometimes graphs are just great for illustrating economic concepts. But the usefulness of graphs is limited. Why? Because, with graphs, it's difficult to work with more than two things at one time—one thing on the x axis, the other thing on the y axis. But mathematical equations aren't limited that way. With equations you can deal with many things—with several products and inputs and other things, all at the same time.

Sometimes we like to be able to deal with many things—not just "consumer goods or industrial goods" but with cars and refrigerators and sofas and raincoats and canned peas and gear grinders and lathes and milling machines and more. If we're growing corn and tomatoes, we might like to see what happens if we simultaneously use less land and more water and less fertilizer and better seed. Economists like to try to figure out what's going to happen if the local plant shuts down—what will happen to employment, wages, grocery sales, new car sales, apartment vacancies, tax collections, highway traffic, school enrollments, welfare payments, and other things. Equations can't give us all the answers, but sometimes they can help.

### You Don't Need Math To Learn Economics

Math can be a powerful tool in economics. But what most people need to know about economics can be learned and understood without any more math than the simple graphs you just learned. In this book you will learn the basic concepts and you will get them integrated into your thought processes, not by working with math, but by working with words and examples that you can understand and relate to. A solid, rigorous treatment of economic concepts does not have to be mathematical. Nor does it have to be dry or heavy or frightening. There's no cause to be apologetic about the non-mathematical approach in learning economics. The "non-math" approach is likely to generate more conceptual depth and a more thorough, more permanent, and more useful knowledge of economics than you would gain from a more mathematical approach. Why so? One reason is that much "relevant economics" just can't be handled with math. Another reason is that very few people can think and conceptualize as well in mathematical terms as they can with familiar words, and with examples they can see and relate to.

As you are learning the economic concepts, if you have a facility with math you may find yourself translating some of the concepts into mathematical terms in your own mind. If you don't have a facility with math, that's no problem at all. If you major in economics, eventually you will need to use some math, but most economics majors manage to get by without using much math. What math you will need depends on what specialties in economics you choose, and how far you plan to go. Be heartened by the fact that if you major in economics, much of the math you will need will sort of "come to you" in your advanced economics courses (and may make good sense for the very first time) as you see the ways math can be used to help fathom "the economic problem."

## ECONOMICS AND MATERIALISM

Some people criticize economics on the grounds that it is not concerned with human values, but with material things. Yes, economics is concerned mostly with material things—with resources and products and energy and effort and with other valuable and scarce things. But this does not mean that economists are "materialists."

### Economics Doesn't Recommend Materialistic Goals

Economics doesn't tell people to go out and be materialistic! Economics only says that if you are using your "scarce things" to best achieve *your* desired objectives, you are making the right choices *for you*. Today, as throughout history, a major objective of most of the people in the world is to have more and better material things—food, clothing, shelter, medicine, transportation, tools, and other things.

It is difficult to pursue any objective, however noble or lofty, without using up some energy and effort and resources—some "material things." In a very poor nation or area (or even family), in which the people have barely enough food and other things to stay alive, a major objective (probably *the* major objective) will be to get more food and other material goods. But in a more affluent society or area or family, more of the scarce resources will be used for other objectives.

For most of us in the advanced nations of the world, there is some abundance of material goods. This makes it much easier for us to go after whatever objectives we desire. Our affluence provides the slack that lets us do our own thing—go to college, buy things we like, waste things, goof around anyway we please.

### Material Things Can Be Used For Noble Objectives

Is economics concerned with material things? Sure. Most of the things people use to try to attain their objectives are material things. Also, most of the efforts of most of the world's people are directed toward satisfying their material wants. Economics is sometimes defined as "the study of how man earns his daily bread." If man's objective is to earn his daily bread, then economics is going to be about how he does that. But if man's objective is to get to the moon or find a cure for cancer or eliminate the slums or get better teachers into the colleges, then economics will be about how the society uses its scarce resources to try to achieve maximum progress toward these objectives. Economics has no business telling people what their objectives *ought to be*. Honest, it doesn't do that.

People do as they do and want what they want for a variety of reasons. No one (certainly no economist!) would ever be so foolish as to suggest that all people will find their greatest satisfaction in life from building up the greatest possible pile of material goods! There are many aspects of human behavior which economics doesn't get into at all. Hopefully, the other social and behavioral sciences are making progress in dealing with some of these difficult questions.

What the economist says is this: "An individual or family or business or society will most rapidly achieve whatever it seeks to achieve if it will economize—that is, if it will conserve, and direct its scarce resources into the uses which will bring maximum progress toward its objectives." The objectives may be bigger pyramids, more food, successful military conquest, universal college education, faster cars, or whatever.

### Most "Good Economics" Is "Good Common Sense"

Now that you're coming to the end of your first chapter in economics, how does it strike you? Are you beginning to get the feeling that much of "good basic economics" is really "good common sense"? I hope so. Because it really is.

Economics is concerned with the kinds of "choice situations" you have been dealing with all your life. Economics will help you to see deeper into the "choice situations" which people and societies are always struggling with. If you put forth the effort, basic economics will become a permanent part of your way of thinking. Economic concepts and principles can really help you to see and decide about the many choices you will have to make throughout your life, both for yourself and for your society. Economics, learned well, will provide lifetime support and illumination for your own good common sense. Don't you think it's worth the effort?

## A WORD ABOUT THE END-OF-CHAPTER REVIEW EXERCISES

Each chapter in this book ends with a section of review exercises. Each section of review exercises contains the following:

1. a list of all the *major* concepts, principles, and terms explained in the chapter;
2. a list of *other* concepts and terms mentioned in the chapter;
3. a list of all the graphs and curves explained in the chapter (if any); and
4. a few discussion questions
   (a) highlighting some of the most frequently misunderstood concepts and principles, and

(b) relating some of the concepts and principles to the "real world."

It is likely that the educational value of each chapter (to you) will be very much determined by the amount of time and effort you put into the review exercises. The chapters are designed for easy reading and understanding. But beware! There's quite a long jump between *reading and understanding* a concept, and *permanently integrating the concept into your way of thinking.*

You won't get much permanent and lasting value out of just reading, from *this* book, or from *any* book. The review exercises are designed to help you to get involved with it—to make this a real educational experience. So what should you do? After reading the chapter, without looking back, try to do this:

1. For the *major* concepts and principles and terms, try to write a paragraph explaining each one, carefully, precisely, thoroughly.
2. For the *other* concepts and terms, try to write a sentence (or a phrase) saying what each one means.
3. For the graphs, try to draw and label each one and write a paragraph or two explaining what it means and why it's important.
4. For the discussion questions, if you have the time you might try writing out complete answers. But if not, at least try to jot down the highlights, or an outline of what a good answer should contain.

Finally, after you have tested yourself and practiced explaining everything, look back into the chapter to see how well you did. Then try to put in more study and practice time on the ones you didn't do too well on. Study your economics and learn it well, and someday (when you get your A in the course) you will feel great! Now, here are the review exercises for Chapter 1. Go get 'em.

---

## REVIEW EXERCISES

**MAJOR CONCEPTS, PRINCIPLES, TERMS (Try to write a paragraph explaining each of these.)**

**economics**
**economizing**
**scarcity**
**the economic problem**
**opportunity cost**
**diminishing marginal utility**
**diminishing marginal rate of transformation**

## OTHER CONCEPTS AND TERMS (Try to write a sentence or phrase explaining what each one means.)

economic value
exchange value
free good
marginal utility
expenditure possibility curve
production possibility curve
transformation curve
budget constraint line
constant outlay line
transformation ratio
trade-off ratio
x axis
y axis
negative slope
positive slope
convex curve
concave curve

## GRAPHS AND CURVES (Try to draw, label, and explain each graph.)

The Expenditure Possibility Curve
The Straight-Line Production Possibility Curve
The Convex—In-Sagging—Production Possibility Curve
The Concave—Out-Bulging—Production Possibility Curve
Production Possibility Curve for a Nation

## QUESTIONS (Try to write out answers, or jot down the highlights.)

1. Suppose all the shoe companies made a mistake and produced so many shoes that they had to lower the price to 10¢ a pair to sell them all. Would shoes still be "scarce," according to the "economics" definition of the word "scarce"? Explain.
2. If you inherited ten million dollars, you sure wouldn't have an "economic problem" then. Right? Discuss.
3. What does it mean when we talk about "transforming" consumer goods into industrial goods? That isn't *really* what happens, is it? Explain.
4. When someone tries to make up an expenditure budget, is he trying to solve his "economic problem"? How about when you work out a time schedule of study hours for each course? How about when the President and Congress try to work out the National budget? Are they working on "the economic problem"? Explain.
5. Can you think of any ways you might shift your own "resource uses" so as to move more rapidly toward *your* objectives? Can you think of any way your *college* might? Your family? Your local city or county government? Your nation? Think about it.

# 2 HOW INDIVIDUALS AND BUSINESSES CHOOSE

*People try to satisfy their wants and businesses try to make enough profits to survive and prosper.*

Everybody wants what he wants. We all try to use the things we have —our money, time, and things—to get as much as we can of what we want. Anyone who succeeds in doing the very best he can with what he has, is optimizing. "Optimizing" is doing the best you can possibly do with the limited things you have to work with. When you are optimizing, you are making as much progress as you can toward your chosen objectives.

## EVERYONE WANTS TO ECONOMIZE, OPTIMIZE, AND MAXIMIZE

What is "optimizing"? It's using each thing you have in the best possible way—in the way that will help you the most to get to your chosen objectives. Suppose you would rather have your cake than eat it. So you keep it. You go around feeling good all day because you know you have that delicious piece of cake waiting for you. You know you have the *opportunity* to eat it any time you want to.

For you, "optimizing" means never running out of cake. You never eat the last piece of one cake until you have another one to take its place. That way you always go around feeling good! See what a personal thing "optimizing" is? Different people "optimize" in different ways.

### Optimizing Is A Very Personal Thing

I'll bet you know people who get greater satisfaction from knowing they *have* something than they could possibly get from using it up. Maybe you are one of those people. Do you always carry a $10 bill tucked away somewhere, just so you will never be broke? Only the most urgent necessity can make you spend that ten! If you are such a person, you always will have "something to fall back on" all your life. Why? Because that is what "optimizing" means to you. Having the $10 gives you a better feeling than you ever could get from anything you might buy with it.

If you are "optimizing," does that mean you are "economizing"? Yes, that's true. So, do the two words mean the same thing? Sort of. But not exactly. The meaning is sort of the same, but the emphasis is different. Economizing emphasizes the *negative* side of choosing. Optimizing emphasizes the *positive* side.

> *Economizing* says: "No! I will *not* use up my money and time and scarce things for purposes that I don't really care about. I will *conserve*, so I can have things that will mean more to me in the long run." That's what economizing means.
>
> *Optimizing* says: "Yes! I *will* use each dollar and each hour and each thing I have in such a way that each will carry me as far and as fast as possible toward my objectives." That's what optimizing means.

Every time you use up or spend up something, you do it for some reason. You have some objective in mind. You would like to maximize your progress toward that objective. In this sense, "economizing" and "optimizing" always mean "maximizing"—that is, using your scarce money, time, and things to get "maximum progress toward your objectives."

Optimizing the use of your study time means maximizing your progress in learning something. If you are using your money, time, and everything else so as to maximize whatever you want most to maximize, then you are economizing and optimizing. For one person, *optimizing* the use of his cake (or pancakes) might mean eating a lot, but to another person it might mean *maximizing* the quantity he has on hand. Whole trunks full of pancakes? Wow!

### Each Individual, Business, And Society Tries To Optimize

As each "economic unit" decides what to do with its scarce things, it tries to maximize its progress toward its desired objectives. One person's objectives will be very different from another's. One quits his job, buys a pickup, and sets out for distant places; another takes an extra job and starts a savings account so he will be able to send his yet-unborn child to college. Both individuals are optimizing, but their objectives differ.

Businesses usually try to make money. Unless they are able to make enough money to cover their costs, they don't survive for very long. Most businesses usually try to maximize their profits. That's the objective businessmen usually have in mind when they're deciding what to do—deciding which choices to make.

What about the society as a whole? What does it try to maximize? The society might try to maximize any number of possible objectives. One society may emphasize producing consumer goods, while another may want to produce more industrial goods so it can have fast economic growth. A society might work for military supremacy, economic stability, economic equality, maximum individual freedom, flush toilets in every home, or whatever else it wants to work for. Ultimately, each society will work for the things wanted most by "the people in charge"—that is, by the people who have the power to influence the "economic choice" decisions. It's pretty important who gets to do that, wouldn't you say? Yes, it sure is.

In some societies, control over the economic resources—that is, the power to decide what to do with what—is diffused among many people. In other societies, this control is concentrated in the hands of one, or a few. Each society has some kind of economic system. It's through the "economic system" that the economic choices are made and carried out.

The next two chapters talk about economic systems—and the organizations and procedures which different societies use to get their "economic problem" solved—to get their choices made and carried out. In Chapters 3 and 4 you'll find out what economic systems are and how they work. The remainder of this chapter explains how the economic choices are made by individuals and families, and by businesses.

## HOW THE INDIVIDUAL SOLVES HIS ECONOMIC PROBLEM

Everyone has wants and objectives. For one individual or family the wants and objectives may be very carefully laid out. Another individual or family may just let things drift along and sort of "happen." Do you have a list of carefully chosen objectives? And do you have a carefully thought out program for using all your money and time and efforts and everything to try to maximize your progress toward your objectives? I doubt it.

Not very many people try to define and list their objectives. But most people are working toward some objectives, just the same. I'll bet you know of some things you're working for. Right? Do you mostly go for maximum immediate satisfaction? Or do you sacrifice a lot of present satisfaction and work for long-run objectives? What are your long-run objectives? These are choices each person must make for himself. That's a part of the individual's economic problem — yours, mine, everybody's.

## Each Family Gets To Choose Its Own Objectives

Perhaps one family's objective is to have the biggest and nicest house in the city. They all work, deny themselves things, never take vacations, save money, and eventually they buy the house of their dreams. After that they spend all their spare time and money painting, decorating, furnishing, and working on the house. That's their chosen objective.

Some other family might have maximum educational progress as their objective. Maybe they spend all their extra time and money on tuition, books, travel, and such things. They read a lot, and they only cut the grass and paint the house just often enough to keep the neighbors from complaining.

You probably know people who really enjoy acquiring things — automobiles and boats, or paintings and furniture and silverware. You probably also know people who don't care at all about acquiring things. They would rather spend their spare time and money night-clubbing and partying and such. Who's right? And who's wrong? Nobody's right or wrong. These are matters of individual taste, attitude, desire. Chacun a son gout! De gustibus non est disputandum! (Let each do his own thing!)

For many individuals and families there is not enough income to achieve any very lofty objectives. Some people have to spend their extra time working on extra jobs trying to increase their incomes, trying to make enough so they can pay the rent and utilities and buy groceries and make the payments on the car and the TV set. Sometimes people seem to just give up on trying to make it. They use each payday as an opportunity to have a little fun at the neighborhood bar, hoping that the finance company won't really repossess the car or the TV set, and that the landlord won't really kick them out on the street.

All these individuals and families are *trying* to use their scarce resources so as to maximize their satisfactions. Everybody wants to get as much as he can of whatever means the most to him. Each person has his own (usually unconscious or partly unconscious) objectives, goals, desires, hungers, wants. Each person (sometimes thoughtfully, sometimes instinctively) uses his scarce resources (his income, his time, his tools, all his things) to try to satisfy his desires. Some people make wise choices and are happy. Some make unwise choices and are miserable.

When people use their incomes and other things in very different ways, this doesn't mean that one person is doing it right and the other is doing it wrong. Each person has his own preferences. Each has his own feelings about how much he is willing to deny himself in the present so he can work for more economic freedom (and the chance to have more choices) in the future. Each knows how much effort he is willing to put into the task of economizing (carefully managing) his time and money and things.

People who want to "get ahead," to have more things to use and to build themselves some future freedom and security, must manage their resources carefully. But those who don't care about improving their future circumstances, don't have to go to all that trouble. In a free society, each person gets to decide for himself how hard he wants to work and what he wants to buy and how much he wants to save and all that.

## Theories Of Consumer Behavior — The Spending Decision

Why do individuals make the choices they make? Various theories in economics deal with this question. One theory says that each time a person gets "an additional something," this "additional" or marginal something adds to the person's total "satisfaction," or utility. The theory says that if the marginal utility (extra satisfaction) I can get from spending ten dollars in the local tavern is greater than the "marginal utility" I can get from having the living-room rug cleaned, then I will choose to spend my ten dollars in the local tavern. That seems logical enough. It's obvious that everyone would like to get as much "extra satisfaction" as he can for each dollar he spends.

Another theory of consumer behavior approaches the question a little differently. It says that when I decide not to have the rug cleaned so that I can spend my ten dollars in the neighborhood tavern, I am "substituting" (giving up) the "opportunity" to have the rug cleaned in exchange for the "opportunity" to spend the evening in the local tavern. The theory says that if I really would rather have the evening in the tavern than to have the rug cleaned (and if both cost the same), then I won't have the rug cleaned. I'll spend the evening in the tavern. I'll get the most extra satisfaction that way, so that's what I'll do.

You will recognize this concept as another appearance of our old friend, opportunity cost. The "opportunity cost" of the evening in the tavern is the "lost opportunity" to get the rug cleaned. The opportunity cost is the opportunity lost. Remember? Since the evening in the tavern was more important to me, I made the right trade-off. I made the right choice for me. (My wife, of course, might disagree violently!)

Both the "marginal utility" and the "marginal rate of substitution" theories of consumer behavior have been developed into detailed complexity by economists. But the logic of the two theories is very similar, and very simple—and it's the simple logic that really matters.

**The Marginal Utility Theory.** The *marginal utility theory* says that if a person is going to spend a dollar, he will spend it in the way he thinks it will bring him the greatest "additional satisfaction." If product x would add only a small amount to his satisfaction, and product y (for the same price) would add more, then he will buy product y.

Each person will try to spend each dollar (and use each "unit" of his time and things) to add as much as possible to his "satisfaction," or "utility," or "progress toward his objectives." If each dollar you spend carries you as far as it possibly can toward your objectives, then by the time you've spent all your money you've done the best you possibly could. If the use of each dollar is optimized, that means each dollar is doing as much as it can do for you. So you must be getting the most you can for your money. You must be achieving the highest level of satisfaction you can. That, in a nutshell, is the idea of the "marginal utility theory" of consumer choice.

**The Marginal Rate Of Substitution Theory.** The *marginal rate of substitution theory* is a way of getting at the trade-offs (opportunity costs) which an individual considers as he is choosing between two things. It says that if you can give up a little of one product and get a little of another product and end up with a higher level of satisfaction than you started with, then you are better off. That was a good move. A person will try to cut down his spending on those things which mean less to him (bring him less satisfaction, less progress towards his objectives), and increase his spending for those things which mean more to him. The idea is that people will be constantly shifting their spending away from the things that mean less to them and toward the things that mean more to them.

Both the marginal utility and the marginal rate of substitution theories describe an optimal position where the individual cannot, by making further changes in his spending pattern, come out with more "total satisfaction" than he is already receiving. This means that he has made the best possible choices for himself about how to spend his money (or about what to do with his time and things). He has achieved the highest "level of satisfaction," or "level of welfare," he can achieve with his limited money and resources. He is "faring as well" as he can, with what he has to work with.

In essence, both the "marginal utility" and the "marginal rate of substitution" theories say this:

> "The individual will spend his money (and use his time and things) doing and getting those things which give him the greatest satisfaction, or pleasure, or move him most rapidly toward whatever objectives he desires. Anytime a person can give up

("trade-off") some of one thing and get something else that he prefers, he will do so. When he does, his "total satisfaction," or "welfare," will go up. He will be better off than before.

### Welfare Economics: How Well The People Fare

Both the "utility" and "substitution" theories can be helpful in understanding consumer behavior. Both are included under that part of economic study usually called welfare economics—the study of what determines *how "well" the people "fare."* At this point it is only necessary for you to realize that what you would have guessed anyway, is really right. Each person's choice about how to use his money or time or things, is a very personal choice. Each person tries to use his money, time, and things to get what he thinks is best for him.

Many people (students, for example) deny themselves things in the present, hoping to build greater future opportunities for more economic choices and a better life. For other people the time horizon is shorter. They want their pleasures in the present, so they follow a "live-it-up-now" philosophy. Both are trying to optimize in their own way.

Perhaps both the "miser" and the "squanderer" are making the right choices for themselves. Only they can tell. The one who best succeeds in solving his economic problem is the one who, after years have gone by, can honestly say: "If I had it all to do over, I would make the same choices again." But no one ever has the chance to "do it all over." So we all should hope that we "luck-out," and make good choices the first time through!

### The Income Decision

As we have been talking about consumer behavior we have been going along as though each person's income is fixed. In reality, we know that each individual can do things to influence the size of his income. Each person has the opportunity to decide how much he is willing to work and sacrifice in order to get more income. With more income, a person gets the opportunity to make more choices.

**More Work Can Bring More Income.** Each person must decide how he wants to use his time and energies. How much of his time does he want to spend earning money? How hard is he willing to work? Each individual can increase his income very quickly if he gets a second job "moonlighting" in the evenings, or on weekends. If a person is willing to move, he may be able to go to another part of the country and get a higher paying job someplace where workers with his skills are getting higher wages.

Another income-increasing opportunity available to the family is for the wife or the older children to take part-time or full-time jobs.

So even in the short run, if employment opportunities are available the family can get an increase in income by increasing the amount of work they do.

**Savings And Investment Can Bring More Income.** Over a longer period of time, most people can increase their incomes by saving money and investing it. The more an individual or family saves and invests, the more future economic choices they will have. The more things they will be able to buy and the higher level of economic welfare they will be able to enjoy.

One way of saving and investing is to put money into a savings account, or bonds, mutual funds, stocks, rental housing, a small business, or some other income-producing investment. Another (more frequent) form of saving and investing occurs when people buy furniture, appliances, an automobile, and (most of all) a house. When people buy these things, they are, in a sense, "saving and investing." As the years go by, they will own more things. They will be able to live better. After all their payments are made they will find that they have extra income. This "extra income" which they no longer need to spend for house payments, car payments, or other payments, will provide them greater freedom and more opportunities in later years.

**Education Is One Kind Of Saving And Investing.** Another important way of saving and investing is by going to college or taking training courses. Investing time and money in getting an education and in developing skills is the most important kind of saving and investing most young people can make. Money and time invested in this way are likely to bring much larger returns than any other investment available.

Further along in this book you will find out that just as the individual can enjoy increased productivity and income from saving and investing, so the nation as a whole can experience *economic growth* (increased capacity to generate output and income) by saving and investing. The "economic growth" of a nation hinges on its willingness to save and invest. But that discussion must wait until later.

In summary, each individual and each family is seeking a unique solution to a unique economic problem. Each must decide how to use up the available time, money, and things.

> How much time and effort to put into earning income? How much in education, training, building job experience? How much in household chores, community improvement, neighborly helpfulness? How much in gossiping and arguing, watching TV, outdoor recreation? Mowing lawns? Sleeping? Other things? How much money to use for food? Clothing? Housing? Education and training? Recreation? Medicine? Religious and charitable contributions? Savings and investments? Other things? Which things to use up? And which to hold on to? Sell the house

and move into a mobile home? Share things? Or be stingy with things? Hide the beer when Uncle Bert comes to visit?

All these choices about how to use up the available time and money and things, are highly individual decisions. De gustibus non est disputandum. Remember? All individuals and families make their own choices based on their own pattern of wishes, desires, and preferences. If a person is not making the choices which bring him maximum progress toward his objectives, then he can improve his situation by changing the way he uses his time or money or things.

### Time Is The Coin Of Your Life

Are you spending your time and money and things in the best ways for you? Most people seem to be more careful with their money and things than with their time. It's easy to waste a lot of time. That's too bad. The poet Carl Sandburg once referred to time as "the coin of your life." Then he went on to suggest that you ought to be careful how you spend your time—that if you aren't careful, other people are likely to spend it for you. That idea is worth thinking about once in awhile. As a student, how you choose to spend your time is likely to be your most important economic problem.

## HOW BUSINESSES SOLVE THEIR ECONOMIC PROBLEM

People have to choose what to do with their scarce things. That's obvious. What about businesses? That's obvious too. Of course they have to decide, to choose what to do with their scarce things.

Each business is constantly facing choices. The businessman has to always be trying to economize and optimize his time, money, and things. He must try to use all the buildings, machines, equipment, raw materials, management and workers' skills and everything else, in the best possible ways. The *best* ways are the ways that help the most to achieve the objectives of the business.

### What Are The Objectives Of A Business?

A few minutes ago when we were talking about how individuals and families make their choices, we sort of dodged the question of "objectives." Each person's wants and objectives and goals are sort of unique—you chase your rainbow and I'll chase mine, chacun a son gout, and all that. Remember? So whatever a person considers to be most important to him, that's the objective which will guide his choices.

Suppose we took the same approach in trying to understand how the businessman makes his choices. Would that be bad? You bet it would be bad! A whole major segment of basic economics would collapse! The thing we call "the theory of the firm" (a fancy way of saying "how businesses choose") would collapse. Then our whole "theoretical model" of how an economic system works, would go down the drain! Let me explain why.

The businessmen are the ones who actually make the decisions about which inputs to use and which products to make. If we can't understand how the businessman chooses which (and how many) inputs to use, and which (and how many) products to make, then we can't understand what's going on in the economic system! Now take it one more step. If we don't know what the *specific objective* of each business is going to be, then there's no way we can figure out what that business is going to do—how it's going to respond to changing conditions and circumstances.

We economists had better be able to fill in the blank in this sentence: "The objective of every business firm is to __________________________." If we can't, a lot of economists are going to have to "go out of business"! So you may be sure that we *can* fill in the blank. What's the answer? What's the objective of every business? To *maximize profits,* that's what! In our "model theory of the business firm" that's what it is, anyway. In Part Six of this book you'll find out quite a lot about the economist's "theory of the firm." But right now we need to go a little further into some of the practical questions about how businesses make their choices —how and why they do as they do.

### Do Businesses Try To Maximize Profits?

Economists usually assume that businesses try to maximize their profits. That's a useful assumption. It lets us know exactly what choices a business will make: which (and how many) inputs it will hire, and which (and how many) products it will produce.

Is maximum profit really the objective of all businesses? Actually, no. Business decisions are based on a variety of different motives. Businesses undertake civic projects, develop playgrounds, install antipollution devices, give guided tours for students, and do other things which add to their costs but which bring them no immediate revenues.

Perhaps we might say that the businesses are trying to maximize their "long-run" profits. Maybe such things as keeping the air and water clean and keeping the employees and their families happy will pay off in the long run. Perhaps so. Still, no matter how we might "nudge it" to try to make it fit our "theory of the firm," most civic-minded business behavior just can't be explained in terms of the profit maximization assumption.

Large modern corporations are owned by thousands of stockholders. Each corporation is operated by a board of directors and a group of executives who respond (more or less) to the wishes of the stockholders. Some of the stockholders may want the business to make the highest possible profits, right now. Others may be more interested in long-run stable growth of the firm, of the industry, and of the economy.

Some stockholders and directors and executives and managers may be more interested in seeing a cleaner environment, or a general improvement in the employment conditions in the city where the plant, or the corporate headquarters is located. Some may even want to maximize the number of Democrats, or relatives, or blacks, or Catholics, or pretty girls working in the local office. Some may even want to maximize the number of business trips to Hawaii!

In the real world the motives which guide business decisions will differ from time to time, and from one business to another. A business on the verge of bankruptcy will be much more interested in making immediate profits than will one which is embarrassed by the fact that its profits are already so high that the labor union is demanding a big wage increase.

Some businesses have more leeway than others, but you may be sure that *no business can afford to ignore the question of whether or not it is going to make a profit*. All businesses have costs. The costs are usually almost as high as (and sometimes higher than) the revenues coming in. This means that the profit margin is usually small. Losses are not unusual. Often the owners who have invested in the business do not receive as much return on their money as they would receive in savings accounts or on government bonds. It isn't unusual for business owners, during bad years, to receive no return at all.

### Business Profits Are Usually Small

Most businesses operate very close to the break-even point. If anything goes wrong, they will be fighting for survival. Small businesses fold up by the thousands every year. All businesses face this problem. The giant Penn-Central Company went into bankruptcy in 1970. Lockheed Aircraft Corporation was hovering close to bankruptcy during 1971. One of Britain's best known and most highly thought of firms—Rolls-Royce—declared bankruptcy in 1971. How can such things happen? Simply by letting costs get too high, or revenues too low.

Making a profit is not the only objective of businesses, true. But the need to make a profit certainly can't be ignored. For the businessman, no other need or objective is as universal, as inescapable, as constantly haunting, as the need to make profits and to avoid losses. Why? The survival of the business depends on it.

Suppose a business is faced with a choice between using or doing or making or buying or selling one thing ("choice x") or another ("choice y"). If "choice x" looks more profitable than "choice y," which do you think the business will choose? Choice x? Probably. Not always, but usually. Although it isn't exactly true, it isn't too unrealistic to assume that the profit motive (the desire for maximum profits) guides all business decisions. It's a very useful assumption, and it's closer to the truth than anything else we can think of.

In the "make-believe, model world" that economists like to play in, every business tries to do everything it can to maximize its profits. In the "model world" each business decision is *always* made on the basis of the expected effect on profits. Suppose a business could expand its output a little and add a little bit to its profits. Would it do that? Sure! Suppose it could hire another worker and make more profits. Would it? Of course. But what if expanding the output or hiring a worker would *reduce* the profits? Would the business do that? Not a chance! (See how the "profit-maximization assumption" lets us say *exactly* what the business will do?)

## The Idea Of "Marginal Profit"

The economist would say it this way: "When faced with a choice, the business will make every move which is expected to bring a marginal profit, and will reject every choice which is expected to bring a marginal loss." So all we have to do is figure out whether a certain move will bring a "marginal profit" or a "marginal loss." Then we can tell right away whether the business will make that move. If the move is expected to bring a marginal profit, it will be made. If it isn't, it won't.

Now you know how businesses make their choices. They choose to do all those things which promise to bring them marginal profits. They choose *not* to do any of those things which promise to bring them marginal losses. *Marginal profits add to total profits* (or subtract from losses). *Marginal losses subtract from total profits* (or add to losses). The more marginal profits the business can get, the larger its total profit will be.

The "choice decisions" of businesses center around such questions as: Which product(s) should we produce? How much of it (of each) should we produce? Which kinds of inputs (labor, resources, machinery, etc.) should we use? How much of each kind? When should we add more inputs? Or cut back some? Should we build a larger factory? Maybe initiate an employee training program? Most of these choices (and many more) are facing most businesses most of the time. How does the businessman decide about these things? How does he choose? In our "profit maximizing, model world" he makes all of the moves which he thinks will be *profitable* and rejects all the others.

## Each Business Tries To Economize And Optimize

Each business that seeks maximum profit will always try to produce more valuable outputs and use less valuable inputs. Each business tries to achieve a high level of efficiency—that is, to produce as much as possible (the most valuable outputs) while using up as little as possible (the least expensive inputs). This will make the total profit (the value of the outputs minus the cost of the inputs) as high as possible.

Each business is constantly trying to optimize the use of its inputs. The most expensive inputs will be used as sparingly as possible. Cheaper, more plentiful inputs will be used whenever possible to replace the more expensive, scarcer inputs. Businesses are always trying to develop better machines and equipment to try to cut down on their costs, and to increase their outputs.

## Marginal "Value Product" And Marginal "Input Cost"

A businessman is always ready to buy any kind of new machine if he thinks it will more than "pay its way." The same is true about hiring labor, and about buying materials and equipment, building buildings, and about every other kind of input. Suppose a businessman is thinking about taking on an extra input (a worker, or a machine, or whatever). If he expects the extra input to "pay off" (in terms of reduced cost or increased output value, or both) then, sure enough, he will take on the extra input. But suppose he thinks the input is going to cost him more than it's worth to him. What then? He won't buy. Common sense, again? You bet.

If the businessman can hire a worker for $20 a day and the worker will increase the output value by $25 a day, that's a good deal. The worker will get the job, all right! Economists say it this way: If the marginal "value product" you get from hiring another man ($25) is greater than the marginal "input cost" you pay to hire him ($20), then you should hire him because he will add to your profits. The new man in this example brings a marginal profit of $5. ($25 extra "value product" minus $20 extra "input cost" = $5 extra profit.) Total profits will be $5 higher (or total losses $5 less) than before.

Why are we talking about "value product"? When a business buys or hires an extra unit of an input, what the extra input really adds is output—more goods—more "physical" product, not more "value" product. Right? Of course. But the businessman doesn't really care about the number of extra units of the "physical" output. What he cares about is how much *value* is going to be added, when the extra unit of input is added. How much will the total revenue go up? That's what he wants to know. So that's why we talk about marginal "value product." *It's the extra "value product" that pushes up the total revenue,* and that's what the profit-maximizing businessman cares about.

Going back to our example, suppose the marginal "value product" had been *less* than the marginal "input cost." Maybe the extra worker would only add a "value product" of $15. Then the marginal input (the extra man) would not be hired. Why? Because he wouldn't "pay his way." He costs more than he's worth. In the economist's language: the marginal "value product" is not great enough to cover the marginal "input cost," so this input would bring a marginal loss. Total profit would be $5 smaller (or total losses $5 larger) than before.

The businessman doesn't mind adding to his costs—not so long as he is adding *more* to his revenues! He doesn't mind having his revenues go down, either—not so long as his costs are going down *more*. It is not how big his total revenue or his total cost is that matters. It's the difference between the two (the *profit*, or *loss*) that counts.

### Marginal Cost And Marginal Revenue

While the business is making resource-use choices, economizing, trying to keep costs down, it is also making choices about how much product to make. It wants to adjust its output to produce the most profitable quantity. Each time it expands its output by one more unit per day, it adds something to its revenue. The cost of expanding the rate of output by one unit is called the marginal cost. The extra revenue gained from the sale of the extra unit of output is called the marginal revenue. If the marginal revenue (say, $15) is greater than the marginal cost (say, $10), the business makes a marginal profit on the extra unit of output ($15 marginal revenue minus $10 marginal cost = $5 marginal profit).

How does a business decide how fast (how much per day) to produce? How long should it keep expanding and adding "marginal output units" to its daily output? The business will keep expanding its daily output as long as each expansion brings a marginal profit. Of course! So long as the extra revenue received from "the sale of the larger flow of output" is *greater than* the extra cost of "producing the larger flow of output" then the increase in the size of the flow of output brings a marginal profit. Total profit goes up.

Whenever marginal profits can be made by expanding the rate of output, then the profit maximizing businessman will be expanding his output. Obviously. But does this mean a business might just keep right on expanding its daily output more and more? Could a business increase its output from its present rate of ten units a day (its "normal daily output"), to maybe a hundred units a day by the end of next week? And to a thousand units a day by the end of the month? Or to a million a day by the first of next May? Not likely! No business could expand its rate of output that much, that fast.

As the output of a factory or a mill or a processing plant (or even a hamburger stand, for that matter) gets larger and larger, pretty soon overcrowding and inefficiencies begin to develop. The inefficiencies cause the "marginal cost" (the cost of expanding the output) to go up. As the marginal cost goes up, what happens to the marginal profit? It goes down.

If the marginal revenue is $15 and the marginal cost is $10, then the marginal profit is $5. Great. Expand the output. But as the daily rate of output expands more and more, inefficiency creeps in more and more. So the marginal cost creeps up more and more. See what's going to happen? Sooner or later (probably pretty soon) the marginal cost is going to inch its way up to where it is as high as the marginal revenue. That means the marginal profit is going to inch its way down to zero. ($15 marginal revenue, minus $15 marginal cost, equals $0 marginal profit.)

When the marginal profit gets down to zero, that's as far as the profit-maximizing business will expand. At that rate of output the profit is maximized. The business has "gathered up" all the marginal profits it can get. It has expanded its output rate to the point where its marginal cost is equal to its marginal revenue, so marginal profit has inched down to zero. The business is producing its most profitable output.

So how does a business decide how much output to produce? It tries to find the output rate where its marginal cost is equal to its marginal revenue (MC=MR). At any smaller output rate (MC<MR), the business would be forfeiting "potential marginal profits," so it would expand to take advantage of all the marginal profits available. At any larger output rate (MC>MR), inefficiencies would be too great, marginal cost would be too high, and marginal losses would be incurred. The business would cut back to eliminate the marginal losses. When *MC=MR, that's the output where marginal profit is zero and total profit is maximized.* (In Part Six of this book you will learn how to use graphs to illustrate these cost and revenue relationships. Then all this will become clearer to you. For now, a basic understanding of the "MC=MR" principle is all you need.)

## The Marginal Input View And The Marginal Output View

I may have confused you a little bit. I was talking about inputs, and then started talking about outputs. What's the relationship between the inputs and the outputs? The more inputs, the more outputs, of course. Nobody is going to add any inputs unless he expects to get more output; nobody is likely to get more output unless he is willing to add more inputs.

You know that a business will add an extra unit of *input* (labor, resources, etc.) if the extra unit is expected to produce more value than it costs (that is, if the marginal "value product" is expected to be greater than the marginal "input cost"). You also know that the business will expand its output rate by another unit if the extra unit is expected to add more to revenue than it adds to costs (that is, if "marginal revenue" is expected to be greater than "marginal cost"). Another thing you know is that a business can't be adding *output* units without adding some *input* units, and it wouldn't be adding *input* units unless it was getting more *output* units. It all ties together.

If you think "the input view" and "the output view" are just two different ways of looking at the same thing, then you're right. Suppose a businessman is thinking about expanding his daily rate of output by hiring another "unit of labor per day" (another worker). Suppose he's pretty sure that his marginal "input cost" will be less than his marginal "value product." Then what can you say about what his "marginal cost" and "marginal revenue" will be? That's easy.

Whenever the marginal "input cost" is less than the marginal "value product," that means that the "marginal cost" must be less than the "marginal revenue." It must be true! These are just two different ways of looking at the same thing. So what's the difference? Simply your point of view.

Sometimes it's more helpful to think in terms of "whether or not to add a unit to the *input* flow." So then it's best to compare marginal input cost with marginal value product. Other times it's more helpful to think in terms of "whether or not to add a unit to the *output* flow." Then it's best to compare "marginal cost" with "marginal revenue."

The only difference between these two views is the *unit* we're looking at. Marginal "input cost" and marginal "value product" always refer to *one input unit*. "Marginal cost" and "marginal revenue" always refer to *one output unit*. That's the only difference.

## The Law Of Diminishing Returns

If a business keeps expanding its output, the marginal cost is going to go up—the marginal product will go down. Remember why? Inefficiency, from trying to produce more than your factory is designed to produce, that's why. Your factory (your fixed input) is too small, too limited, to let your other inputs (your variable inputs) work efficiently. What we are talking about is what economists call the law of diminishing returns.

This "law of diminishing returns" begins by assuming you have a factory or a plot of ground or some other "fixed input" to work with. Then it says that if you keep trying to expand the output from this fixed input (by adding more "variable inputs"—labor, raw materials, equipment, etc.), you can do so. *But* after awhile, the marginal product will start getting smaller and smaller. The marginal cost will start getting higher and higher. If anybody was stupid enough, he could keep on adding variable inputs until it would foul up the whole operation. (I don't suppose anyone would be that stupid.)

The "law of diminishing returns," simple and obvious though it is, is a very basic, important concept. This "law" is always there, always working, exerting important influences on the decisions (the choices) of everyone. It's constantly influencing all who organize and undertake productive activities—manufacturers, farmers, wholesalers, retailers, grocers, bankers, everyone. Even the productive activities of households cannot escape it. Everyone wants a "big enough but not too big" place. Everyone wants enough "fixed input capacity," but no one wants wasteful "excess capacity."

What determines how high (or low) the marginal cost will be at different rates of output? The size of the plant (the size of the "fixed inputs"), of course. If you expect to produce 100 units a day you will build a "100-unit plant." But then if you try to use that plant to produce 200 units a day, your marginal cost is going to be very high. If a businessman guesses wrong about his plant size, he can always make adjustments by moving, or building a bigger or smaller place. That may be expensive, and it may take quite awhile. It's much better (but not always easy) to do it right the first time.

### The Businessman Needs To Understand Economic Concepts

It takes more than an understanding of basic economic concepts to be a successful businessman. On the other hand, no one is likely to be very successful in business unless he does understand the basic economic concepts—either through his own good common sense, or from education, training, or experience. Usually it takes some combination of most, or all of these.

Each businessman will work to be successful. The one who succeeds in producing the right quantities of the right products while keeping his costs low enough will succeed. The successful, profitable business will grow, hire more people, build more plant space, buy more machines, and add to the total growth of the economy. The unprofitable business will fire employees and reduce output. Unless the businessman finds out what's wrong and does something about it, he will go broke.

There are many reasons for success in business, just as there are many reasons for success in life. But in both cases the answer rests to a very large extent on the way the economic problem is solved. Success for each of us (whatever it means to each of us) depends on our making the right choices about what to do with what we have to work with. The most successful people are the ones who do the best job of economizing and optimizing. They are the ones who make the wisest choices about how to "spend up" their scarce time and money and things.

## MARGINAL THINKING IS ALWAYS BEST

As you look back over the previous sections of this chapter it may surprise you to find that all the "maximizing decisions," the choices which bring the most progress toward the chosen objectives (for individuals, families, and businesses) are *marginal* decisions. The same would be true for all economic units—all people, all organizations, all societies, all nations.

What's so important about making the right "marginal" choices? Suppose each little choice, each little move, each little bit of something used up, is used so that it does as much good as possible. If each choice leads to the best possible move in the best possible direction, then all these little "best moves" must take us to the best possible place. See it? Sure.

If you really optimize each little bit of your time and money and things, then you're doing the best you can possibly do. Wherever you want to get to, you're getting there as fast as you can go. See the importance of making the right "little choices"?

### Take Care Of Your Little Choices

The wisdom of thinking marginally—of breaking down the big choices into lots of little choices—is not new. Some 200 years ago Benjamin Franklin told us that we should take care of our pennies—that if we did, our dollars would take care of themselves. You can see that the whole idea of "marginal thinking" is just good common sense, anyway. Still, it's a very important concept. Every day small businesses fail and people do foolish things just because they don't know how (or don't bother) to "think marginally."

### The Businessman Must "Think Marginally"

Making "marginal adjustments" is "fine tuning," sort of. The idea is that you start from where you are and then consider making little adjustments to see if you might come out better. Suppose you are running a factory. You ask yourself: "If I expand my output by one unit per day how much will that add to my daily cost? And how much to my daily revenue? If you think the marginal revenue will be greater than the marginal cost, you will expand your output. No matter what your situation was *before*, you are bound to be better off (make more profit) after you expand your output.

Suppose Mr. Bellino, your friendly neighborhood grocer, is thinking about spending $100 a month on advertising. How does he decide? If he thinks he will get enough extra revenue to more than cover the cost (the $100 advertising expense), then he will spend the money. If he doesn't think the advertising will pay off, he won't spend the money.

At any moment, most of a businessman's choices have already been made, at least for the time being. Mr. Bellino has already paid for his store. Most of his equipment is bought and paid for. He hasn't paid his taxes yet, but he's going to. (He doesn't have much choice about that.) What choices can he make on a day-to-day basis? Marginal choices. He makes them every day. Every hour. Every minute, even. These little moment-to-moment choices are the decisions which will determine whether his business will make money or lose money—whether he will prosper, just barely hang on, or go broke.

What are these little choices he makes? He must decide:

> whether to use a little more shelf space for Pepsi and a little less for Coke; whether to run a "special" this week on chicken or on beef; whether to run a little ad or a big one in tomorrow's paper;

whether to charge more for cold beer than for warm beer; whether to extend credit to Mrs. Capozza; whether to have the delivery boy work every evening or only on Thursday, Friday, and Saturday; whether to stock the new "Bella Maria" brand of bakery products, and if so, what to move out to make room; whether to put a little sign or a big sign in the window announcing the "special" on canned peas; whether to continue to close at 11 p.m., or to stay open until midnight; and dozens of other little decisions.

All of Mr. Bellino's marginal choices are important. Some are critical. His most critical choices are likely to be the same as yours and mine. What choices? The little marginal choices about how to "spend up" each little unit of time.

Should he spend his next five minutes sweeping the sidewalk? Catching up on his bookkeeping? Being friendly and helpful to the new customer who just came in? Painting a sign to go in the window? Planning next week's promotion? Calling several wholesalers to try to get a better price for corned beef? Making up "Italian grinders" for the noon rush? Restocking the shelves? Or how?

If Mr. Bellino makes the right choices on most of these little marginal decisions, he is likely to have a successful, profitable business. He will serve his neighborhood well, and for that he will be rewarded with profits. But if he makes too many of the wrong marginal choices, soon his store will be up for sale. Truly, in the success of a business, it's the little things that count.

## Good Enough Is Best

It's easy to see how thinking marginally applies to the businessman. It's just about as easy to see how it applies to almost any other "choice situation" you can imagine. Let's take some examples of things you're familiar with.

When you are washing the car, or the dishes, or the kitchen floor, or when you're mowing the lawn, or ironing a shirt or skirt, or studying economics, how much time and effort should you put into each task? That's up to you. But you can be quite sure that you will optimize by thinking marginally. That means *doing a job well enough to meet your own personal objectives*. (And it means *not* doing it any *better* than that.)

For one person an absolutely clean, shining, spotless, dust- and grease-free car, inside and out and under the hood, is just about the most important thing in the world. To him, "good enough" means perfection. Since perfection is impossible, he will never do anything in his whole life except work for and work on his beautiful car. You can't say he's wrong. But you may have a different set of objectives for yourself.

You may be highly interested in studying economics and making an "A" in the course. You may also be embarrassed about driving a dirty car. So if you have an hour to burn, you will spend fifteen minutes quickly scrubbing down your car, then you will spend the next forty-five minutes studying economics. Why? Because the "marginal satisfaction" you got from washing the car during the first fifteen minutes was highly important. The "marginal satisfaction" you would have received from spending another forty-five minutes washing the car would be very low—much lower than the marginal satisfaction you can get by spending that time studying econ.

After a fellow spends a certain amount of time washing his car, the returns begin to diminish. The same holds true for almost anything you can think of. You could spend all night studying this page of this economics book. You could memorize it so you could quote it forward and backward. You could count the words and the letters and the commas and periods, and then diagram each sentence. You could scratch off some of the letters and run a chemical analysis to find out what kind of ink South-Western Publishing Co. uses in its economics books. With your fertile imagination you could think of ways to spend the rest of the semester on this one page! Why don't you? Because your good common sense has already told you that "good enough is best." You're thinking marginally.

How much time should you spend on each page, section, or chapter in a book? Enough to learn each "just well enough." Spend too little time and you don't learn it, so the time is wasted. Spend too much time and you're wasting time.

At first, you can get increasing returns as you study a chapter. At the beginning, it doesn't make much sense. Then later you understand it, but you don't really have a solid grasp of it. Finally, you get it all put together. You know how to think about and explain and relate to the concepts. That's good enough. Quit. You will get a greater marginal return by spending your next "marginal units of time and energy" on some other subject or some other chapter or some other concept.

How much time and effort should you put into anything you do? As long as "the marginal return" you're getting is worth more to you than "the marginal effort" you're spending, you're doing just fine. As long as the additional "value" or "satisfaction" you're getting from the time and effort you're spending is *greater* than the additional satisfaction you could get from spending that time and effort in any other way, you are optimizing. That's the best you can do. You're making the best marginal choices, for you.

### All Intelligent People Think Marginally

Intuitively, all intelligent people think marginally, more or less. It's just good common sense. But once in awhile it's a good idea to stop and remind ourselves that we really do have a choice about how to spend each five minutes during each day, how to spend each dollar we have, how to use each thing we have—and that success in college (and in life) depends very much on how wisely we make these little choices.

To "think marginally" means to make each little choice so it will do you the most good. It means each time you spend each extra bit of effort or time or money or anything, make it do the best it can, for you. It means to optimize the use of everything you have to work with. The importance of making each decision "at the margin" and of concentrating on this "marginal trade off" is easy to see, once you start to think about it.

In a business enterprise, "marginal decision-making" is crucial. The successful businessman thinks marginally. You can be certain of that! A person had better not try running his own business if he doesn't think marginally. He'd better get a job working for some businessman who does!

### The Marginal Approach Doesn't Ensure The Best Choices—But It Can Help

Perhaps Benjamin Franklin wouldn't mind if we borrow his idea and say it this way: "Take care of your little choices, and your big objectives and goals will take care of themselves." That's good economics. But recognize that thinking marginally will not always make things come out the way you expected. The marginal approach is essential in making intelligent choices, but it can't insure you (or a businessman, or your college president, or the President of the United States) against making wrong choices.

Throughout this chapter we have been talking about choosing. That's what economics is all about. Remember? We have been talking about choice-making by *people,* and by *businesses.* That's important. The issue of choice-making by *the society as a whole* may be even more important. That's what the next chapter is about. First, though, take long enough to be sure you have a good understanding and a comfortable feel for the simple but very basic concepts which underlie the choice-decisions of individuals and businesses. It might help for you to spend some time with the questions and exercises. Do it if it helps. Don't, if it doesn't. Think marginally.

---

## REVIEW EXERCISES

**MAJOR CONCEPTS, PRINCIPLES, TERMS (Try to write a paragraph explaining each of these.)**

the "marginal utility" theory of consumer behavior
the "marginal rate of substitution" theory of consumer behavior

the output rate where MC = MR brings maximum profit
the law of diminishing returns
"thinking marginally"
the profit-maximization assumption

**OTHER CONCEPTS AND TERMS (Try to write a sentence or phrase explaining what each one means.)**

optimize
economize
maximize
welfare economics
theory of the firm
marginal profit
marginal loss
marginal "value product"
marginal "input cost"
marginal cost
marginal revenue
fixed inputs
variable inputs

**QUESTIONS (Try to write out answers, or jot down the highlights.)**

1. For many people, as the years go by life seems to get a little easier. Do you suppose it might be because they have been "saving and investing"? Or why? Explain.
2. If a businessman keeps on expanding the daily rate of output of his factory, sooner or later the per-unit cost of his product will go up. The "law of diminishing returns" says so. But suppose, to expand his output the businessman buys another factory, and another, and another. Would the "law of diminishing returns" push up the cost then, too? Explain.
3. A producer, thinking marginally (as all good businessmen do), is always thinking about cutting back or speeding up his daily rate of output.
   a. Suppose the marginal cost of expanding his present rate of output by one unit would be *less than* the marginal revenue. What should he do? Why?
   b. Suppose the marginal cost would be *greater than* the marginal revenue. What should he do? Why?
4. One of the things about a "free society" is that each individual has the right to make the *wrong choices* for himself. Are you making any of the wrong choices for yourself? Is there any way that you might "think marginally," and shift some of your own time and energy from one direction to another and come out better? Think about it.

# 3 HOW DIFFERENT SOCIETIES CHOOSE

*The production and distribution choices can be made by social custom, government control, or the market process.*

The economic concept "you can't have your cake and eat it too," is as true for the society as a whole as it is for each individual or business. In each society people are busy growing food and making things. These same people at the same time are eating up, using up, and wearing out these things. Which things to grow and make? And then, who gets to have them and use them up? These are the questions every society somehow must answer.

**Every Society Must Solve Its Economic Problem**

Every society which has ever existed has had to solve its economic problem—has had to face these resource-use questions and come up with some answers. The approach differs from one society to another, and from one time to another. But before we talk about the different ways societies solve their economic problem, let's talk about the three basic questions which every society must answer—the three basic kinds of choices which every society somehow must make. When these three basic questions are answered, then for that society, that gives the answer to their "economic problem." It may not be the "best" answer, but it is *that society's* answer.

## EACH SOCIETY MUST MAKE THREE BASIC ECONOMIC CHOICES

The "economic problem" of the society can be broken down into three basic questions. First: the output question—what to produce? Second: the input question—which resources to use? And third: the distribution question—who gets to have what share of the output? Each of these questions needs to be explained, and then we need to talk about ways of getting the answers. That's what this chapter is about.

### The Output Question: What Are We Going To Produce?

What are we going to produce? Of all the possible things we might spend our time and resources making, which ones will we choose to make? Will we put more effort into building houses, or into growing food? What kind of food? grain? or beef? Or perhaps we should erect monuments to our ancestors. Or paint beautiful pictures and make beautiful music and write books. Will we produce military tanks or farm tractors? automobiles or trucks? if automobiles, big cars or little cars? school buildings or hospitals?

The list of possible products could be endless. Yet, these choices must be made for each society. The question is: Of all the thousands (or millions) of things we *might* use our scarce resources to produce, which things and how much of each *will* we produce? The quantities of some things—canned Mississippi mud—will be zero. Other things—perhaps bread—will be assigned higher priority. Somehow this question "what to produce?" must be answered.

### The Input Question: Which Resources Will We Use?

Which resources will we use in making our chosen products? Almost anything can be produced using different combinations of inputs. Roads could be paved with straw, wood chips, boards, gravel, leather, bricks, concrete, asphalt, steel, copper, gold, economics books, or almost anything you could think of. Ridiculous? Well, for some of the examples, yes. Yet it illustrates the fact that almost everything you can think of could be produced in several different ways. Somehow it must be decided which of the society's resources will be used to produce which things.

Just as we can use different materials, we can use different sources of power and different techniques of doing each job—different production methods. For power we can use electricity, steam, draft animals, internal combustion engines, manpower, the wind, water, sunshine, or some other source of power. If we want to make steam we can use wood, coal, oil, gas, nuclear energy, or buffalo chips. As tools we can use steam shovels, or hand shovels, conveyor belts or wheelbarrows, paint brushes or paint sprayers, farm tractors or hand plows, adding machines or computers. To travel or to move things we can use trucks or boats or railroads or pipelines or airplanes or horses or pogo-sticks. Again, the list of possibilities is endless. And again, some of the alternatives are a little bit ridiculous. No one is going to pave roads with gold and no one is going to use a steam shovel to plant tulip bulbs.

No society would have any trouble making the obvious choices—for example, the choice of whether to pave the roads with gold or gravel. But most choices aren't this easy. Suppose we're trying to decide if we should pave the roads with crushed granite, or natural gravel from the gravel pit, or tailings from the local copper mine—or perhaps some mixture of all three. Now the best choice is not so obvious. Right? These close choices—where the little marginal "fine-tuning" decisions must be made—are difficult. Should we use a little more land and a little less fertilizer to produce our grain? Should we produce a little more corn and a little less wheat? Should we use migrant labor or harvesting machines? More account clerks or a computer?

If the people are going to enjoy the most efficient society—if the uses of all the resources are to be optimized toward the society's objectives—then somehow the society must get the best possible answers to these questions. We must produce just the right amounts (proportions) of just the right *output products*. We must use just the right amounts (proportions) of just the right *inputs* in making each product. You can see that in a modern society these choice questions can get very complex! But before we go into that we need to talk about the third basic choice.

### The Distribution Question: How Much Will Each Person Get?

How much of the total output does each person get to have, or to use? We are using up our resources, producing all these things, but then what happens? What are we going to do with all this output? Who's going to get all those products that are being made?

How much of the output are *you* going to get? And how much of it will belong to me? I'm sure I wouldn't want all of the output, but I think I would like to have a little more of it than I'm going to get. I think you probably would like to have a little more of it than you are going to get, too. I think most people are going to want more. But we can't *all* have more. There's just no way!

How will we decide who gets how much? Will the men get the most and the women have to beg the men for a share? Maybe we will share it all equally among everyone who is over 21, and let all the young people beg for their shares. Or why not let a larger share go to everyone who voted for the winner in the last election, and give everyone else just enough to keep from starving?

There are lots of ways we could divide up the output. A smaller share for short people and a larger share for tall ones? A larger share for educated people and a smaller share for ignorant ones? A larger share for native-born citizens and a smaller share for immigrants? Larger shares for the diligent ones? the tricky ones? the ones whose fathers got large shares? Larger shares for the ones who produce the most? For the ones who make the most noise and threats? For everyone who has a monopoly in something? Who? And *how much larger* will be the shares of the favored ones? How are we going to decide?

You can see that the question of "who is going to get how much" can get to be pretty complex! But complex or not, the answers have to be found. Somehow the society must devise some kind of system for getting each person's share figured out.

## The Production Questions Involve Opportunity Costs

Now you know the three basic kinds of choices which must be made. Sometimes the first two (the output question and the input question) are lumped together and called the production question. In its broadest meaning, the "production question" is concerned with: (a) choosing the specific products to be produced and deciding how much of each will be produced; (b) choosing the specific resources to be used and deciding how much of each will go into making each product; and then (c) stimulating and directing the people and the other inputs to go to the necessary places and to do the necessary things to carry out the society's production choices.

There are thousands of little "choice decisions" which must be made to answer the society's production question. Will the men build the roads? or the women? Will we haul the rocks and gravel in wagons? or in wheelbarrows? If we decide to use wagons to haul the rocks and gravel, somebody is going to have to decide if the wagons will have wooden wheels or bronze wheels or steel wheels. And someone is going to have to build the wagons. With the lumber we use to build the wagons, we can't also build a house for someone to live in, or a boat to use for catching fish. Opportunity costs are staring us in the face. Which resources will we try to conserve? Which will we use up freely?

The "production question" can get about as complex as the distribution question. But, one way or another, every society must get the production question answered. There's just no way around it.

## There Are Never Enough Resources To Do Everything

There will never be enough resources and products to give everyone enough to fulfill all his wants and objectives and also to permit the society to fulfill all its objectives and goals. Sometimes we like to make-believe about this. We like to think that the reason there are so many factories and power plants polluting the air and the water, and the reason people must work so hard and the reason we use up resources so fast is because "those other people" want too many things.

"Those other people" are too materialistic. They want bigger cars, bigger houses, bigger everything. And electric everything—blenders and can openers and TV's and hair driers and frying pans and air conditioners and you name it. If "those other people" didn't want so much we could all slow down and escape from this rush-rush-rush—produce-produce-produce hangup. Then we would all be better off. But would we?

## Most People Disagree With Some Of Their Society's Choices

Isn't there *something* each of us wants more of, for ourselves? or for the society? Maybe not an electric can opener or a bigger car—but maybe a new tire and enough gasoline for a trip to the beach? Perhaps more books? Or maybe more books for the college library. Or more college libraries. Or more colleges. Or more libraries. A trip to Singapore? Or France? Maybe better sewage systems in the small towns and villages throughout the nation and the world so we can stop polluting the streams and rivers. Maybe better health and medical care and better food and housing and education and transportation for the urban and rural poor in our country—and in other countries, too. How about a major effort to develop and install antipollution equipment? Or to make automobiles safer? Or to cure muscular dystrophy?

How long is this list? As long as the sheet of paper you have to write it on. One who argues that the economy is producing too much probably doesn't really mean that at all. What he probably means is: "I don't think the economy is producing the right things. I would rather see our resources and energies used less toward some of these objectives, and more toward some other objectives."

Almost everyone sees some desired objectives going unfilled. If you wish, you can disagree with the society's answers to the production question. You can argue that we are producing the wrong things, or that we are using up the wrong resources in the process. Probably anyone, in any society (anyone who takes time to think about it), will disagree with some of the production choices of his society. So long as individuals have different wants, ideas, beliefs, objectives, and different degrees of information and misinformation, how could it be otherwise?

Yes, the production question is always difficult for a society to answer. But what about the other basic question—the distribution question. Is that difficult, too? You bet.

### Everyone Seems To Want A Larger Share Of The Output

You know that the "distribution question" concerns the decision about who (that is, what specific people, which individuals) will get what shares (how much) of the things which are being produced. Now that all these cakes have been baked, who gets to have them? or eat them? Will everyone share equally? Or will everyone who has royal blood, or is kin to the chief have first claim on the cakes? and on the fish that are caught? the vegetables that are grown? the houses that are built? the horses that are raised? the services of the best physicians?

Most of the people are always going to get a somewhat smaller share of the society's output than they want. Each choice to let one person *have* something is a choice that someone else will *not* have it. These are the very difficult choices posed by the distribution question.* There is no way of working out each person's distributive share so that all the people will be completely satisfied. Most people are going to want at least a little more than they get. From the time you first argued about the size of your weekly allowance, you probably have been trying in one way or another to increase your "distributive share." Throughout your life you probably will go to all kinds of trouble to try to keep your share rising.

### Your Money Income Brings You Your "Distributive Share"

In all the modern countries of the world, each person initially receives his income as *money*. The money he receives is his claim to his "distributive share" of the output. Each person's money income lets him "go into the market and claim his share" of the things produced. The bigger your money income, the bigger share of the output you can buy. Anything which changes your money income changes your "distributive share."

In every modern economy the distribution question is essentially the question of how much money—how much *income*—each person will get. No matter how the question is answered, you may be absolutely sure that nobody will be completely satisfied with the answer.

### There Are Only Three Ways The Society Can Choose

By now you probably are impressed with the fact that society's economic problem is not going to be easy to solve. But somehow it must be solved. There are several different ways a society might go about getting

---

*The word "distribution" has two different meanings. As used here it is concerned with: "How much of the output each person will receive." A completely different meaning is used by the businessman to refer to the "marketing channel" through which he "distributes" his products to consumers—that is, the movement of products from manufacturer to wholesaler to retailer to final consumer. Be careful not to confuse these two meanings of the word "distribution."

the answers to the production and distribution questions. But all the different ways can be grouped into three different categories, according to the *process* the society uses to get the choices worked out. These are: (1) the social process, (2) the political process, and (3) the market process.

All three of these processes are at work in every society. In one society, one process may be dominant. In another society, another may be dominant. But all three processes will be in there, influencing the choices (at least to some extent) in every society, all the time. So now let's talk about these three processes—social, political, and market—these three approaches which all societies use for solving "the economic problem."

## HOW THE SOCIAL PROCESS MAKES THE BASIC ECONOMIC CHOICES

Many of the production and distribution questions in every society are answered by the social process—that is, by the customs and the traditions, the "usual way of doing things" in that society. If you ask why Dad mows the lawn and Mom washes the dishes, the answer is: "That's the way it is usually done in our society." Why does the husband usually work outside the home to earn income, while the wife usually stays home and uses up her "scarce productive energies" (her labor) in domestic duties? It's customary. It's traditional.

Why does the son wash the car and the daughter mop the kitchen floor? Why do the children in the family receive allowances? Why do the children sometimes get more allowance as they get older? Again, because this is the way it is usually done in our society. But wait. Are we talking about the basic production and distribution questions? Of course we are! Mopping floors, washing dishes, mowing lawns and washing cars are all *productive* activities. And children's allowances certainly have an influence on the distribution question—on "who gets how much"!

In our society and throughout most of the world, the *men* usually lay bricks, fly planes, collect garbage, fix pipes, fight wars, pass laws, and build highways. Women usually run typewriters, vacuum rugs, teach kindergarten, nurse the sick, work as airline stewardesses, cook, clean house, and raise kids. Why? Because it's traditional. Even in the most modern nations people aren't yet free of traditional bonds. Tradition still tries to force people into the customary roles passed down from ancient and medieval times. In the "less modernized" societies the bonds of tradition are much stronger. *Most* of the production and distribution questions are answered by social custom—by tradition.

### The Social Process Is Strongest In Less Developed Countries

Even today, the less developed countries of the world have very strong social pressures which influence the production and distribution choices.

In almost all tradition-bound societies there is a hierarchy of royalty, or chiefdom, with the highest-ranking chiefs having the largest influence on how the society's resources will be used. The high chiefs receive more of the output than do the others.

In a society which operates mostly on custom and tradition, each individual's life—economic, social, and personal—is largely predetermined for him by tradition. He follows the paths of his forefathers. The position and activities of each person in the society, including the work he will do and the share of the output he will receive, are largely determined by customs and traditions from the past. The same patterns have been repeated by each new generation, century after century.

### In The Traditional Society, Everyone Shares

In most traditional societies the people share the output on the basis of kinship and bloodlines, but usually there also is some special reward for good work. The fisherman who catches the most and biggest fish usually gets a larger share of the catch than the one who brings in no fish at all. But the one who brings in no fish still gets a share. The situation usually is: "Most of what you produce belongs to the other members of your family, extended family, clan, and society. But something extra of what you produce will go to you as your reward for producing it."

If the fisherman catches a big fish and eats it all himself he is actually stealing from the others. Why? Because according to the rules of his society it isn't *his* fish. It's *their* fish! (It's the same as if you are working on an Aransas Pass shrimp trawler, and you steal and sell boxes of the shrimp you catch. They aren't your shrimp!)

### The Traditional Society Discourages Saving

The traditional society's way of answering the distribution question keeps everyone fed, but it discourages saving. It's almost impossible for anyone to work hard and build up some savings for himself. Whenever a person saves, his savings "rightfully belong" to everybody—so why save?

In recent years some of the ambitious young men from the traditional societies have moved to other places where they can earn money and keep it for themselves. Those who get away, and save, know that if they go back home, most of their savings will have to be shared with others. Yet, often the feeling of "their just and rightful obligation" is so strong that either they return home, or they send most of their money back to be shared. Nobody likes to feel that he's stealing from his family, or from his relatives and friends. Would you? Of course not.

You can see how a society organized primarily on the principle of tradition would discourage savings and investments. There is not much incentive for the individual to try to get ahead. In many traditional societies, "getting ahead" is discouraged. But even if it isn't, anyone who wants to get ahead must pull up his entire family and clan (and maybe the whole society) with him. Not very many people are either able or willing to undertake such a formidable task!

### The Traditional Societies Are Eroding (And Exploding!) Away

Much of the political unrest throughout the world today reflects the erosion and breakdown of the traditional societies. Their customs and traditions have maintained social and political stability, solved the economic problem, and held together and maintained these societies over the centuries. But the breakdown is inevitable. The traditional societies are not designed to encourage—or even to permit—the kinds of rapid changes required to bring about the higher productivity and increased standards of living which all people now are demanding. The day of the traditional society is rapidly passing into history. Sad, perhaps. Anthropologists like to observe them; all of us can learn much from the study of such societies. But not very many of the traditional society's young people (who have glimpsed the "free outside world") seem to want to live in such a society anymore. Most of the people seem to want to have more personal freedom, and more things—better tools, clean water, modern medicine, outboard motors, cars, transistor radios, and cokes, beer, and cigarettes. But such "standards of living" for the masses simply are not compatible with the traditional society.

### Anthropology And Sociology Explain The Social Process

Economics is not the place to look if you want to understand how the basic economic choices are made through the social process—through custom and tradition. Anyone who is deeply interested in such questions should study anthropology and sociology. It is in understanding the social process—how the society functions—that you will find the key to understanding how the economic choices are made in the traditional society.

Even though the traditional societies are eroding (and exploding!) away, the influences of social processes on the economic choices will continue to be important in all societies. The diminishing but continuing importance of the family as a "producing, distributing, consuming economic unit" is one good example. The diminishing but continuing economic inequality of women is another.

## HOW THE POLITICAL PROCESS MAKES THE BASIC ECONOMIC CHOICES

Billions of dollars worth of our resources are being used to produce the U.S. interstate highway network. Space exploration, military activities, education programs are all using up billions of dollars worth of resources. How did the society decide that all these resources would be used to produce these things? Governments decided. The federal government, the state governments, and the local governments made the choices.

### Governments Make Many Of The Economic Choices In Every Country

All countries rely on the political process for many of their economic choices. In the United States, governmental decisions play a very important role in answering both the production and the distribution questions. You can see the government's influence on the production question everywhere you look—streets and highways, post offices and public buildings, parks and recreation areas, schools and prisons, and B-52's flying overhead.

What about the distribution question? Can you see the government's influence there, too? Sure. The government requires people to forego some of the goods they could buy, and to pay taxes with that money instead. When people pay taxes their "distributive shares" of the society's output are reduced. Then when the government gives money to unemployed people, to families of dependent children, disabled veterans, old people, and to students, that increases their "distributive shares." That way, people with no income of their own can have a share of the output of the society.

These examples of government influence (and there are many, many more) show that even in a "free economy" such as that of the United States, the political process (government) makes a lot of the production and distribution choices. Yet the United States is one of the countries in which the government has the least influence on the economic choices. Most of the production and distribution choices in the United States are *not* decided through the political process; political (governmental) control of the economic choices is the exception rather than the rule.

In some countries—for example the Soviet Union and Communist China—political control over the production and distribution choices is the usual thing. In those countries the choices which are *not* made by government are the exceptions. Even in the countries of Western Europe, more of the economic questions are decided by government than in the United States. You will be hearing more about that in the next chapter.

## Governments Direct The Economy By "Command"

Whenever a production or distribution choice is made by governmental decision (through the political processes of the society), this overrules any other choice which might be made—either by tradition, or by "the free individual who wants to do his own thing." When the government levies a tax, the people pay. There is no choice. When the State Highway Department says a new highway is going to occupy your front yard, sure enough, that's what happens. (You get paid for the land, of course, but you don't have much choice in the matter.) When the city zones your vacant lot as "open space," not to be used for buildings, then you won't get a building permit, and that's that! (Unless you can get the zoning board to change the decision, or get the city council to change the zoning law.)

When the resource-use choices are made by the political process, this sometimes is called the command method for making the choices. You can understand why. Notice that the word "command" does not have to mean that there is a dictator making the decisions. Your own local city council and county board are making some of these choices every day.

## The Command Method May Be Direct Or Indirect

Sometimes the command method involves direct allocation: "This piece of land will be used for a naval base; that one for a recreation area. That man will serve in the armed forces." Sometimes the "command" (political process) method uses the *indirect* approach. Resources are induced to move into the uses the government wishes, but not by direct order. The government simply offers attractive prices for those resources which it wishes to control. Thus the people and resources are persuaded (rather than ordered) to do the government's bidding. In the Communist countries, many of the choices are made by direct allocation. But in most modern countries, most of the government-directed economic choices are carried out by the use of price and profit incentives.

If you are thinking about being a school teacher, chances are you are planning to work for some state or local government. School teachers are teaching school because the government pays them enough to induce them (or to let them) keep doing that. If the government stopped paying them, the school teachers would have to quit teaching and find some other kind of work to do.

Contractors build government buildings because they are offered payments to do so. Engineers construct dams, aerospace firms make Apollo moon-landing ships, and policemen try to keep the peace, all because the political process induces them—pays them—to do these things. The political process (the government) makes choices about how it wants some of the resources (human, and other) to be used. Then the government carries out the choices by offering income (that is, a distributive share of society's output) to those who will do what it wants done.

### Political Science And History Explain The Political Process

To understand how the choices are made by government, you need to understand the political processes of the society. You need to understand how the government functions—how it makes its decisions. Why does the government decide to expand the interstate highway system, to cut back the space program, to raise the tuition at the state university, to put a ceiling on professors' salaries, to build a bridge across Oregon Inlet, or to reduce payments for "aid to families of dependent children"? Such decisions are all made through the political process. How does it work?

In a country like the United States, each governmental choice usually involves many individual decisions. The decisions are made by elected representatives and by administrative officials and subofficials and assistants and clerks and secretaries and maybe others. No one can ever be quite sure about the reasons why any legislator decides and votes as he does. Sometimes his decision may be based on a careful analysis of the issues; sometimes perhaps on personal whim. One legislator may vote "yes" because "the party" is for it; another, in response to pressures from his local constituents.

One thing we can be sure of. Unless each legislator responds reasonably well to the wishes of a goodly number of his constituents, he will one day cease to be a legislator. In countries with less democratic forms of government, the political process is less responsive to the wishes of the people. But even in the United States where we enjoy some of the world's most democratic governments, the responsiveness of the political process on economic matters is far from perfect.

### Economics Explains The Market Process

Just as it is necessary to study anthropology and sociology to understand how the social process makes the economic choices for the society, so is it necessary to study political science, history and the workings of governments to understand how these choices are made through the political process. It is only for the third and final "choice-making process" that you must study economics to understand how it works.

The third choice-making process is called the market process, or simply, "the market." The study of "the market process"—what it is and how it works—is a major part of economics. Almost everything in this book—most of what you have already seen and almost everything coming up—is concerned in one way or another with the nature and operation of "the market."

## HOW THE MARKET PROCESS MAKES THE BASIC ECONOMIC CHOICES

Suppose the government isn't getting involved in the production and distribution choices. And suppose there are no traditions directing the economic activities and choices in the society. Then neither the political process nor the social process is going to solve the economic problem. But somehow things have to get produced and distributed. Otherwise everyone will starve. So what happens?

Without governmental or social control over the economic choices, everyone is on his own. Anyone who wants something had better make it for himself. That seems to be the only way. But wait! Perhaps somehow he can *induce someone else to make it for him.* Great idea! But how might he do that? It's easy. Just offer to pay the other fellow enough, or to trade him something he wants. Simple? Logical? Natural? Yes. And that, in a nutshell, is how the market process works.

### How "The Market" Answers The Production Question

In its barest essentials, this is the way the market process works: Someone is producing the things you want because you are buying those things. You produce something *others* want, because that's the way you get the money to buy what *you* want. The other person wants your money (as his income), and the only way he can get it is to produce something for you. So, just like that, the society chooses "what to produce." It's automatic! Neither "tradition" nor "command" need to be involved at all. "The market" answers the question "what to produce," automatically.

### How "The Market" Answers The Distribution Question

What about the distribution question? How does the market process solve that? That's the other side of the coin. The fellow who produces a lot of what I want gets a lot of my money. That's his income. With his income he can buy a big "distributive share" of the society's output.

Where do I get all my money? By doing and making things other people are willing to pay for. Of course! The more I produce of the things others want, the more money I get to buy the things I want. That's how I get to claim my "distributive share." See how the distribution question is automatically decided by the market process? Each person's production—the value of what he produces—determines his income, and (therefore) the size of his share of the society's output. This is called the productivity principle of distribution.

Now you know how "the market" answers two of the basic questions: (1) What to produce? Produce those things the people want and are willing to pay for. Produce the things the people *demand.* (2) Who gets to have how much of the product? The ones who produce the most will get the largest incomes and can buy the largest shares. Those who produce the least will receive the least. But there's one more question. Which resources will be used for what?

### How "The Market" Conserves Society's Resources

When I'm making the things other people want, which of society's resources do I use? The cheapest ones I can get to do the job. Of course! I never use moon rocks to pave people's driveways. I help society by conserving its moon rocks. Who tells me to conserve the moon rocks? Or is it just that I am a good guy and am doing my bit for the good of mankind? Neither. The market induces me (forces me, really) to conserve the moon rocks. Here's how.

Great scarcity puts a high price on the moon rocks. The high price convinces me to conserve them—not to use them to pave driveways. Suppose I did buy and use moon rocks to pave driveways. What would happen then? I would be severely punished by the society. I would lose a great deal of money on the job, and go broke. My "distributive share" would drop to zero because I used up more of society's valuable things than I produced for society in return. See what a high price can do? It forces people to conserve the society's most valued things. And, just as with everything else about the market process, it works *automatically.*

### The Driving Force Is Self-Interest

What is the driving force of the market process? The desires of people. Each person is looking out for himself, but the only way anyone can get what he wants is to produce something the other people want. I make something for you and you reward me with money. Then I can buy what I want from someone else, and I will reward him with money. On and on it goes.

Everybody is working for everybody else, and each is rewarded for how much he does and how efficiently he does it. All three of the "basic economic questions" are being answered all the time, automatically. As each individual, family, and business works to solve its own economic problem, the society's economic problem is automatically solved.

People buy the things they want most. This stimulates other people to go into business and make more of those things. So the "output question" is answered. The business is interested in making profit, so it uses the inputs which will do the job at least cost. The businessman, to be successful, must carefully conserve and optimize the use of society's resources. So the "input question" is answered.

The owners of the most profitable businesses will get the best incomes. Their most productive employees will get the best wages. The people who supply the best resources and equipment and products to the businesses will receive the highest incomes. Everyone who gets a good income will enjoy a good distributive share of society's output. But no income goes to those who don't work. Those who don't produce anything don't get anything. The output goes to the productive ones. In this way the distribution question is answered.

The market process is a very natural sort of thing. Each person decides what he wants to do to get enough income to buy the things he wants. As a result, each person winds up doing or making something that someone else needs or wants. How naturally and easily the market process makes the three basic choices! Let's summarize it once more. But first, a word of warning. We aren't talking about "the American Economy" or about *any* real-world economy. Far from it! We're talking about the *pure market process*—which does not exist, has never existed, and will never exist in the real world in anything approaching its "pure" form. More on that later, but now, the summary.

### Summary Highlights Of The Market Process

**What To Produce?** Produce those things the people are buying. The more being bought, the more will be produced.

**Which Resources To Use?** Use the cheapest ones to do the job adequately, and optimize the use of each one. The cheap resources are the ones society has a lot of. It's better to use these than to use the more expensive, more scarce, more valuable ones which really ought to be conserved.

**Who Gets How Much?** The people who produce the most of what other people want the most (and use up the least valuable amounts of society's resources in the process) will get the most income. Each person's income will reflect the value which society places on what he (either he himself, or something he owns) produces. Each person's income lets him claim his distributive share of the output.

The market process is really only an extension of the way things work in nature. In nature, each living thing must produce to stay alive. Each animal gets to consume whatever he produces. If he produces much, he lives good. If he produces nothing, he starves. That's the way it is in nature, with all the animals, and birds—and even plants!

A tree in the forest solves its economic problem by sending its roots deep and wide to "produce" moisture and minerals and things. However much it produces, that's how much it gets to consume—no more, and no less. If the roots don't go wide and deep enough, the tree doesn't produce and consume enough to stay alive. So it dies. That's the way it is in nature with all living things. The productive ones get to consume and live. The unproductive ones die.

The market process modifies this natural "produce and consume cycle" in only one way: It gives each person the opportunity to produce *one* thing and then consume a *different* thing. Each person produces *indirectly* for himself, by producing *directly* for others. The market process lets each person produce "for himself," and at the same time, to specialize in producing something he's good at.

### The Market Permits Individual Freedom Of Choice

The "pure market process" permits each person to be free to choose what and how much he wants to produce, and what things to buy with the money he earns. Each person is free to influence the size of his income by deciding how productive he wants to be. Each is free to decide for himself what products he wants to buy. As all the individuals make their choices, they automatically determine the choices for the society. The market process just lets nature take its course.

## ALL THREE PROCESSES ARE AT WORK IN EVERY SOCIETY

The three processes—social, political, and market—are at work throughout the world today, exerting their influences on the production and distribution choices. All three processes are at work in every society, in every nation. You know that now. But you don't yet know much about how these intermixed processes are actually working today in the real world. You know that to understand how the choices are made by the social process (tradition) you need to understand the functioning of the society itself. You know that to understand how a society's choices are made by the political process (command) you need to understand how the government of that society functions—how the governmental decisions are made and carried out. And you know that to understand how the market process makes the choices you need to understand economics.

### The Social Process Reflects The Kind Of Society

If the economic choices in a society are going to be made by the social process, then it's pretty important what kind of a society (what kind of "social process") exists there. If it is a very rigid society, controlled by taboos and omens and superstitions, economic conditions of most of the people may be less than pleasant. On the other hand, if the society is controlled by love and mutual sharing, the production and distribution choices will look a lot different.

### The Political Process Reflects The Kind Of Government

What about the political process? Does it make much difference what kind of government exists? Of course! Suppose a nation has traditions of high morality and democracy, and a high degree of responsiveness of the political process to the wishes of the people. Then the "political process" (command) choices will reflect the wishes of the people. But what if the political process is not responsive to the wishes of the people? Then the economic choices are likely to be very different.

### The Market Process Works Independently

With the market process, it doesn't really make too much difference what kind of society or government exists. So long as the market is permitted to make the economic choices the results come out the same way, regardless of the kind of social or political system that exists. This fact has been a very strong argument in favor of the market process as a way of organizing an economic system. As you know, this idea is strong in the "philosophical heritage" of the United States—free enterprise, freedom of the individual, minimum influence by government, and all that.

Do not be worried that you don't yet understand much about the details of how the market process works. It's really all very logical and easy to understand, and it will be coming to you little by little, off and on, throughout most of this book. For now, all you should really understand is what the market process is, and the bare essentials of how it works—how it makes the three basic choices for the society.

You should know that the production problem is solved automatically as the businesses produce the things the people are buying; that businesses try to produce these things at the least possible cost and this conserves society's scarce resources; and that each individual's share of the output is determined by the value of what he produces. If you really understand this much about the market process, that's enough for now.

The next chapter discusses some of the different *economic systems* which exist in the world today. We will look at the ways in which social, political, and market processes are working to make the economic choices in the economic systems of *capitalism, socialism,* and *communism.* But before we go on, let's ask one more question about the market process.

## DOES THE MARKET PROCESS ALWAYS MAKE "GOOD" CHOICES?

Are the choices made by the pure market process always "good" choices? Does the market process always produce the "best" combination of outputs, using the "best" combinations of inputs? Does it always distribute the outputs so that everyone gets the "best," most "rightful" share? the share he "should" get? No, it really doesn't always do all these things. It can't.

Many times throughout this book you will see examples of the inability of the market process to fulfill the wishes of the society. One of the reasons why no "pure market" economic system exists in the world is because (in its pure form) no society would put up with it! Neither in the "pure model" of the market process, nor in the real world, is the market process (acting alone) capable of making acceptable social choices about everything. About lots of things, yes. But about some things, no.

In education, anti-pollution, social security, and many other times the market process is "overruled" by the society. The *political process* takes over and tries to do what the society thinks "should be" done. How do we know what "should" be done? There's no way to be sure. Whatever is decided, some people will disagree.

### Positive Economics And Normative Economics

For more than a century economists have been arguing with each other about whether or not economics and economists ought to get involved in questions of "what should be." Should economics try to be a "positive science"? to deal only with scientific laws of cause and effect? with questions of "what is and what would happen if"? and stay away from such questions as "what *should* happen"? and "what *would be best* for the society"? Or should economics be a "normative" field? Should it deal with some of the "less scientific" (more philosophical) questions of *what should be*? Should economics deal with "value judgments"? and try to figure out how *good* or *bad* one choice or another might be, for the society?

### Positive Economics Deals With Economic "Laws"— With How Things "Are."

"If you don't pay a man to work for you he will not work for you." "If the price of fried chicken doubles and the price of hamburger goes down, people will buy less fried chicken and more hamburgers." These are statements of positive economics. Positive economics doesn't say it's good or bad—it just tells "what would happen if. . . ."

### Normative Economics Deals With Questions Of How Things "Should Be"

Normative economics deals with "value" issues—the issues of good or bad, right or wrong, better or worse—and of how to make better economic choices for the society.

When economists recommend such things as progressive income taxes and social security programs as "just and equitable," then their recommendations are based on normative economics—on an idea, or philosophy of "social justice." But *the economist's knowledge of positive economics is what enables him to make realistic and feasible normative recommendations.* Anyone who really knows the "laws of positive economics" knows the most basic law: you can't have your cake and eat it too. That's a good thing to keep in mind (and so easy to forget!) when we start making "normative" recommendations.

In the past few years, more and more economists have been getting into the normative, "value judgment" issues of economics. Normative economics is much less definite, much less "sure and scientific" than positive economics. For example, it's much easier to say what will happen to a family's food budget if the father loses his job (positive economics), than it is to say what the minimum income of a family of four "should" be (normative economics), or to decide who "should be" required to help to support the family when the father loses his job (normative economics). Yet, even in the face of uncertainties, many economists are willing to get involved in the many pressing normative economic issues of our day.

Realistic and effective involvement in the *normative* economic issues requires a basic understanding of the concepts (the "laws") of positive economics. That's why it's good that you are studying economics. When you get to the end of this book, you will have a good basic understanding of the concepts (the "laws") of positive economics. So you will be able to make better, more realistic normative recommendations.

Are you ready now to look at some real-world economic systems? As you know, no real-world economic system relies entirely on the market process to make the production and distribution choices—that is, no "pure-market economic system" exists in the real world. You also know that no "pure-tradition" or "pure-command" economic system exists, either.

What kinds of economic systems *really do exist* in the real world of the 1970's? That interesting question is what we're going to be talking about in the next chapter. As soon as you're sure you understand and will remember all we have been talking about in this chapter, then go ahead. You'll be ready to learn about "comparative economic systems."

## REVIEW EXERCISES

### MAJOR CONCEPTS, PRINCIPLES, TERMS (Try to write a paragraph explaining each of these.)

the three basic questions
the three ways of making the choices
economics of the "tradition-oriented" society
how the political process makes the choices
how "the market" answers the production question
how "the market" answers the distribution question

### OTHER CONCEPTS AND TERMS (Try to write a sentence or phrase explaining what each one means.)

the output question
the input question
the distribution question
the production question
distributive share
the social process
the political process
the market process
command: the direct approach
command: the indirect approach
productivity principle of distribution

### QUESTIONS (Try to write out answers, or jot down the highlights.)

1. What are some of the ways you are dissatisfied with the *production* choices in your society? What about the *distribution* choices? Discuss.
2. You know that there is no country in the world in which *all* of the economic choices are made by either the social process, or the political process, or the market process, but do you think there ever *could be*? Discuss.
3. What are some examples you have run into *today* (since you got out of bed this morning) of economic choice-making by tradition? by command? by the market?
4. Can you think of any choices that are now being made (in your family, at your college, in your society, in your *world*), where you think the "choice-making process" ought to be changed (from tradition, to command, to market, or vice versa)? You'll know a lot more about this issue when you finish the next chapter, but it would be a good idea for you to take a few minutes to think about it, right now.

# 4 THE ECONOMIC SYSTEM DIRECTS THE CHOICES

*Capitalism, socialism, and communism are different arrangements for making and carrying out the economic choices of the society.*

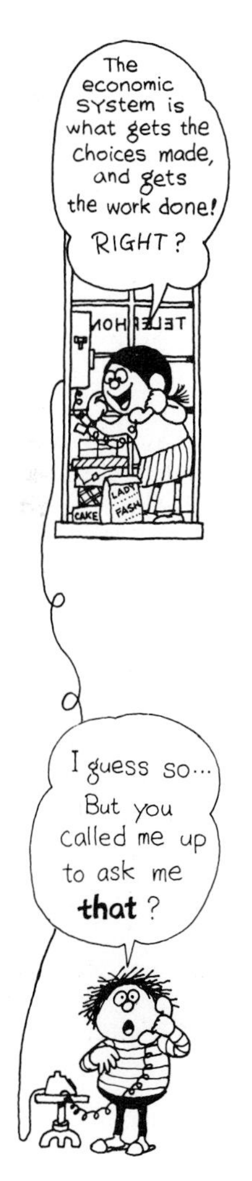


Each society must somehow get its economic problem solved. That's obvious. There must be some process, some mechanism, some systematic way to get the basic production and distribution questions answered. Every society has some kind of organization and some procedures for getting these questions answered. This organized set of procedures is called the economic system.

A society's economic system determines how the choices will be made. The economic system is nothing more or less than the normal procedure by which, or through which, the questions are answered and the economic decisions are carried out. The economic system gets the choices made, and gets the chosen things produced and distributed.

You know that there are three processes which are embodied in, and which make up, a society's economic system: the social process, the political process, and the market process. It's much easier to talk about these three processes than it is to talk about economic systems. The three processes are easy to understand. Each one has a precise meaning. But when we get into economic systems in the real world, nothing is very neat or definite anymore.

**Each Economic System Is Unique**

Each real-world economic system is unique. All three processes—social, political, and market—are at work in all systems, all the time. Each economic system is changing, evolving, all the time—sometimes slowly, sometimes dramatically, but constantly.

The only truly accurate way to refer to a nation's economic system is to refer to it as "that nation's economic system at that moment in time." For example, you could talk about "the American economic system of the mid-1970's" and compare it with "the American economic system of the mid-1960's." These two "comparative systems" are similar in many ways, but they're different in a number of ways, too. Or you might compare the American economic system of the mid-1970's with the system of the 1940's, or the 1920's, or the 1880's. Soon you will see the differences beginning to outnumber the similarities! What you would be observing, of course, is the evolution of the American economic system.

There is no modern nation in which the economic system has not undergone major changes, just during your lifetime. Rapid change keeps going on and on, in the American economic system and in all other countries. Yet we keep calling each of these changing, evolving systems by the same name. For example, we still refer to the American economic system as "capitalism." How can that be? It's just that each of the "system names" has a very broad (and not very precise) meaning. Don't get hung up on the names. How we "label" an economic system really doesn't tell you very much about it.

### The Three Major Systems: Capitalism, Socialism, Communism

We use the words "capitalism," "socialism," and "communism" to refer to the major forms of economic systems in the world today. You already know that an economic system labeled *capitalism* would allow individuals and businesses much freedom to make their own economic choices. The market process would play an important role, but certainly not the only role. You also probably know that in an economic system labeled *socialism*, the government plays an important role in making and carrying out the economic choices, and that in *communism*, the government makes almost all of the economic decisions. But all these terms leave broad latitude for different degrees of emphasis, from one country to another, and from one time to another.

We call the economy of the United States "capitalism." In reality, it's a mixed economy in which governmental decisions influence many of the choices. We refer to the economic systems of the Soviet Union, Cuba, Red China and others as "communism." But even in these countries the market process plays a role.

The economic systems of Western Europe and Great Britain and of many of the other countries in the world are called "socialist." But they could be called either "capitalism" or "socialism," depending on what you want to emphasize. Perhaps if you want to emphasize the fact that a system contains many "socialistic departures" from capitalism, you would call it "socialism." But if you wanted to emphasize the many "capitalistic departures" from a "pure socialist" system, probably you would call it "capitalism."

In the so-called "socialist" countries, some of the major industries are government owned and operated, but by far the greater number of businesses are privately owned and operated. These private firms operate in response to the market process. Even the government-owned enterprises respond to the market process—production is adjusted according to what the people are buying. Otherwise there would be surpluses of some things and shortages of other things.

All three of the major forms of economic systems—capitalism, socialism, communism—contain similarities. But it would be a serious error for you to assume that they are all about the same. There are some very important differences—like night and day—between some of these systems. Now, before you get completely confused, let's take a closer look at each system.

## CAPITALISM

Capitalism is a form of economic system in which most of the means of production—the factories, tools, equipment, coal mines, oil wells, railroads, and others—are owned by private individuals, not by the government. The owners are free to use their factories, mines, and all their other things in any way they wish. In general, people and businesses will seek to increase their profits. They will use their resources, energies, and tools in response to the market process. The workers and resource-owners who respond best—by making the things that society wants most—will enjoy the biggest incomes and profits. Those who do *not* do, or make, something to meet the wants of others, won't get anything.

### Free Private Enterprise

Capitalism has many names. Sometimes it is called the *free enterprise system.* This term emphasizes the fact that each person is free to go into whatever business enterprise he wishes. Another term for capitalism is the *laissez-faire system.* Laissez faire is a French term meaning that the government lets the people be free to make or to do whatever they want to make or to do with their economic assets—their resources, tools, and time and efforts. Capitalism is also called the *private enterprise system.* This term emphasizes the fact that it is private individuals and businesses—not the government—who own and control the society's factories and natural resources.

### Competition

Capitalism is sometimes referred to as the "competitive" system, or as the system of *competitive free enterprise.* This term emphasizes the fact that each person can produce anything he wants to, and sell it in any market he wants to, in competition with other sellers who also will be trying to sell things in that market. It also means any buyer can enter any market and compete against other buyers for the available goods.

If you see a service station that looks like it's making big profits, what do you do? You build a station on the opposite corner and compete, to try to get some of the profits. This is "free competition." Capitalism encourages competition. Competition is essential if the market process is going to work. Think about it for a minute and you probably will see why. Try to figure it out, but if you can't, don't worry. It will be explained later.

If we put together all the names describing it, capitalism is a system of "laissez faire-competitive-free-private enterprise." This is really a pretty good description of what capitalism is. It is a system in which the people are in control of the factories and tools and resources, and everyone is free to follow the influences of the market process. Those who do as society wishes will prosper and grow. Those who do not will not.

### Pure Capitalism Would Be Intolerable

Where would you look to find a pure system of "laissez faire-competitive-free-private enterprise capitalism"? In the economics books, of course! No such pure system ever existed in the real world. Some of the effects of such a system would be intolerable. For example, suppose a man borrows money—mortgages his house, car, and furniture—and goes into business, and loses it all. In "pure capitalism," all his little children are going to starve. That isn't too good, so the government takes some of the income and profits away from someone else, and gives some to "the loser" and his children. With "pure capitalism" this wouldn't happen. Government wouldn't bail out the losers.

With "pure capitalism" we wouldn't have any national, state, or local parks or recreation areas. There would be no public schools, no public libraries, no public health and welfare programs, no "public" anything! Unmodified capitalism would be too harsh to be socially acceptable. When capitalism is modified to make it more "socially acceptable," it moves toward socialism. Capitalism in the real world has undergone continual and very significant change. Throughout its past history it has been altered continually by the political process.

Real-world capitalism, in the American economy or wherever it exists, is always some mixture of capitalism and socialism. Really, it's mixed socio-capitalism. The mixture is constantly changing. You will hear more about that later. But first we need to talk about socialism.

## SOCIALISM

Just as we can talk about a system of "pure capitalism" and then admit that no such system exists, we also can talk about "pure socialism." The idea of socialism is that the economic choices are made and carried out through the political processes of the society. If the political processes are democratic, then you have "democratic socialism." If the political processes are autocratic or dictatorial, then you have "autocratic socialism."

### Utopian Socialism

If the society could work out a way of selflessly sharing both the work and the product, this would be "utopian socialism." This kind of "socialism" would be the same thing as "utopian communism." The words "socialism" and "communism" are derived from the same thought —socialism says "society-ism," and communism says "community-ism." But in real world economic systems, that isn't what these words mean.

### Real-World Socialism

When used to label a real-world economic system, "socialism" usually means that the government owns and controls some of the major industries and provides important "welfare programs" to meet some of the economic needs of the people. The government may own and operate such industries as electric power, coal, oil and gas, transportation, communications, perhaps steel and chemicals, and perhaps others.

The economic system of each "socialist" country is different from that of any other. One country will have more and different industries under government ownership than another; the welfare programs will be different from one country to another. These variations, from one system to another, are ignored when we try to put different societies, governments, or economic systems into categories—like "socialism," and "capitalism."

### Is The U.S. Economy Capitalism? Or Socialism?

How many of what kinds of choices must the government be making— what things must the government be doing—for us to call an economic system "socialism"? There is no rule to go by. Most people don't think of the U.S. economy as socialism. Yet you know that all three levels of government in the United States make many economic choices. All are engaged in many kinds of economic activities. The federal government dredges rivers, builds dams and bridges, produces and sells electric power and chemicals, manages many kinds of natural resources, operates recreation areas, and (since 1971) even operates a nationwide railroad passenger service (Amtrak). In addition it regulates and controls transportation, communications, electric power, natural gas and several other industries.

The state governments in the United States are in the insurance business, the liquor business, and various others; local governments are involved in transportation service, water and sewer service, electric power production, and various other kinds of enterprises. And that isn't all. The influence of government touches each business and individual in several ways and exerts some influence on most of their economic decisions. For example, no business in the U.S. economy can afford to make any important decision without first considering the effect of the decision on the taxes it will have to pay. Licenses, zoning regulations, building codes, pollution regulations, minimum wage laws, and many other government influences, mean that "free enterprise" in the United States is really not so free after all. Yet we still refer to the U.S. economy as a "free enterprise" economy—as an economic system of "capitalism."

### Capitalism And Socialism May Be Indistinguishable

If the U.S. economic system (with so much governmental influence and control) can be called capitalism, then what is socialism? The truth is that in most real-world cases it isn't very different from the American economic system of "modified capitalism." Take the British system, for example. A few more of the major industries and resources are owned and controlled by the government—coal, transportation, communication, for example;—and the British government is somewhat more involved in using resources for "social welfare" purposes—in medical care for example. But when you really look at what is happening and how, you realize that the British and U.S. economies are really very much alike. Yet we refer to the U.S. system as "capitalism," and to the British system as "socialism." Why?

In Britain, and in the other "socialist" countries of Western Europe, the political process does make more of the choices than in the United States. Not greatly more, but more. But there is another reason we don't call the U.S. economy "socialism." Many people in the United States have been taught that socialism is "bad" and capitalism is "good." The Democratic Party, the Republican Party, and the people in general are constantly working to bring about changes in our economic system—changes which, in Western Europe, would be called "socialist reforms." But you seldom hear anyone calling the changes "socialist reforms," not in the United States!

### Socialist Reforms

We are constantly making socialist reforms. Extensions of unemployment and social security benefits, medicaid and medicare programs, public welfare and public assistance programs, college fellowships with family support, some government subsidies for industry, minimum wage laws, and many other "government interventions in the economic process" are aimed at getting rid of some of the socially undesirable effects of raw capitalism. Such changes are really *socialist reforms*—reforms involving government action to influence the economic choices, aimed toward the objective of a more "socially acceptable" society. But the word "socialism" has been turned into such an emotional "scare word" in the United States that no matter how many "socialist reforms" are undertaken in this country it isn't likely that many people will call them by that name.

If you should happen to make a trip to Western Europe and find the "good old American spirit of capitalism" everywhere you go, don't be too surprised. The difference between the "capitalistic socialism" of the Western European countries and the "socialistic capitalism" of the United States really is not so easy to see! There are some differences, but the systems are really much more alike than different. In both instances the demand of the people, working through the market process, is a very important force in influencing the production and distribution choices for the society; in both instances the political process of "governmental command" also plays a very important role in making and carrying out the economic choices. What I'm saying is this: the U.S. economy and all these other economies are really systems of "mixed socio-capitalism." It probably would eliminate a lot of confusion if we would start calling all of them by that name.

## ECONOMIC SYSTEM? OR POLITICAL SYSTEM?

One cause of great confusion about economic systems is the tendency of many people to confuse the political (governmental) system with the economic system. The two are always related, but they're essentially different things. It will be of great help to you in understanding different economic systems if you will carefully separate in your mind the economic system from the political (governmental) system.

Democracy and dictatorship are forms of government; capitalism and socialism are forms of economic systems. Either capitalism or socialism could exist under either a democratic or dictatorial government. There is sometimes a tendency to tie democracy with the concept of "laissez faire-competitive-free-private-enterprise capitalism" and to equate socialism with dictatorship. To see how absurd this is, let's look at some examples.

There is no question that the British government is democratic. In some ways the British government is more directly responsive to the people than is the government of the United States. Yet we call the British economic system "socialism" and the American system "capitalism." The word "socialism," by itself, really doesn't say anything about the form or processes of the government. It only says that the government admits the responsibility of playing an important role in deciding and carrying out the production and distribution choices for the society.

Take another example: In the United States today there is much more government control over the economic choices than there was 100 years ago. Yet it should be difficult to produce evidence that the government has become any less *democratic*. All the evidence points the other way. Even people without property, blacks, women, and eighteen-year-olds now can vote!

### Could Capitalism Exist Under Dictatorship?

Could an economic system of capitalism—even modified capitalism—operate under a dictatorship? Yes, it could. But unless there was a wise and just dictator it is not likely that private ownership of the major industries would long survive. The dictator and his supporters usually want to get and hold as much power over the country as they can. One way of holding and exercising power is by controlling the productive resources of the nation.

One of the frequent purposes of revolution is to get control of the productive assets of the nation, and then to use these assets for the purposes and objectives of the dictator and his supporters. To say it differently, usually one of the things the dictatorial government wants *most* to do is to hold and exercise the power to make the economic choices—the production and distribution choices—for the society. If capitalism exists, that means that the *market process* will be in control of most of these choices. Most dictators and their friends aren't the type to put up with that sort of thing for very long.

### Most Revolutionaries Want To Change The Economic System

A major reason for the unrest, wars, and revolutions of the past quarter century has been the desire of one group or another to gain and exercise control over the society's productive resources. Each group wants the power to have its own way in making the production and distribution choices. Certainly this was the most important objective of the Castro revolution in Cuba. It has been a major factor in the unrest and instability in the Latin American countries. The communist revolutions in both the Soviet Union and Red China were inspired by the desire of the revolutionaries to gain and exercise control over the land, factories, and other economic resources of the nation.

The general problem of political instability in Southeast Asia (which generated the Vietnam War) is very largely an issue of economics. It is a question of who (what individuals or groups) will have the economic power—who will get to control the use of the productive assets and economic resources of the nation. The revolutionaries are driven by dissatisfaction with the existing arrangement. They want to force the economic assets and the "choice-making power" to change hands—from someone else's, to theirs.

The aim of the Communist revolutionaries is to change the economic setup. The objective of the communists is to get the factories, farms, land, and all the other things we call "private property" away from the present owners—individuals, families, businesses, other organizations. They want to take over the government, and then "transfer" the ownership and control of virtually everything (and everybody) to the government. Sounds like the communists believe in a lot of socialism, doesn't it? You're right. Extreme socialism, without democracy. Let's talk about communism.

## COMMUNISM

Communism is a difficult term to understand. For one thing, it is a political term as well as an economic term. It describes, in one breath, both a political system and an economic system. Another problem is that there are so many different meanings of the term.

### The Communist Party

First, "Communist" is the name of a political party. The Communist party controls the government and runs things in the Soviet Union, in the East European "satellite" countries, in Cuba, in Red China, in North Vietnam, and in North Korea. Communist political parties also exist in the other countries of the world, but usually they are a small minority and have not been able to get control. The objective of each Communist party is to get and then exercise control over the economy—to make the economic choices for the society.

### The Communist Philosophy And Predictions

A second meaning of "communism" is "the philosophy and predictions of the communist philosophers"—of Marx, Engels, Lenin, Mao-Tse-tung, and the other philosophers of the ideas of communism. This meaning of the word refers to the belief, the conviction, that the communist doctrine is true and that the predictions are really going to happen. So one definition of "a communist" would be "one who believes in communism"—one who believes in the doctrine of communism. The real communist believer takes these philosophies as a kind of religion, to be lived by and preached to others.

The communist philosophy generally holds that the market process really isn't going to work as it is supposed to. A few monopolists will gain control and enrich themselves, while everyone else suffers. With such a bad system, everybody but the few "fat cats" will be miserable. Soon the miserable masses will revolt, kill off all the monopolists, and take control. In this way, capitalism will be destroyed. Thereafter the government will be in the hands of the Communist party leaders, who will be in control of all the productive resources—of everything in the nation.

Sometime later there will evolve "an economy of abundance" in which there will be no economic problem of material goods for anyone. A beautiful world will evolve in which people will no longer be selfish; each person will really have all he wants. People will continue to produce for society because of their love for society. The government will "wither away" because in such a society of mutual love and trust and respect for each other, where there are no unfulfilled desires or wishes, who needs a government? The idea of the coming of this utopian, ideal world is where the word "communism" comes from. The people would live as one big happy community.

You can see how the philosophy of communism could have an almost religious appeal to those who are willing to believe. It describes a "heaven-type" situation, and then offers it here on earth to those who will believe and follow. But you get into "communist heaven," not by being peaceful and loving thy neighbor, but by being destructive and maybe killing him. The most fanatical communists believe that all the predictions are really going to come true. They feel duty-bound to help the process along by doing everything they can to weaken or destroy capitalism, or to cause disruption and revolution in any country in which anything that looks like capitalism, exists.

### Utopian Communism

A third meaning of "communism" is the idea of living in a "communal society" with a group of other people. The clan in some traditional societies, and even the family in modern societies, operates from the economic point of view, as a sort of "commune." The idea of the commune is that each person works, not for himself, but for the group. Each plays his role in helping the communal unit to prosper. Each gets to share in the prosperity of the group. Each does his bit because he wants to help the group; each gets his share because the others want him to have what he needs. What each gets is not related to how much he produces.

This kind of communism is often called "utopian communism." It means the same thing as "utopian socialism." It has never existed except on a very small scale, as tried by various groups of a few people. In your studies of real-world economic systems you can forget about utopian communism. It doesn't exist.

The collective farms set up by the communists in the Soviet Union and Red China are sometimes called "communes." But in fact these collective farms more closely resemble forced labor camps. The people live where and produce what the government says. Then the government takes much of the output.

### The Communist Government

A fourth meaning of "communism" is the form of government which is set up when the Communist party gains control. The form differs in some ways from one communist country to another, but in all instances the governmental power is centralized in the hands of one or a very few people—those who are in control of the party. The government has complete power over everything, including the lives and activities of the individuals.

The government is not responsive to the immediate desires of the people—that is, it is not democratic. The only way the great mass of the people could ever change their leaders or force a change in policy would be to revolt and overthrow the government. This is the political, or governmental, meaning of the word communism. It is a form of dictatorship.

### The Communist Economic System

Fifth and finally, we come to the economic meaning of communism. It is the form of economic system which is set up when the Communist party takes over. You already know who is going to make the economic choices. The government, of course! The economic system of communism is really an extreme form of socialism. Not just a few major industries, but *all* industries and farms are government-owned and operated.

The natural forces of the market are reduced to the point where they hardly exert any influence on the economic choices. The government owns and directs virtually all of the resources. All the production and distribution decisions are made by the government. Since the Communist party is in control of the government, the party controls the economic choices. Since the party is controlled by a few people, these few people are really the ones who make the choices for the society.

The kinds and quantities of outputs and inputs of the farms and factories are all determined by the government's economic plan. Industries operate as the government directs. Whether or not the economy will produce more heavy machinery and less consumer goods is not determined by private investors. There aren't any private investors! The government decides.

The number of economic decisions which individuals control is very small. Each person is free to spend his small income among the few kinds of consumer goods which the "economic plan" calls for. In general, each person's income is determined by the work he does and how productive he is. But it is the important people in the party (in the government) who get the largest shares.

In the communist system, all the society's inputs—workers and resources and everything—move according to the government's economic plan. But it isn't only the communist systems which use economic plans. All economic systems include some *economic planning* by the government. Before we talk any more about the communist system, we need to talk about economic planning.

## ECONOMIC PLANNING

In any kind of economic system, whenever the political process makes the economic choices, it is necessary for the government to do some kind of "economic planning." Economic planning means about what you would think. It means deciding how to use (how to economize and optimize) the productive resources of the society. But it also includes the idea of implementing, or carrying out the plans.

### Designing The Plan

The Soviet Union and most of the other communist countries in the world today operate on a "five-year plan." Of course there are shorter range and longer-range plans, and sub plans, and all that. But the idea of the five-year plan is that a set of objectives is laid out in detail, to be achieved during the coming five years. Then all the resource uses are aimed toward fulfilling the objectives of the plan.

For example, the plan would include (as objectives) the completion of certain additions to the total highway network, construction of new buildings, houses, and all that; opening up new mines for coal, iron ore, and other things; new railroads which might be constructed; added capacity in the production of steel, chemicals, and other industrial products; total amounts of consumer goods, in each category (clothing, food, etc.) for each year (each month, week, and day, really) within the planning period. The plan would include construction and operation of schools and hospitals, numbers of people to be educated in various ways, numbers of motor vehicles and kinds of vehicles to be produced; amounts of output to be produced for export in order to get the dollars, francs, marks, and other currencies necessary to buy the things wanted from other countries; numbers of people to be moved from one industrial center or activity to another, and on, and on, and on.

The five-year plan attempts to figure out just exactly how much of each of the various kinds of resources and products the economy will be capable of producing, and then plans exactly how each of these should be used—for consumption, or for further production; and if for further production, for the further production of which things. Can you see that this is like a big "production possibility," or "transformation" curve operation? Only instead of having only two choices, the choices are almost infinite; and instead of knowing for sure exactly how much we can produce with the available resources, these things can only be estimated. How are all these things decided? Through the political process, of course! The government experts (economists) work with the political leaders to get the plans all worked out.

## All Economic Systems Use Some Planning

Do you get the idea of how economic planning works? Of course you know that all economic systems use some economic planning. Without economic planning, do you suppose there would be any schools or colleges in your state? or a highway system? or any policemen or firemen in your town? Or do you suppose there would be a United States flag on the moon? or an intracoastal waterway running from Maine to Mexico? or a Saint Lawrence Seaway? or national parks and recreation areas? None of these would exist without economic planning. There wouldn't be an Army or a Navy, either, or an Air Force or a Marine Corps.

The amount of "economic planning" going on in a society indicates the extent to which the political process is making the economic choices. If the government makes most of the choices, then there is a lot of economic planning. The government decides about things more than the people, as buyers, do. But when the choices are left up to the market process, economic planning isn't necessary. Resources and products and workers automatically flow to where they are demanded.

## Economic Planning Reduces The Individual's Choices

Economic planning can make a lot of difference in the lives of the people. It matters whether there is a little or a lot of planning, and it matters how the plans are made. How much democracy is used in designing the plan? Who gets to have an influence on the objectives? Who decides which resources will be used in which way? If there's a lot of planning, everyone's life is patterned by the plan, so these questions become very important.

If there is very little economic planning, then the market process will be in control of most of the choices. People will respond to the market. The ones who are able to get a lot of money will have a lot of influence. Each person can try to get more money so he can have more influence, and have more things.

In an economy controlled by economic planning, individuals have less "freedom of choice." If the government of the United States decides to use more resources for medicare (and to tax you to pay for it), then the society is *not* going to be able to produce that sporty little car you had wanted to buy, but which, because of the high taxes, you aren't going to be able to pay for, now. In the Soviet Union, if you wanted some new gloves and a warmer coat but the government decided to put more of its resources into the production of a new steel mill, then the new gloves and warmer coat are not going to be available to you, no matter how much you would like to have them.

See how economic planning thwarts the individual's desires for things? This is not to say that economic planning is always undesirable. Of course not! No intelligent person would suggest that "maximum individual freedom of choice, in all things" is the paramount objective! But the farther we go with economic planning, the more we cut down on the individual's freedom of choice. That's a fact, and we should recognize it.

### Implementing The Plan

It makes a lot of difference how the plan is carried out. There are several different ways to induce the resources to move to where and to do what the plan requires. But essentially, there are only two kinds of approaches: (1) the resources can be allocated directly—that is, "ordered," or "directed" to go where and do what the plan requires; or (2) the resources can be "enticed" to do as the plan requires, by offering rewards (money).

The first approach can be called "allocation by direct order," or simply "direct allocation." The second might be called "allocation by incentive" and it usually means using the market process—that is, offering to pay people more to do the things the plan calls for.

### Using The Market Mechanism

In an economic system which relies mostly on the market process, it's just natural for the government's economic plans to be carried out through the "market mechanism." If the government wants a canal dug, a highway built, or someone to teach in the local school, what does it do? It doesn't *order people* to do these things. Instead, the government *offers to pay* to get the job done. In the case of the highway, the government can pay a contractor. Then the contractor will hire the labor and buy all the things he needs, then build the highway. Or the government might hire the people, buy the rocks and gravel and steel and cement and graders and all the other things, and build the highway itself. In each of these cases you can see that the market mechanism is being used to carry out the plan. People are automatically responding to the "price and profit incentives" offered by the government.

Most of the economic planning decisions in the United States, and in the other modern non-communist countries, are carried out by the market mechanism. This means that if you look at the total amounts of money being spent by the national, state, and various local (county, township, city, village) governments, you will get some idea of the total amount of resource use which is being directed by the political process—by governmental "economic plans."

What does it look like in the United States? Would you believe that during the average day, week, or month, total government spending amounts to about one-third of the value of all goods and services being produced? The total value of everything produced in the country is running well over a trillion dollars per year. Total government spending amounts to about one-third of this total.

This "government spending" figure is not a precisely accurate indication of the importance of the political process in making the economic choices in the United States. Why not? Well, because some of the spending goes to individuals who are unemployed, or on welfare, or retired, and they take the money and then go out and buy the things they want. Also, some of the government spending is not really for what we would call "final goods." But don't worry about these details. Just remember that this "one-third" figure is not a "precise measure" of anything. It just gives us a fairly good indication of how much of the society's resources are being directed by government budgets—by governmental "economic plans."

### Direct Allocation

There's another reason why you shouldn't rely too much on the "government spending figures" to show the importance of "political process choices" in a country like the United States. Many "political process choices" or "command decisions" are carried out by direct allocation. Many of the "direct allocations" consist of limitations on the private uses of resources. For example, the uses of almost all of the land in the urban areas of the United States (and in many rural areas too) are restricted by some kind of "zoning," or "land use restrictions." This means the local government tells you what you can do (and what you cannot do) with "your" (the society's) land.

Government restrictions go much further than just limiting land uses. Governments limit the kinds of activities you can undertake in various places, and the kinds of resources you can use for what purposes. For some things, even the prices you can charge (or that you must pay) are set by government. Such regulations and restrictions are influencing the economic decisions—the resource-use choices—of *all* the people, all the time.

As population expands and as technology develops, we all become more interdependent. Everything anyone does, affects everyone (sort of). People keep using up and messing up more and more of the society's natural resources and natural environment. How can the society protect its natural and environmental resources? By restricting our individual freedom. As time goes on you will see the political process adding more and more controls and restrictions.

In the United States and in most other nations, the government doesn't usually require a company to expand its plant to a given size, or to relocate in some other area, or to use one kind of production process or another. But some of this kind of direct government regulation *is* done, even in the United States. It is done all the time in the governmentally regulated industries—that is, in the public utilities (such as the electric, gas, and phone companies).

As more and more people live closer and closer together, as the actions of each person have an increasing influence on the conditions and lives of everyone else, the government can be expected to come in and play a larger role. That's the way it has been happening. It seems likely to keep on moving in that direction. Many people will lose some of their individual freedom. Our "private property rights" will be limited more and more.

Maybe lots of people will be unhappy about the drift of things. But the real question is this; all things considered, what are the alternatives? Given the hard facts (economic and political) of the real world, expanding the role of the political process (that is, more economic planning) may be the only realistic approach available to us.

## Economic Planning In The Soviet Union

In the Soviet Union and in the other communist countries, economic planning gets into every aspect of life. The plan includes not only the choices as to which final products are to be produced. It also includes the choices of which resources and which techniques will be used. It determines the way industry and agriculture will be organized, the kinds of machinery to be used, and all the other production details.

The plan includes decisions about consumption levels in general and about how much of which kinds of consumer goods will be produced. All these decisions are made by the government, and included in the plan. But this is not to say that how much each person will get—that is, each person's income—is predetermined by the plan. In the Soviet Union and in the other communist countries, the productivity principle of distribution is used just as it is in the United States and in the other countries of "mixed socio-capitalism."

In recent years the Soviet Union has been expanding the production of consumer goods. The people are being permitted to have a few more things. The rapid growth which the Soviet economy has experienced under the consumption restrictions (forced savings) and heavy investments of the five-year plans, is now paying off (in greater output). But don't make the mistake of thinking that the Soviet Union and the other communist countries are permitting their people to enjoy the high levels of living which most of the people in the other developed countries are able to enjoy. Some day that may come about. But it has yet a long way to go.

## ECONOMIC PLANNING AND THE MARKET PROCESS IN THE WORLD TODAY

It might make a lot of sense for everybody to stop talking about "capitalism, socialism, and communism" when we are trying to understand and compare real-world economic systems. Perhaps we really ought to talk about the extent to which *the market process* influences the resource choices as compared with the extent to which economic planning is in charge of things—and which kinds of things. Then, for the "economic planning" kinds of choices, we might ask questions about how the plans are made, and *what techniques* are used for carrying out the plans.

If we can get reasonably good answers to these questions, we can get a reasonably good look at what the economic system is like—what it's doing, how it functions. If we look at the world today, using this approach, what do we see? First, we see that in *every* nation, many of the major resource-use decisions are made by the political process—are made by government plan. In most of the "mixed economies" these decisions are made through some kind of (more or less) "democratic" process and are carried out through the "market mechanism"—by using wage, price, and profit incentives to get the job done.

Among these "mixed economies of socio-capitalism," there are several differences. But really, these countries are much more alike than different. The market process plays a major role in all of them. No *one* of them goes very much farther with economic planning than the others. If you take a close look at the economies of such different countries as Britain, France, Japan, Sweden, Italy, West Germany, Mexico, Australia, Thailand, Canada, Brazil, Argentina, Ethiopia, the United States (or almost any other noncommunist country), in all of them you will find a basic reliance on the market process for making most of the economic choices. But always the market process is forced to operate within various restrictions established by the government (and, in many cases, by tradition).

## The Communist Systems Really Are Different

The economies of the Soviet Union, Red China, and the other communist countries are different from the "mixed economies" of the world, on all three counts:

1. The communist countries do *not* rely on the market process to make any of the significant economic choices. That is, the choices are made by the political process—by economic planning.
2. In the communist countries, the economic choices (the plans) are *not* designed in response to the wishes of the people. That is, the economic planning process is not in any sense "democratic." The few who are in power decide on the objectives, then make the economic choices and design the plans.
3. The plans are not usually carried out through a system of "market-process" incentives and rewards. Profit incentives are not usually used to get the resources to move to where they are needed. Instead, resources are usually "directed," or "administered" (although this has been changing some, in recent years).

I'm sure you have no trouble understanding items (1) and (2)—that the communist countries don't rely on the market process to make their economic choices, and that their economic planning process is not democratic. But the third item—that resources are "administered"—may not be quite clear. It's important that you understand this concept. Let's talk about it and use some examples.

## How Resources Are Administered

You already have had a lot of experience with resource administration: your college administrators decide how each classroom will be used at each hour of the day; who will live in which dorm; how much land space will be used for athletic fields and how much for parking lots, and who will get to play in which fields and who will get to park in which lots; how much floor space will be used for the library, how much for study areas, for student lounges and for faculty lounges; which professors will teach which courses with how many students in which rooms at which times; which students will get to use which of the college's resources, to take which courses . . .—and on and on the list could go. These are examples of resource administration.

At your college, all the resource-use choices may be made through very democratic processes. The students and faculty may be the ones who have the ultimate say on all these issues. If so, then everything may be going just fine. But, likely as not, many of these decisions will be made undemocratically—will be imposed by someone else who "knows what's best for the good of your college and is going to give it to you whether you like it or not."

Sometimes your own college administration may be the "dictator." Or it may be your state legislature. Or maybe both. But whoever does it, whenever "autocratic resource-use planning and administration" happens, unless it's done with great wisdom—with great awareness of and sensitivity to the people and issues involved—then problems will arise. Things won't go smoothly. Morale will be low.

The power to order people around is an essential part of any system of "resource administration." If the administrators are wise and just, and if they use effective leadership techniques and all that, it can work out just fine. But unfortunately, there just don't seem to be enough wise, just, and sensitive administrators to go around. So it isn't unusual for problems to arise.

### How "Market-Mechanism-Type" Incentives Could Be Used

Do you begin to get a "feel" for the idea of what "resource administration" is? When your dad hands you a bucket of paint and tells you to paint the bathroom, or when your mom hands you a broom and tells you to sweep the steps—and if you feel that you *must* do it (or else face unpleasant consequences)—that's resource administration. How might "market-mechanism-type" incentives be used instead? Your dad might offer to pay to get the bathroom painted. He could keep raising the "price" until you (or someone) will gladly grab the brush and start slopping paint all over. Or your mom could pay to get the steps swept. Obviously.

But what about your college? You know that your college is sort of a little "administered economic system." How could it use "market-mechanism-type" incentives, instead of "direct allocation"? instead of resource administration? For many things, it couldn't. For some things, it does. Raises and promotions go to the professors who do a good job (and/or to the ones who please the administrators). Sometimes parking fees for the distant lots are low, and for the close-in ones, high. But these examples don't really get into the basic resource-use choices for the college. Let's take one that does.

### A "Price System" For Your College's "Planned Economy"

Our colleges need more facilities—more and better buildings, classrooms, listening booths, study areas, and many other facilities. But did you ever stop to think of the amount of wasted, excess capacity that exists on the college campus at three a.m.? Wow! Why don't our resource administrators start scheduling classes all night long? Because the faculty and students would revolt, that's why! And because administrators don't like to work nights, either.

Suppose someone began to think about the millions of dollars of savings which might result from running the college all night. Is there any way that people might be *induced* to go to college on the "midnite shift"? Some of the savings might be offered as "bonuses" to the professors and as "scholarships" to the students. Do you suppose that could work? You bet it could! All it would require is that administrators set the right "price incentives"—the right "bonus" and "scholarship" payments.

### Price Adjustments To "Fine-Tune" The Resource Flows

If too many professors and students volunteered for the night shift, the bonuses and scholarships could be reduced. If not enough volunteered, the payments could be raised. Maybe it really ought to be tried! It might work!

The administrators could assign "penalty charges" to professors and students who chose the most desirable class times (like 10:00 a.m., M. W. F.), and offer "bonus payments" to those who chose the less desirable times. Eight a.m. and four p.m. classes might carry a two percent bonus; 9:00 a.m., noon, and 3:00 p.m. classes, no bonus and no penalty; 10:00 and 11:00 a.m. and 1:00 and 2:00 p.m. classes might carry a five percent penalty charge. Then evening and night classes would all carry bonuses of various sizes.

The bonuses and penalties would have to be adjusted to get just the right number of professors and students to volunteer at each time. For some hours—like midnight on Saturday night, or 8:00 a.m. Sunday—perhaps the bonus payments would need to be so very high that those hours would have to be left idle. Do you think a system like this could be made to work on your campus? How about all over the country? It's interesting to think about. The savings might run into several millions! But the purpose here is not to recommend it. It's just to show how "market-type" incentives might be used.

Once we get going on this "market-mechanism-type" incentive program, we might just as well keep going. We could build in bonus payments and penalty charges for professors on the basis of their productivity—how many students they deal with, how much they publish, or whatever measure of productivity is decided on. Each student could make "large-class or small-class" choices, and "high-paid prof or low-paid prof" choices and he could get rebates or pay extra, according to what choices he made.

See how it might work? There would be some problems, of course. This approach might not meet our ideas of "equity and justice," but it sure would take care of the problem of your being ordered around by your college administrators! Many professors and students and administrators would perform quite differently under such a system.

Now you know how "market-mechanism-type" incentives ("the price mechanism") might be used to carry out economic plans. The payments ("prices") are adjusted to get the right people to go to the right places and do the right things. This makes it unnecessary to order people around, so *policing* (watching to see that everyone follows orders) becomes unnecessary. The "price mechanism" is very efficient in getting the resources to do what the plan calls for. Everything just moves automatically!

### Do Communist Countries Use The Price Mechanism?

Now that you see what an efficient "resource director" the "price mechanism" can be, you may wonder why the communist countries use resource administration (direct allocation) instead. One reason is philosophical. They associate wage and profit incentives with their philosophical adversary, capitalism. Anything that looks like capitalism is bad. Anyone who suggests using profit incentives to get resources to move around, soon may find that he is going on a long trip—and not coming back. A second reason is that the communist planners haven't been fully aware of the advantages. Understanding the price mechanism and how it works isn't exactly the communist party leader's cup of tea!

But now that all this has been said, let's go on and admit that the communist countries *are* using market-mechanism-type incentives (the price mechanism) more and more. The communist countries are beginning to appreciate the great efficiency of the market mechanism as a tool for carrying out economic plans. They're finding out that an efficient way to get a hole dug is to offer money to anyone who will dig it—and let the digger make a "profit" on the job if he can; a good way to get a factory or a hotel or a restaurant of anything else operated *efficiently* is to let the manager work to try to make a profit.

## ARE ECONOMIC SYSTEMS BECOMING MORE ALIKE?

All economic systems are constantly changing, responding to new circumstances, reflecting changing ideas. It's happening in the United States, Canada, Western Europe, Japan, the Soviet Union, the Soviet satellite countries, Red China, and everywhere else.

There have been several significant changes in the U.S. economic system since you were born. There have been changes in the government's role in directing and developing and conserving human and natural resources. There are many new programs for economic welfare, and many new governmental influences on people's incomes, and on prices and wages and profits. The U.S. economic system of the 1970's really is significantly different from the system of the 1950's. It's changing so fast it's hard to keep up with it!

What about the communist countries? The Soviet Union, for example? After the Bolshevik (communist) revolution in 1917 the Soviet system sort of fell apart. The communist leaders imposed rigid controls, but they didn't know how to get the system put together again. So in 1921, Lenin announced his "New Economic Policy"—a "temporary step backward toward capitalism"—and urged the people to produce and try to make profits. The purpose was to overcome the complete economic collapse following the communist take-over. Remember in the book (or the movie) *Dr. Zhivago,* how tough things were?

Beginning in 1928 with the first five-year plan, tough and rigid economic controls were imposed again. That was less than half a century ago. Today the Soviet economy is still rigidly controlled. No doubt about that! But it's considerably different. Consumers get to have more things. People are ordered around less, and wage and price incentives are used more. The economic plan now responds more to the people's demands for consumer goods.

### Communist Countries Are Beginning To Use Profit Incentives

The Soviet Union and its satellite countries are discovering that enterprises which cannot be operated efficiently by government resource-administrators, can sometimes be operated very efficiently by a private individual seeking profit. What's the difference? The profit seeker has a very strong incentive to economize and optimize! The November 15, 1971 issue of *U.S. News and World Report* carries the article: "Why Reds Are Turning to Capitalism."

This article describes several instances in which serious inefficiencies had existed in hotels and other service establishments. Then individual managers were allowed to take over the establishments, run them as their own businesses, and try to make profits. In most cases the improvements in service and the reductions in cost were dramatic!

### The Influence Of Pragmatism

Are the "Reds turning to capitalism" as the title of the *U.S. News* article says? No, not really. But there seems to be a movement away from the philosophy that "anything that looks like capitalism must be despised and shunned." The communist countries aren't becoming "capitalist." They are simply indicating their ability to use an approach which has been a hallmark of American life—in economics, in politics, in everything. What approach? Pragmatism.

The pragmatic approach aims for results. Try something. If it works, that's good. Do it some more. If it doesn't work, stop doing it. Try something else. Pragmatism recognizes only *effectiveness*.

Pragmatism has been leading the Soviet Union and other communist countries to use the market mechanism. Why? For philosophical reasons? Of course not! Why, then? Because for some things, it's effective. It's efficient. It works too well to be ignored.

What about the United States, with its strong laissez-faire, anti-government philosophy? We have been hastily introducing economic planning techniques which the communist (and other) countries have been developing. Why? For philosophical reasons? Are we swinging over to communism? Of course not! Then why are we using more and more economic planning? Because *it works*, that's why!

So can we say that the economic systems in the real world are becoming more alike? It seems sort of logical to assume that they would. If one system starts out with a very extreme position—using all dictatorial economic planning and direct-order resource administration—and then begins to change, which way *can* it change? Or if a system starts out "laissez faire and pure market process all the way," which way *can* it change? If the two systems are as far apart as they can get, then as they change we might expect them to move closer together—to become more alike. That seems to be what's happening.

What's the ultimate result of all these "economic system changes" going to be? Will all the nations eventually wind up with the same kind of economic system? A system with some blend of the market process and government planning, perhaps as modified by local customs and traditions? I don't know. Nobody knows. It's an interesting idea, though.

Changing economic and political systems is a fact of life. People always seem to like to steal each other's secrets. Over the next several years I think it's a safe bet that the United States and the other economies of "mixed socio-capitalism" will be doing more and more economic planning, while the rigidly planned (communist) economies will be using the market mechanism more and more. How far will it all go? Your guess is as good as mine.

No society has ever devised any "completely satisfactory" way of making the economic choices. No system of choices quite reflects the wishes of *all* the people. It isn't likely that one ever will. But the market process (with some modifications to temper its harshness) has a lot going for it.

In the final chapter of this book—after you know a lot more economics than you know now—you'll see this issue again. There, you'll find out more about the continuing evolution of the world's economic systems, and about the increasing role of government planning in the mixed economies of socio-capitalism. But first, you need to know more economics. Especially, you need to know more about the market process—about what it is and how it works. We'll be going farther into that, in the next chapter.

Before you go on, take some time to think about and try to remember all the new things you've been reading about in this chapter. Maybe the review exercises will help you to do that.

---

## REVIEW EXERCISES

**MAJOR CONCEPTS, PRINCIPLES, TERMS (Try to write a paragraph explaining each of these.)**

the economic system
the meanings of capitalism
the meanings of socialism
the meanings of communism
mixed socio-capitalism
economic planning

**OTHER CONCEPTS AND TERMS (Try to write a sentence or phrase explaining what each one means.)**

free enterprise system
laissez-faire system
private enterprise system
competitive system
utopian socialism
socialist reforms
direct allocation
allocation by incentive
resource administration
price mechanism
pragmatism

**QUESTIONS (Try to write out answers, or jot down the highlights.)**

1. Think about an economic system of "pure capitalism." What are some of the ways in which such a system would be intolerable, from *your* point of view?
2. Think of some things which are now being "administered" at your college, but which might be handled by the "price mechanism" instead. Would you like to see any of these things handled by some kind of "price, incentive, reward system"? Discuss.
3. Suppose you were to rank different real-world economic systems on a scale (from 1 to 10) to indicate how much the "political process," and how much the "market process" was responsible for the choices in that economic system. (Let 1 = pure *political* process, and 10 = pure *market* process.) What number do you think you would assign to the present economic system of the United States? Great Britain? Japan? West Germany? East Germany? the USSR? What number would you assign to the economy of each country, 20 years ago? What do you think the number for each country will be, 20 years from now? Think about it.

---

# 5 SPECIALIZATION GIVES US MORE TO CHOOSE FROM

*The market process fosters specialization and trade, and everyone gets to have more cake and eat more too.*

Suppose a society's economic system is operating *entirely* on the "pure market process." (None ever did, but just suppose.) Why do people produce? Because they would get pretty hungry if they didn't. They produce because they want to consume.

Some people may be producing their own food and clothes and shelter and all, but most people will be producing just one thing (or maybe a few things). Why? Because it's more efficient to specialize in producing something you're good at and then trade to get the other things you want. It's "the market" that lets you do that. It lets you specialize, and trade.

If you couldn't trade, then when you wanted something you would have to produce it for yourself. What a tough life that would be! What the market lets you do, is this: It lets you work for the things you want, not *directly* (by making the things yourself) but *indirectly*. The market lets you make something you're really *good* at making, and then *trade* to get the things you want.

All this may seem pretty obvious to you right now, but after awhile you're going to see that there's a lot more to it than you might think. I'm not going to go lurching ahead and leave you confused, and that's why we're taking each step nice and easy. Our next "nice and easy" step is to show how the market process might evolve. When you finish this step you'll see how all this "specializing and trading" helps the people to live much better. You'll really understand it.

## HOW THE MARKET PROCESS MIGHT EVOLVE

Let's begin with a "Robinson Crusoe-type world." Suppose there is only one person living on a big tropical island in the middle of the Pacific Ocean. Obviously the one person will make all the economic decisions for himself. If he is hungry he will try to "produce" (find, catch, or gather) something to eat. If he wants protection from the sun and rain he will try to "produce" (find, or make) something to use for clothing, and shelter. Whatever he is successful in producing, that is what he gets to have as food, clothing, and shelter. If he doesn't produce very much he doesn't get very much.

### A "Robinson Crusoe" Makes His Own Choices

When a person is all alone, his production and distribution choices are automatically decided. He solves his production problem by (1) deciding what he wants most, and then (2) producing it in the most efficient way he can. The distribution question is so automatically solved that it's a little ridiculous even to mention it. Obviously he gets a "100 percent share"—all the output goes to him. Whatever he produces, that's what he gets. There is no question of "who gets what share." There's no one to share with. (I'll bet he wishes there were.)

It's all so simple and easy to see, when there's only one person. But when we introduce more people, the picture gets a little more complex. Still, it operates essentially in the same way. Let's introduce more people and see what happens.

Suppose that now there are four large families (clans, maybe) living on the island. Each family is located some distance from the others. One family lives on the east side of the island and one on the west side; one on the north side and one on the south side. Each family is entirely self-sufficient. Each produces only for itself—there is no trading between one family and another.

### With More People The Choices Are More Complex

Even within each island-family the production and distribution choices are somewhat more complex than for one person alone. In one family, father and the boys may be responsible for growing and gathering vegetables and fruits while mother and the girls catch fish from the lagoon. Father may get the biggest and best fish and fruits, mother second best, and the children, whatever is left. In another family, father and the boys may fish while mother and the girls grow and gather vegetables and fruits, and all share equally at suppertime. The third family may have the women doing all the work and the best of everything going to the father and the oldest son. All these different kinds of arrangements (and many others) have actually existed in various societies at various times and places throughout the world. What we're talking about, of course, is "the production and distribution choices as decided by tradition."

Now let's suppose that the family living on the east side of the island finds fishing to be very good. But the family on the west side has very poor fishing. Ah, but the westside family has very fertile soil. Coconuts, breadfruit, and other tropical vegetables and fruits are plentiful. You know what's going to happen. Right?

The westside family will become the farmers, the eastside family will become the fishermen, and the two families will trade. How do you know? It's just obvious. Nobody is going to keep trying to catch fish where there aren't any fish! No one is going to keep trying to gather breadfruit where it doesn't grow! So the eastside islanders will catch fish and trade for breadfruit; the westside islanders will gather breadfruit and trade for fish. Everyone will benefit. Soon you will see exactly how this "specialization and trade" gets started. But first, maybe you should study the map of our little island. You'll find it on the next page.

### Trade Adds To The Complexity

When trade arises, the economy is going to get more complex. Production will become more specialized and much more efficient. The distribution question—the question of which family gets how much fish and how much breadfruit—will get more complex, too. But the most productive families will get most of the output.

Suppose the people living on the south side of the island are not very productive at anything. Either because of poor fishing and growing conditions, or their own inability or laziness or for some other reason, the southside family doesn't produce very much of anything. Since it doesn't produce much, it doesn't have much to consume. After trade arises it still isn't going to have much to consume unless, with specialization, it can produce a lot more. If it doesn't have much to trade, it won't receive much in return. The "productivity principle of distribution" really works; each family's share of the output depends on how much that family produces. That's the way it is if they don't trade; that's still the way it is, if they do.

### The Northside Family Starts Producing Things For Sale

Now let's suppose the people on the north side of the island—an inventive and industrious family—start producing bows and arrows, and

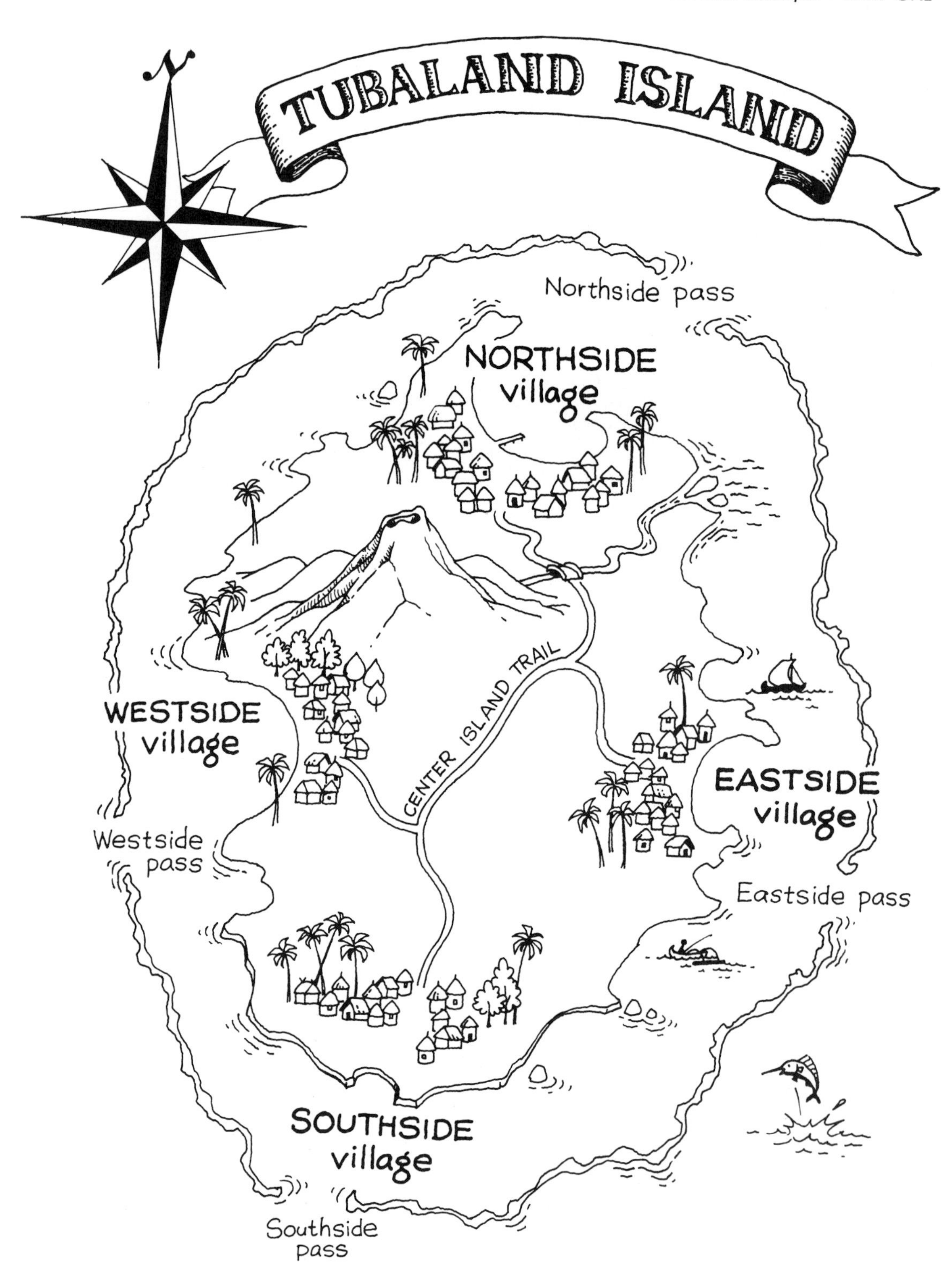

TUBALAND ISLAND
N
Northside pass
NORTHSIDE
village
WESTSIDE
village
CENTER ISLAND TRAIL
EASTSIDE
village
Westside
pass
Eastside pass
SOUTHSIDE
village
Southside
pass

"tuba" (an alcoholic beverage made from the sap of the coconut palm). Early one Saturday they begin beating the drums, inviting all the other island families to come for a visit, to enjoy refreshments, and to try out (and perhaps to buy) some of the new products. All the families come. Before noon all the tuba is sold. For the tuba, they paid handsomely, with tropical vegetables, fruits, and dried fishsticks. But no one bought any bows and arrows.

It doesn't take a genius to figure out what the northside islanders will produce next week. More bows and arrows? Of course not. They are going to make tuba. Why? Because that is what people are buying. The market process is beginning to answer the production question. The northside islanders are seeking profit so they produce tuba. That's what their society wants them to produce. The market tells them so!

### "Tuba" Is A Profitable Drink — To Sell

The way things are going it seems that the northsiders are going to get very rich. All the island people (the society) place high value on tuba. The northside islanders are the only ones who know how to make tuba. It seems that these inventive and industrious people, by producing and selling tuba, are going to get more fish, vegetables, and fruits, than those who catch the fish and grow and gather the vegetables and fruits.

From this simple illustration it's obvious what will happen if the members of the society are free to "let nature take its course." People will make those products which other people want to buy. Through the market process, the individual motive, "I will make what I need to satisfy my wants," changes to "I will get more of what I want by making what *you* want, and trading." The market process lets each specialize in what he chooses to do. It gives each person an incentive — a reward for making what society wants him to make.

The northside islanders are producing tuba and becoming wealthy. The southside islanders aren't producing very much of anything. They are staying poor. With the market working, it's almost as though each family is producing directly for itself. Yet with the market working each family can be much more productive. All can live much better by producing for the market, because the market lets each family specialize in whatever it's good at. Still, if a family doesn't produce anything, it isn't going to be helped by the market. They who produce nothing, get nothing.

Now you have seen a simple "market system" arise. But so far, not much trading is going on. The northside islanders are producing and "selling" (trading) tuba, but everything else is going just as it was before. Wouldn't it be a good idea for the other families to specialize in something, and trade? It really would. That's exactly what's going to happen. Just watch.

## THE GAINS FROM TRADE

You know that almost any "economic unit" could produce almost anything if it was willing to work at it hard enough. Any individual, family, business, farm, or nation could use its resources to make a hundred different kinds of things. An Iowa farmer could produce corn or tomatoes, chickens or eggs, coconuts or bananas, or polar bear skins—or any combination of these. How will he decide which to produce?

### The Producer Specializes In His Most Profitable Product

If the market process is working, it's easy. He figures how much it would cost to make each product, and how much revenue he would get from it. Then he chooses to produce the most profitable product, or the most profitable combination of products. He doesn't produce bananas (or coconuts or polar bear skins) because the cost would be ten times as high as the value of the product! He might be able to break even on chickens and eggs, or tomatoes. But for *his* farm, *corn* is the thing. He owns some of the best corn land in the state. So what will he grow? Polar bears?

If the market process is working as it's supposed to, each producer will automatically make the most efficient, most productive (the optimum) use of his (the society's) resources. He will produce the highest-priced outputs (the ones his society values most) and use up the lowest-priced inputs (the ones his society values least). It's a neat system. But from this kind of example you can't see how much good comes from all this. To really see the gains from trade, we need to start back at the beginning.

### Different "Opportunity Cost Ratios" Generate Trade

Let's go back to our island. Remember that fish are much more plentiful on the east side of the island and breadfruit grows much better on the west side. But both the eastside islanders and the westside islanders like to eat both fish, and breadfruit. Suppose the eastside islanders can work all day and catch five baskets of fish. Or if they wanted breadfruit and worked all day at that, they could gather one basket of breadfruit. So what is the opportunity cost of a basket of breadfruit? Five baskets of fish! So the value of breadfruit is high: one basketful of breadfruit is worth five baskets of fish.

**To The Eastside Islanders Breadfruit Is Very Scarce And Valuable.** Breadfruit is five times as valuable as fish, to the eastside islanders. How do we know? Because the only time they will go after breadfruit (and give up five times as much fish) is when the one basket of breadfruit is at least as valuable to them as the five baskets of fish they are giving up. Otherwise, they would be fishing instead. They will conserve breadfruit, and eat mostly fish. At suppertime, whenever one of the eastside children reaches for a second slice of breadfruit he gets his hand slapped.

**To The Westside Islanders Fish Is Very Scarce And Valuable.** Now let's look at the westside islanders where the situation is exactly reversed. In one day's work they can produce five baskets of breadfruit or one basket of fish. When they spend a day fishing, the one basket of fish they catch "costs" them five baskets of breadfruit. Fish are *five times* as valuable to them as breadfruit. They eat mostly breadfruit. Woe be unto the westside child who reaches for his second piece of char-broiled fish!

### Eastside Fish Will Be Traded For Westside Breadfruit

Let's suppose that both families spend half their productive time catching fish and the other half gathering breadfruit. In their normal twenty-workday month, the eastside family will produce 50 baskets of fish (10 days of fishing, five baskets a day = 50), and 10 baskets of breadfruit (10 days gathering breadfruit, 1 basket a day = 10).

Over on the west side, the situation is exactly reversed. They wind up with 50 baskets of breadfruit and only 10 baskets of fish. They also get a total of 60 baskets of food per month. Figure 5-1 uses transformation (production possibility) curves to show this situation. You should study Figure 5-1 for a few minutes, now.

### Trade Permits Specialization And Increased Output For All

There is really a great opportunity for the eastside islanders and the westside islanders to gain from trading fish and breadfruit. The eastsiders forfeit five baskets of fish every time they give up a day's fishing and go after breadfruit. And they only get one basket of breadfruit. If they could trade *less than five baskets* of fish and get *one basket* of breadfruit in exchange, they would be better off.

Suppose they could trade four baskets of fish for one basket of breadfruit. They would be ahead by one basket of fish. If they could trade *one basket* of fish for *one basket* of breadfruit they would wind up with *five times* as much breadfruit as they are now getting. With a one-to-one trade ratio, breadfruit would become as easy to get and as cheap as fish!

Even if the eastside islanders had to pay four and a half baskets of fish for one basket of breadfruit, they still would be one-half basket of fish better off than if they didn't trade. They could spend a day producing five baskets of fish (instead of producing one basket of breadfruit). Then they could trade four and a half baskets of fish for one basket of breadfruit, and wind up with one basket of breadfruit and one-half basket of

Fig. 5-1 **PRODUCTION POSSIBILITIES FOR SEPARATE ECONOMIC UNITS**

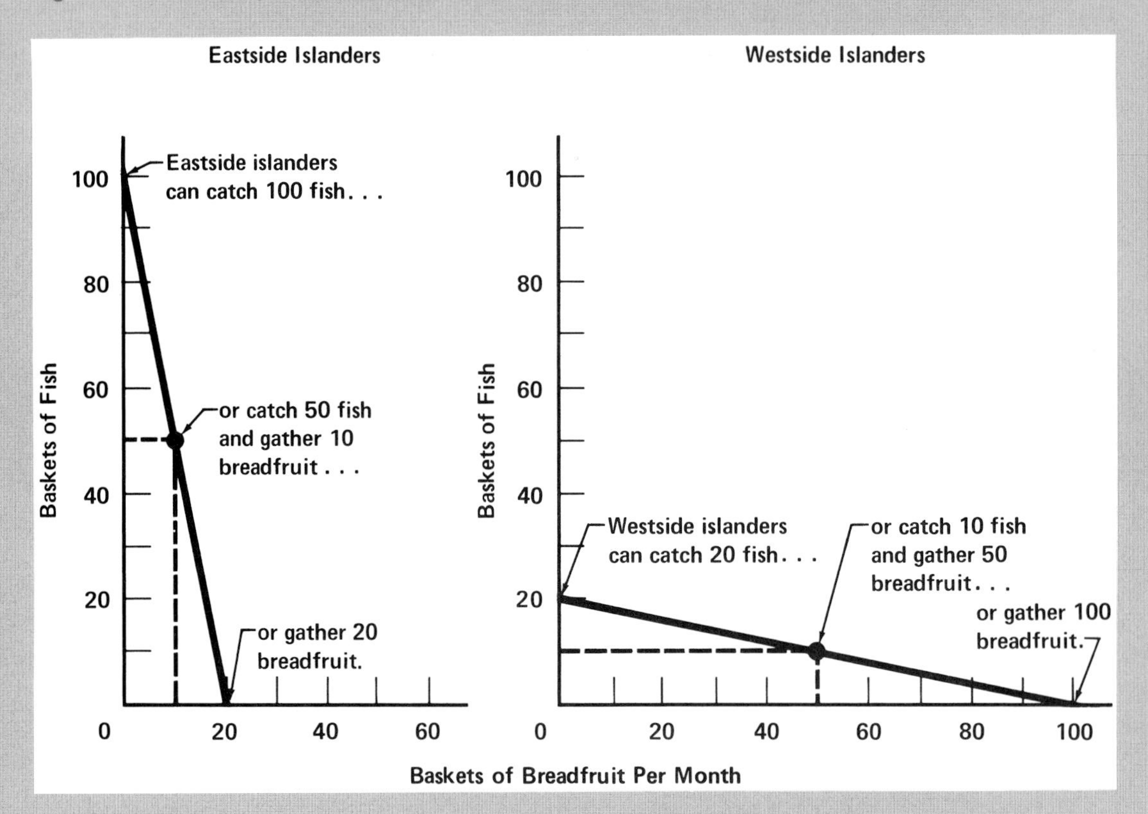


The Eastside Islanders Are Great Fishermen And The Westside Islanders Are Great Breadfruit Gatherers.

The "opportunity cost ratio," or "transformation ratio" for the eastsiders is 1 basket of breadfruit for 5 baskets of fish, or a 1 to 5 ratio of breadfruit to fish.

The transformation ratio for the westsiders is 5 baskets of breadfruit for one basket of fish, or a 1 to 1/5 ratio of breadfruit to fish.

To the eastside islanders, a basket of breadfruit is very valuable. It is worth 5 baskets of fish; to the westsiders a basket of breadfruit is worth only 1/5 of a basket of fish. To say it the other way, to the eastsiders, fish is not very valuable. It is only worth 1/5 basket of breadfruit. But to the westsiders, fish is very valuable. It is worth 5 baskets of breadfruit. What a great opportunity to gain from trade!

fish. That's better than one basket of breadfruit and no fish at all! (which is what they would get if they produced the breadfruit themselves).

### The Production Ratio Is Different From The Trade Ratio

For the eastside islanders, the "trade-off ratio" in production (between fish and breadfruit) is 5 for 1. If the trade-off ratio *in trade* is anything *less than 5* for 1, then they should get their breadfruit by trading—not by producing it. Isn't it obvious? When they *produce* to get a basket of breadfruit the trade-off cost is 5 baskets of fish. If they can *trade* and get a basket of breadfruit for *less than* 5 baskets of fish, then they will come out with more by trading.

The eastside islanders would like to get the products they want at the lowest possible cost. So if the cost of breadfruit (in terms of the fish they have to give up) *in trade* is less than the cost *in production,* they will want to get their breadfruit *in trade*. Why? Because the cost is lower. They will specialize in producing fish and then trade for breadfruit. Sure.

Any individual, family, business, or society will get more of what they want if they will follow this general principle: If the opportunity cost is higher *in production* than it is *in trade,* then you should trade to get what you want. You should get what you want at the lowest opportunity cost (the lowest "trade-off" cost) you can arrange. That way you'll get the most of what you want, with what you have to work with.

Now take a look at the westside islanders. They are facing the same kind of situation the eastside islanders are facing, except that the "trade-off ratio in production" (the transformation ratio) between the two products is exactly reversed. To the westside islanders the opportunity cost of *one basket* of fish, is *five baskets* of breadfruit. If they could trade *less than five baskets* of breadfruit and get back *one basket* of fish, they would be better off. So we have a beautiful setup, just waiting for trade to begin.

### The Island Chiefs Discover The Gains From Trade

Suppose that one day the chief of the eastside islanders is walking down Center Island Trail and meets the chief of the westside islanders. They stop to talk. Soon the eastside chief begins to brag about the good fishing. The westside chief responds with stories about the fabulous breadfruit harvest. Each challenges the other to prove his story. Soon they have worked out an agreement to meet regularly, and to trade. They agree that they will trade one for one—one basket of fish for one basket of breadfruit.

Once the bargain is made, the eastside chief hurries home as fast as he can, hardly able to contain his glee. What a great trick he has pulled off! He knows breadfruit is *five times* as valuable as fish, yet he arranged for a one-for-one trade. How lucky his people are to have such a wise chief, to bring them such good fortune! Such a bargain! And at the expense of the westside islanders! What stupid people those westsiders must be, to have such a stupid chief! He rushes on home to tell his people.

At the same time, guess what the westside chief is doing and thinking? The same things, of course! He knows that fish are *five times* as valuable as breadfruit. But he has talked the eastside chief into trading each basket of fish for only *one* basket of breadfruit! What a dopey bunch those eastsiders must be, to have such a stupid chief! Each chief keeps thinking, "What a steal! What a steal!"

## The Gains From Trade: More Output For Everybody

The beauty of this situation is that both chiefs are right. It is "a steal." But they're both wrong about where all the extra fish and breadfruit are going to come from. The gains of one obviously aren't coming from the other. Both are gaining equally. Nobody is losing. The "steal" that each chief has made, is coming from the great increase in total output.

The "steal" is coming from the gains from trade. A great increase in output of both fish and breadfruit is going to result when each family specializes in the production of the thing they can produce best. "The gains from trade" come from *increased productivity*. The increased productivity comes from specialization. Trade is what makes it possible.

Once the trade arrangement is set up, the eastside islanders produce only fish. They produce a total of 100 baskets of fish per month, and trade 50 baskets of fish to the westsiders for 50 baskets of breadfruit. The westsiders produce 100 baskets of breadfruit, and trade 50 baskets of breadfruit to the eastsiders for 50 baskets of fish. Now both families have much more food than before. All the eastside kids and the westside kids can reach for all the second helpings they want. Even the chickens and pigs are finding that life is easier these days. See the great gains from trade? The entire society benefits. Even the chickens and pigs, too.

## "Economic Integration" Benefits Almost Everybody

Figure 5-2 puts together the production possibility curves of the two families. The "combined" production possibility curve shows the great gains in output which result from the economic integration of the two formerly separate economies. In the production of fish and breadfruit, these two formerly separate units—the eastside economy and the westside economy—are now integrated into *one* economic unit. The transformation curve in Figure 5-2 shows the production possibilities for this "newly integrated" economic unit. Figure 5-2 builds on Figure 5-1. Perhaps you should review Figure 5-1 first, and then study Figure 5-2.

Fig. 5-2 **THE PRODUCTION POSSIBILITY CURVES ADDED TOGETHER**

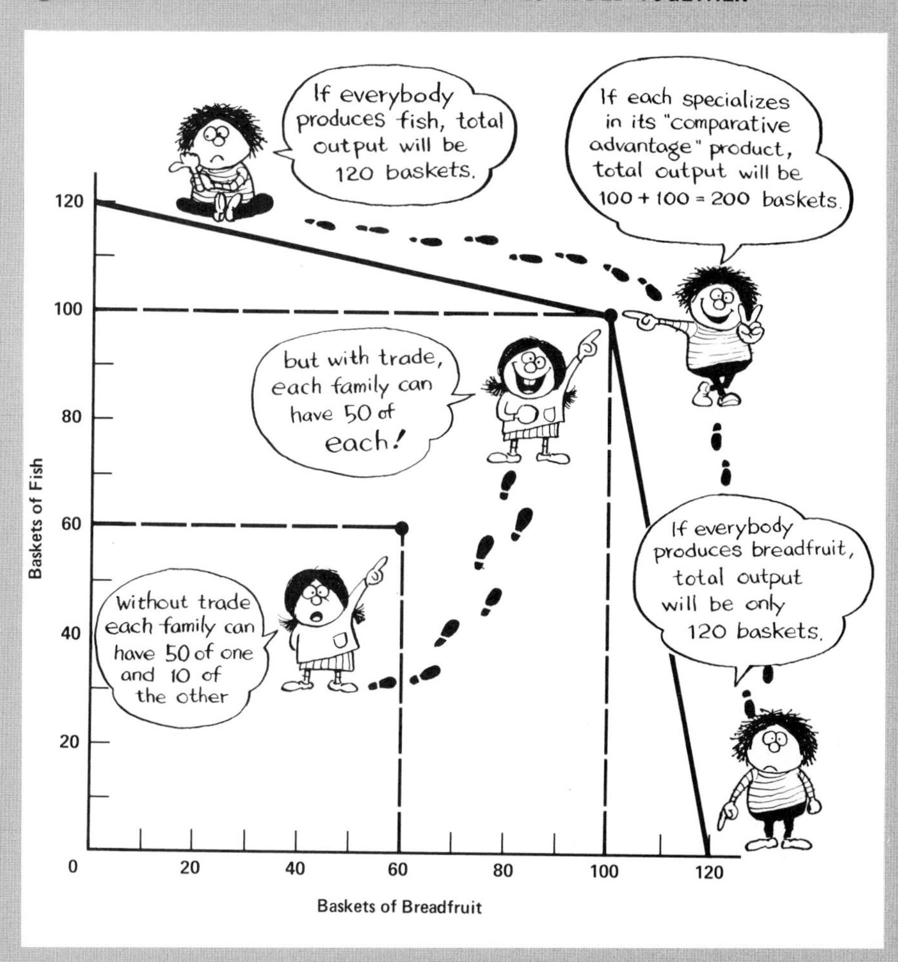


Much More Can Be Produced When The Eastsiders And Westsiders Specialize And Trade.

This concave (out-bulging) production possibility curve puts together the two curves (eastsiders and westsiders), as shown in Figure 5-1. The great outward bulge in the curve shows how much more product it is possible to get by letting each specialize, and then trade.

This integrated east-west economic unit can produce 100 baskets of fish *and* 100 baskets of breadfruit per month—a total product of 200 baskets a month. Without trade, each could produce only 50 of their specialty and 10 of the other, for a total of 60 for each family—120 for the two families combined. The specialization and trade doesn't quite double their food output, but it almost does!

"Economic integration" permits the free movement of goods and resources (of inputs and outputs) among economic units. The benefits can be really great. In our island example, no one is working any harder than before. Yet the total product of the society is almost doubled. Perhaps they will decide to go on a four-day work week, or a six-hour day. Or perhaps they will spend more time building better houses, or developing a better water supply. Or maybe they will build a harbor so they can export dried fish, and copra (dried coconut meat). Then they will be able to gain even more, by trading with other islands and other countries.

### The Price System Stimulates Trade

In a society where the market process is functioning, the *price system* automatically gets people to specialize and trade. In our example, if the market process with its automatic price system had been working on the island, the trade would have developed automatically. Here's how it would happen.

The eastside islanders (the good fishermen) value breadfruit highly. It is very scarce, so its *price* (on their side of the island) is very high. But to the westside islanders (the good breadfruit growers) breadfruit is *low-priced.* Fish is the highly valued, very scarce, *high-priced* item. What's going to happen? It's obvious.

One day, the eastside fishermen will hear about the high westside price of fish. They will start catching more fish and selling them to the westsiders. Big profits! But as fish become more available to the westsiders, what happens? The more available the fish become, the less scarce they become. So the less valuable they become. The price goes down.

What about breadfruit? It works the same way. The westside breadfruit growers will hear about the high eastside price of breadfruit. They will start producing more baskets of breadfruit and selling them on the east side of the island. As more breadfruit is supplied to the east side, some of the scarcity is relieved, so the eastside price of breadfruit will go down.

### Trade Brings The Prices Into Balance

Without trade, eastside fish would stay low-priced and breadfruit, high-priced. Westside fish would stay high-priced and breadfruit, low-priced. But the market process will *generate* trade and bring the prices into balance.

If the market process is working, the ultimate result will be the same as what happened when the sly old chiefs outfoxed each other. The *price mechanism* will induce the eastside islanders to specialize in fish and the westsiders to specialize in breadfruit. Then each will sell to the other.

See how efficiently the "price mechanism" can get the right production choices made? Automatically!

## ABSOLUTE ADVANTAGE AND COMPARATIVE ADVANTAGE

In our example we have given both the eastside islanders and the westside islanders an absolute advantage in something. The eastsiders have an "absolute advantage" (are definitely more productive) in producing fish. The westsiders have an absolute advantage in producing breadfruit. But suppose one family doesn't have an absolute advantage in *anything*. What then? Can they trade, and gain? To find the answer to that, let's go on with our example.

### It's The Comparative Advantage That Counts

Remember the southside islanders? The unproductive ones? Suppose it takes them a *whole week* (5 days) to produce a basket of fish. In one day they can only produce 1/5 of a basket of fish. With breadfruit, things are bad too — but not quite as bad. They can produce three baskets of breadfruit a week (3/5ths of a basket a day). Let's summarize the southsiders' situation.

In one day the southsiders can produce 1/5th of a basket of fish, or 3/5ths of a basket of breadfruit. If they fish all week they wind up with one basket of fish (1/5th basket a day, for 5 days = 1 basket). That's all they get to eat that week. If they spend the week gathering breadfruit, they get three basketfuls (3/5ths basket a day, for 5 days = 3 baskets). The southside islanders aren't nearly as productive as either the eastsiders or the westsiders. They have an absolute disadvantage in producing both products. Can they possibly gain anything from entering the going trade between east and west? If you've really been thinking about it, you won't be too surprised to find out that the answer is: Yes, they certainly can!

### Opportunity Cost Ratios Determine Comparative Advantage

The southside islanders' opportunity cost ratio between breadfruit and fish is three for one. Each time the southsiders decide to *not* produce breadfruit for a week so that they can produce fish, how much breadfruit do they lose? Three baskets. How much fish do they get? One basket. So whenever they produce their own fish, each *one basket* of fish is costing them three baskets of breadfruit.

Suppose the islanders' "going exchange rate" between fish and breadfruit is still one-for-one. Then if the southsiders produce breadfruit and

**Fig. 5-3 PRODUCTION POSSIBILITIES AND TRADE POSSIBILITIES, COMPARED**

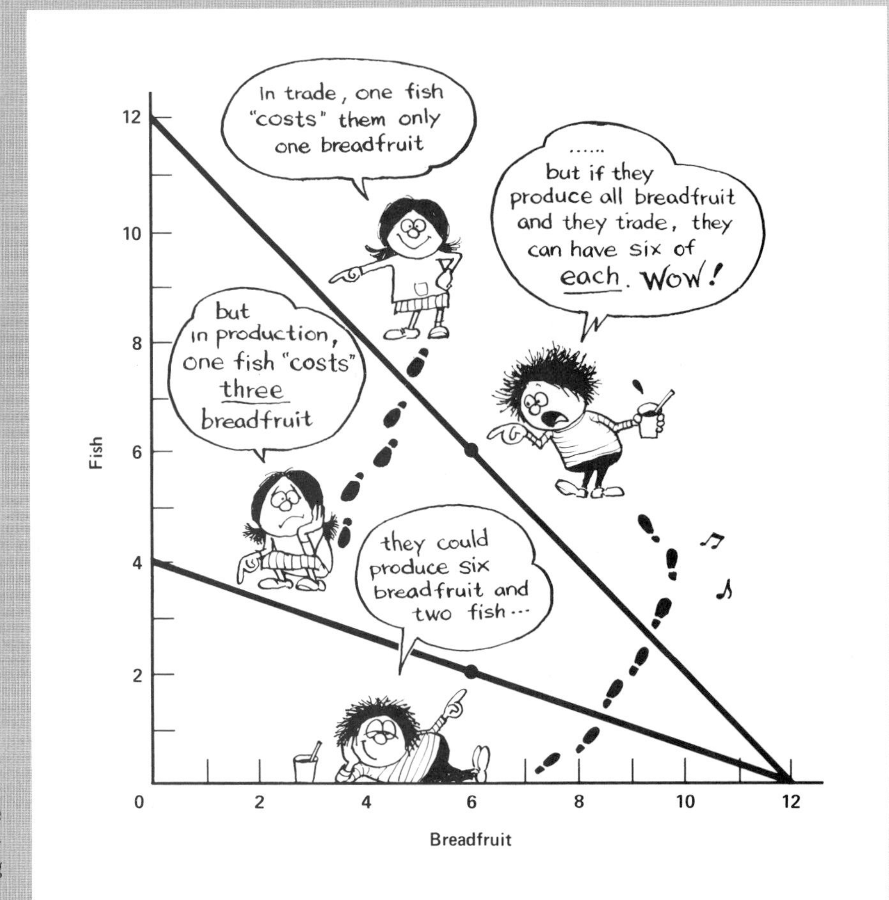


The Southside Islanders Have Absolute Disadvantage In Everything, But Still They Have Comparative Advantage In Something.

Their Comparative Advantage Is In The Product In Which They Are "Least Worse Off" Relative To Their Trading Partners.

*In production*, the opportunity cost (transformation) ratio between fish and breadfruit for the southside islanders is 4 to 12, or one to three. If the southsiders produce their own, the opportunity cost of one basket of fish is three baskets of breadfruit.

Since the "going exchange rate" on the island is one for one, the southsiders could trade one basket of breadfruit and get back one basket of fish. So the opportunity cost of one basket of fish *in trade* is only one basket of breadfruit. So by specializing in the production of breadfruit and trading, the southsiders can get three times as much fish as if they had produced their own.

The effect of the southsider's food supply is the same as if they had discovered some new technique to make them three times as productive in catching fish. All hail the gains from trade!—(even when—or *especially* when—you have absolute disadvantage in everything).

But maybe it isn't all as bad as it seems. Maybe the southside islanders are a very small family and the eastsiders and westsiders are very big families.

trade to get fish, how much breadfruit will they have to give up to get a basket of fish? Only one! So can they gain from trading? Of course! When they *trade* to get a basket of fish, the cost is only *one basket* of breadfruit. But when they *produce* the basket of fish, the cost is *three baskets* of breadfruit. When they trade, the cost of fish is only one-third as much as when they produce the fish themselves! By producing breadfruit and trading for fish, they can get three times as much fish.

Figure 5-3 illustrates the choices available to the southside islanders. You should study that figure now.

## Absolute Advantage? Or Comparative Advantage?

The southside islanders don't have an absolute advantage in anything. They are very unproductive as compared with the eastside and westside islanders. Still, the southside islanders have a comparative advantage in the production of breadfruit. *"Comparative advantage" means "the opportunity cost in production" is less than "the opportunity cost in trade."* When you have a comparative advantage in something (like breadfruit) you don't trade to get it. You produce it yourself. Then you trade to get other things (like fish). You trade to get the things in which you have a comparative disadvantage.

"Comparative disadvantage," quite obviously, means the opposite of "comparative advantage." It means *"the opportunity cost in trade" is less than "the opportunity cost in production."* You don't produce the things in which you have "comparative disadvantage" (like fish). You produce something else (like breadfruit) and then trade.

Let's say it another way. Can you get more of product x by producing product y and then trading to get x? If so, you have a comparative advantage in y. You have a comparative disadvantage in x. You can get more x by producing y and trading than you could get by producing x yourself.

Whenever one trading partner (individual, business, or nation) has a comparative advantage in one thing, the other party automatically has a comparative advantage in the other thing. Anytime it's possible for one "trader" to come out better, that means the other "trader" has the opposite opportunity to come out better. Anytime the "trade-off ratio" or "opportunity cost ratio" or "transformation ratio" between two products is different for any two economic units—individuals, businesses, or nations—then each unit will have a comparative advantage in one of the products. One will have a comparative advantage in one of the products, and the other will have *exactly the same amount* of comparative advantage in the other product. Your comparative advantage is my comparative disadvantage; my comparative advantage is your comparative disadvantage. It's the same "comparative situation," looked at from two different points of view.

If the southsiders produce all breadfruit and trade for fish, they can get three times as much fish as they could get by producing fish themselves. You can see that the question of whether or not they have an *absolute advantage* makes no difference whatsoever in determining whether or not they can gain from trade. It is not the *absolute* efficiency in producing any one thing that counts—it is the *relative* efficiency among the various things that might be produced that tells you whether or not to specialize and trade. Perhaps the southside islanders will always be poor. But they will be *less poor* if they will specialize and trade.

In the real world, every economic unit, every country (no matter how rich or poor) has "comparative advantage" in some things and "comparative disadvantage" in others. Each unit will get more output if it concentrates on producing the things in which it has "comparative advantage," and then trades for the other things it wants. Economists call this idea the law of comparative advantage.

## International Trade

Suppose there are "national boundaries" separating these island families into different "nations." Then all this trade we have been talking about would be international trade. Would that make any difference? No. Not from the point of view of the economics involved.

With "international trade," there might be some difficulty working out the exchange because of the different kinds of money used in the different countries. Also, the Westside Fishermen's Association and the Eastside Breadfruit Growers Association may scream for protection against "foreign competition." Or the eastside chief may not want to become dependent on westside breadfruit because he's planning a war against the westsiders. Or there might be some other complications. But from the point of view of the opportunity for both to benefit, it makes no difference whether the trading partners are in the same city, in different states, or on opposite sides of the world.

The gains from trade are shared by both parties. It isn't that one wins and the other loses. Both always benefit (or at least they always expect to). Otherwise, why would they trade? Trade restrictions always reduce or eliminate the benefits of trade. If we stop trading, *both* parties lose "the gains from trade." Of course.

## The Law Of Comparative Advantage

The "law of comparative advantage" emphasizes the potential benefits of trade. This "law" was mentioned a minute ago. It says that any economic unit—individual, business, nation—will achieve the greatest production possibility, the highest total output, the highest level of economic well-being, by concentrating its energies on the production of those things in which it has *comparative advantage*, and then trading for the other things it wants. This law says that the plumber should work as a plumber, collect his money, then buy his shoes from the shoemaker. The shoemaker should concentrate his energies on making shoes, sell the shoes for money, then use the money to hire someone to install his plumbing. Through this arrangement, the society will have more and better shoes and more and better plumbing. Everyone will benefit.

## THE ADVANTAGES OF SPECIALIZATION

People who concentrate on doing the things they can do best, usually produce more, receive higher incomes, and enjoy higher standards of living than those who don't have any "specialty." But specialization requires trade. One of the major factors that distinguishes the poverty-striken economies of the "underdeveloped world" from the affluent societies of the "modern world," is the degree of specialization.

### We All Depend On Specialization And Technology

You already know that the market process permits each individual to produce for others and thereby to get the things he wants. It is a simple extension of the Robinson-Crusoe world where each person produces to satisfy his own wants. But why bother with the market? Why not let each individual produce for himself and not be concerned about prices and exchange and all that? Why not? Because without the market we couldn't specialize, that's why not.

Think for a minute. What would you be eating today if all you had to eat was what you produced for yourself? What would you be wearing? In what kind of house would you be living? How would you keep your house warm? What kind of medical care do you suppose you could succeed in giving yourself? What sort of transportation would you use to carry you from one place to another? Things would be sort of primitive. Right?

The only way it is possible for people to have more than just the bare necessities of life, is through specialization and trade. People concentrate on doing the thing they can do best. They work at becoming much better at their jobs. But in order for people to specialize, some people must want to buy what the others are producing. If you're going to specialize in medicine, then you must depend on many people to buy your services. Unless "the market" is large enough and the "market demand" great enough to buy up what you are producing, you cannot specialize.

If a 20-acre field is best for producing cotton, its owner will let it "specialize" in cotton. But suppose the "market" is too small. People don't want to buy all that cotton. Then he can't specialize in cotton. Suppose this is a "subsistence type" economy and not much trade is going on, in anything. Then the owner of the 20-acre field must use the field to grow himself some food to eat. See how impossible it is to specialize, without markets, and trade?

Business firms are always trying to design better, more specialized machines and tools. The most *efficient machines* are the *highly specialized* ones. Highly specialized machines produce only one thing and they produce it with great efficiency. For example, compare the value of output per hour you could make with a hacksaw, to that which you could make running the machine that bores the holes in an engine block—or running the machine that puts the caps on beer bottles. Specialization is really great! But unless there's a big market for engines, and for bottled beer, we can't afford such highly specialized machines.

## Specialization And Division Of Labor

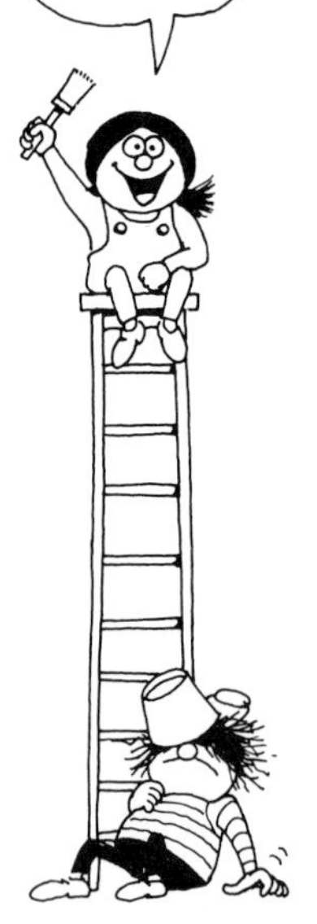


The word "specialization" is frequently used together with *division of labor*. Both expressions mean essentially the same thing. Both mean that each unit of productive input—each person, each piece of land, each machine—does only a part of the total production job. The total production operation is divided up so that each task can become specialized.

Different people have different skills and abilities. By specializing, most people can become even better, more productive at their specialties. With division of labor comes the opportunity to use *more advanced technology*—that is, better, more efficient ways of doing things. As new technology is discovered, new machines are built to embody the new technology. Workers are trained to use the new technology. Productivity increases. Output gets larger and larger. The total productive capacity of the economy increases. This is the process of economic growth.

The society gets more output as technology develops. The different parts of the production process become more specialized. It's a remarkable process, and it tends to come about all by itself. If you just let people alone in a society where the market process is working, these people will figure out ways of specializing. They will figure out how to trade the things they are producing, to get the things they want. Through these individual efforts, specialization and trade evolve and expand. Output per person increases. The society moves to higher and higher standards of living.

The effects of specialization and trade are truly remarkable—so remarkable, in fact, that it's almost impossible for anyone in the modern world to visualize a world without specialization and trade. It's an important aspect of the very fabric of modern society. It influences every moment of the lives of each of us. And the results—in terms of fewer hours, easier work, and more and better things of all kinds—are truly phenomenal. Yet we are so accustomed to it that we just take it for granted.

Specialization is great; still, there is a cost. What's the cost? Interdependency. Vulnerability. We all become dependent on each other.

### With Specialization, Everyone Is Dependent On Everyone Else

One man alone on an island does not need to worry about what his neighbor does. But when there are many people, all producing and trading with each other, no man stands alone anymore. Each person is dependent on, and his life is very much influenced by what all the other people (other buyers and sellers) decide to do. Once these complex interdependencies develop, each person, each business, and each part of the economic system becomes dependent on all the others. Whatever happens anywhere has an effect everywhere. There's not much anyone can do to protect himself.

The more specialized and interdependent the individual units in the economy become, the more productive they can be, the higher standards of living they can enjoy, and the greater "freedom from work" they can experience. Also, the more speedily the society can achieve whatever objectives it seeks. Specialization is really great. But with interdependence, we are all vulnerable to each other. If I produce eggs and you egg-buyers decide you don't like eggs, it hurts me. If I am depending on you to produce the gasoline I use in my car and suddenly you decide to go on strike and not produce gasoline, it hurts me. If the workers in my steel mill demand a wage increase and I can't raise my price to cover it, it hurts me. If I do raise my price and then my customers start buying their steel from foreign producers, I go out of business and fire my workers. It hurts me. It hurts my workers. Before it's all over it's going to hurt the grocers and the bakers and the bartenders and a lot of other people in my town.

With specialization and exchange we all live much better. But we forfeit our independence. That's the cost, but we're willing to pay it.

Without specialization we would all be struggling along at bare subsistence. Are we willing to do that? No. We aren't willing to give up the easy work and the free time, or the warm houses and vacuum cleaners and fast cars, or the eyeglasses and dentures and allergy shots. Specialization and division of labor "cost us" our economic independence, but it's quite clear that this is a cost we are willing to pay.

Specialization and trade arise spontaneously. The market process and the magical "price mechanism" see to that. But in order for this to work, several essential requirements must be met. Part Two of this book talks about those requirements. Also, it carries some of the economic concepts a step closer to the real world. Before we get into that, though, here are a few words of "farewell to Part One."

## A SUMMARY COMMENT ON PART ONE

Economics, on the surface, is as simple as anything can be. We're all faced with making choices. We, as individuals, as members of families, of businesses, of other organizations, and of society, are constantly involved in (and are feeling the results of) the economic choices. Different people, businesses, societies, approach the questions in different ways. The "economic results" are very different from one place to another and from one time to another. Different societies have different economic systems. But the basic concepts, the basic issues, are universal. The same ones are constantly at work, everywhere.

There is no "right" set of answers to the production and distribution questions. For any economic unit—any individual, organization, or society—the best choices are those which will bring the most progress toward the sought objectives. The optimum position of any economic unit will be achieved only when the unit takes advantage of all the available "marginal improvements."

The production choices reflect the opportunities for specialization and trade. It would be rare indeed for an economic unit to find it impossible to gain from specializing and trading! The law of comparative advantage is undeniable. Every economic unit has a comparative advantage in something.

The market process is a miraculous thing. It permits the people to be free of direct social or political controls, yet it gets the people to do what society wants done. It gets people and businesses to follow the law of comparative advantage. It gets the little "marginal decisions" made very efficiently. But there are some problems with letting the market process make all the choices. In fact, several problems. You already know about some. You'll find out about more, later.

You're now only about one-sixth of the way through the book, and already you know much more about economics than most people will ever learn. You could quit now and be ahead. But think of how much farther ahead you will be when you finish the book!

Part Two follows right along in the same vein. You will find out more about economic choices and about the market process. You will find out more about the important force of human selfishness, and about how the force of competition keeps selfishness harnessed and working for the good of society. You'll be surprised how much you're going to find out about money. When you finish Part Two you will know a lot more economics.

Part Two is going to be just as easy as Part One. In a way, easier. I think you will enjoy some of the new things. But don't be in a hurry to go on. First, why not go back and spend some time thinking about all the new things you've seen here? Part Two builds on Part One. The better you know the basic concepts of Part One, the better Part Two will go for you.

---

## REVIEW EXERCISES

### MAJOR CONCEPTS, PRINCIPLES, TERMS (Try to write a paragraph explaining each of these.)

how the "market process" might arise
the gains from trade
the advantages of specialization
trade requires different "trade-off ratios"
how the "price mechanism" generates trade

### OTHER CONCEPTS AND TERMS (Try to write a sentence or phrase explaining what each one means.)

economic integration
absolute advantage
absolute disadvantage
comparative advantage
comparative disadvantage
the law of comparative advantage
international trade
economic growth
division of labor
specialization and technology
specialization and interdependence

### GRAPHS AND CURVES (Try to draw, label, and explain each graph.)

Production Possibilities for Separate Economic Units
The Production Possibility Curves Added Together
Production Possibilities and Trade Possibilities Compared

### QUESTIONS (Try to write out answers, or jot down the highlights.)

1. The westside islanders *really* shouldn't be buying their fish from the eastsiders. They should "buy locally," from the Westside Fishermen's Association. That would be good for the local economy. Support your local businesses! Keep the money at home! Right? Discuss.

2. One impediment to the economic development of the poor nations of the world is that the poor nations can't benefit from the gains from trade. As compared with the advanced nations, a poor, underdeveloped nation couldn't possibly have a comparative advantage in *anything*. Discuss.
3. Most of the European countries have joined the "European Economic Community" (the Common Market) to eliminate trade (and other) restrictions on things (and people) moving among the member countries. If it's such a good idea for Europe, why not the whole world? Discuss.
4. Suppose stringent restrictions were imposed on trade between all the states in the United States. What difference do you suppose that would make? Do you think it would have any effect on *you personally*, in your daily life? Think about it.

# PART 2

IF YOU TAKE CARE OF SOCIETY'S THINGS,
SOCIETY WILL TAKE CARE OF YOU!

The people own things and produce things and buy and sell things, and the ones who make the best uses of the things they have, will get the most of what they want.

# 6 ESSENTIAL CONCEPTS: THE MARKET, PRIVATE PROPERTY, THE FACTORS OF PRODUCTION

*"Capitalism" could not exist without free markets, private property, and responsive labor, land, and capital.*

This chapter continues the flow of thought from Part One. Part Two tries to give you a solid understanding of some of the conditions and institutions needed for the market process to function. Without these conditions the economic system called "capitalism" could not exist.

This chapter talks about free markets, private property, and "factors of production." You will see how all these work together in the "pure market model." Then later you will see why it doesn't always work out in the real world exactly like it does in the model.

## WHAT IS "THE MARKET"?

You've read a lot, now, about the market process and how it works. But so far we haven't stopped to talk about what *the market* is! You have an idea of what it is, of course. I doubt if learning some definition of "the market" would help you to understand it much better. If you just read about it for awhile I think you will develop a "feel" for what it is. Here are some thoughts about "the market."

### "The Market" Is A Concept

Is the market a place? or a thing? Neither, really. It's a concept. If you are growing tomatoes in your backyard "for sale," you are producing

"for the market." You might sell some to your neighbor and some in your little stand by the roadside and some to the manager of the local supermarket. But in either case, you are producing "for the market." Your efforts are being directed by the market. If people stop buying tomatoes, you will stop producing them.

If you mow lawns to earn money, you are producing a service "for the market." If your father is a steelworker or a bricklayer or a truck driver or a dentist or a grocer, he is "producing" goods or services "for the market." Probably he is selling his labor services in the "labor market." When you spend your income, you are buying things from "the market." You may spend money in several stores, supermarkets, gas stations, and restaurants. Still you are buying from "the market." When the local grocer hires you to drive the delivery truck, he is buying your labor in the labor market.

The market is a concept. To say that you are selling your services or your products "in the market" simply means that whoever wants to buy and is willing to pay the price can have what you are selling. If you are "buying in the market" you are buying those things you want from whatever seller has them available. If you are a seller of tomatoes or of lawn mowing services, or of your skills at running a bulldozer, you see the market as *all those buyers who might want to buy what you have to sell.* If you are a buyer of anything—a businessman buying a bulldozer operator's services, or a housewife buying tomatoes—you see the market as *all those sellers who might offer to sell what you want to buy.*

The market may seem to be a fuzzy sort of thing. But for each person (or business) who is making and selling something, it's very real. If the market doesn't want what the seller has to offer, the seller gets the message, pronto! If nobody will buy your tomatoes, it won't take you long to get the message. The market is telling you something. It's telling you that you are using your energies and other resources to do something "the market" doesn't want you to do with those resources.

If the market system is working, and if you want an income, you must do something that the market wants you to do. The market is difficult to visualize, "in general." But whenever it needs to send you a message, you may be sure that sooner or later you will get that message. The sooner you do, the better for you!

### The "Market Structure": How Many Buyers And Sellers?

Even though "the market" doesn't usually exist in any one place, it is still useful to think of it as being "structured," or "built up" of *buyers* on one side and *sellers* on the other. When economists talk about the market structure they're talking about *the number, and the relative sizes of the buyers and sellers* in that market. Are some buyers or sellers big and powerful while others are small and insignificant? Or are the buyers and sellers all small? These are the questions you would ask if you wanted to find out about the "market structure."

If there is only one seller, we call that kind of market structure monopoly. It can make a lot of difference whether there are many buyers and sellers of something, or only a few, or perhaps only one. If there are only a few buyers or sellers, each will have a considerable amount of market power (which means the same thing as monopoly power). If there are many small buyers and sellers, no one will have any "market power," or "monopoly power." *With many small buyers or sellers, each will simply respond to market conditions.* Since no one is big enough to have any "market power," each buyer or seller can only "respond." No one buyer or seller is big enough to have *any noticeable influence* on market prices, or on the quantities of things being produced and exchanged.

### Market Structure, And Competition

Sometimes the market for something is made up of a great number of small buyers and sellers. Other times there may be only one seller and many buyers (for example, the local electric company selling electricity). Or there may be only one buyer and many sellers (the local electric company buying the services of—that is, hiring—electrical workers). You can see that the kind of competition existing in any market will depend a lot on the kind of "market structure." When there's only one seller there isn't going to be much competition on the seller's side of that market! A market which has only one buyer certainly won't have much competition on the buyer's side, either!

Later on, when we get into the issue of competition and monopoly, you will gain a greater appreciation for the importance of "market structure." But for now, it is enough for you to be sure that you have a clear understanding of the meaning of the term. Now that you have that, you're ready to go on into the discussion of private property.

## PRIVATE PROPERTY

The market process couldn't work unless people had the right to keep the things they earned. That right is called the private property right. *The "private property right" is a "social institution" which exists to some extent in every society, but which does not exist to an "absolute," or "complete" extent in any society.*

The "concept of private property"—the idea that people have a governmentally protected right to own things—is very basic. I'm going to try to show you some different, helpful ways to look at and think about this "social institution" called "private property."

### Private Property Rights

The *concept* of private property is simply the idea that the private individual has the right to own and to do whatever he pleases with anything which is "his"—his house, his land, his car, his guitar, his money, his anything. His right to have and to use "his things" is supported by law. Anyone who interferes with another person's *property rights* is breaking the law.

In order for "private property rights" to exist in the society, there must be some strong, stable, dependable force in the society—either government, or tradition, taboos, or something—to protect the "property rights" of each individual. Otherwise the things owned by some people would be taken over by other more greedy and powerful ones.

### Your Wealth Is Your Private Property

When we talk about private property, we aren't just thinking about houses and land—that is, real estate. The concept includes everything an individual can own. If you were to add up the value of everything you own (that is, all your private property), you would arrive at a figure showing the total value of your wealth. As you make more income, you add to this "wealth." As you use up your income, or wear out your car, or your shoes, or whatever, you subtract from this wealth. Your "wealth," at any moment, is the total value of your private property at that moment. Your *private property rights* are the rights you have to keep the things you own and to do anything you please with them.

### The Market Process Requires Private Property

The market process could not work without private property. The reason people work for income, or make things for themselves, or plant trees in their backyards or undertake any other productive effort, is usually because these efforts add to the total value of their private property—their money, and things—their wealth. Unless the individual is going to be able to keep the money he earns, or live in the house he builds, or eat the fish he catches, or drink the tuba he makes, or keep the money he gets when he sells his tuba, then he probably would rather not work to produce these things.

People work and produce things so they can have more private property. When you have property you can enjoy having it, or enjoy using it up. People work so they can get more cake to have, or to eat. The government decides how "complete" your property rights will be. Then it guarantees and protects your property rights against encroachment by others. Stealing is punished in every society!

Anything which interferes very much with a person's private property rights is likely also to interfere with the person's incentive to work. When government collects taxes from people, this reduces the "private property" (income) the people get from working. This may reduce their incentive to work. It is sometimes argued that if income taxes in the United States (and in other nations) are increased much more, the high taxes may reduce the people's work-incentives so much that the functioning of the economy will be impaired.

### Property Brings Perpetual Income To Its Owner

The "private property right" lets a person own something and enjoy using it, or rent it out, or invest it and receive income from it. It may be difficult to save and acquire wealth, but once a person has acquired some wealth, he can invest this wealth and receive income from the investment. This gives him a *continuing, perpetual income* that he does not have to keep working for! If you keep your money invested in bonds, or in a savings account, *you will keep getting income from it forever!*

If a person invests in a house to live in, or a car to drive, he is still making an "investment" from which he will receive a continuing "income," for several years—not more "money income," but his "real income" will go up because he owns the house or the car.

A person usually gets his real income—the goods and services he gets to use and enjoy—by spending his money income for the things he wants. But once you own your house or your car "free and clear," you get the "real income" of using it without having to make any payments to anyone. You no longer have to pay out your "money income" as rent, or mortgage payments, or car payments. You can use your "money income" to buy more of some other kinds of "real income"—more steaks, or lawn furniture, or a new stereo system, or a trip to Daytona or Lahaina or Acapulco.

Yes, your income *really is* higher when you own useful, productive property. This is true if you get money from someone else for the use of your property; it is also true if you use your property yourself.

The basic "private property right" can be thought of in many ways. One interesting and useful way is to think of it as "the right to receive real income (today) without having to work for it (today)." Most people save up for vacations. Some people save all of their working lives and make investments so they can enjoy a high level of "real income" after they retire. What they are doing is storing up some "private property" (money, or bonds, or real estate, or whatever) so that later, they can enjoy some real income *without having to work for it.*

### The More You Own, The Less Work You Must Do

If a person gets wealthy enough—that is, enough private property—he will never have to work again. Some people are born wealthy. They could go through life living on the income from their property if they wanted to. Some do. But most of them wind up working *harder* because of all their important economic choices—all the important decisions which most wealthy people must make. They must decide:

> Which businesses to expand? to contract? Which manager to promote? to fire? Which investment to shift? Which worthy cause to support? Which political candidate to endorse? and hundreds of other choices.

I suppose we all dream of how nice it might be to be a millionaire. But a careful look at "the average day in the life of the average millionaire" would probably convince most of us that we certainly wouldn't want to have to put up with all that! And those who are willing to put up with it and live that kind of life—to work and study and think that hard, that constantly—will probably become millionaires anyway if that's what they want most to achieve in life.

Want to be a millionaire? No. But want to have enough property to be able to take a good long vacation and travel next summer? Or to have an extra bedroom, or an extra bath? an extra car? a boat? Yes! Most people have a strong incentive to acquire just a little more private property—*just a little more* "real income"—than they already have. Don't you? *This is the driving force that makes the market process work.*

### Some Private Property Exists In All Economic Systems

The desire to get and have a few more things is so strong an incentive for most people that even in the governmentally controlled societies of Communist China and the Soviet Union, individuals are permitted to own some personal, private property. They usually receive income and products as rewards for their productive efforts.

Throughout history, every successful economic system, every successful society has had some arrangement for giving people "as their own private property," at least some share of the value of the output they produce. If the "pure market system" was working, the person who produced something would get to keep it all. But no society has ever worked quite that way.

The productivity principle of distribution is a vital part of the market process. You can see how the productivity principle depends on the existence of "private property." If each person is going to receive a share according to the value of what he produces, then whatever he receives as his share must become his "private property"—to keep, and to use.

### Private Property Rights Are Always Limited

If you think about it for a minute it becomes obvious that no society could give individuals "complete, absolute, unlimited private property rights." For example, suppose the man running the local junkyard has a big pile of old rubber tires in the corner of his lot. He needs the space for junk cars, so he decides to dispose of the pile of old tires. That is his "private property right." The cheapest way to get rid of the tires is to burn them, so that's what he does.

The stinking smoke from the burning tires blows in the windows of the school down the road. All the children turn green. Before the day is over, many scientifically prepared, well-balanced lunches get flushed. By seven p.m. that day, we find that the County Board is meeting in special session. (The daughter of the Board Chairman was one of the victims.) By 7:15 p.m., a new law has been passed removing the "property right" of a person to dispose of his rubber tires in this county, by burning. A property right has been removed "by due process of law."

Maybe the junk man should not have had the "right" to burn the tires in the first place. Certainly as more people get crowded closer and closer together it becomes more and more necessary to limit the freedom of each individual to do whatever he pleases with his "property."

### Property Can Be Taken By "Due Process"

The U.S. Constitution says that your property rights cannot be taken away except by due process of law. But what this also means is that your property rights *can* be taken away—so long as it is done by due process of law! When the State Highway Department decides that a new highway is going to go across the corner of your front lawn, the fact that the land is your private property does not stop the highway. The land is taken by "due process," and "just compensation" is paid to you for your loss. Each time the government collects taxes from you, this takes some of your private property. When a wealthy person dies and leaves all his private property to his heirs, the government takes a sizable share. All these are limitations on the individual's "rights of private property."

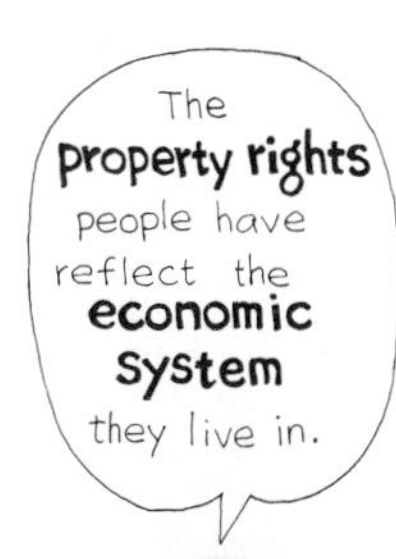

Many limits on private property are imposed by the government, to protect the community. People cannot own certain firearms without governmental consent. No unauthorized person is supposed to own or to transport certain drugs. No one can build anything on his property unless the structure is in keeping with the local zoning laws. We are all told what we must and must not do with our automobiles. On and on the list could go.

### Property Rights Are Becoming More Limited

We all live within a complex network of restrictions on our private property rights. In the coming years, with larger population, increasing technology, increasing demand for goods, the crowding of more people into the cities, increasing pressures on the environment, and all the other complex conditions which are advancing toward us over the horizon, you may be sure that more restrictions and limitations on our private property rights are on the way. This result seems inevitable.

How far can the local, state and national governments go in limiting private property rights before the functioning of the market process is impaired? No one knows for sure. The important idea is the obvious one: if the market process is going to work, the people must be permitted to have, to keep, and to use at least *some* of what they produce. But just *how much*? No one knows.

## THE INCOME-SEEKING FACTORS OF PRODUCTION

Whenever something is going to be made, some "inputs" are going to be needed. Obviously. To really see how the market process operates, you need to know more about "inputs"—that is, about the factors of production.

There are hundreds—even thousands of different kinds of factors of production—everything from the electricity that runs the machines to the paper that packages the products. I'm sure you could list a dozen different kinds of inputs without even thinking about it. But with all these different kinds of inputs, these factors of production buzzing around, how do we ever make sense out of what's going on? Easy. We combine and simplify.

When we economists talk about the "factors of production," we have something very specific and precise in mind. You will need to grasp these special meanings so that what comes later will make sense to you. So let's talk about the "factors of production": labor, land, and capital.

### Labor, And Wages

A man is not a factor of production. He is a man. But the effort or manpower he exerts to make something, is labor. "Labor" is a factor of production. A man sells his labor to a business in exchange for income. Or perhaps he uses his labor himself to grow corn and tomatoes in his backyard field. In a market-directed economy, each person is free to sell or to use his labor in any way he chooses. He owns his labor. He can sell it to the highest bidder, use it himself, or just rest—not use his labor to produce anything at all. Labor, then, is one of the factors of production. Some of the factor "labor" is required as an input, for almost any kind of production you can think of.

Wages are the payments people receive when they sell their labor. The wage rate is "the *price* of a unit of labor." As you know, there are many different kinds of labor. Some kinds sell for low prices, others sell for high prices. Doctors and lawyers sell highly skilled labor and receive high "wages." The efforts of unskilled mill workers and migrant farm workers are not valued nearly so highly. Their labor sells for a much lower price (wage rate). But both are selling labor—their own time, energies, and efforts—in the "labor markets."

### Land, And Rents

A second factor of production is land. But, like labor, "land" isn't exactly what you would think it is. Actually, it's what you would think it is, plus a lot more. In order to simplify things and limit the number of different "factors of production," economists usually consider "land" to include all the natural resources—all the "free gifts of nature." When we say "land," then, we include the minerals in the land, the trees in the natural forests, the natural lakes and streams, and even the fish swimming in these lakes and streams, the deer in the forest, and the wild buffalo roaming on the plains. (I don't know why it is that the only thing buffalo ever seem to do is "roam." I suppose it's because "roam" rhymes with "home." But roaming or not, wild buffalo are "land.")

Why do economists include the roaming buffalo and all the other "free gifts of nature" in our definition of "land"? Only because it's useful to do it that way. More on that, later.

Most of the land in the world is scarce. It has economic value and is owned by somebody—individuals, businesses, other organizations, or governments. Just as people can sell their time and efforts (labor) for a price (wage), so also can the use of the land be sold for a price. Some kinds of land have high value and bring big incomes to their owners. Other land has low value. Owners of the low-value land won't receive much income from their land.

The price paid for the use of land (that is, the income received by the owner of the land) is called rent. If you own a very productive oil well or gold mine, or half an acre just off Times Square, the "rent" you receive will put you on easy street. But if you own a barren plot in some secluded spot, better not quit your job and plan to live it up on your "rent" income!

**Capital, And Interest**

The third and final factor of production is capital. Like land and labor, the chances are this term doesn't mean exactly what you thought it did. Capital is the only one of the three factors left. So this term must include all the inputs *except* labor and land. And so it does!

If "land" includes all the natural resources (the "free gifts of nature") and "labor" includes all human effort, then what's capital?

"Capital" must include all the inputs *except* "human effort" (labor) and "natural things" (land). What's left? All the "things" other than the "natural things." The man-made things, of course!

*All the "man-made inputs" are called "capital."* There are thousands of different kinds of "produced" things which are used as inputs—ranging from factories and machines and trucks, to electricity and note pads and paper clips. All these different things are lumped together and called "capital."

We define "capital" as *produced things which are going to be used as inputs for further production.* Capital is anything which has been produced and which is not available for the immediate satisfaction of someone's desires. A machine, a building, a shovel is capital. Even a can of peas on the grocer's shelf, is "capital," to the grocer. But to the housewife who buys it, the can of peas is a "consumer good."

Someone must *invest* to make "capital." If you "invest" your income or your efforts in building "capital," you can receive more *future* income because your capital is productive. It will help you to produce more things. If you start using more efficient machines and tools, you will produce more product. Then when you sell this marginal product, you will get more revenue. See? Your capital has a "marginal revenue product," just as your labor does!

When you save (you don't consume) and put your money (or time and effort) into building capital, you are *investing*. When you invest and then receive income from the capital you invested in, this income is called *interest*. The "interest" is the money you receive because your capital is productive, just as your wage or your rent is the money you receive because your labor or your land is productive.

The economist's use of the term "capital" will be a bit strange to you at first. *When a businessman uses the word "capital," he usually means the "money" he must have to operate his business*—to buy the machines and raw materials and other things he needs. But when the economist says "capital," he doesn't mean the money. He means the factory, machines, equipment, materials, and other things the businessman will buy with the money. To avoid confusion, economists frequently say "capital goods," or "capital equipment" to make it obvious that they are not talking about the money. It isn't really the money that's productive, that produces interest—it's the machines and equipment and all those things—the *real* inputs—the capital.

Some capital is essential for almost any kind of production. Even if you are going into the business of selling hot dogs on the beach, you need some capital. You need a little hot dog stand, a grill, a box of rolls, an oven to heat the rolls, a package of wieners, some mustard and relish and a few other things. All these things are your capital.

If you are going to grow corn or tomatoes, you need a shovel or a hoe or a plow or a garden tractor or a sharp stick or some kind of tool (capital), and some seeds (capital), fertilizer (capital), some bug spray (capital), and a basket (capital) to use in gathering your harvest. It wouldn't hurt if you had a couple of fellows to help you (labor) and a naturally fertile garden plot (land). But some capital is essential.

## Capital Good? Or Consumer Good?

Various kinds of capital are being produced all the time. All the things that are being produced, if they are going to be used as inputs for further production, are capital. All the things being produced are being produced for one of two reasons: either to be consumed (used to satisfy people's wants), or to be used to help to produce something else (used as capital).

*"Consumption" is the ultimate objective of all production.* "Final goods" are the things that come out at the *very end* of the "production pipeline." Final goods are used by "final consumers" for the ultimate purpose—to provide "satisfaction," or "pleasure," or "utility," whatever that means to each consumer.

Everything that is being produced is aimed, either directly or indirectly, toward the ultimate objective of "final consumption." If a good is aimed *directly* toward final consumption, it's a consumer good. If it's aimed *indirectly* toward final consumption (if it's going to be used to produce something else) then it's a capital good. All along the "production chain," (like from iron ore, to an automobile) everything being produced is capital, *except* the good bought by the final consumer (the automobile).

Whether something is a capital good or a consumer good often depends on how it will be used. A small private twin-jet aircraft owned by a millionaire movie star so he can have fun zooming around the country visiting his girl friends, is a consumer good. An identical aircraft owned by a businessman so that he can more efficiently serve his customers, is a capital good.

The tank full of oil a person is going to use to heat his house this winter is a consumer good; the tank of oil the power company is going to use to fire its boilers is a capital good. If your "aggie agent" (your county agricultural agent) convinces you to feed your chocolate cake to your cow so she will give chocolate milk, then your chocolate cake becomes capital!

### A Fourth Factor Of Production?

Now you know about the three factors of production — labor, land, and capital. And you know about the three kinds of income (distributive shares) the owners of these factors can earn — wages, rent, and interest. You also know that we have defined each factor in such a way that everything is covered.

All human effort is "labor." All *things* used in production are either man-made, or they aren't man-made. If they *are* man-made they're capital. If they aren't man-made, they're land. So everything is covered.

A fourth factor of production? How could that be? Maybe there can't. But then again, maybe there can. Why? Because something seems to be missing. People earn wages, rent, and interest by selling the services of their labor, land, and capital. Fine. But what about *profits*? What factor of production gets the profits?

### What About Profits?

A very successful (and lucky) businessman in a good year may earn a very high profit. What is he getting paid for? Not for the use of his land, or his capital. If it was for that, we economists would say: "What he's earning is really rent and interest and he only *thinks* it's profit." It's not for the use of his labor, either. If it was, we would say: "He's really only earning wages and he thinks it's profit." No, it's something else.

Over the years, economists have had trouble trying to decide what profit really is. Or, to say it differently, we've had trouble deciding the *most useful* way to conceptualize profit. What is the true nature and role of profit? How does it operate? What does it do in the "pure-model market process"? What does it do in the real-world economic systems of "mixed socio-capitalism."?

The market system couldn't work without profit. (If it isn't obvious to you now, you'll see why, later.) But is profit a payment to some factor of production? If so, what factor? And if not, why is profit so necessary to the functioning of the system? What is it a payment for? You can see the confusion that arises when we let profit just "hang loose," without any factor of production to attach itself to.

### The Entrepreneur, And Profits

Economists, neat- and orderly-minded as we are, don't like for things to be "hanging loose." So we invented a factor of production to go with profit. What factor? The entrepreneur — the farsighted resource-manager who brings together the other three factors, gets them organized, and directs them into socially desired production.

The entrepreneur is the one who decides what to produce and how much, which inputs to use, and all that. He hires and pays the owners of the labor, land, and capital, and he hopes that when he sells the product he will get all his money back, *plus* some profit. If he produces the right amounts of the right things and does it efficiently, he will make a profit. If he doesn't, he won't. So *we can think of profit as being the payment which goes to the fourth factor of production: the entrepreneur.*

This "entrepreneur" approach to explaining profit is easy enough to understand. Right? Here's another approach, just as easy to understand.

## Risk And Uncertainty, And Profit

Another way is to explain profit as the "incentive payment" which is necessary to induce people (factor-owners) to take risks. Uncertainty is everywhere. Unless a lot of people were willing (somehow induced) to take some chances, the market system couldn't work.

Without the profit incentive, who would quit his steady job at a steady salary and start an auto-repair shop? Who would take his money out of his savings account and invest it in a McDonald's drive-in? or in an organic food store? or in a reverb amp for Friday night "gigs"? Who would spend money drilling for oil? Who would go hungry and work in a basement trying to invent a better antipollution device? or TV tube? or vaulting pole? or pool cue? Probably nobody.

*Profit can be looked at as the reward you get for taking a chance for society, and winning.* If you hire some factors of production and make something the people want, and if you do it efficiently, you will be rewarded. Your profit is your reward. But if you guess wrong, you get no reward. You lose. Your loss is your "punishment" for using society's resources in ways the society didn't want its resources to be used.

Either way you look at it, you can easily see how important profit (and loss) is. It is essential to the functioning of the market process. It can be regulated, taxed, and manipulated without destroying the system —but only within limits. Just as private property, wages, rent, and interest are essential, so too profit is essential for the market process to work. In some ways, profit might be considered the most essential of all.

Now can you answer the question about the "fourth factor of production"? Is there one? Is the entrepreneur a separate factor of production? Or not? You can have your own way, on this one. You decide.

Sometimes it may be helpful (and neat and orderly) to think of four factors: labor, land, capital, and the entrepreneur; and four sources of income: wages, rent, interest, and profit. There are times when it is helpful to break down the factors even further. The labor factor can become several different factors: unskilled, semi-skilled, skilled, service, clerical, professional, young, old, male, female, and so on. And capital and land can each be broken into many separate factors.

### You Can Define As Many Factors As You Wish

How many factors are there? As many as you want to define. Use the most useful breakdown—the one that will help you most to observe what you want to observe. For a basic understanding of the market process you can think of either three factors (with profit going to any factor-owner who takes a risk), or four factors (with the risk-taker, resource manager thought of as a separate, profit-seeking factor). You can see the market process working just fine, either way. Personally, I like the three-factor breakdown, so that's the one you will see in this book.

Now we need to leave this issue and get back to the market process. How does the market process move the factors of production to the chosen places and into the production of the chosen things? How does it get the factors to do what it wants them to do?

### The Factors Of Production Respond To "The Market"

The factors of production—labor, land, and capital—are the "essential ingredients" in the production process. These factors provide the productive base of every society. These input factors are what the society tries to optimize. The production question for the society is really the question of choosing the way each one of these factors will be used. Which land, which labor, and which capital will we use in which ways to produce which things? Will we use our factors to produce more consumer goods? Or more capital goods? Whichever ones we produce, will we use a lot of labor and a little capital and land? or much land and capital and only a little labor? Which input combinations? How do we decide? When all these questions are answered, the production question for the society is answered.

### Owners Move Their Factors To The High-Paying Jobs

If the market process is in charge, the answers to all these questions will emerge automatically. The owners of the factors of production will be free to use their factors in any way they wish. Each person will move his factors into the use which offers the highest income.

Each kind of labor tends to flow towards its highest-priced (most valued) use. The same is true for land and for capital. As each factor-owner tries to increase his own income, he automatically moves his factors into the uses society values most.

If the specialized labor needed to do some important task is very scarce, that means the society has a great need for more of that kind of labor. The price (wage) offered for it will be very high. The high wage will induce some people in other occupations to study or retrain so they can get one of these specialized, high-paying jobs. As more people respond to the high wage, the great scarcity will be relieved. The wage will move down. See how the market automatically moves the factors of production to where the society most wants them to be?

What a neat system it is! Each factor is moved (by its owner) into that activity which society most wants it to perform. This happens automatically. Why? Because the owner of each factor is trying to get more income. The highly efficient "factor market" accomplishes these three objectives:

1. it moves the factors of production from one activity into another;
2. it stimulates the development of new labor skills and new kinds of capital in the areas and activities in which society's demands are greatest; and
3. it discourages people from using their labor, land, or capital in the kinds of work society *doesn't* want done.

#### The Market System Is Really Efficient!

The great efficiency of the market system in directing the factors of production results from the fact that everything happens so automatically. No one needs to run a survey to find out what things need to be done, and where. No one needs to be assigned the job of going around contacting the individual workers, or the owners of land or capital, trying to induce or force them to move their factors of production from one job or one place to another. The market process handles it all automatically.

Also, this process simultaneously and automatically solves the distribution problem. Each person gets his "income share"—his wages, rent, interest, or profit—on the basis of how well he responds to the society's wishes—that is, how successful he is in using his factors to do what the society most wants those factors to do.

### DEMAND DIRECTS THE FACTORS OF PRODUCTION

As the factor owners are moving their factors around seeking the highest wages, rents, interest, and profits, the factors are being moved automatically to where the people who are spending money, buying things, *want* the factors to go. A factor of production—labor, land, capital—will only bring high income to its owner if it is used to produce something people want to buy. Why did the northside family shift their efforts (labor), natural resources (land), and tools (capital) out of production of bows and arrows and into the production of tuba? Because people were buying tuba! Nobody was buying bows and arrows. This is the way demand works through the market process to direct the input factors to produce the things most wanted by the society.

### Factors Respond By Making What People Are Buying

As people buy more of a product, more factors flow into the production of that product. The factors flow away from the production of the things people are buying less of. The factor owners who keep on using their factors to produce things for which the demand is going down, are not going to enjoy very high incomes. Those factors which are moved rapidly into the areas of expanding demand, will bring more income to their owners. It is the high incomes available in the "expanding demand markets" and the low incomes in the "declining demand markets" that cause the factor-owners to move their factors (labor, land, capital) from the declining markets to the expanding markets.

It is the movement of factors from the declining to the expanding industries which relieves the great scarcity of factors and products in the expanding industries. This influx of factors also keeps factor prices and product prices in the expanding industries from going sky high. When the demand increases, if output can be expanded quickly, the higher demand can be satisfied quickly. Then, no serious shortages will persist. Factor and product prices will not go up very much.

Suppose the factors do not move quickly in response to the expanding demand. What happens? Serious shortages develop. The price goes higher and higher.

### Competition Is The Name Of The Game

The essential role of *competition* in making the whole market process work is pretty obvious. Right? Competing businesses move into the industries where demand is increasing. This pulls more factors of production into the areas where society wants them to be. That's the way the market system works.

Suppose all the businesses in a profitable industry could get together and *prevent anyone else from entering* their industry. Could they make more profits? You bet! Anything which prevents competition can increase the opportunity for profits. But if other businesses or factors of production aren't free to move into the more profitable industry, then the market system can't work right. Anything that interferes with competition strikes at the heart of the market process. *The market system cannot operate without competition.*

The vital role of competition, and also the important issues of inequality, and economic growth, are explained in the following chapter.

But before you go on, take time to do the exercises, think about the questions, and review the essential concepts of market structure, private property, the factors of production, and the role of demand—of how demand directs the inputs to go where the society wants them to go, and do the things the society wants them to do. After you finish reviewing and doing the exercises, you will be ready for Chapter 7.

---

## REVIEW EXERCISES

### MAJOR CONCEPTS, PRINCIPLES, TERMS (Try to write a paragraph explaining each of these.)

the market
market structure and competition
the role of "private property"
the factors of production
the role of profits
demand directs the factors
the role of competition

### OTHER CONCEPTS AND TERMS (Try to write a sentence or phrase explaining what each one means.)

market structure
monopoly
market power
monopoly power
the concept of private property
private property rights
wealth
rent income
money income
labor
wage
land
rent
capital
interest
profit
capital good
consumer good
the entrepreneur

### QUESTIONS (Try to write out answers, or jot down the highlights.)

1. Suppose there's only one dealer in your city who can repair your small foreign car. What kind of a market structure would you say that is? Suppose you don't like his repair work, or his prices. What can you do? Discuss.
2. What are some of the ways in which the rights of private property have become more restricted over the past several decades? (If you can't think of anything, ask your parents, or grandparents. They'll

tell you!) What additional restrictions do you expect will be imposed over the next decade or two? Discuss.

3. What's so *efficient* about letting the market process direct the factors of production?
4. The *production question* for every society is really a question of what to do with the available inputs—with the available factors of production. Explain how the *market process* gets the question answered and gets the society's wishes carried out.

---

# 7 MORE CONCEPTS: COMPETITION, INEQUALITY, ECONOMIC GROWTH

*Everyone competes, produces more, tries to get a larger share, and the economy grows.*

The market process can't work right unless several special conditions exist. The last chapter talked about the need for private property and for factors of production which will respond to the demands of the society—that is, factors which will respond by doing the things for which they get paid most.

This chapter talks about the essential role of competition. It explains how the lack of competition short-circuits the system and dumps big rewards in the lap of the monopolist. Then this chapter explains why inequality is necessary, and finally it describes the process of economic growth. The next chapter will talk about money—about what it is, where it comes from and how it works to tie all the parts of the market system together. When you finish the next chapter you will understand how money lubricates the "market mechanism" so it can work with speed and efficiency. But let's slow down now and take it all in order, one step at a time. Right now it's time to talk about competition.

## COMPETITION IS WHAT KEEPS EVERYONE IN LINE

To show the essential role of competition, let's go back to the beginning days of our make-believe island economy. Remember the first "market day"? Everyone traded things to get tuba, and they bought all the tuba in a hurry. Suppose that on the next morning the northside islanders are sitting around, happily admiring their loot and planning what to do next. Their first idea is to produce all the tuba they can. They figure: "Make more tuba, get more loot!" This is what the market process is supposed to get them to do. But maybe they decide not to do this.

Suppose the shrewd old northside chief decides they can make more profit, *not* by increasing the output, but by raising the price instead. He says to himself: "Even if we double the price of tuba, they probably will buy almost as much as before. The demand for alcoholic beverages isn't very *elastic*." Well, he probably didn't say it exactly that way. But that's what he had in mind. He means he could raise the price and people would still buy almost as much as before—the "quantity bought" would not change much in response to the price change. That means not much "elasticity of demand." (You'll hear more about elasticity of demand—that is, about the responsiveness of buyers to price changes—later.)

### The Effect Of Monopoly

Suppose the northside chief decides to hold the output at the same level, but to double the price. Soon the northside islanders will be taking in about twice as much "loot" in trade for their tuba. But they aren't working any harder or producing any more tuba than before! (Actually they're producing a little less, because at the higher price some people decide not to buy as much.) A good deal for the northsiders? Sure! But not so good for the other islanders.

What has happened to our neat and orderly market process? It's supposed to get production and distribution to respond automatically to the wishes of the people. But it isn't working right. The people want more tuba, but the tuba doesn't come! Instead, the price just goes up higher and higher. The "production choice" is not being made by the society, but by the northside chief. But even worse: the northsiders are getting a very large distributive share (income) as their reward for *not* producing the extra tuba the society wants!

All of a sudden everything has turned upside down. The market process is not bringing forth the right products. And it is rewarding the wrong people. Neither the production nor the distribution choice is responding to the wishes of the society. What's wrong? Monopoly, that's what. The northside chief has complete control of the supply. He is using his monopoly position to restrict output and raise the price—to enrich himself at the expense of everyone else in the society.

If the market process is working right, people are supposed to be rewarded for producing more of those things society wants most. But just look what happens when someone gets monopoly power—that is, the power to influence or control the amount of the product flowing across the market from sellers to buyers. Sometimes a market may have *a few sellers*. Then each seller will have a small amount of "monopoly power."

When there is only *one* seller, that seller's "monopoly power" will be much greater than if there were *two*, or *three*, or *more* sellers.

The northside chief doesn't have "just a little" monopoly power. He has *complete* monopoly power in the "tuba market." He has control over the entire supply. He can limit the output, keep the price high, and make big profits.

Any producer or seller who has "monopoly power" can do the exact opposite of what the market process says he's supposed to do, and he can get rewarded for it! People are supposed to produce more of the most scarce things, to help relieve the intense scarcity. But the monopolist can restrict his output and thus maintain (or further intensify) the scarcity. Then he can raise the price and be rewarded for taking this harmful action against the people. Can you see what a threat monopoly can be, to the proper operation of the market process?

### The Remedies For Monopoly

What will happen next? Will the other islanders let the northsiders keep their monopoly on tuba? What else can they do? Perhaps the other chiefs will get together, organize an army, and attack the northsiders to force them to share their secrets about how to make tuba. Or maybe the problem can be handled peacefully. Maybe an "all-island government" can be established, and the government can control the price. Or maybe the government can pass some laws or make some rules about monopoly and competition. If they did this, the island would have its first "government regulation" or "antimonopoly" or "antitrust" laws.

Perhaps the government's new rules will require the northsiders to share their secrets with at least one other family, so that there can be two producers of tuba. The two producers could then compete with each other. This might result in greater output and lower prices. Another possibility is that the government might decide to regulate the output and prices of tuba.

### Natural Monopolies Must Be Regulated

The government might decide that since the northsiders are the only ones who can make tuba efficiently, healthy competition is not possible in this industry. Therefore, the free-market process could *never* work right. The tuba industry would be called a natural monopoly. Competition would be inefficient, so the monopoly must be permitted to continue to exist. But to protect the people, a "natural monopoly" must be regulated by the government. The government would decide what quantity and quality of tuba should be produced, and the price at which it would be sold.

This is what happens with "natural monopolies" in all the market-directed economies. Telephone service, electric and water services, airline, rail, and bus service, and several other services are provided by governmentally regulated "natural monopolies." Natural monopolies are industries where competition would be too inefficient (too wasteful) to be permitted.

### Competition May Arise Automatically

It might be possible for the island's monopoly problem to be solved without either government control, or war. The southside chief might be jealous of the great profits being made by the northsiders. The lure of profit might stimulate him enough to overcome his natural laziness. Maybe one day he will sneak up and hide in a palm tree and spy on the northsiders' tuba-making operations. After he learns how to make tuba, he can sneak back home and go into the tuba business himself.

The new competition from the southsiders will add to the tuba output, reduce the great scarcity, and force the price to move down. In this case, competition has arisen by itself. Whenever this happens, it isn't necessary for the government to come in and take any action at all. Of course the northside chief and the southside chief might then get together and work out a cartel arrangement—an agreement to limit production and "stabilize" the price and not compete with each other. This "cartel" agreement would eliminate the competition between the two sellers. A cartel creates monopoly. The islanders might have to pass an "antitrust law" to prevent the northsiders and southsiders from entering into any such "conspiracy to restrain trade."

When competition can arise by itself to overcome or eliminate monopoly, governmental action is not necessary. But whenever monopoly exists and competition cannot or does not arise to curb the market power of the monopolist, governmental action becomes essential—either (1) to generate some new competition, or (2) to break up the monopoly into several smaller businesses, or (3) to regulate and control the outputs and prices of the monopolized products or services. Approaches (1) and (2) will "fix the market structure" so that the market process can work as it should. The third approach (3) recognizes that the market structure can't be "fixed," and lets *the government* (the political process) make the price and output choices.

### Competition Protects The Consumers

What's wrong with monopoly? It simply prevents the market process from working as it should. Competition forces producers to respond to consumer demand. With competition, producers must produce and sell more when the demand increases. If they don't, they will lose their customers to competitors.

When the demand for something increases, the present producers will produce more. Also, new producers will come in and start to produce the profitable product. More producers always try to enter the industries where demand is strong—where prices and profits are high. The market system is supposed to work that way.

**Competition Guarantees That An Increase In Demand Will Bring More Output.** Competition keeps each producer on his toes. It causes him to respond as quickly and fully as possible to changes in demand. The market process could not operate without the force of competition to keep all the producers and sellers in line.

Anytime a seller of anything is free of competition, he is free of the "natural control mechanism" of the market. He is free to increase his price and restrict his output. Then he gets rewarded—not for producing what society wants most, but the opposite—for *restricting* the output of the things society wants more of. Can you see that monopoly power could be a very profitable thing to have? The antitrust (antimonopoly) laws, in the United States and in other countries, try to prevent firms from creating or expanding their market power. These laws try to prevent firms from freeing themselves from the natural control mechanism of the market—that is, from competition.

## The U.S. Antitrust Laws

In the United States, the Sherman Antitrust Act (1890) and the Clayton Act (1914) are two of the basic antimonopoly laws. The Federal Trade Commission (FTC) Act (1914) which set up the Federal Trade Commission to keep an eye on business behavior, is another. These and other antitrust laws try to prevent firms from eliminating competition. It is illegal for firms *to get together* with their competitors, or *to destroy* them, or *to shut them out* of markets. The antitrust laws try to prevent any action which would interfere with free competition in any market for any product. These laws have been far from completely successful in preventing the existence and growth of monopoly power, but they seem to have helped some. You'll hear more on this, later.

If the society is going to permit "free business enterprises" to set their own prices and produce whatever they wish, then competition is essential. Only when we have competition to protect the consumers can free enterprise be justified. Without competition, "free enterprise" would become a system in which most of the economic choices for the society would be made by the monopolists. The few would get very rich while most of the people would live in poverty. This is the kind of situation Karl Marx predicted.

Marx said that monopoly power would grow worse and worse. Eventually a very few monopolists would gain control of all of society's capital and land. Then one day the poverty-stricken masses of downtrodden people would revolt, kill the monopolists, and take over. Luckily, Marx was mistaken. "The masses" in the industrialized countries have become very much better off — not worse off. Governments have increased their responsiveness to "the masses," and have increased their restrictions and controls on monopoly power.

What you really need to understand, at this point, is why competition is so essential to the operation of the free-market process. Also, you should be aware that governments do try to maintain competition, and that governments control the prices and outputs of the "natural monopolies," such as the utilities. But before we leave this subject for now, you should also be aware that governments sometimes take action to eliminate competition — to create, to perpetuate, or to encourage the growth of monopoly power.

### Governments Sometimes Encourage Monopoly

In the United States, some of the most important examples of government-created or government-supported monopoly (market) power are found in the markets for labor, and for agricultural products. In the labor market the government sets the minimum price (wage), and labor unions are given "monopoly power" so they can bargain for higher prices for the sale of labor services. In the markets for some agricultural products the government sets minimum prices, restricts outputs, and buys up surpluses.

The patent and copyright laws are designed to give monopoly to inventors, singers, writers, and others. There are several other examples, but that's enough on this subject for now. We will come back to the issue of monopoly and competition, later. Right now it's time to take a look at another important issue: economic inequality.

## THE ESSENTIAL ROLE OF INEQUALITY

Did you ever stop to think about how "reward-oriented" we all are? Why does a person struggle out of bed in the morning and hurry to get to work on time? What gets the taxi driver to keep fighting the rush-hour traffic? What gets farmers up at dawn to plow, sow, harvest? What gets people to save and invest? Why are people willing to work overtime? nights? weekends? Why would a family pull up stakes and move to some distant city where better jobs are available? You know the answer. The promise of rewards. The promise of higher incomes—a larger share of the output—more and better things. Of course.

## We Are All "Reward-Oriented"

Even our approach to education is highly "reward-oriented." We professors offer (and most students seem to want) academic rewards, almost on a day-to-day basis. Schools give rewards in the form of grades, quality points, course credits, semester hours, Dean's list, scholarships, degrees, cum laude, and other such "contrived" rewards. Most of us admit that emphasis on these rewards often diverts attention away from the real purposes of being in college. Yet most people seem to insist on them.

Some people suggest that working for money, trying to get ahead, to pay off the mortgage, to gain some financial security—that these objectives divert attention from the real purposes of life. Perhaps so. Nonetheless, this is the way most people are. This is the way the world is organized.

Perhaps all this emphasis on "seeking rewards" appeals to the basic nature, the survival instinct, of all living things. It isn't just true of people. All animals and all plants are seeking, reaching for rewards. The most successful ones get the biggest rewards. The least successful ones may get no rewards at all. The result? Inequality.

Inequality is essential in any system which offers incentives, or rewards. The "market system," in its pure form, operates entirely on the basis of incentives. Rewards. Inequality is essential to the functioning of the system.

## A Reward-System Requires Inequality

If you are going to be rewarded for performing some task, that means that you must *not* receive the reward unless you perform the task. Obviously. If you are going to get the reward anyway (task, or not), then it isn't a reward anymore!

There aren't very many ways to get the people of the world to do the work that is necessary to keep society fed, clothed, housed, doctored, transported, taught, protected, entertained, etc. One way to get people to work is to reward them. Another way is to order them to do it and punish them if they don't. If rewards are used, inequality results—some people may get a lot more than others. But if rewards aren't used, any other system results in a loss of individual freedom of choice. So which system are we going to use? That's the kind of tough dilemma every society must face.

Either we put up with the inequality and use rewards as incentives, or we use something like the military draft to force people to do what the ones in charge tell us to do. Under the draft laws, all "able-bodied men" are equal. If the system is working as it should, there's absolutely no inequality. But notice also that no man is free. Even the "direct command" economic systems are likely to exhibit a high degree of inequality. Unproductive people don't get very much income in those systems, either. And the ones in charge somehow seem to wind up with quite a lot.

The market process absolutely requires that economic inequality be permitted to exist. Neither the political process nor the social process *absolutely requires* economic inequality in order to get the economic problem taken care of. But in every past or present real-world economic system, a high degree of economic inequality has always existed. Capitalism requires it. Conceivably, a system of pure socialism or communism could function without it—but none ever has.

Capitalism is sometimes criticized because inequality is an essential part of the system. Certainly, on the face of it, inequality is not the most desirable human condition. No one who has a feeling for humanity likes to think of wealthy ladies feeding milk to their cats, while on the other side of town children are suffering of malnutrition because they don't have enough milk. But unfortunately, in every economic system today and throughout history, we can find examples of great waste on the part of some people, while others suffer from malnutrition.

### Dramatic Examples Don't Help Much

Probably such dramatic contrasts will always be discoverable in every society. But examples of such dramatic economic injustices serve very little purpose—except to fuel the emotional fires which demagogues try to build to hypnotize otherwise intelligent people. Rare and dramatic examples only becloud the real issues. The real issues hinge on the actual extent of deprivation of the poor and waste by the rich—and on finding *practical* solutions to the problem—solutions which will do more good than harm.

We know that inequality is a fact of life in every economic system, and that in capitalism some inequality is essential. But *how much* inequality is essential to permit the market process to work? People argue about that question all the time. The truth is that nobody really knows the answer.

In recent years, all market-directed systems have been undergoing constant changes, to reduce inequality. The political process has been making more and more of the distribution choices. Income taxes, business taxes, inheritance taxes and many other taxes are collected from people who are productive (or who own capital which is productive). Then the government gives money to the unemployed, sick, aged, dependent children, and others who cannot support themselves. Some such income redistribution program is now working in all modern societies, both in the market-oriented countries and in the communist countries too.

### How Much Income Redistribution Is Possible?

In every market-oriented economy, the governments take property "by due process of law" (taxes) from the people who have "earned" it as income (as rewards for their production), and then "redistribute" this property (as money, goods, services) to other people—people who did not earn it. At what point does this "redistribution" process destroy people's incentives to work? At what point do the productive people say, "Why work more (or risk more), when I will get to keep so little of the extra income I earn?" At what point do the unproductive ones say, "Why try to get a job? I am already receiving about as much as I would get if I went to work." No one knows the answers to these questions. No one knows how much the existing redistribution programs are interfering with incentives.

If people think marginally (and of course people do think marginally—marginal thinking is just common sense!), any redistribution of income will have *some* effect on *some* people's behavior.

Suppose a person is "right on the fence." He just can't make up his mind about whether or not to work overtime, or maybe to move somewhere else to get a higher-paying (more productive) job, or maybe to take a training course to improve his income. Is the income tax likely to influence his decision? Of course. If he's going to get to keep all the extra money he makes, he is more likely to do the extra work. If he's going to have to pay half of it to the government in taxes, he may decide to forget the whole thing.

Some people are always on the "threshold of reaction" (right on the fence), so surely our present-day income redistribution programs must be having some effect on some people's choices—to work or not to work, to invest or not to invest, and all that. But how much effect? On how many? We don't know. We just don't know.

### People Disagree About The Effects Of Income Redistribution

Some people say that the income redistribution programs are destroying the free enterprise economy. They say that reduced incentives are slowing down economic growth; that people are less willing to save and invest; that people and businesses are no longer so interested in working harder and becoming more productive—so the market process is losing its ability to work for the society. Others argue that no real harm has resulted, that much more of the output of the society now is going to the people who need it, that there is less waste by the rich, and that the income redistribution programs are good and need to be extended.

This issue will be one of the important economic policy dilemmas you will hear people arguing about throughout your lifetime. Let's hope we soon learn to understand it better. In Part Eight of this book you'll be taking a closer look at it.

We know that if the market process is going to work, incentives are essential. Rewards must be used. So inequality can never be eliminated. To eliminate inequality would require that rewards be eliminated. And to do that would be to "kill the goose that lays the golden egg." For truly, it is the *incentive of reward* which works in the market process to make the total output high enough so that there is enough for some people to waste things. It is the *incentive of reward* which generates enough total output so that some of it can be redistributed from the original producer-earners to the unproductive ones. As society evolves, maybe rewards will become much less important, much less necessary as a technique for motivating people. But that hasn't happened yet.

### The Great Dilemma Of The Market System

Waste by some while others go hungry? That isn't good. So we have a dilemma. Right? Yes. There's a bad flaw in the system. But as yet, no method anyone has ever devised for getting the world's work done, for stimulating people to produce the needed output—has come out any better. Actually, most of the tried alternatives seem to have turned out worse.

All this doesn't mean that we should stop seeking better answers to the distribution question. But it does suggest that maybe we shouldn't completely destroy the "incentive and reward" approach until after we have figured out something else that might work. That's the tough part. It's a lot easier to protest or support things than it is to understand them! But you've already found that out. Right?

### Market Incentives Really Do Work

Incentives, rewards, and the resulting inequality are essential in the operation of the market process. These are the conditions which stimulate and capture and harness and channel the energies of the society, and aim them toward the desired objectives. The incentive of reward stimulates and directs the economy and keeps it going and growing.

Would you be just as interested in a college degree if you knew it would have no effect on your lifetime income? Some people would, but many wouldn't. The market process stimulates and directs most of us. As we try to get ahead—to achieve some "personal economic growth"—

automatically we are working for the growth of the entire economy—and of the world, for that matter.

The market process exerts a powerful force for the growth of the economy. It stimulates people and businesses to expand their capacity to produce. Each time an individual's productive ability increases, the society's productive capacity expands. It's only because of this growth (and the specialization and trade which it embodies and permits) that we have the time to sit around and worry about these issues of economics, and inequality, and all that. If it wasn't for economic growth, we'd all be out grubbing for food! (At least the few survivors among us would be.) Do you get the idea that "economic growth" is a very important subject? It sure is. That's what the rest of this chapter will be talking about.

## ECONOMIC GROWTH COMES FROM MORE AND BETTER FACTORS OF PRODUCTION

The concept of economic growth can be applied to an individual, a business, an organization, an area or region, a society, a nation, or even to the entire world. Usually it is used to refer to a country—a nation. The simplest meaning of economic growth is "increasing capacity to produce," or "increasing productive ability."

The term "economic growth" is often used to mean "*per capita* increase in productive ability"—the ability to produce more output, *per person*. This is the kind of growth which can provide an increasing standard of living for the people. As economic growth begins, often the population begins to expand too. When that happens, the output *per person* winds up no bigger than it was before. The "average person" is no better off than before. We'll talk more about this problem later. First, you need to understand *the process of economic growth*. The next several pages provide a step-by-step description of the process. You will see how each economic unit—individual, business, or nation—acquires the "more and better factors of production" which lead to economic growth.

### Economic Growth Requires Saving And Investing

Suppose you have been growing corn and tomatoes in your backyard plot for the past few years. You have worked very hard, but your output has been small. This year you want to use your labor and land more efficiently. You decide you need more and better capital. You want to get a garden tractor and buy better fertilizer and better seed. How can you do that? You must save. Or you might borrow from someone else who has saved. But *in order for new capital to become available, someone must save*—that is, *someone must produce more output than he consumes*. Can you see that this must be true?

If everyone consumes exactly as much as he produces, what's left over? Nothing. Capital, remember, is anything that we produce and keep, to use in future production. If there are no *savings of output* anywhere in the country, that means we are eating up all our output. Output is like cake. We can't have it and eat it too!

There cannot be any new capital to use to help us to produce more next year, unless we produce more than we consume this year. This is another one of those little "obviously true but often forgotten" concepts in economics. If we want some left over to use tomorrow, then we can't eat it all today. Truly, this is the key to economic growth.

You can increase the "productive capacity" of your backyard farm, by spending less of your income for consumer goods (saving money), and then buying capital goods instead. Anyone can do that—any farmer, any businessman, any large corporation, any individual. When someone saves some of his income and then spends his savings to get more capital, his capacity to produce increases. That results in economic growth.

The more you save and invest, the more "economic growth" you can have. The more your economic growth, the more output you can produce. For a society, if all the factors of production are used to satisfy "consumer objectives," then productive capacity cannot expand. Growth can't occur. But if many of the factors are used to make capital goods instead of consumer goods, productive capacity will increase. Economic growth will occur. As the years go by, the society's output can get larger and larger. So if the population doesn't expand too much, standards of living can go higher and higher.

**Economic Growth Means Increasing Capacity To Produce**

If we are going to increase the capacity to produce—that is, increase the total amounts of things we are able to produce in a day, a week, or a year—we must have either *more*, or *better* factors of production (labor, land, capital) or else we must learn to make better use of what we have. Obviously! What else is there for us to work with, but the factors of production? Nothing.

If an individual wants to experience "economic growth," he can improve the productivity of his labor (by education, training, observation, trial and error, experience, etc.), or he can save his income and buy (invest in) more and better land, or more and better capital. If a nation wants to experience economic growth, it must do the same thing an individual (or a business) must do. It must get more and better factors of production (or it must get things better organized, so it can make better use of what it has).

A nation usually can't get more land. Land area and the other free gifts of nature are just about all given out already. So the nation which wants economic growth must concentrate on the other two factors—labor and capital. Essentially, economic growth for any nation hinges on the organization and development of a better labor force (better utilization, education, attitudes, skills, etc.) and somehow getting more and better capital (building more power plants, factories, railroads, and producing or importing more machines, equipment, etc.).

Every nation is constantly working to improve its labor skills, and to get more and better capital. Everyone wants the kind of capital with the latest technology, the highest efficiency, the greatest productivity. With improved labor and more and better capital, productive capacity increases. Total output can increase. The real income of the people—that is, the goods and services which the people get to have and enjoy—can increase.

With economic growth, people can have more and better things—food, transportation, medical care, libraries, refrigerators, telephones, air conditioners, fish hatcheries, colleges, hamburger drive-ins, antipollution devices or whatever the society chooses to produce with its increased productive capacity. But the people can't enjoy any of these extra things if the population expands fast enough to eat up, use up, and wear out all of the increased output!

## Some Advantages Of Economic Growth

Economic growth can be a great thing for a society. It lets the society produce enough output so that people can live more humanely—more like people, and less like animals. It provides some "surplus" above what is required for subsistence. It lets the productive ones share with the unproductive ones—the sick, the crippled, the very old, the very young.

Economic growth lets the society feed and care for the young people for the first 20 years of their lives so they can go to school—or maybe just have fun. Usually nobody has to go hungry to make up for this "extravagance." Growth lets each person have free time to think—to "do his own thing"—to relax and enjoy life.

Yes, economic growth can be well worth striving for. All the less-developed countries place great emphasis on stimulating economic growth. The very lives of many of their people depend on it. But when we choose the path leading to economic growth, just as with every other choice we make, there is a cost.

## The Cost Of Economic Growth Is Present Consumption

What is the opportunity cost of economic growth? When you save—when you produce things you don't consume—when you use your factors of production to make capital instead of consumer goods—what happens? You give up the opportunity to consume more, today. You are giving up your chance to eat cake today—or even to *have* cake today—when you are using your resources to build a better oven (capital) for baking cakes in the future.

The worker who is training to improve his skills, or the student who is building his education for the future, is not producing anything. At least, not now. Later he will be more productive. But today he produces nothing. Who produces the food to feed him? Someone else must. This means that some other people must have less so the trainee or student can eat.

You can see that the opportunity cost of rapid economic growth could be very high. Those factors of production we are using to produce gear grinders can't also be used to produce canned peas. Nobody eats gear grinders. When a poor society produces more gear grinders (less peas), someone goes hungry. Of course, *someday* the gears made by the gear grinders are going into tractors. With the tractors, the poor society can produce many more cans of peas. But until then, the people must force themselves (or be forced) to get by with less food. That's the opportunity cost of economic growth. Is it worth it? That's a tough question.

Should some people be forced to go hungry now, to bring a better future for the children and grandchildren of tomorrow? What a difficult choice for a society to have to make! But many of the less-developed countries are having to make this choice.

### Economic Growth Usually Requires Forced Savings

The problems of economic growth are difficult and serious. Most of the difficulties stem from the "opportunity cost" aspect of growth—the fact that some of the current output must go into improving labor and technology and into building capital for the future. It is not easy to do this, when all the output is needed now, for current consumption.

The important thing to understand is that growth requires saving—*producing but not consuming*. Saving is never easy. Poor people (and poor nations) find it almost impossible to save. Usually a society can't get started on the path of economic growth unless the people are somehow forced to save.

The modern nations have experienced very rapid economic growth over the past century. Large savings were required to bring this about. Most of these savings were forced on most of the people. How? By unequal income distribution! Most of the people never got the chance to use up much of the output. If they had had a choice, they would have used up more. They would not have saved so much of the society's output. Most of the economic growth would not have occurred.

In every society, most of the people are prevented from using up as many things as they would like to use up. Either the social process or the political process or the market process limits their consumption. The limited consumption (forced saving) releases the factors of production from making consumer goods, so more capital can be produced. The result: economic growth. The issue of economic growth takes us right back to the "basic economic choices of every society." If growth is going to occur, the society must choose to produce fewer consumer goods, so that some capital goods can be produced. Somehow the society must prevent "the masses" from receiving too much income. In an advanced, wealthy society, it's no problem. Everyone can save some. But if everybody is poor, people won't save. If they are allowed to, they will use up all the output of the society.

In a less-developed country, if all the output is equally distributed among the people, nobody's share will be very large. Everybody will use up all of his share of the output and still want more. The level of savings will be zero. No growth can occur.

## Wealthy People Generate Economic Growth

In an economy where the market process is answering the production and distribution questions, who does most of the saving? And then who uses those savings to demand more capital, factories and machines, research, new technology, and the other things necessary for economic growth? The rich people and the big corporations, of course!

From the early 1800's to the time of the Great Depression of the 1930's, the great surge of economic growth in the United States was financed mostly by the high incomes of the wealthy people and big businesses. These "capitalists" and big businesses developed some very productive capital which brought them large incomes. They spent their large incomes to demand (and thereby to cause the society to produce) more and better capital. As they acquired more capital their incomes went up even more, so they built even more capital. As this process continued, the economic conditions of the masses improved a little, but not much. Mostly, they stayed poor. But the economy grew by leaps and bounds.

During the latter 1800's and early 1900's, some very wealthy capitalists had personal incomes (tax free!) of several million dollars a year. You can't spend several million dollars a year, year after year, on consumer goods. You just can't, no matter how many mansions and yachts you buy! The only thing you can do with that much money is to use it to build more capital. And that brings you even more income. You get even richer. The thing keeps on snowballing, getting bigger and bigger, and there seems to be no way to shut it off! Once your personal income gets up to, say, five million dollars a year, the only way you can keep from getting more and more wealthy is to spend up, or give away, or somehow get rid of about $15,000 a day! How would you like to try to do that for a few days?



## ECONOMIC GROWTH IN DIFFERENT SOCIETIES

The "economic process" of growth is essentially the same, no matter what country or society or economic system you happen to be looking at. Saving is essential. Resources and energies must be used to produce more and better capital, and to improve the labor force. But different societies, different economic systems, accomplish these results in different ways.

### Growth By The Market Process

In the United States and in most other countries, the *market process* brought about the rapid economic growth. The high incomes of the "capitalists" (and the low incomes of most of the other people) "forced" the society to save. This permitted the growth to occur. It was painful for the poor people. They suffered, but now we are all sharing the benefits of their sacrifices. It all resulted from "forced savings"—the low incomes of the workers—and the high incomes and capital-building of the rich.

### Growth In The Traditional Society

In societies controlled by tradition there usually isn't much growth. If any economic growth occurs, it is accomplished in the traditional ways of doing things. For example, it might be "understood" that all able-bodied men must work for three days each month clearing and developing roads and digging channels and clearing the harbor. This social requirement for capital-building, by labor, would cut down on the time the people could spend catching fish and gathering breadfruit and brewing tuba and producing other consumer goods—or just resting. The amount of the consumer goods (or leisure time) would be reduced, and the society's investment in capital would be increased. The society is "transforming" consumer goods into capital goods. The society is saving and investing.

Suppose the traditions required that the workers go from one family farm to another, clearing brush, planting young palm trees, expanding the taro patches and building copra-drying sheds. All these efforts would be "capital-building." The more this sort of thing is done, the more productive the society (economy) will be in the future. In underdeveloped, tradition-oriented societies, this is often the way the capital-building actually takes place. But usually there isn't much growth. The amount of effort usually is only sufficient to maintain the existing capital, and to help keep up with the needs of the expanding population.

### Growth In The Communist Countries

In societies in which government control over the production and distribution choices is dominant—that is, in the command-oriented, communist societies—the government decides how much capital will be produced. The decision becomes a part of the "economic plan" (perhaps the "five-year plan") of the country. If much capital and few consumer goods are included in the plan, then the level of consumption in the society will be low, and the growth rate will be high. If the amount of consumer goods produced is too small, some people will have to go hungry.

Rapid economic growth has been a major objective of the communist countries. In the Soviet Union, thousands of people actually died during the 1920's because of the lack of consumer goods. Both in the Soviet Union and in Communist China there has been a very determined effort by the government to hold down consumption and to build capital rapidly.

The growth of the Soviet Union over the past fifty years has been very rapid, especially considering the poor conditions which existed when the Communists took over, in 1917. But the most rapid growth rate of any nation in modern times has occurred in Japan, just in the last few years. In Japan, the relatively low wage rates have held down consumption, and profits have been high. This has provided funds for investment in more capital. Also, the good profit picture has caused technology and capital to move from the United States and from other nations, into Japan.

### As Wages Rise, The Growth Rate May Slow Down

The market process can be a powerful force for economic growth. But in the beginning it works only if wages are low enough and profits high enough to generate savings and investment. That's what is necessary to direct the factors of production into making more and better capital. But here's the catch. These days, people don't like to live in poverty while a few wealthy capitalists live like kings. That's the way it was in the last century, but these days—in the 1970's—most people wouldn't put up with that much inequality.

Everyone always seems to want a larger share of the output than he is getting. That has always been true. But these days, many people have the necessary economic and political power to be able to get more. No longer are people content to "suffer in silence." In the United States and other nations, workers have gained much more power. So have the farmers and fishermen and other economic groups. Also, the voice of the poor has become quite loud and powerful.

These days, the society takes a more humane attitude toward people. The government takes money from the rich, and gives to the poor. As all this happens, more of society's resources go into consumer goods—TV sets, cars, living room furniture, air conditioners, movies, canned peas, pork chops, TV dinners, outboard motors, disposable diapers, airline travel, electric can openers, and all the other things the "average people" will choose if they have the money. So now the economy produces more consumer goods instead of capital. The growth rate is not so fast anymore.

### The Highly Developed Economy Just Keeps On Growing

In the United States, today, we see an economy of high mass consumption and low profit levels. Even so, the U.S. economy can continue to grow through the market process. How can this be? It's just that *so much* growth has already occurred—the economy is now so highly productive—that almost everyone can live like a king, and still there are enough factors of production left over so that new capital can be produced and growth can continue.

Remember the snowballing effect of having a five million dollar annual income? If you ever reach that level, you can't help getting richer and richer. It's sort of that way for the economy as a whole. In the United States the total productive capacity is great enough to provide high standards of living for the people, and with enough factors of production left over to make *even more capital*!

In an economically advanced society like the United States, even the "ordinary people" have incomes high enough to allow them to do some saving if they want to. They can invest in capital if they want to. Or they can just put their savings in the bank and let others do the investing. In the United States millions of people save some of their incomes. Many hundreds of corporations earn income (profits) but don't pay all the income out to the stockholders. They keep the income and invest it in more capital. See how the market process automatically keeps the growth process going in an advanced country? But what about in an "unadvanced" country? It doesn't work that way.

### The Growth Problem In The Less-Developed Countries

In the advanced nations the people can enjoy high standards of living, while growth continues. But not so for the "unadvanced" nations. If the less developed countries (LDC's) are going to generate much economic growth out of their own resources, some of the people are going to have to go hungry some of the time. But many of the people are already getting by with less than they need, so there isn't much "slack" to get the growth process started.

Another serious problem in the LDC's is that things are so bad that there aren't very many promising opportunities for investment. Almost all LDC's have their few wealthy people who earn big incomes. But where do the wealthy people invest their money? They trade their money for dollars (or pounds, or yen, of some other foreign money) and then make investments in other countries. Of course! Why should they invest in some shaky enterprise in their shaky country? The lack of promising investment opportunities is a real and serious problem in the LDC's. It's no different, really, from the situation of a wealthy person whose home may be in a depressed, poverty-stricken county in Appalachia. Will he invest his money in his local county? Would you?

So how can the growth process get started in the LDC's? The governments of the highly developed nations might make grants or loans of capital to the governments or businesses in the LDC's. Or some businesses in the United States and other advanced nations might go into the LDC's and start modern farms and factories and processing plants and other modern activities. But neither of these possibilities seems likely to occur on a large enough scale to really get things going.

**There Is No Easy Solution To The Problem Of The LDC's**

Can more capital be brought in from outside, while the local government keeps consumption at a bare minimum and places severe limits on population expansion? And perhaps pulls the surplus workers from the farms, trains them, and puts them to work building roads and railroads and docks and harbors and power plants and things? Perhaps. But only "perhaps."

One thing seems sure. Only carefully made plans, with the force of the government carrying out the plans—and with some generous help from outside—seem to offer any real promise of breaking this vicious circle of poverty in most of the underdeveloped countries of the world. We'll talk more about this in Chapter 30, which deals with current international problems and issues. When you get to Chapter 30, you will understand a lot more economics, and you will be able to understand this development problem much better. But even there, you'll still find that there's no easy solution—maybe, in some cases no visible solution at all.

## A SUMMARY OF THE CONCEPTS OF ECONOMIC GROWTH

The basic concepts involved in the issue of economic growth are really very simple. If we are going to be more productive tomorrow than we are today, then somehow we must do something today that will result in our increased productivity, tomorrow. Obviously. But whenever we use our efforts today to help us *tomorrow*, we are giving up the chance to use our efforts today to get what we want *today*. We are giving up today's pleasure for the sake of a better tomorrow.

Anything anyone does today to provide a better way to produce things tomorrow, will mean that he will have less to consume (or less leisure time) today. Economic growth requires choosing *not* to produce consumer goods, but to produce capital goods instead; *not* to spend your time earning money so you can live better today, but go to school or learn new skills instead; *not* to spend your income for the new car you would like to have, but to spend it for a better set of tools instead. The opportunity cost of economic growth is current consumption? Sure.

### The Distribution Choices Influence The Growth Rate Indirectly

It's easy to see how the rate of growth is influenced by how the distribution question is answered in the society. If everyone gets an equal share, most of the output will be consumed. But if a few people get a great amount (or if the government takes a great amount) and the majority of the people get very little, then growth can be rapid. The distribution choice influences the growth rate indirectly. Unequal distribution causes "forced savings" and generates more growth.

### The Production Choices Influence The Growth Rate Directly

The production choice determines the growth rate *directly*. If the production question is answered by deciding to produce many consumer goods and few capital goods, the growth rate will be low. If the decision is made to produce fewer consumer goods and more capital goods, the growth rate will be higher.

The more economic growth a society experiences, the easier it is for that society to grow even faster. As growth occurs, the society gains *more* and *better* factors of production. This lets the people live better while the economy grows even more.

Those of us in the advanced nations, where we work so little and have so much, are truly indebted to our ancestors—not just to our rich ancestors who built the capital and ensured the growth, but also to our poor ancestors who, through no choice of their own, worked long hours, lived in poverty, and made all this "industrialization," all this "capital accumulation," all this "economic growth," possible.

The major concepts discussed in this chapter and in the previous one—the market, private property, the factors of production, competition, inequality, and economic growth—carry you a long way toward developing an understanding of the market process and how it works. But one important link is missing. You don't yet know about money.

One essential requirement for the market process to work is this: *There must be some kind of money.* Truly, money serves as the lifeblood of the market system. We have already talked about the other required conditions. So now it is time to talk about that surprising, remarkable thing—that thing about which (as Will Rogers observed) people think and talk so much, yet know so little. It's time to talk about *money*. That's the subject of Chapter 8, coming up next.

---

## REVIEW EXERCISES

### MAJOR CONCEPTS, PRINCIPLES, TERMS (Try to write a paragraph explaining each of these.)

the role of competition
the role of inequality
the great dilemma of the market system (inequality)
the role of savings and investment
the process of economic growth

### OTHER CONCEPTS AND TERMS (Try to write a sentence or phrase explaining what each one means.)

elasticity of demand
monopoly power
natural monopoly
anti-trust laws
the Sherman Act
the Clayton Act
the FTC Act
income redistribution
economic growth
savings of money
savings of output
forced savings
saving and investing
capital accumulation

### QUESTIONS (Try to write out answers, or jot down the highlights.)

1. Businesses that favor the "laissez-faire" approach by government should be very much in favor of strong government programs to prevent monopoly and to strengthen competition. Right? Discuss.
2. Some colleges now let students take some courses on a "Pass-Fail" basis.
   (a) Do you think the "pass-fail option" is a move toward greater "equality," and less emphasis on the "reward approach" as a method of getting students to learn?
   (b) Do you think you would be learning more economics, or less, if you were taking this course "pass-fail"? How about if there were no grade and no credit and no college degree? Nothing to

stimulate you but your own "yearning for learning." What about that?

(c) How far do you think colleges can go in eliminating their artificial "incentive reward" systems? And give "equal rewards" (or no rewards) to everybody?

(d) How far do you think a modern economic system can go toward eliminating pay incentives, and giving equal incomes to everybody? Nobody knows, really, but what do you think?

3. Trying to get economic growth is sort of like trying to borrow money from a bank. Usually, the more severely you *need* it, the harder it is to get! Explain.
4. One way of thinking about economic growth is to think of it as "increased production potential *per capita*." If that's what economic growth is, then one way to get it would be to *get the population to decline*. Right? Discuss.

# 8 MONEY: THE LIFE BLOOD OF THE MARKET PROCESS

*Something will arise automatically to serve as the medium of exchange so the market mechanism can work.*

Money is one of the most important, yet most thoroughly misunderstood things in modern society. This chapter moves along very slowly and carefully to try to dispel the myths about money and to give you a really honest understanding of what money is and how it works. Some of the things you will read on the following pages will be very different from what you have always thought about money. Some of the things may not be easy for you to believe. But it's all true. Take your time and try to really understand it. So much of what's going on in the modern world is incomprehensible to most people simply because they don't understand about money. Here's your chance to escape forever from "the ranks of the unaware." Here's your introduction to the magic world of money.

## THE MEDIUM OF EXCHANGE

Everyone has heard that money is a medium of exchange. Exactly what does that mean? Let's look at some examples. A newspaper is a medium for spreading the news. We talk about newspapers and radio and TV as being "news media." A truck is a medium for transporting things from one place to another. We talk about trucks, railroads, airlines, pipelines as "transportation media." You have heard of the spiritualist who acts as a medium for contacting spirits. So what is a "medium"?

A medium is anything which serves as a go-between. It permits or makes it easier for something to happen. A medium lets you accomplish some objective which could not otherwise be accomplished, or which could be accomplished only with great difficulty. When we say that money is a "medium of exchange," we mean that money makes it easier for people to exchange things—to trade things.

Most of the trade that goes on in the modern world could not be accomplished without something to serve as the medium of exchange—that is, it could not be accomplished without something to serve as money. *Whatever serves as money, is money.* That's the way money is defined.

Money is anything which serves as the medium of exchange. It is not "the medium of exchange" because it is money; it is money because it is serving as "the medium of exchange." To say it differently: Anything which performs the function of medium of exchange, is money. No matter what it is, so long as it is serving as the exchange medium, it is money. After I show you some examples you will understand what all this means.

## The Need For A Medium Of Exchange

I am a fisherman. I just caught a boat load of fish and I want to exchange all these fish for a new suit. Without an "exchange medium," this is very difficult to do. Try to find a suit-seller who wants a boat load of fish! But with an exchange medium, it's easy.

What will I accept as a medium, to help me exchange my fish for a new suit? I will accept anything for my fish—I don't care what it is—so long as I am sure the suit-seller will accept it in exchange for a suit. If there is any doubt about whether or not the suit-seller will accept it for a suit, then I will not accept it for my fish. If I think he *might not* accept it then I *will not* accept it.

Now, let's be more realistic. A fisherman really doesn't want to exchange his boat load of fish just for a suit. He wants groceries, electricity and water, housing (rent), and a few glasses of beer and some jukebox music at the local tavern. So whatever he accepts in exchange for his fish cannot be something acceptable *only* to the suit-seller. It must be something that is acceptable to all those other people, too. Why might all those other people accept it? Only because they are confident that everyone they do business with will accept it too. See how it all ties together? When everyone is confident that everyone else will accept something as money, then that something becomes *generally acceptable*. Then—and only then—it is money.

## Money Is Money Because People Believe It Is Money

Now you know what *money* is. Money is anything which is *generally acceptable* in exchange for things. People will accept it for whatever they have to sell—products, natural resources, labor services, or whatever. The fisherman will accept it for his fish because he is sure other people will accept it for the things he wants to buy. It's as simple as that.

In any society, at any given place or time, *anything* which performs the function of exchange medium, is money. It is money because everyone is confident that it is "good money." If some people begin to question the acceptability of whatever is being used as money, then they will refuse to accept it in exchange for the things they have to sell. Once this happens, everyone will stop accepting it. It will cease to be money.

If you heard that there was a great flood of near-perfect counterfeit twenty-dollar bills in your neighborhood, you might decide not to accept any twenty-dollar bills. Others also refuse to accept twenty-dollar bills. So twenty-dollar bills cease to be money in your neighborhood. How quickly something can cease to be money when the people lose confidence in its general acceptability!

## The Functions Of Money

As soon as something begins to perform the role of medium of exchange—that is, as soon as it becomes generally acceptable in exchange for things—then that something becomes money. Once it becomes money it automatically begins to perform all the functions of money. There are four functions of money. The first and indispensable function of money is to serve as the exchange medium. You already know that. But there are three more functions.

Money serves as the standard for valuing things. Once money exists as a medium of exchange, people automatically begin to compare the values of things in terms of how much "money" each thing is worth. Is my boat worth more than your car? How can we tell? Translate the boat into its "dollar value" and do the same for the car. Then you know immediately which is worth more. A $3,000 car is worth more than a $1,500 boat. Right? Of course.

People in every society need some way to compare the values of things. Money makes this easy to do. So money serves as the standard of value, or unit of value for the society. The "value unit" function and the "medium-of-exchange" function are sometimes called the two *primary functions of money*. The "primary functions" are the functions which money performs at the present time. You might call them the "current functions" of money—the functions of valuing things and exchanging things *now*.

The two other functions of money serve the purpose of "transferring value" from one period of time to another. These are called the *secondary functions of money*. You might call them the "future functions" of money.

Whenever a person or a business takes on an obligation and agrees to make some future payment, the obligation (debt) is usually stated in money. The money unit (for example, dollars) is usually the most convenient way to express the amount of the debt. Everyone understands exactly what the other person has in mind. When money is used as the measure of a debt, we say that it is functioning as the standard of deferred payments.

In its other secondary function, money serves as a convenient thing to save—a convenient form in which to hold savings. It is a thing which can be used to "store up value" for the future. When you save money, what function is the money performing for you? It is serving as a store of value. That's the fourth and final "function."

If a person wants to save, he must have some way to store up the value he is saving. He could buy and hold land, or he might buy and store steel ingots, or mahogany lumber, or antique furniture or anything else he thinks will hold its value. Money is a very convenient form for storing up value. If a person saves money (instead of land, or steel, or whatever), he can always exchange it for other things in a hurry. A person knows that he can use the money anytime, for whatever he might want to buy. No other kind of asset is as liquid (as easily and rapidly exchangeable into other things) as money.

The four functions which money performs for the society—medium of exchange, unit of value, standard of deferred payments, and store of value—are *vital* in the operation of the market process. Even in tradition or command-oriented societies, these functions must be performed. It would be difficult even for a primitive society to operate without something to perform the functions of money. Modern society couldn't exist without money.

## HOW MONEY MIGHT EVOLVE

To illustrate how money might evolve, let's go back to our island economy. Remember when the northside islanders started selling tuba? They were accepting all sorts of things in exchange—dried fishsticks, breadfruit, taro, and anything else the other islanders wanted to trade. But much time has passed since that day. Now the northside islanders no longer accept "any old thing" in exchange for their tuba. Here's what happened.

### Fishsticks Become Money

One day the westside islanders took several baskets of breadfruit and taro to the northside islanders to trade for tuba. But it happened that on that day the northside islanders already had all the breadfruit and taro they could use, so they refused to accept the breadfruit and taro in exchange for tuba. They explained that the breadfruit and taro would all be spoiled before they could eat it. Then the northside chief made a suggestion.

"Why don't you trade your breadfruit and taro to someone for dried fishsticks? We will sell you all the tuba you want to exchange for dried fishsticks. Dried fishsticks last forever. If we get too many fishsticks, we can always ship them and sell them in Japan. The Japanese call them "katsuobushi" and value them highly. They pay us with Japanese money (yen) and we can use the yen to buy all sorts of things from Japan."

So the westside islanders go around trying to find somebody who will give them dried fishsticks in exchange for their breadfruit and taro. Sure enough, the eastside islanders are willing to trade fishsticks for the breadfruit and taro. So the westsiders make the trade, then go back to the northsiders and exchange the fishsticks for the jugs of tuba they wanted. In this trade, what function did the fishsticks perform? Medium of exchange? Of course.

Soon the word gets around that the northside islanders will always accept fishsticks in exchange for tuba. Because of this, and because tuba is something everyone seems to want from time to time, other people around the island begin to accept fishsticks in exchange for the things they want to sell. As soon as everyone realizes that everyone else will accept fishsticks in exchange for things, then no one hesitates to accept fishsticks in exchange for whatever he wants to sell. Fishsticks have become money? Just automatically? Right!

Anyone who wants to trade anything can first sell it for fishsticks, then use the fishsticks to buy whatever he wants. As long as fishsticks continue to be generally acceptable in exchange for goods—that is, as long as everyone is confident that everyone else will accept fishsticks—then fishsticks will continue to be money.

See how easy and simple it is for money to arise? The gains from trade are so great, and trade without money is so difficult, that some kind of money just naturally evolves. It has actually happened this way in every society—small or large, primitive or advanced. Something always arises to perform the functions of (and therefore, to become) money.

### Specially Printed Paper Becomes Money

Now suppose that over the years the northside islanders become very wealthy from making tuba and selling it for dried fishsticks. They have a great warehouse which they built from hewn mahogany logs, just to hold the fishsticks. One member of the northside family stands guard every night so no one will steal any of their fishstick money. When the warehouse gets too full, they export some fishsticks (katsuobushi) to

Japan in exchange for better tuba-making equipment, and for all kinds of luxury items to enjoy.

These days, whenever the members of the wealthy northside family travel around the island buying whatever they want, they don't carry sacks of fishstick money with them. They give out small pieces of paper saying: "The northside family will pay to the bearer, on demand, one fishstick." These neatly cut, rectangular pieces of paper are beautifully designed in green and gold. The northside chief had them printed in Japan.

The "paper money" idea was thought of by the northside chief as a way to make trade more convenient. He thought: "Why should we travel all over the island buying things and paying with fishsticks, when a few days later the people always bring all the fishsticks back to us and buy tuba? Why not keep the fishsticks in our warehouse all the time, where they are safe and dry? We can just make payments with slips of paper saying 'the northside islanders owe you one fishstick.' Then when the people come to us to trade the fishsticks for tuba, they can give us back these IOU's." This seemed like a good idea, so they decided to try it.

At first some of the islanders were wary about accepting pieces of paper in exchange for their fruits and vegetables. But a few tried it, and it worked so well that soon everyone wanted to be paid in "paper money" instead of "fishstick money." Now, the people have complete faith in the paper money. They are sure that everyone else will accept it, and they know it's fully backed by fishsticks—they could always turn it in to the northsiders and get fishsticks if they wanted to.

Now that modern times are here, when the eastside islanders want to buy something from the westsiders they never carry fishsticks with them. They simply use the little pieces of printed paper as money. Occasionally someone on the island will present a "fishstick bill" to the northside family in exchange for a real fishstick, but this doesn't happen very often. Usually it goes the other way. Whenever someone makes several bags of dried fishsticks, if he doesn't want to eat them soon, he takes them to the northside islanders and exchanges them for paper money. The pieces of paper are so much easier to keep, to hide, or to use in exchange, than are the fishsticks. How easily and naturally and automatically the society has moved from a less efficient kind of money (fishsticks) to a more efficient kind—paper! But even paper is not efficient enough. Watch what happens next.

## The Eastside Islanders "Deposit" Their Money

Over the years the eastside family has been very industrious. They have been producing many fishsticks and turning them in to the northside islanders for paper money. Also they have been making other products and selling them to other people in exchange for paper money. The savings of the eastside family now amount to many thousands of the fishstick-backed paper bills. The family begins to get really worried about the possibility of a fire or a robbery which would wipe out their savings.

The northside family has a large stone house with a big fireproof vault built in the basement. The eastside chief asks the northside chief to please store the paper money in the vault. The northside chief agrees. So the eastside chief delivers the paper money and receives in exchange, a receipt. You might say the eastside islanders have put their money into a "bank account" for safe keeping.

### Bank Accounts Become Money

A week or two later, the eastside family wants to buy some breadfruit tree logs from the westside family so they can make outrigger canoes. First they go to the northside chief and get (withdraw) some of their paper money from the vault. Then they buy the logs and pay the money to the westsiders. The westsiders don't want to keep all this cash lying around, so they take it to the northside chief and ask him to keep it in the vault for them. They "deposit" their money and, in exchange, get a receipt.

A week or two later, the eastsiders need to withdraw more money to buy taro from the westsiders. As soon as the westsiders get the money, they take it right back to the northside chief and ask him to put it back in the vault. They "deposit" it again. This sort of thing continues to happen every week or two. One day the northside chief has an idea. He says to the eastside chief: "Every week or two, you eastsiders come and get some of your money from your box in the vault, and you go and spend it to buy things from the westsiders. Then the westsiders bring the money right back, and we have to open the vault and let them put the money into *their* vault box. That seems to me like a lot of trouble for nothing. Let me make a suggestion. When you want to buy something, why don't you just write a note to the westsiders authorizing me to take the money out of your vault box and put it into their vault box? That way, it will save both of you a lot of trouble, and I will only have to open the big heavy door to the vault one time for each transaction. Here. Write your authorization on these special pieces of paper I had printed up in Japan. These are called 'checks.' All you do is write who gets the money (whose vault box I should put it in), and how much; and then sign your name. I'll take care of transferring the money, and I'll keep a record. I'll send you a statement once a month to let you know how much has been deposited, how much has been paid out, and how much is left in your vault box. Want to try it?"

The eastside chief thinks this is a little unusual, but he decides to give it a try. Then when he tries to buy something the westsiders are a little wary about this flimsy way of doing business, but soon they decide to take a chance and accept the eastside chief's "check." Lo and behold, it turns out that the system works fine! So, whenever the eastsiders want to buy from the westsiders, they write checks. It works so well that soon all the people on the island start keeping their paper money in the northside vault. When they want to buy things they write a note (check) telling the northside chief to transfer the money from one vault box to the other.

### An Efficient Money System Has Evolved

Notice how far the island's monetary system has evolved now. How efficient it is! Seldom does anyone exchange checks for paper money. Almost all of the paper money just stays in the vault. The most efficient way of doing business is by writing checks to each other. Paper money is only used for small, day-to-day transactions. No one ever uses the actual fishsticks to make payments. Not anymore. How medieval that would be! The islanders now have a modern, efficient monetary system. But it isn't through evolving yet. Just watch.

As the years go by, trade picks up on the island. In a few years it gets so busy that the northside chief has to spend much of his time going into the vault and moving paper money from one owner's box to another. One day he says to himself: "This is ridiculous. All these fishstick bills are exactly the same. Why should I keep moving them from one box to another? Why not count the number of bills in each box and make a record. Then I will know how much belongs to each 'depositor.' After that, I can dump all the bills in the same pile and not worry about which ones belong to which people. Then I won't have to be going into the vault all the time and moving cash from box to box.

I will keep the account books upstairs in my office. When the westside chief comes in with a check from the eastside chief, I will just add the amount to the westsiders' account, and subtract it from the eastsiders' account. How easy that will be! I will also keep some fishstick bills up there in the office safe, so that if anyone wants any cash I'll have it right there. I may *never* have to open this big, heavy door to the vault, ever again. (How stupid I was not to have thought of this sooner!)"

### Account Figures Begin To Be Used As Money

The northside chief puts his plan into action. It works beautifully. Sometimes someone comes in and wants cash (fishstick bills); but before the wall safe gets empty, someone else comes in and deposits cash and replenishes the supply. Last year, just before Christmas, many people came in to get cash because they were making lots of small gift purchases. The wall safe was emptied and the northside chief had to open the vault and bring more cash upstairs. But right after Christmas, all the cash came back and was deposited in people's accounts. The wall safe got overstuffed with cash and the chief had to take some of the currency to the basement and put it back in the vault.

The monetary system is working just great. The dried fishsticks never leave the warehouse. Most of the cash never leaves the vault. People do business mostly by writing checks. Whatever you "spend" is subtracted from your account and added to someone else's account. Whatever checks you receive are added to your account and subtracted from someone else's account.

The account figures are being transferred back and forth. So what's serving as money? Would you believe, the account figures?

How neatly it all works. It's working just like a modern monetary system. And it all came about so naturally. Why? Because the benefits of an efficient monetary system are so great and so obvious, that's why. The islanders couldn't enjoy the full advantages of specialization and trade without an efficient medium of exchange. So one evolved. But look out! A disaster is about to happen.

### The "Backing" For The Money Is Destroyed

Let's suppose that one spring the tuna from which the fishsticks are made, all migrate away so that no more fishsticks can be produced. Also, that same spring, a midnight fire breaks out in the warehouse. The flames leaping skyward can be seen all over the island. Everyone awakens and hurries anxiously, apprehensively, to see what is burning. When they arrive the entire warehouse is ablaze. They stand dumbfounded, watching the flickering reflection of the flames lighting the circle of sad, tearful faces of their friends from all over the island. The fire is beyond control. Everyone knows that their money—the island's store of treasure—is gone. And they know it can never be replaced.

Almost overcome by grief, the eastside chief speaks: "Woe be unto all of us. Our money is gone. Our savings of paper money now are only paper. Our checking accounts mean nothing. We are all poor again. We must all go back to the hard, primitive life we had to lead before our fishstick money freed us from all that. Truly, this is the saddest day in our lives, and the saddest day in the entire history of this island."

Then the wise old northside chief who earlier had withdrawn from the crowd to think, came forth and began to speak. Everyone's attention was captured by his deep, resonant, reassuring voice: "Do not despair, my friends. Truly, nothing is lost. The kindly gods have relieved us of a burden. No longer will we need to waste our energies guarding and repairing the warehouse, or producing fishsticks which no one will ever eat. You have my word that your paper money and your checking accounts are as good as they ever were. You can't exchange them for fishsticks, but you never wanted to do that anyway. You *can* exchange them for tuba anytime you want to, and you can still spend your fishstick money for anything else you want to buy from us. We will all miss the delicious flavor of dried fishsticks. But except for that, nothing is lost. Your money is still just as good for buying tuba and other things as it was before."

At first the others can't see how "fishstick money" can be good if there are no fishsticks to back it up. What good is paper money if the *backing* is gone? But then the westside chief announces: "As long as the northside chief will continue to accept this money in exchange for tuba, we will accept it in exchange for breadfruit and taro."

The eastside chief announces: "As long as we are sure we can use this money to buy tuba and breadfruit and taro, we will accept it for smoked lagoon fish, turtle meat, and anything else we have to sell."

### The Monetary System Doesn't Need Any "Backing"

The monetary system has been saved by the wisdom and quick, reassuring action of the northside chief. He was wise enough to know that it is not the "backing," or what "backs up" the money that gives it its value—it is what you can *buy* with it. It is "faith in its general acceptability" which makes money, money. The northside chief knew that if he could assure everyone that the currency and the checking accounts were still generally acceptable, then the monetary system would continue to work just as before. But if the people lost confidence in the general acceptability of the money, the monetary system would collapse.

It is always true in every society that the monetary system operates on confidence. If the American people, for any reason, lost confidence in the future general acceptability of the U.S. dollar, the U.S. monetary system would collapse. The dollar would cease to be money. Unless something was done quickly to reestablish some kind of money, production would stop. Drastic action would be required to prevent most of the people from starving to death.

The wise old northside chief saved the monetary system and the island's economy by guaranteeing that the money was still acceptable in exchange—it still had purchasing power. Before he spoke, the monetary system had been destroyed because the people *thought* it had. After he spoke the money was good again. Why? Because everyone was confident that it was.

The only thing the island's fishstick money can now be used for, is to buy things. But that is what money is always wanted for anyway! People don't want money so they can "turn it in" for whatever "backs it up." Of course not. They want it so they can buy things.

As long as everyone is confident that other people will accept it, money continues to serve as money. The value of money is determined, not by the number of fishsticks or grains of gold you could get for it, but by what you can buy with it. This is the true value of money in every society. The value of a dollar is simply determined by what you can buy with a dollar. You accept dollars, not because they are backed by gold (which they are not), but because other people will accept them in exchange for the things you want to buy.

## Most Of The Paper Money Is Destroyed

Even though the "backing" behind the fishstick money is gone, the monetary system functions just as well as it did before. Everyone continues to have complete confidence in the money. Then as time goes on, the northside chief occasionally looks into the big vault and sees the great pile of paper money lying there. One day he picks up one of the aging pieces of paper money and reads what it says: "The northside chief will pay to the bearer on demand, one fishstick." Suddenly it hits him. "These are nothing but my old IOU's! Each one only says that I owe the bearer one fishstick. Why should I have to keep and store and guard all these old IOU's? If I could get rid of them we could use the vault space as a wine cellar, for aging our tuba. Why shouldn't I just burn up all this old money? If I ever need more of these bills (because people want to carry more cash, like at Christmas time), I can always get a new batch printed up in Japan. So why keep this pile of old paper money?" The chief ponders over this for quite awhile. But, try as he will, he can think of no good reason to keep the paper money. So one day he clears out the vault and burns the money. He is very careful not to let anyone see him do it. He doesn't think they would understand.

Now the island has no fishsticks "backing up" the money, and almost no paper money backing up the checking accounts. Then, one day before long the eastside chief comes and wants to see his money, just to be sure everything is all right. The northside chief opens the account book and shows the eastside chief his balance. But the eastside chief says: "I want to see the *money*, not a *number* in your account book!"

"The number in the account book *is* your money, my friend," says the northside chief. "It's perfectly good money. You can spend it by writing checks. But you can't change it into fishsticks. And you can't change it into paper money either, unless you will wait for two weeks while I have the paper money printed up."

## The Eastside Chief Demands Cash

The eastside chief is really upset. He has heard of many strange things, but this takes the cake! The idea that the *number* showing his *checking account balance* is his money—the idea that *the number in the account book is the only form in which his money exists*—that is just too much for him to comprehend. He says, "I'll be back in two weeks to get my cash and you'd better have it ready for me when I get here. Otherwise, be ready for war!"

The northside chief orders the bills printed in Japan, and two weeks later the eastside chief returns, gets his cash, closes out his checking account, and goes home with a big basketful of new paper money. He really doesn't have any more money than he had before. He has just changed the form of his money—from an accounting figure, to printed pieces of paper. Then as the weeks go by the eastside family finds out that using cash is not nearly as convenient as using checks. Also, they get more and more worried about losing their money by theft or fire.

One day the eastside chief finally realizes he did the wrong thing. He takes his basketful of cash back to the northside chief, apologizes, reopens his checking account, and deposits the cash. Now he has changed the form of his money from paper bills back into an accounting figure. The paper bills really aren't money any longer. They aren't serving as a medium of exchange anymore. They are old IOU's of the northside chief, and as long as they are in his vault they are not serving as money. That night the northside chief discreetly takes the cash out and burns it up. The eastside chief has regained his confidence in "checking account" money. Once again everyone is happy and all goes well.

### The Island Now Has A Modern Monetary System

The island finally has arrived at a modern monetary system. Most of the money exists only in the account book of the northside chief. The accounts are not fully "backed" by paper money (currency), and the currency is not "backed" by anything. People make payments by writing checks. Each check results in a subtraction from one account and an addition to another. That's all. No currency changes hands. Currency representing these accounts doesn't even exist.

What do you think of the island's monetary system? Would you believe that's the way it works in every modern society? This example really does fairly well describe the money system in the United States and in every other modern nation in the world. You will hear more about this later. But first we need to look at the interesting effect which borrowing can have on the money supply.

## LENDING MONEY SOMETIMES CREATES MONEY

Suppose the eastsiders want to buy a plot of land from the westsiders, but they don't have enough money in their account. The eastside chief goes to see the northside chief who, as you already know by now, is operating as the island's banker. The eastside chief asks if he can borrow 10,000 fishsticks (FS 10,000) to buy some land from the westsiders. The northside chief realizes that the eastside chief doesn't want cash—he only wants the amount added to his checking account so he can write a check to buy the land. Then the westsiders will deposit the check to their account, and the "loaned money" (FS 10,000) will be subtracted from the eastsiders' account and added to the westsiders' account.

The northside chief, knowing the eastside chief's credit is good, says, "Certainly. We will lend you as much as you would like." The northside chief adds FS 10,000 to the account of the eastside chief, and the eastside chief signs a note promising to repay the money, plus interest.

What has happened to the supply of money on the island? It's FS 10,000 bigger than before! The eastside family now has FS 10,000 which it did not have before. Nobody had it before. It didn't exist! The northside chief, acting as the banker, simply created that FS 10,000 when he added that amount to the eastsiders' account. Is it *really* money that has been *created*? Yes! It really is.

Look at it this way. Suppose *before* the FS 10,000 loan, you went around the island and found out how much money each person had (in cash and in his checking account). If you added up all these figures, what would you come out with? The total size of the island's money supply? Of course! Then *after* the FS 10,000 loan, suppose you did it again. You would come up with a larger total than before. How much larger? FS 10,000 larger. Where did the extra FS 10,000 come from? It was *created*? Right!

The eastside family will now buy the land from the westside family and pay them by check. Then the new FS 10,000 will be subtracted from the eastsiders' account and added to the westsiders' account. The new FS 10,000 will become the westsiders' money. Before the eastside chief borrowed it, whose money was it? Nobody's. Where was it? Nowhere. It didn't exist. When the northside chief added the FS 10,000 to the eastsiders' bank account, that really did create money.

### All The Money On The Island Is Really "Debt"

All the money on the island consists of debt. It's the debt of the northside chief. The "balance" in each account is the amount the northside chief owes to the owner of the account. And it's those "account balances" which make up most of the island's money supply.

Since the debt of the northside chief serves as money, any increase in that debt is an increase in the money supply. When the eastside chief borrows, he gives the northside chief a promissory note saying, "I owe you FS 10,000." What does the eastside chief get in return? He gets a

FS 10,000 addition to his checking account. He gets a "deposit slip" from the northside chief which says (in effect), "I owe *you* FS 10,000."

See what has happened? Each has given the other a debt, in exchange for a debt! Each owes the other an additional FS 10,000! But the important thing is that the northside chief's debt is monetized debt—that is, *his debt serves as money.* This transaction creates money because the debt of the northside chief *is* money. It's the only kind of money that exists on the island.

### Increasing The Checking Account Balances Creates Money

Whenever the northside chief is lending money to people by adding to their checking accounts, he is adding to the island's total money supply. The people are getting more money to spend. But whenever a person pays back the northside chief by writing a check and using up his checking account balance, this destroys money. It reduces the size of the total money supply of the island.

The total money supply is made up of the checking account balances of the people, plus the little bit of paper money they like to carry around. Anything which increases the size of the checking account balances increases the money supply; anything which reduces the size of the checking account balances reduces the money supply. If anything happened to cause the northside chief's bank to collapse, all of the checking account balances would be gone, so all of the "checking account money" would be gone. That's about the way it is in the real world, too. If the bank where you keep your money suddenly collapsed, your "checking account money" would be gone. (But the Federal Depositors Insurance Corporation [FDIC]—the government agency which "guarantees" the safety of your bank account—probably would reimburse you.)

Later you will see how this simple island example of "creating money by making loans" is more true to life than you might think. Most of the money in the United States and in other modern nations is actually created by bank loans. It doesn't happen exactly as in the island example, but almost! Later you will find out exactly how it works. If you understand how it happens on the island you won't have any trouble seeing how it works in the real world.

## MONEY IN THE UNITED STATES

To see just how similar the monetary systems of the modern world are to our island example, let's take a quick look at money in the United States. In this country most of the money is made up of all the checking account balances in the banks. Each individual depositor and each business depositor has an account balance. Each depositor's account balance is his money. Paper money ("currency" or "cash") doesn't exist to back up these checking accounts. If the currency did exist, all the vaults of all the banks in the country would be stuffed full and there still wouldn't be enough space to store all the "currency"!

People don't like to hold large amounts of paper money. They don't like to use paper money except for small transactions. They would rather have most of their money in the form of an account figure at the bank than to have it in the form of paper bills. If there was enough paper money to "pay off" all the checking account balances in the country, no one would want to hold all this cash. Nobody wants several thousand dollars in cash lying around the house. The banks certainly don't want several billions in cash lying idle in their vaults! If all this cash existed, someone would have to arrange for it to be taken out and burned, to get rid of it. That's just what would happen. Just like on the island.

### The Federal Reserve Banks Issue Paper Money

In the United States, the central banking system is the *Federal Reserve System*. The twelve Federal Reserve Banks in this system are the ones that issue the *currency* (paper money). If more people go to their local banks and cash checks to get paper money, the banks pay out the cash they have on hand. If they run short, they get more from the Federal Banks, which get it from the Bureau of Printing and Engraving (U.S. Treasury).

The paper money in the United States consists of *Federal Reserve Notes*—simply the IOU's of the Federal Reserve Banks. Just as the northside chief could have a little piece of paper printed up to say: "This piece of paper represents one fishstick," so the Federal Reserve Bank of New York (or St. Louis or Chicago or Minneapolis or Denver or Dallas or any other) can have a piece of paper printed up to say: "This is one dollar." Reach in your wallet and pull out a "Federal Reserve Note." The black numbers on the face of the bill, and the letter and the printing on the black seal tell you which Federal Reserve Bank issued the note. The number "1" and the letter "A" indicate Boston; 2 and B-New York; 3 and C-Philadelphia; 4 and D-Cleveland; 5 and E-Richmond; 6 and F-Atlanta; and so on to 12 and L for San Francisco.

### Currency Is Printed Up As Needed

The number and sizes (denominations) of the notes issued by each Federal Reserve Bank are determined by the currency needs in their district. That depends on the desires of the people and businesses to hold and use "cash money" instead of "checking account money." If people want to hold more cash (perhaps at Christmas time), they cash checks. Then their banks ask for (buy) more currency from the Federal Reserve Banks. The Federal Reserve Banks get more currency printed up, as needed. If the people decide to change their form of money from "currency" back to "checking account money," they simply deposit the Federal Reserve Notes (currency) in their bank accounts. The banks send the surplus bills back to the Federal Reserve Banks, where some (the newest) are kept for future use and the others are destroyed.

Be careful to say "money" when you mean money, and "currency" when you mean only "the paper kind" of money, and coins. You know that money is anything people will generally accept in exchange for things. And you know that "the paper kind" is only a small percentage of the total. So be careful to say it that way. In the United States and other modern countries, money consists primarily of people's and businesses' checking account balances on the books (or in the "computer memories") of the banks.

### The Present U.S. Money Supply

In the United States, in 1973 the total amount of money in existence (the money supply) amounted to something more than $250 billion. Of this total, some $200 billion (about 80%) consisted of demand deposits—that is, the checking account balances of people and businesses. So you see that more than three-fourths of the U.S. money supply is really nothing more than little numbers written down on the account books of banks! Total currency (paper money and coins) added up to about $50 billion—about 20 percent of the total money supply. The value of all the paper money amounted to about $44 billion, and the other $6 billion consisted of coins. (Many of the coins are not being used as money but are being held by collectors and speculators.)

None of the money actually being used in the United States is backed in any way by gold or silver, or by anything else. It has been about forty years since a person could turn in paper money and get gold for it; it has been a few years since a person could turn in any of his paper money and get silver coins. But no matter. Money is valuable because it will buy things. The more it will buy, the more valuable it is. As long as people will accept it in exchange for things, it is good money—as good as gold.

## THE EVOLUTION OF MONEY IN THE REAL WORLD

Money is so useful and it evolves so naturally that in every society something has evolved into money. Seashells, stones, various kinds of metals, diamonds, cigarettes, fish hooks, grain, bullets, and almost every other imaginable thing has been "money" at one place or another or at one time or another throughout history.

### At First, Money Must Be Full-Bodied

How does something get to the exalted position of general acceptability? How does it become money? Usually it starts out as something useful—some commodity or tool, something used and wanted by most of the people. At first, people will not be sure whether other people will accept it as money. Still, they are willing to accept it because it is "full-bodied." If others will not accept it as money it can always be used for its original purpose anyway. For many years in the evolution of money, the "money thing" must continue to be full-bodied—that is, it must have as much value as a "thing" as it has as "money." Otherwise people would not trust it and would not accept it.

Over the centuries, precious metals have been valued for their own sake. It was almost inevitable that gold and silver would evolve into money—so convenient, so attractive, so permanent, so easily made into bigger or smaller sizes, and such a little bit can be worth so much! So gold winds up performing the functions of money. People go around with gold dust or gold nuggets in a little pouch. Before they can trade they must find someone who has a scale to weigh the bits of gold.

### Full-Bodied Coins Emerge

One day someone gets the idea that if the little pieces of gold were melted down into carefully measured sizes, and if each piece was stamped to indicate the weight, trade would be much easier. No longer would everyone need a scale for weighing gold. Gold would become a better medium of exchange—better money. This idea lead to the development of uniform little pieces of metal, each stamped to indicate its weight. But who can you trust to stamp the right weight on the pieces of gold? Not everybody! So before long, coins are being made and certified as to weight and value, by the government.

Once the government gets involved, the next step is to reduce the amount of precious metal in each coin. So long as people continue to accept the smaller or less pure coin at face value, the "un-full-bodied" coin (token coin) works just as well as a full-bodied one. Smaller coins are more convenient, anyway. (Did you ever see one of the old full-bodied copper pennies which were once used as part of the U.S. money supply? How would you like to lug around a pocket full of those?) The next step in the evolution of money comes with the introduction of paper money, backed by (convertible into) the gold or silver coins.

Then as time goes on, people stop carrying and exchanging metal coins and begin to use paper money almost entirely.

### The Valuable Substance Is Removed From Money

The two final steps in the evolution of a modern monetary system are taken (1) when the precious metals are withdrawn from the monetary system entirely, and (2) when the people begin carrying out most of their exchanges by writing checks. Most of the money supply becomes the checking account balances of the people and businesses. Why? Because the government says so? No. Because this is the form in which the people choose to hold their money. It is the most convenient, most efficient form of money for most transactions.

So now you know the way the monetary systems have evolved in all the nations of the modern world. Interesting, isn't it? Hard to believe? Maybe. But it really did happen that way.

### Money Is Essential In Every Society

It would be impossible to overstate the importance of money in the operation of the market process, or in the operation of any modern economy. The economy simply could not function without money. The market process just naturally happens. But it couldn't happen without money. Money is the medium which permits the market process to function. Truly, money is the life-blood of the market process. There is no society in which something has not arisen to play this vital role—the role of money.

Exchange, specialization, division of labor, technological development—these are essential for *any* society which seeks to achieve high standards of living for the people. You know about the gains from specialization and trade. You know that the advantages can't come about without exchange. And you know that the many opportunities for exchange can't be fulfilled without money.

## THE MIRACULOUS MARKET PROCESS

Now that you're at the end of Part Two, you may be beginning to develop some real appreciation for the neatness and efficiency of the market process. The market process evolves naturally. Specialization and trade develop. Money appears. Output expands more and more. This whole thing is sort of miraculous. Yet it all works so naturally that there doesn't seem to be anything so very impressive about it. But think about it. Each individual is free to make his own choices, free to follow his own wishes and desires. Yet, most of the right choices seem to be made. Most of the right things seem to happen. Society thrives. People live better and better, year after year. Isn't that something to be impressed about?

Looking at the functioning of the market process is like looking at the functioning of the human body. It operates so miraculously that it's almost unbelievable; yet everything works so easily and naturally that we sort of ignore it all. It happens every day. It's just "the usual thing," so we see no need to stand back and look at it, and marvel. But when you stop for a minute and take a close look, it's really impressive.

At this point you know a lot of very basic economic concepts and principles. Not all you need to know, but a lot. Most of the concepts have been explained with simple, make-believe examples. I have used that approach because I think it's vital for you to understand each concept without "the complexities of the real world" to confuse you. If you've been working at it steadily—really thinking about the questions and doing the exercises—you've come a long way. You're probably making an A in your "econ" course, too. Great! But even greater, you are beginning to gain a lasting understanding, an awareness of and an appreciation for the economic things going on in your life and world. That's what you're really after. Right?

You're now at the end of what might be called "the first major step" in your study of economics. You know many basic concepts and principles, and you have a clear overview of how the market process functions. I hope you feel that you have learned some interesting and worthwhile things.

Part three, coming up now, will take you on some completely new and different excursions in economics. I think you'll find some interesting (and surprising) things there, too. But before you go ahead, maybe it would be a good idea for you to spend some time reviewing Parts One and Two. Make sure you have a good grasp of all these basic concepts. Nail them down securely. You'll need them later in this book. And in life.

---

## REVIEW EXERCISES

**MAJOR CONCEPTS, PRINCIPLES, TERMS (Try to write a paragraph explaining each of these.)**

the meaning and importance of money
how "full bodied" money might evolve
how paper money might evolve
how "checking account" money might evolve
how lending can create money
how repaying loans can destroy money

**OTHER CONCEPTS AND TERMS (Try to write a sentence or phrase explaining what each one means.)**

medium of exchange
standard (or unit) of value
primary functions of money
secondary functions of money
standard of deferred payments
store of value
liquid
monetary "backing"
currency
monetized debt
FDIC
Federal Reserve System
Federal Reserve Bank
Federal Reserve Notes
demand deposits
full-bodied money

**QUESTIONS (Try to write out answers, or jot down the highlights.)**

1. Do you think *tuba* (instead of fishsticks) might have evolved into money, on the island? Explain in detail.
2. During the great U.S. and worldwide depression of the 1930's, not very many people had very much money. A lot of banks had collapsed—gone bankrupt. Where do you suppose most of the money had gone? Did the rich people have it squirreled away somewhere? Or what? Discuss.
3. If you lend your friend five dollars so he can buy himself a few glasses of beer, that doesn't create any money. But suppose you were so rich and well known that everybody knew your signature and would accept your IOU. If you write your friend an IOU for five dollars, and he goes and "spends" the IOU at the local bar and the barkeeper uses it to pay the waitress and the waitress uses it to pay the taxi driver and the taxi driver uses it to buy gas. . . . Does that mean that your loan has *created money*? Discuss.

# PART 3

THE MORE WE ALL SPEND
THE MORE WE ALL EARN!

The people earn and spend and earn and spend; the more they spend the more they earn and the more they earn the more they spend and the more they get to have.

# 9 HOW TOTAL SPENDING SUPPORTS THE NATION'S ECONOMY

*The size of the national output or income depends on how much the people and businesses are spending.*

What determines whether you will be rich or poor? Is it up to you? Partly, yes. But only partly. In the modern world all our destinies are tied together. Everything you do has an influence on somebody else. And how well-off *you* are depends a lot on what *others* are doing. This is perhaps true in every aspect of life. It is certainly true in economics.

## THE ECONOMIC SYSTEM IS ALL TIED TOGETHER

The more specialized the economic system and the more "well-off" we are, the more interdependent we are—the more dependent on each other we become, and the more dependent on the economic system we become. In the modern economy, very few of us could even survive if the system quit working. In a more primitive economy, survival would not be so difficult. For example, in the island economy, if the eastside islanders and the westside islanders somehow experienced a trade breakdown—if money collapsed because of loss of confidence, or if trade stopped for some other reason—this would lower the standard of living. But the eastsiders would still be able to eat plenty of fish and the westsiders would have their breadfruit. No one would starve. How different this is from the situation in a big, modern city!

### We Depend On The Economic System For Survival

How many people in the modern city could survive if the economic system broke down completely? There would be no water, no electricity, no transportation, no source of food. The people just couldn't live. Our means of survival would be gone. Except for those few who could find some way to become self-sufficient, we would all just die. It isn't really important for you to go around worrying about the coming of this kind of a catastrophe. This kind of a "total depression" is not at all likely. If something like this threatened to happen, something would be done to prevent it.

But sometimes there really are "little breakdowns" in the system. These "little breakdowns" don't usually cause anyone to starve to death, but they do hurt many people. Suppose you are working at the Chicago stockyards and you are told that because of a declining demand for beef you are no longer needed. Many of your friends receive the same notice. Maybe if the drop in demand for meat products is great enough, the stockyards will go on "part-time operation," or maybe even shut down entirely.

If the stockyards lay you off and shut down, does this mean there is a "little breakdown" of the economic system? Maybe yes, maybe no. Maybe people are buying less beef and pork, but buying more of other things—maybe poultry. Maybe there is a boom in the egg business and maybe you can get a job there. You start watching the want ads. But you don't see any jobs available. Suppose one day you are walking down the street and bump into one of your poultry-farming friends. He's looking for a job at the stockyards. He tells you that the poultry farm workers are having the same problem as the stockyard workers. Layoffs. No jobs. So what's going on?

You figure that maybe people are buying less food so they can buy more cars and nicer clothes. So you hitchhike to Detroit to get a job in an automobile factory. Then you find that the automobile workers are being laid off too. So you hitchhike down to Burlington, North Carolina and try to get a job in the textile mills. But there you find the same problem. What's going on?

## A New Kind Of "Economic Problem"

What you are seeing now is a new kind of "economic problem"—the problem of depression—the problem of a slowdown of the whole economic system. This chapter and several of the ones that follow will explain about speedups and slowdowns in the economic system. The economy never seems to want to run at just the right speed. Why not? And what can be done about it? That's what we're getting into now.

Parts One and Two of this book were concerned with economic concepts of a somewhat different kind than the ones we're getting into now. We were concerned with the question of choosing *which way* to use our resources. Which things to produce? Which resources to use in the process? How to share the output? Now, suddenly we are looking at a different kind of problem. We see idle factors of production everywhere. People are trying to sell their labor, but no one wants to buy it. Businesses are trying to sell beef and pork, and chickens and eggs, but nobody is trying to buy them. The automobile companies would like to produce and sell more cars, but there just aren't many buyers. The same is true for new clothes and shoes and furniture and airline tickets and for almost everything else.

As you hitchhike around the country looking for a job you begin to get hungry. You certainly would like to have some of that beef to eat! The meatcutters certainly would like to have jobs cutting sirloin strip steaks for you to eat. The cattle ranchers certainly would like to find buyers for their beef cattle. The stockyards certainly would like to get back into full-time production again. You certainly would like to get your job back, at the stockyards. And you certainly would like to have that steak!

All the needed inputs—all the factors of production—are available, itching to go to work. You, and many other people, are hungering for steak. So why isn't the steak being produced? Simply this: The economic system isn't working right. The market process isn't doing what it's supposed to do. The system is suffering from a "partial breakdown."

## When One Thing Slows Down, Other Things Slow Down

The economic system is not producing steak for you, because you aren't offering any money to buy steak. You aren't offering any money because you don't have any money. You don't have any money because you lost your job at the stockyards. You lost your job at the stockyards because people weren't buying steaks. It's all a big circle!

If people stop buying things, businesses stop producing things. Then factors of production get laid off. So people's incomes get cut off. So they don't buy as much. So businesses produce even less. They fire more workers. Things get worse and worse. The economy goes into a recession and then on down into a depression.

If people shift their spending from steaks to eggs, there is no problem. The factors of production move from where they are less wanted to where they are more wanted. But if people stop spending for steaks and for eggs and for everything else, then the people who make these things—that means all of us, as workers, as businessmen, as clerks, as investors, as service and repair people, and so on—all of us will lose our sources of income. Then we will be forced to stop spending. Can you see how different this new kind of economic problem is from the "choice" (the "opportunity cost") problem we were talking about in Parts One and Two?

We are talking about the problem of a *partial breakdown* of the economic system. The economic system is the "thing" that gets our factors organized and directs them into doing or making the things society chooses. When the factor owners want their factors to be used, and the society wants the output, but still the factors are *not* being used, this means that the method of organizing and directing these factors—the economic system—is not doing what it's supposed to. It isn't working as it should.

Now it's time to start talking about *what determines the speed at which the economic system will run*. How much will we produce this week? This month? This year? How fast will we use our productive factors? Will we operate our factories and farms and stores, and use our labor force at 100 percent of capacity? At 90 percent of capacity? Or at only 50 percent of capacity? How are these questions decided? These are the issues we will be concerned with in this chapter and throughout the remainder of Part Three.

### The Circular Flow of Production and Consumption

The economic process consists of producing things and consuming things. If the "model market process" was working in its *pure* sense, everybody would be doing some of both—producing *and* consuming. Who would produce? Everybody. That's the only way you could get to have your income and to claim a share of the output. Who would consume? Everybody. Either in the "pure model" or the real world, everybody must consume to stay alive!

The "production-consumption cycle" can be thought of as a *circular flow*. Let's look at this "circular flow" on our make-believe island and watch it develop as the economy gets more complex. First, think of a "Robinson Crusoe world."

One man alone on an island would be producing and consuming at the same time throughout much of the day. He climbs a palm tree and gets a coconut, then drinks the juice and eats the coconut meat on the spot. He catches a fish and tosses it on the coals. As soon as it is cooked he peels off the charred skin and eats the tender, char-broiled meat inside. By the end of the day he has eaten all the food he has produced. His "circular flow" of production-consumption is a tight little circle! If he wants more output (food, or shelter, or fishing spears, or whatever) he must speed up his "economic system." He can start working longer and harder and produce more. Then he will get more output and (simultaneously) more things to have and to consume—more real income.

In the Robinson Crusoe world it's very easy to see how the production-consumption circle (the output-income circle) works. Although it isn't quite as obvious, soon you will see that the same kind of output-income circle also exists in a complex, real world economic system.

### Total Output Equals Total Income

In the Robinson Crusoe world, the one man produces in response to his own wishes and desires—for food, clothing, shelter, and perhaps to build himself some capital. His "level of economic activity"—how much he works and how much he produces—is determined by his own wants and desires. Whatever total amount he produces, that is the amount he gets to have—to consume, or to "save" and use as capital. His total output is equal to his total income (real income). When there's only one person, all these things are easy to see. But now let's complicate the picture by bringing in more people.

Suppose there are two families on the island, the eastside islanders and the westside islanders. The eastside islanders produce fish and the westside islanders produce breadfruit. They meet in the middle of the island every afternoon at 4 o'clock and trade baskets of fish for baskets of breakfruit, one for one. The eastside islanders take half of their fish catch and the westside islanders take half of their breadfruit harvest to the "market place," and trade. The total output of the island is equal to the total "real income" the people are receiving.

"Total island product" each day, equals "total island income" for that day. The total amount of fish and breadfruit produced by the islanders is equal to the total amount of fish and breadfruit received by the islanders. Obvious, isn't it? It's a neat little circular flow. They produce from sunrise to 4 p.m., then they trade, and then consume until bedtime. If they sometimes produce more, they receive more. That is, if output goes up, real income goes up. If they produce less, they receive less. Real income goes down.

As time goes on, two new families move to the island from neighboring islands. Then we have the northside islanders and the southside islanders. The new islanders soon begin to participate in the trade. At 4 p.m. each day they take their fish and breadfruit and bananas and tuba and other things to the "market place." But now things begin to get confusing. The eastside islanders want to trade some fish for bananas, but the southside islanders who have bananas don't want fish today. They want tuba (coconut palm booze) instead. So the eastsiders first trade fish for tuba, then trade tuba for bananas. The island economy is beginning to need a medium of exchange. They need something which will serve as (and which, therefore, will be) money.

## THE PRODUCTION-CONSUMPTION CIRCLE CAN BE BROKEN BY SAVING MONEY

As the years go by, through a series of rather miraculous accidents, money arises in the form of dried fishsticks. Then it evolves into paper money backed by fishsticks. Finally, the money evolves into "bank account money" which is transferred from one person to another by writing checks. The development of this highly efficient form of money solves the "medium of exchange" problem. But wait! The existence of this medium of exchange (money) leads to a *new* kind of problem. Watch what happens.

Things sure are a lot more convenient with money! No longer is it necessary for everybody to go to the "market spot" at 4 o'clock every afternoon. Now, when somebody wants fish, or breadfruit, bananas, tuba, or anything else, he can just go to the producers and buy it. Each person can now specialize in exactly what he wants to, and then sell his output in the market for the going price. Money lets it all work. So what's this "new kind of problem" that money brings?

Last October when some of the eastside islanders were out fishing they were passing the time, talking about Christmas. Soon they were all bragging about all the Christmas presents they were going to buy for their friends and loved ones. The more they talked the more they realized that if they really were going to buy all those things they had better start saving some money.

That day, after they sailed their outrigger canoes to shore, the fishermen sold their fish to the waiting buyers on the beach, just as always. But instead of going around the island and spending all their money for breadfruit, bananas, tuba, and other things, each one decided that he would spend only *half* of his money. He would save the other half—just deposit it and leave it in his bank account. These good, solid, thoughtful, thrifty eastside islanders are certainly doing a fine thing, saving for Christmas. Right? But that's the start of a chain reaction which is going to develop into a serious problem. Watch.

**Savings May Leave Surplus Products In The Market**

At sundown that day the westside, northside, and southside islanders couldn't understand why they had so much left over—breadfruit, bananas, coconuts, tuba, everything. Why all the surpluses? You know why. The thrifty eastside islanders didn't spend all their money, so the other producers can't sell all their products. So what's going to happen next? Maybe the surpluses of breadfruit, bananas, tuba and all, will force the prices to go down. At the lower prices, maybe all the surpluses will be bought up, and everything will be all right again. Do you think it really will happen that way? Maybe, but probably not.

Over the years the fishermen and breadfruit gatherers and banana growers and tuba makers have established trade associations and unions. They now insist on "fair and just wages and prices." Each chief has made it quite clear that the "going price" for his product is the rock-bottom price he will accept. If the demand for his product happens to drop, he will not permit the price to go down. He will cut back production instead.

You can see that all the sellers on the island have enough "monopoly power" or "market power" to keep their prices from falling. If demand drops they will hold their prices up and just cut back production. They would rather *dump the surpluses in the ocean* than to let their prices go down!

So here we are, with the eastside islanders not spending all the money they are getting from the other islanders. They are saving for Christmas. But as a result of their saving (not respending the money they receive), surpluses develop in the markets for breadfruit, bananas, tuba, and all the other things. Then the westside, northside, and southside islanders cut back their production rather than let the price go down. While all this is going on, what do you think is happening to the demand for fish?

### Increased Saving Sets Off A Chain Reaction

Yesterday when the fishermen decided to save half their income, the demand for fish was as great as it always had been. But today, what about it? When the outrigger canoes come sailing in to the beach this afternoon there aren't as many buyers as before. Can you guess why?

Where do the westside, northside, and southside islanders get the money to buy fish? By selling their own products, of course. When they can't sell all their products, they don't get as much money. So they can't buy as much fish.

Today the eastside islanders find that they can't sell all their fish. Several baskets of fish are left over. The fishermen can't understand what happened. Today the eastsiders face the same problem the others faced yesterday. And today the eastside chief decrees that tomorrow some of the fishermen will not go out fishing. No sense producing more fish than can be sold for the going price!

What's going on, on our happy little island? We have unemployment! The incomes of all the island people are now lower than before. Both their money incomes and their real incomes are lower. What about the eastsiders' Christmas savings plan? That's out the window, too! Their incomes are now so low that *they can't save*. The economy is depressed. Total economic activity—total production, output, and income in the economy—has declined. The eastside islanders didn't want to cause trouble. They only wanted to save for Christmas. But look at the trouble they caused! Why?

### Money Makes It Easy To Defer Spending—To Save

To save means to "not-consume" some of what you produce—or, in money terms, it means to "not-spend for consumer goods" some of the income you earn. When you decide to "not-spend," this means you are selling more products and services to others than you are buying back from them. When the eastside islanders started saving ("not-spending") they cut off some of the money which had been going to the other islanders. They stopped giving the others the money needed to buy all the fish!

Can you see that the economy operates as a big circle? What you have just seen is what happens when somebody doesn't do his part to keep the money moving around the circle. Whenever someone withdraws some of the money from the spending circle, soon everyone winds up getting less. Any withdrawal from the spending circle is felt all the way around the circle.

Back when the eastsiders and westsiders were meeting at 4 o'clock and trading fish for breadfruit, there was no problem. When the fish sellers "sold" (traded) their fish, they automatically "bought" (received in trade) an equal value of breadfruit and took it home with them. There was no way a person could "sell" his fish and then "not-buy" an equivalent amount of breadfruit. It always came out even. Obviously!

But once *money* is introduced, this permits a lot of *slack* in the system. When money is used, the people can take their fish to the market, leave their fish, get their money and go home. They don't have to take home the breadfruit. They can leave all the baskets of breadfruit just sitting there in the market. That's the problem!

### Money Is A "Claim Check"

Money is really a sort of "claim check," or "credit slip." It's your "right to claim the goods which society owes you." When you place goods in the market or perform services for the market, you get "claim checks." You can use these "claim checks" to claim whatever other good or services you want from the market. When you claim the things you want, you pass along your "claim checks" to someone else, who can then claim the things he wants. But when someone doesn't use his "claim checks" (money), this means that some of the goods in the market will be unclaimed, "surplus" goods.

The owner of the surplus (unsold) goods will not be able to get any "claim checks." So he can't claim the goods he wants from other sellers. So the other sellers soon will have surpluses too. The chain reaction just keeps going on and on, and the economy slows down more and more. See how it all fits together?

Before money is introduced, the total value of products a person takes into the market must be equal to the total value of products he takes back out of the market. So long as *things* are being traded for *things*, it must work this way. But with a medium of exchange (money) a person can put more products *into the market* than he takes back *out of the market*. He may decide to hold his "claim checks" and claim his goods later. That's when the trouble starts.

## Sometimes The Amount Supplied Determines The Amount Demanded: Say's Law

The more a person sells in the market the more he can buy. When people produce goods and take them to the market, they receive money. With this money they can buy an equal value of goods from the market and go home. A Frenchman, Jean Baptiste Say (in the early 1800's) explained this concept, saying that the more a person *brings to* the market, the more he will *take from* the market. The more he *supplies*, the more he *demands*. This principle is sometimes called "Say's Law of Markets," or just, *Say's Law*. "Say's Law" simply says that the more a person supplies to the market the more he will demand from the market. So, in effect, "supply creates its own demand." Obviously Say's Law is not precisely true; otherwise, there would never be surpluses in the market. The market would always be "cleared."

If everybody who sells something in the market turns right around and buys something of equal value, the total amount taken to the market will be equal to the total amount taken home. There will be no surpluses and no shortages. Everything works out exactly right. But when people take things and leave them in the market and get their money and go home, this breaks the production-consumption circle. Some producers will not be able to sell all their output. This is what you saw happening on our little island when the thrifty eastsiders started saving for Christmas.

## The "Secondary Functions Of Money" Allow Some Slack

I suppose if you wanted to get fancy, you could say that the island's economy was "wrecked by one of the secondary functions of money." The eastside islanders were using some of their money, *not* as a medium of exchange, but as a *store of value*, instead. That's what created the problem.

It's those *secondary functions*, those "future functions" of money that cause all the trouble. That's what brings the slack in the system and keeps Say's Law from working all the time. The *store of value* function lets people save—to spend *less* than they earn. The *standard of deferred payments* function lets people go into debt—to spend *more* than they earn.

The two "secondary functions" have offsetting effects on the economy, of course. When people spend *less* than they receive, that *slows things down*. When people spend *more* than they receive, that *speeds things up*. If everyone would just use each dollar he receives, right away, as a medium of exchange, then Say's Law would always hold true. But people don't. So we have speedups and slowdowns in the economy. You'll hear much more about all this, very soon.

### The "Paradox Of Thrift"—Sometimes Saving May Be Self-Defeating

The island people are facing hard times. Why? Because the eastside islanders wanted to be thrifty and save for Christmas. They were putting more goods into the market than they were taking back out of the market. They were *saving*. Saving must be a very bad thing. Look at the trouble it brought to our happy little island!

How can *saving* be *bad*? You already know that economic growth depends on saving. You know that if people are consuming as much as they are producing, there is nothing left over for economic growth. Saving is absolutely essential for economic growth. And economic growth provides opportunities for everybody to get more things and not have to work so hard. It sets more people free to go fishing weekends, or to go to college, or to "do their own thing." Economic growth seems to be good, so how can saving be bad?

Just like most of the other answers in economics, the solution to this dilemma is really very simple. Your good common sense probably could figure it out for you. But I'm going to tell you, anyway: The savings must be *invested* to bring economic growth. If the savings are *invested*, then there's no problem of surpluses, or unemployment! See how simple?

## INVESTMENT IS ESSENTIAL TO TRANSLATE SAVINGS INTO GROWTH

"Saving" means "not consuming." That's all it means. It means *production* is greater than *consumption*. It means people aren't spending all their incomes for *consumer goods*. It means leaving some of the current output free, in the market. "Economic growth" occurs when this "left-over output" is then used for the purpose of increasing future production. For growth to occur, the surplus which is left because people save (don't consume as much as they produce) must be bought and used as *capital*. The unconsumed surplus can't be left lying around in the market. (That's what causes depressions!) The surplus must be bought and used to help in future production.

To have growth, there must be saving. But just saving is not enough. In order to link savings to growth, there must be *investment*. Savings cannot be translated into growth, without "investment."

*People can save as much as they want to, just as long as investors buy up everything the savers decide to "not-buy."* As long as investors invest as much money as the savers save, no slowdown will occur. There will be no problem of surpluses or unemployment. Everything will be fine, and the economy will be growing all the time.

Just think of "saving" as "producing more than you consume." That means "leaving some output in the market." "Investing" means "buying up this left over output." If the amount of investing is equal to the amount of saving, then there's no problem. All the markets are "cleared."

I'm sure you can see why. Also, notice that in order for the investors to be able to buy things, somebody *must* be saving. If there isn't anything left over to be used as capital, then the investors won't be able to buy anything. Obviously!

Let's see what would have happened on our island if investment spending had increased enough to offset the savings of the eastside fishermen. Suppose, on the day the eastsiders decided to save their money, the son of the northside chief decided he would like to go into the business of producing pearl necklaces. He thinks he could sell a lot of necklaces at Christmas time. He needs to buy some canoes, underwater goggles, other diving equipment, oyster knives, and some other things. But he doesn't have the money to buy all these things. So (just to make it all work out very neatly) let's suppose he goes to the eastside islanders to borrow some money. He promises to pay them back (with interest) one week before Christmas. They agree to lend him their savings. They arrange it so that each day they will turn over to him all the money they save, that day.

### Investment Demand Stimulates Capital Goods Production

When our business entrepreneur (the son of the northside chief) takes his borrowed money and goes into the market to buy things, he doesn't buy up the surplus breadfruit and bananas and all. He wants pearl-diving equipment and canoes and things like that. He starts spending money for canoes and other capital goods. His spending for capital goods is just great enough to exactly offset the eastsiders' reduced spending for consumer goods. The shift in spending convinces some of the people who previously were producing consumer goods—breadfruit, bananas, tuba—to start producing capital goods—canoes, diving masks, oyster knives, and other necklace-making equipment.

See how the factors of production will shift from the production of *consumer goods* to the production of *capital goods*? This is a temporary problem which the "market process" can take care of. There is no problem of unemployment, no drop in outputs or incomes—no hard times at all! Why? Because the increase in investment spending (for capital goods) is exactly the right size to offset the decrease in consumer spending (for consumer goods). That is, the new rate of investment spending is exactly equal to (and exactly offsets) the new rate of savings. See how neatly it works out?

You know that goods are produced for two reasons: One is to satisfy present consumption; the other is for use in future production. If we are producing goods to be used in future production, this means we are producing things which will not be bought by consumers. The consumers are saving some income (are "not-spending" for consumer goods) and are leaving some factors of production in the market to be used to produce goods for the investors.

If the total value of the things people *don't buy* as consumer goods is exactly equal to the total value of things that investors *do buy* as capital goods, then the total size, or "speed" of the circular flow, just stays the same. Total outputs and total incomes do not change. Total economic activity continues at the same level. But the minute total investment spending for capital goods gets *smaller* than the total amount the people are leaving in the market (saving) then there will be surplus goods in the market. Whenever this happens output will be cut back. Unemployment will develop. Total economic activity will slow down. Outputs and incomes will decline.

The only way the surplus goods left by the savers can be bought up and cleared out of the market is for someone to get the money the savers took home (or get an equal amount of money from somewhere) and go into the market and buy up those surplus goods. Whenever this happens there isn't a problem. Whenever it doesn't, there is.

### When Investment Spending Offsets Savings There Are No Surpluses Or Shortages

As long as the investors are spending as much as the savers are saving, the market will be cleared and the economy will keep on operating at the same level. No surpluses will pile up in the market because total spending, or aggregate demand is great enough to "clear the market." The investors are deciding to spend exactly enough to offset the amount the savers are deciding to "not-spend." All goes well.

### Too Much Investment Spending Brings Inflation

Now suppose the investors want to start spending *more* than the savers are saving. What happens? The investors will still be in the market looking for goods after the goods are all gone! What would happen then? Maybe producers would try to produce more, to meet the demands of the investors. But suppose we already have a "full employment" economy. The producers are producing all they can produce. Then, no matter how you slice it, that's all they can produce!

Given enough time, the producers could build more and better capital to help them produce more—that is, they could invest in capital and bring more economic growth. But isn't that just what they're *trying to do now?* They're trying to spend more money to invest in more capital, but no more comes forth. Growth takes time. We can't invest any more in growth than the people are saving—that is, producing but not consuming. So what happens in the meanwhile? We have shortages.

The investors are trying to buy more capital goods than the economy is producing. They start bidding against each other trying to get the capital goods they want. They try to pull the factors of production away from making consumer goods and into making capital goods. Then the consumers start facing shortages.

All the people are employed in good jobs these days, and they want what they want—new cars, new refrigerators, new houses, you name it. They're spending their money to try to get the things they want. Shortages are everywhere. Prices begin moving up. You know what we call this: Inflation. The investors and the consumers are all bidding against each other, trying to get the available goods and factors of production. Prices keep moving up, faster and faster.

## Too Little Investment Spending Brings Recession

Now let's take another look at the opposite case. Suppose savings are high. There is a large amount of output that the consumers are not buying. Lots of goods and factors of production are available for the investors. Now, if they want to, the investors can build new factories, buy new machines, order fleets of new trucks—they can get plenty of everything they need for rapid economic growth. But suppose they decide they don't want to invest very much right now. Then what? I'll bet you know what.

If the businessmen don't think this is a very good time to expand, then we have a problem. The amount the consumers are leaving in the market by *saving* (not-spending) is greater than the amount the businesses want to take out of the market by *investing*. So there lie all those goods, waiting, unbought in the market. The consumers don't buy them. The investors don't buy them. What happens?

The producers who produced the unbought goods will just hold on to them. Not because they want to, but because they don't have much choice. They are caught with unwanted surpluses. They will cut back production and lay off workers. Some producers will shut down their plants. We have an economic slowdown, unemployment—a recession. If it continues and gets really bad, we will call it a depression.

## With Unsold Surpluses, Total Output Still Equals Total Income

A little while ago you saw Robinson Crusoe producing and consuming. His "total real output" was equal to his "total real income," all the time. His real output *was* his real income. Remember?

Next, you saw the eastside islanders catching and selling fish and getting income. The more "dollars worth" of fish they produced, the more "dollars worth" of income they got. The total value of their output was equal to the total income they received when they sold it. The dollar value of their output was equal to their dollars of income.

Suppose one day the eastside islanders couldn't sell all their fish. Then their *money incomes* would be smaller. They would have to take a *part* of their income in the same way Robinson Crusoe always took *all* of his income—that is, in goods (fish), instead of in money.

Their *money income* is equal to the value of the output they produce and *sell for money*. The rest of their income is made up of "real income" —the unsold fish. Whatever the value of the unsold fish happens to be, that's the value of that part of the output, and that's the value of that part of their incomes. Can you see it? (Really, I'm not playing little word games with you. There's an important concept unfolding here and you will need it later. You'll see.)

### Realized Investment Is Always Equal To Savings

Suppose the eastside islanders are trying to sell all their fish, but the westside islanders won't buy all their fish. The westsiders are saving. So what happens? *The eastsiders are forced to take a part of their income in fish.* In fact, the eastsiders are being *forced to "invest" in fish!*

Anytime a producer hires labor, buys materials, and builds something, (maybe boats) he is *investing* in those boats. He hopes to sell them. He doesn't want to keep his money invested in boats. But suppose the boat-buyers decide to save—to "not-buy" the boats? The boatbuilder is forced to take his "income" not in *money*, but in *boats*. He is forced to *invest* his income in boats!

As the boat-buyers *save* more, the boatbuilder is forced to *invest* more (to take more of his income) in boats. But if this sort of thing goes on for very long, the boatbuilder will stop producing boats. He'll fire his workers, stop buying lumber and screws and paint and things. He'll stop spending for investments in boats, and just shut down. If the same thing is happening to lots of other businesses too, soon the whole economy will slow down. Outputs and incomes will go down. Unemployment will be everywhere. Soon, those boat-buyers who decided to save (to "not-buy" boats) will lose their jobs. No work, no income. No money to spend and no money to save, either.

Now think about what you just saw happening. Savers *saved more* and *forced investors to invest more*. Then, pretty soon the investors cut back, *invested less* and *forced savers to save less*. How about that? (Read this section again and watch it happen. You'll be seeing this important principle again, later.)

### Consumer Spending Plus Investment Spending: A Summary

Now let's summarize the effects of consumer spending and investment spending. You know that total spending is what is needed to pull the total output across the market, from sellers to buyers. For production to keep on at the same level, total spending must be great enough to "clear the market."

Total spending is made up of consumer spending (C) and investment spending (I).

1. If consumer spending (C) plus investment spending (I) is *large enough* to buy up (to pull across the market) the "full employment output" of the economy, then we will have full employment.
2. If the total rate of spending (C+I) is *too great* for the "full employment output," then shortages will develop and prices will rise. We will have "inflation."
3. If the total rate of spending (C+I) is *too small* to buy up the full employment output, surpluses will be left in the market. Production will be cut back and unemployment will develop. We will have "recession," or "depression."

### When Total Spending Decreases, Prices Don't Fall

When total spending is too small to buy up all the output, this leaves surpluses in the market. Why don't prices just go down until all the surpluses are cleared? You already know the answer, but here it is again.

Prices will not fall because sellers don't want them to fall, and most sellers have enough market power to keep them from falling. Both the sellers of products and the sellers of factors of production fight very hard to prevent price reductions for the things they have to sell. They would rather see less output and more unemployment, than lower prices. Workers don't want lower wages, land owners don't want lower rents, and the owners of capital and the businessmen don't want lower interest and profits. How bad would the economy have to get before the steelworkers or the autoworkers or the teamsters or the communications workers or the school teachers or anyone else would accept a major wage cut? Pretty bad. Right?

When businesses are faced with declining demands they can cut back output and still survive—sometimes even make profits. But to keep producing and let surpluses pile up to be sold for "whatever the market will bring"—that would be industrial suicide. In the pure market model, declining demand would have an immediate effect on prices and soon the market would be cleared. (Of course several people might starve in the process.) But in the "controlled markets" of the real world, declining demands do not exert much downward influence on prices. In almost all real-world markets, reduced spending results in production cuts and unemployment—not declining prices.

### Employment And Unemployment

You know that sometimes the society's economic system does not *fully employ* all its "able and willing" factors of production. This means that sometimes it might be possible for us to produce more cake to have, or to eat, without having to give up anything. No opportunity costs? How could that be possible?

Suppose we are doing nothing with some of our factors of production. Then we start using these "surplus" factors to bake more cakes. What do we lose? Nothing! We lose only the "nothing" we were producing with these factors before. What a bargain for the society! Why should the society ever let its "able and willing" productive factors go unemployed? You know some of the answer already. But let's take it from the beginning again, and go further with it.

If we were all producing just for ourselves, we could always use our productive factors as much as we wanted to. But when we start producing for "the market" this isn't true any more. If the market doesn't buy our output, we stop producing. If we stop producing, we stop getting income. If we stop getting income, we stop buying from the market. Then other sellers stop producing. So their incomes stop. So they stop buying. Soon, all throughout the society there are unemployed factors wanting to produce, and there are hungry people wanting to consume. But the system isn't working to put these two things together. Hard times are here for all of us!

### Spending Flows Support Employment And Production

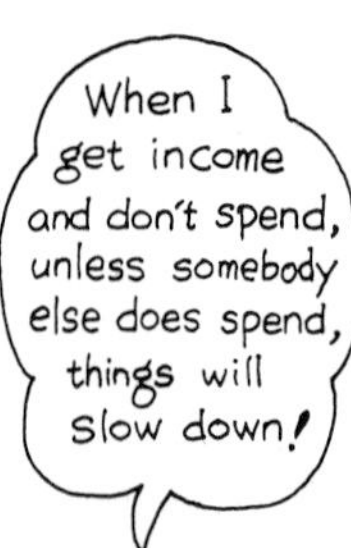

How does the market process get the factors to produce things? By offering them income, of course. Where do people get the money to offer other people incomes? From their own incomes—from selling whatever they have to sell. It's all very tightly tied together. If anything happens to change one person's spending, that affects everybody else's income and everybody else's spending. A *very interdependent system*!

If anything happens to reduce the total rate of spending, the total output will not be bought, so then not as much will be produced. The factors of production will not be demanded as much. Unemployment will develop. We will have unemployed labor, land, capital. People's incomes will fall. Those whose factors are unemployed will get no income at all. People with no income don't spend much. Total spending drops even more. This results in even more unbought output, more layoffs, more unemployed factors, more spending cuts. On and on goes the downward spiral of recession, on down into depression.

In this interdependent system each person is both a producer and a consumer. As a producer you sell your services or your products. With the money you receive you buy the things you want. When you come to the market as a seller (to sell your labor services, or your products) people buy from you and pay you money. Then you go to the "buyer's side" of the market and spend your money for the things you want. When you are on the buyer's side of the market buying, other people are on the seller's side selling; when you are on the seller's side selling, other people are on the buyer's side buying.

Actually you don't go into the market and do all your selling and get your money and then do all your buying. We are all moving back and forth between the buyer's side and the seller's side of the market every day. When we are at our jobs working we are selling in the market; when we stop on the way home and pick up a loaf of bread and a jar of peanut butter, we are on the buyer's side of the market. If you lose your job, you lose your income. You are no longer a seller in the market. So unless you have some savings (some unused "claim checks") or can borrow someone else's "claim checks," you lose your chance to be a buyer.

Are you beginning to get an overview concept of how the total economic system works? Of what keeps it moving? Of what speeds it up and slows it down? If so, you now are ready to understand the meaning of a new term—*macroeconomics*.

## AN OVERVIEW OF MACROECONOMICS

Ever since you started reading this chapter you have been aware that we're now in a "different sort of ball game." We haven't been concerned with *which things* to produce. We have been concerned with such questions as: Will we operate our economic system at full capacity? Will our actual production reach our full potential? That is, will we produce an output which is on the "production possibility curve"? Or will we have unemployed resources and produce less than we are capable of producing? I'm sure you realize that these questions take us into a new area in the study of economics.

You might say that the area of economics we are now talking about is "the economics of full employment and unemployment." Or you could call it "the economics of full production and underproduction." Or you could call it "the economics of over-spending and under-spending." Or you could call it "the economics of inflation and recession." Or you could call it "the study of the overall level of economic activity." Or you could call it "the study of the causes of speedups and slowdowns of the economic system." You could call it any or all of these things and that would be all right. All these things have essentially the same meanings. But we have a special name for this part of the study of economics. We call it *macroeconomics*.

## What Is Macroeconomics?

When we talk about macroeconomics we are thinking about the overall "level" or "rate" of economic activity (employment, production, output, income) and about what might cause it to change. Just like everything else in economics, macroeconomics is concerned with questions about *how* and *why* things happen. Why is there so much unemployed labor this spring? Why are several of the major business corporations reporting losses instead of profits? Why are department store sales running below the level for this time last year? Why did my boss say unless business picks up within the next two weeks, he will have to let me go? And why did one of my best customers come by yesterday to tell me he can't buy any more corn or tomatoes until he gets his job back? These are "unemployment" problems—problems of an economic slowdown—of the underutilization of the society's factors of production. "Macroeconomics" is concerned with the question of what might be causing things like this to happen.

There is another kind of problem in "macroeconomics." Sometimes it goes the other way. Sometimes the problem is one of "overemployment" and "inflation." People try to buy more than the factories and farms can produce. Suppose demand for eggs is up. The price moves up. Egg producers try to hire more factors of production. They try to get some of the stockyard workers to shift to egg production. But people are also demanding more beef and pork. The Chicago stockyards not only are trying to keep their workers—they are also trying to pull in more workers from the poultry business! The automobile companies are looking for extra workers. So are the textile mills and the airlines and the local taxi companies and everybody else.

The lady next door comes by my roadside stand and finds that I have already sold all my corn and tomatoes. She says she will pay an extra 50¢ if I will be sure to save some for her tomorrow. See what's happening? Shortages are developing throughout the economic system. All the resources are fully employed, yet people are still trying to get more output. But the economic system can't possibly produce any more than it is capable of producing! (That's a safe statement.) So what happens?

People keep trying to buy more. Demand increases. Prices rise. What prices? Prices for all the things the people are trying to get more of—almost everything—all products, all factors of production. What do we call this situation? *Inflation*. Unemployment is one of the problems of macroeconomics. Inflation is the other.

## The Key To Macroeconomics Is Total Spending

Macroeconomics is concerned with problems of recession, depression, and inflation. It is concerned with the question of how fast, how fully we operate our economic system. It is not concerned with the question of whether we use our factors to produce eggs or steaks. It is only concerned with the question of whether or not we use our factors to produce *something*. What's the key to keeping production levels high? You already know. Keeping *spending* high. It's the level of total spending that determines whether or not we will keep our factors fully employed, producing things.

If enough people are spending enough money to buy all the things that all of us want to produce, then all of us will produce. If not, then we will not. If enough money is being spent for beef and pork to keep the stockyards and all their employees and cattle cars and everything else operating at full capacity, then the stockyards will operate at full capacity. If people are spending enough to buy my complete output of corn and tomatoes every year, then I will keep on using my factors to produce corn and tomatoes. The same holds true for all producers, throughout the economic system.

If anybody ever asks you why so many automobile workers are unemployed, you know how to answer them. They are unemployed because the people are not spending enough money to buy up the "full employment output" of automobiles. If somebody asks you why the stockyard workers are unemployed, you can answer them quick as a wink: because people aren't spending enough money to buy up the "full employment output" of the stockyards. Then someone asks you a very broad question: "Why are so many people, and factories, and railroad cars, and machine tools, and processing plants 'unemployed'?" Your answer: "Because the people and businesses throughout the country are not spending enough to buy up the 'full employment output' of the economic system."

Then your question-asking friend asks you the ultimate question—the really tough one: "Tell me *why* the people aren't spending enough to buy up the 'full employment output' of the economic system. That's *really* what I want to know!" This question stumps you. You admit that you don't know the answer to that one.

You have just discovered another one of those tricky little things that economics has so many of—answers which really aren't answers at all. You've been answering questions by making obvious statements—by stating truisms. When we say that producers don't produce because people don't buy, this doesn't explain very much. But it gives us a start. At least we know the next question to ask: *"Why don't people buy?"* Macroeconomics is concerned with trying to get at the answer to this question—to *find out* and *explain* why people do, or don't buy.

Any number of things might influence the "level," or "rate" of total spending. One important influence is money. How easy is it to borrow money to build a new factory? To buy a new car? How high will the interest cost be? Another important influence is the general outlook of the people—their expectations. If people think times are getting better, they are likely to spend more freely than if they expect hard times ahead. We will be digging into these and other influences, soon enough.

If we can understand what determines the level of total spending, we can understand why the level of economic activity is high or low. We can understand *why* we have surpluses and unemployment, or shortages and inflation. That is what macroeconomics is all about. Macroeconomics tries to explain what determines the level of total spending (the *size* of the "total spending flow") in the economic system. Why? Because *in a market-directed economy, total spending is the key to understanding total output and total income.* All of Parts Three, Four, and Five of this book are concerned with macroeconomics. But before we go on in macroeconomics, perhaps it's time now to talk about the opposite term: *microeconomics.*

### What Is Microeconomics?

"Microeconomics" is what you were reading about in the first two parts of this book. Microeconomics is concerned with the question of *how the basic economic choices are made.* Mostly, microeconomics deals with the market process and how it works to make the basic choices for society.

Microeconomics explains how the market process brings improved welfare to the people through specialization, trade, optimizing the use of all of society's factors of production, stimulating the production of the most wanted goods, and all that. Microeconomics is the study of trade-offs, of opportunity costs, of substitution, of transformation, of choosing between this and that. Microeconomics just *assumes* that there is full employment. It assumes that the society will be producing someplace *on* its production possibility curve—not *beneath* it.

Microeconomics is concerned with the choice of whether to have your cake, or eat it. Whether to have chocolate cake or coconut cake. Whether to bake the cake in an electric oven, a gas oven, or in a little pit of hot rocks in the ground. Microeconomics is concerned with choosing which things to do with the scarce resources available to us. Microeconomics is the study of choice. You know a lot about microeconomics. In fact almost all the economics you have learned so far, has been microeconomics.

### Microeconomics And Macroeconomics

The terms "microeconomics" and "macroeconomics" are still new. During the past two decades these terms have come into very popular use. The terms are very helpful in separating the two basic segments or approaches to the study of economics. But, as you might suppose, for words which have burst into prominence in an old and established profession, not all economists are in complete agreement as to exactly what the terms mean. You may run across several slightly different meanings for both of these terms, but all the definitions are fairly close to the meanings as explained here. Now, a final statement about the meanings of microeconomics and macroeconomics. Then you'll be ready to go farther into the concepts, principles, and issues of macroeconomics.

Microeconomics is concerned with choices among alternatives. It concentrates on "the economic unit"—the individual consumer, the individual producer, the market for an individual product. If we are thinking about the egg market, why consumers demand more eggs, how the price moves in response to that demand, and how resources then shift into the production of eggs—all these issues are a part of microeconomics. If Willie Wonka is deciding how fast to operate his candy factory, how many boxes of each kind to produce per day, how many workers to employ, whether to pack the boxes of candy in cardboard or wooden cartons for shipment, whether or not to ship by truck, rail, or airlines—all these are questions of microeconomics.

In the market system, each consumer and each producer makes his choices for himself. Then it works out so that society's economic choices reflect the choices of the people. How all this happens is the subject of microeconomics.

Macroeconomics, on the other hand, is concerned with the *total level of economic activity—the rate of employment, production, output, income*. You know that all these are tied together, and they all hinge on the rate of total spending. Macroeconomics, then, is the study of what determines the level, or rate of total spending. Anything which influences the overall level of spending (anything which causes people or businesses to spend more, or less) is a part of the study of macroeconomics. If there is not enough spending to keep the economy fully employed, then we will have surpluses and unemployment. But if people are spending to try to buy more than the economy can produce, then we have the opposite macroeconomic problem—the problem of shortages, and inflation.

No society likes either unemployment, or inflation. If the causes of the level of total spending can be understood, perhaps steps can be taken to stabilize total spending. Perhaps recession and inflation can be overcome; full employment and stable prices can be maintained. As soon as you understand more about the basic principles of total spending, output, income, and employment—that is, about macroeconomics—we will get into these issues of "economic stabilization policy."

The next chapter builds on this one. Be sure you have a good grasp of this one before you go ahead.

## REVIEW EXERCISES

### MAJOR CONCEPTS, PRINCIPLES, TERMS (Try to write a paragraph explaining each of these.)

the interdependence of the modern economy
"partial breakdown" of the economic system
the "secondary functions" of money can cause trouble
investment translates savings into growth
how more saving forces more (real) investment
how less investment spending forces less saving
how C+I supports the economy
macroeconomics
microeconomics

### OTHER CONCEPTS AND TERMS (Try to write a sentence or phrase explaining what each one means.)

real income
Say's Law
saving
investing
aggregate demand
inflation
recession
depression

### QUESTIONS (Try to write out answers, or jot down the highlights.)

1. When the eastsiders saved for Christmas that caused a depression. So how about this: When the government collects taxes from people's paychecks, that's like forcing the people to save (not spend) some of their incomes. Why doesn't that cause the economy to slow down and go into a depression? Discuss.
2. If people save a lot and the investors want to spend just as much as the savers are saving, then what the savers are doing is *good* for the society. But suppose the investors don't want to spend as much as the savers are trying to save. Then what the savers are doing is *bad* for the society. Discuss.
3. If Say's Law was really true, then there wouldn't be any study of macroeconomics—only microeconomics. Right? Discuss.
4. Suppose all prices were free to go down. Then if total spending started to slow down, prices would go down, people would buy more of the lower-priced goods, the surpluses would be all bought up, and there wouldn't be any unemployment and no recession or depression. Discuss.
5. Suppose you are a taxi driver and you skimp on consumer spending so you can spend the money for a new set of tires for your cab. You are *saving* and *investing*, both at the same time. Suppose *all* savers immediately spent their savings for investment (capital) goods. Could the economy ever have a depression? Would Say's Law always be true? Discuss.

# 10 CONCEPTS OF NATIONAL OUTPUT AND INCOME

*The circular flows of income and output, and the basic questions of macroeconomics.*

Suppose you could somehow rise way up above and look down on a "transparent economic system." Maybe you could go up in a balloon, then lean out and look down. What would you see? Lots of activity, right? You bet.

You'd see people doing things, making things, and getting paid with money. You'd see the same people hurrying around, spending the money, buying the things other people were making. Everybody making things and getting money, then buying things and spending the money. All this "receiving and spending and receiving and spending" going on and on and on.

*Things* keep going from the makers to the buyers. *Dollars* seem to just keep going back and forth. A fellow gets some dollars then he spends them. Then he gets them again, then spends them again. Over and over this happens. It's all sort of like a big circle. For a few minutes let's pretend that the "transparent economic system" you're looking down on, *really is* a big circle. Let's look at it that way.

## THE CIRCULAR FLOW OF THE ECONOMY

You can think of the total amount of money being spent by everybody—by all the people and all the businesses—as the total spending flow. The "total spending flow" is flowing for one reason—to buy things. The "total spending flow" is the total amount of spending being "pushed into" the market by all the buyers in the society.

As the "total spending flow" pours into the markets, the money doesn't just puddle up in the streets. Of course not! It goes to people. It is this money, pouring into the hands of the producers and workers and everybody, which gets them to keep on producing and working and making the things the spenders are buying. Nobody keeps making things for sale unless others keep buying. Right?

The people who are working and producing and selling things are the ones who are the *receivers* of the spending flow. But as they are receiving the money with one hand, guess what they are doing with the other hand. *Spending* it, of course. Each person who receives money from the spending flow, runs right around to the buyer's side of the market and spends it again. This is the way the spending flow keeps on flowing.

**Total Amounts Spent Equal Total Amounts Received**

How much do all the people and businesses spend in a week? And how much do they all receive? Can you see that the two amounts—the amount spent and the amount received—must be equal? The total amount that all the people and businesses are receiving must be equal to the total amount that all the people and businesses are spending. Think about it for a minute and you will see that it has to be true.

When a handful of dollars is being spent by somebody, then that handful of dollars is also being received by somebody. Obviously! Otherwise the dollars *would* be puddling up in the streets! Every spender must have a receiver. So total amounts spent must equal total amounts received. No doubt about it.

All the people and businesses are receiving money for what they are selling. At the same time they are spending money to buy things. Each person and business is just passing the money along, from the ones he gets it from, to the ones he gives it to. The "total spending flow" keeps flowing, from buyers to sellers.

The producer or the worker sells his product (or his labor) and gets money. Then he becomes the buyer and spends the money again. After the buyer spends his money, he then becomes a producer or worker again and gets more money. Around and around the money goes. Each person who receives money just turns around and spends it again.

You may begin to get a little bit bored looking down from your balloon, watching this "transparent economic system" in action. It's just the same old thing, over and over. As long as everyone keeps on spending all the money he receives, the total spending flow will continue flowing at the same rate of speed. The economy will keep running along at the same speed, month after month, year after year. The total spending flow changes only when people decide *not* to spend as much (or to spend more than) they receive. That's when things get interesting (and sometimes, tough)! We'll talk more about that, later. But now it's time to stop looking at the money flow for awhile and to look instead at the flows of things—real goods. It's time to look at *real* output and *real* income.

## Real Output Equals Real Income

It's obvious that total spending equals total receipts. Each time someone spends a dollar, someone must receive it. The same is true for real output and real income. It's just as obvious, too, if you think about it.

Look at it this way. Every time something is produced, somebody gets to have it. If a thousand new Ford Galaxie 500's were produced last week, then at the end of the week *somebody* owns a thousand new Ford Galaxie 500's (unless some of them got dumped in Lake St. Clair or destroyed by earthquakes or washed away by a flood—which I doubt.) The total amount of goods made this week must be equal to the total amount of goods received this week. Every new thing must have an *owner*. No mystery about that, is there?

If nobody makes anything, nobody receives anything. If I produce a thousand dollars worth of corn and tomatoes, then somebody is going to receive a thousand dollars worth of corn and tomatoes. If I sell my output of corn and tomatoes to you then you will receive the thousand dollars worth of corn and tomatoes. (And I will receive a thousand dollars worth of money.) If I don't sell my output to anybody, then it's still mine. I'm the one who will receive it as "real income." Maybe I receive it or maybe you receive it. But *somebody* must receive it. It doesn't just pile up in the streets!

Everything produced during any day, week, month, or year, has an owner. The owner of each new product has something that didn't exist before. He has some "added something." He has some "real income." The more new things produced, the more new things (real income) the people will get.

**The Real Output Of The Society Is The Real Income Of The Society.** If we look at it from the point of view of those who produce it, it's real output. If we look at it from the point of view of those who receive it, it's real income. But we are looking at the same flow of goods, no matter which way we look at it. We can look at it as the flow of real output from the producers or as the flow of real income to the receivers. Real output equals real income? More than that. Real output *is* real income.

## Flows Of Money Equal Flows Of Goods

Now let's tie the *money* flow (total spending equals total receipts) together with the *product* flow (real output equals real income). Your common sense would tell you that the spending flow and the output flow must be closely related. Why do people put money into the *spending* flow? To get some of the things in the *output* flow, of course! Why do people put goods into the *output* flow? To get some of the money in the *spending* flow. Of course!

Do you begin to get the feeling that all these flows are going to wind up being equal? Can't you just see everyone pushing in goods and taking out money and then turning around and pushing the money back in and taking out goods? Really, that's the way it works. For every dollar's worth of goods flowing, there's a dollar's worth of money flowing. The two amounts must be equal.

Why do people push goods into the flow and take out money and then turn right around and put the money back in and take out goods? That seems to be so much wasted energy. But no! They're putting in a lot of one thing—their speciality—and taking out small amounts of many kinds of things—they're taking out the great variety of things they want.

Specialization and trade is the name of the game. Remember? That's what the whole "market system" is all about. It's a system to stimulate people to produce something they're good at and then to permit them to exchange what they make for the many other things they want. It's so obvious and so simple that people sometimes forget what the market system is all about!

People are selling goods and factors of production. In exchange for these goods and factors they are on the "receiving side" of the spending flow. These same people, when they are buying goods and services from others, are on the "spending side" of the spending flow. At any moment, the total amount of new product being produced is equal to the total amount of new product being received. Obviously. Just as obviously, the total amount being sold is equal to the total amount being bought; and the total amount being spent is equal to the total amount being received.

Now you can see the equalities (identities, really) between real output and real income, goods sold and goods bought, and money spent and money received. You can see all these things going on simultaneously as you look down from your balloon and observe this "transparent economic system." So now that you have it all clearly visualized, here's a different way of looking at it. I think it will be more helpful than confusing. I hope so.

### The Circular Flow Between Households And Businesses

Pretend that "businesses" buy all the factors of production and then use those factors to produce all the outputs. The "households" own, and sell all the factors of production to the businesses, and then buy all the outputs from the businesses. Businesses receive money when they sell products to households. Then they spend the money to pay the households for the factors. Then the households spend the money again to buy more products from the businesses, then the households get the money back again as factor payments. This circle keeps on going and the money keeps on flowing, from households to businesses to households to businesses, around and around and around.

The households own all the factors of production. They sell factors to the businesses. When the businesses pay for the factors the income goes to the households. Then, when the households spend the money to buy the products from the businesses, this completes the circle. The households wind up getting the goods and the businesses get the money back.

Isn't this a neat way to look at any economy? Really, it's a pretty good picture of what actually is going on all the time: People selling factors to businesses and using the money they earn to buy the outputs of the businesses; businesses paying out money to the factor owners, the factors making things, and then the businesses selling the things to the factor owners, to get the money back.

In this illustration, don't think of "the business" as being "somebody" —some wealthy person—some "businessman." That isn't the idea. The business is sort of "a place where everybody goes to sell factors and receive incomes." Even the man who owns the business—the buildings and machines and all that—is selling those factors to the business in exchange for an income from the business. The business owner doesn't actually "live in the business." He lives in a household just like everybody else. He just sells his capital and managerial skills to the business. Then he receives income from the business as his "household income," just like everybody else. The only difference is that the owner of the business is the last man in the "income-receiving line." He gets what's left over after all the other factors get paid off. You might say he is the "residual claimant." His household gets whatever is left over after all the other households get paid.

Actually, when we say "businesses" what we really mean is "the *buyers' side* of the *input* markets, and the *sellers' side* of the *output* markets." Businesses buy inputs and sell outputs. When we say "households" what we really mean is "the *sellers' side* of the *input* markets, and the *buyers' side* of the *output* markets." Households sell inputs and buy outputs. Of course! All the people who are earning incomes are selling in the input markets. And just about everyone is buying in the output markets. We all have to eat!

Figure 10-1 shows a diagram which may help you to visualize this circular flow between households and businesses. You should study that diagram now.

**Fig. 10-1 THE CIRCULAR FLOW BETWEEN HOUSEHOLDS AND BUSINESSES**

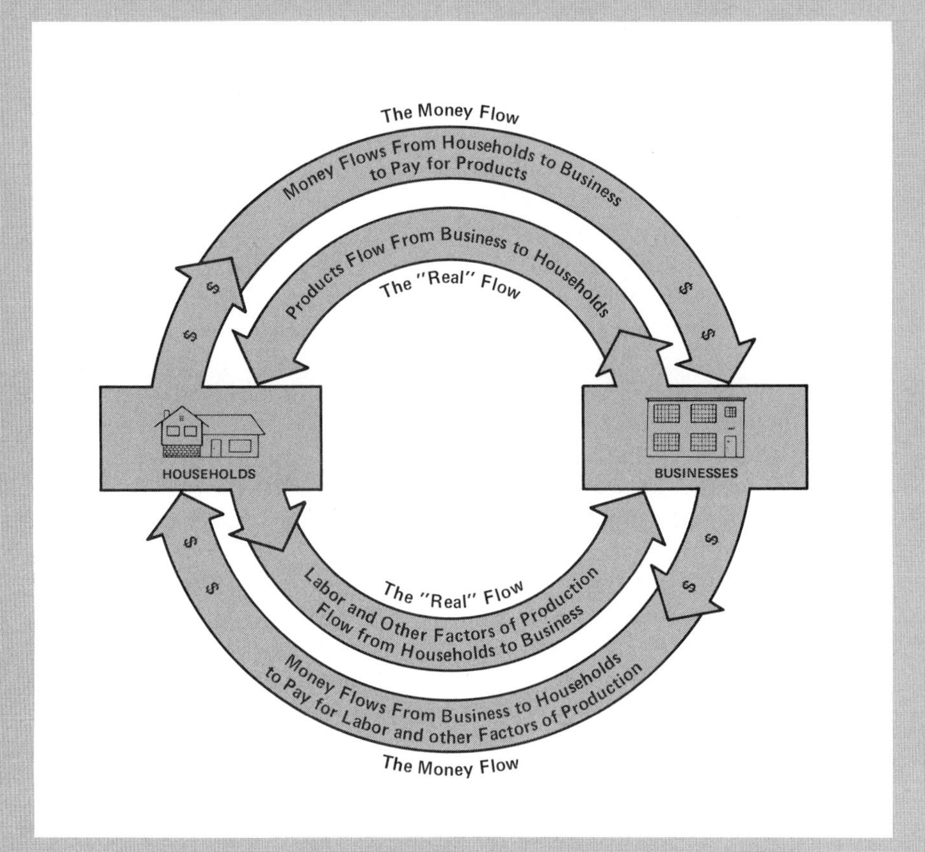


Around and around it goes. Money flowing one way, things flowing the other.

You can see that if anything makes the money flow smaller, the "real flow" will get smaller also. The two flows are completely dependent on each other and are (quite obviously) equal.

## The Interdependence Of The Circular Flow

The circular flow concept really shows how much we are all dependent on each other. If the flow of money being paid to the business gets smaller, then the businesses can't pay out as much money to the households. Employment is cut back and output is reduced. The money flow gets smaller and the real flow gets smaller.

The circle fits together so tightly that if any part of the system speeds up or slows down, all the other parts of the system—both the money flows and the real flows—will be forced to speed up or slow down, too. Once it starts to speed up or slow down, it's likely to go on changing more and more in the same direction. How interdependent the system is, and how tightly it all fits together!

## It's The Total Spending For New Output That Matters

We have been talking about "the spending flow" as though all spending was spent for *new output*, and as though all spending created *new income* for someone. This is not quite true. Here's an example that will explain why not.

If I produce a bushel of corn and sell it to you for $10, then that "sale" is a part of total spending for output. It creates income for me. Then, after you have bought the corn, suppose somebody offers you $11 for it and you sell it to him. The $11 you get does not represent another $11 worth of output, and income. Obviously not. Only one dollar of it would be income for you. That would be your payment for increasing the value of the corn by "buying, keeping, and carrying" the corn to a time and place where it was more valuable—$1 more valuable.

You have increased the value of the output of the economic system by $1, and your income has gone up by $1. But *not* by $11! The value of *your* "output" is the value *you* have added. That's what you get to keep as income. One dollar. Your "value added" is *your* output.

Throughout this book, whenever we're talking about *total spending* we're *really* talking about total spending *for new output*. That means all the money flowing in the total spending stream will be flowing to the receivers as *income*. We don't count the spending and re-spending, over and over, for the same old things. So when we say "total spending," we really mean "total spending for new output."

## Each "Economic Unit" Influences The Total Economy

Think about the "modern economic system." What a complex piece of machinery it is! Like the inside of a watch. Only instead of dozens of little wheels there are millions of little wheels. Each little wheel is "locked in" and turning with the others. Yet each wheel has its own little engine which it can speed up or slow down if it wants to. But as each tries to speed up or slow down, it influences the speed of all the other little wheels "locked in" around it. How fast (or how slowly) the whole machine—the economic system—actually will be going depends on how fast (or how slowly) all the little wheels—the individual economic units—are trying to go.

The more you work and produce and earn and spend, the more you help to speed up the "total mechanism"—the economic system. If you produce a lot, earn a lot, spend a lot, then you push the system to speed up—and it does! When a businessman buys more materials, hires more people, and increases his output, he pushes the system to speed up. But when people or businesses produce less, earn less, spend less, they slow down the system. Suppose we are trying to decide how fast we think this "total machine" ought to be going. Should it be running at maximum possible speed? At maximum physical capacity? Of course not. People don't like to put in their absolute maximum effort: Sixteen hours a day? Seven days a week? No holidays? No vacations? That's no fun!

No economic system has ever succeeded in running at its absolute maximum physical capacity over long periods of time. Very few people would want it to. So, except perhaps when there is some sort of national emergency, the question is not whether or not the economic machine is running at maximum possible speed. That isn't the desired objective. We only want it to be running "fast enough"—that is, as fast as we want it to run. And this brings us to *the basic questions of macroeconomics.*

## THE BASIC QUESTIONS OF MACROECONOMICS

The real macroeconomic questions are these: "Is the economic machine running as fast as the people *want* it to run? And no faster? And if so, why? And if not, why not? Does every person who *wants* to work a 40-hour week have a chance to do so? Without being pressured into working weekends too? Do all the factory owners who want to operate their factories on an 8-hour day, 5-day week have a chance to do so? And without being induced to run a night shift?

Is there enough *aggregate demand* (total spending) to buy up everything the people and businesses are trying to sell? And no more than that? And if so, why? And if not, why not?

Is the aggregate demand (total spending) of households great enough to buy all the output the businesses want to produce? And no greater than that? Is the aggregate demand (total spending) of businesses great enough to buy all the factors of production (labor, land, capital) the households want to sell? And no greater than that? If so, the economy will have full employment, and there will be no "excess demand pressures."

Whenever there are surpluses, this means that some of the outputs and some of the labor and other factors are remaining unbought in the markets. There is unemployment. Whenever this happens, the economic system is running too slowly. The "owners" of the unemployed labor and other unemployed factors are not very happy about this situation. If you couldn't get a job, how would you feel? You'd say: "Let's do something to speed up the economy so I can get a job!" Right?

Now look at the opposite situation. Suppose the economic system is trying to run too fast. If aggregate demand (total spending for goods, and factors) is too great, shortages will appear. Consumers will be trying to buy more goods than are available. Businessmen will be trying to get more factors than are available. Prices will rise. As businesses offer more money to get the factors to work harder, faster, longer hours, outputs will increase some. But the economic system is trying to run too fast. And it's riding for a fall! Later on, you will see why.

## THE NORMATIVE AND THE POSITIVE MACRO-QUESTIONS

You can see that "the basic questions of macroeconomics" can be looked at in several different ways. But all the macroeconomic questions are concerned with the same issue—the issue of *how fast the economic system will run*. For a minute, let's take this basic question apart. Let's look at it from two different points of view. First, we can look at the *normative* question: "How fast *should* the economy run?" Then we can look at the question economists usually talk about—the *positive* question: "What determines how fast the economy actually *will* run? (If you aren't sure about the meanings of the terms "normative" and "positive," perhaps you should go back to Chapter 3 and review the section headed "Positive Economics and Normative Economics.")

### The Normative Macro-Question

The normative macroeconomic question is: "How fast do we *want* the system to run?" "What's the best, the most desirable speed of the economic system, to bring the most good to the society?" "What speed is 'fast enough but not too fast'?"

Each society, somehow, explicitly or implicitly, must answer this question. When the Bible says: "And on the seventh day He rested"—it's talking about the normative macro-question. When the society decides to have an eight-hour workday, and a five-day workweek, and vacation time each year, it is deciding the "normative macro-question" for the society. When we decide to (or not to) use schools for productive purposes at night and on weekends, we are making a "normative macro-decision."

### Custom Influences The Normative Macro-Question

In each society, custom and tradition are very strong in influencing the answer to this normative question of how hard the economy *should* work—how fast it *should* run. You can see that in the *pure market* economy, the socially desirable rate of output would be reflected in the amounts the people were trying to buy, and in the amounts the producers and workers and farmers and everyone were willing to produce and sell.

In the model pure market system, whatever rate of output the society wants, that's the rate of output the society gets. The rate of employment and production automatically reflects the customs, traditions, wishes, and desires of the society. Just automatically!

Even in a completely controlled economy, the government planners must build their employment and output goals within the limits of "what the people are likely to be willing to put up with." These limits are mostly determined by the customs and traditions of the society. But even so, if it wants to, an autocratic government can change people's ideas about such things as working on the Sabbath, or working long hours, or on the night shift.

### The Government Can Force More Output

Economic plans, enforced with the power of law, sometimes can result in very large increases in the total rate of employment and output—that is, can push the economy to speed up to a rate much faster than the society "naturally" would seem to want it to go. Very few people seem to want to work a ten-hour day, seven day week, and forfeit all vacations. But if the government decides to enforce this kind of maximum productive effort, it can do so.

### The U.S. World War II Experience

During World War II, under tight government planning, the American economy went on a 24-hour day, seven-day week. Vacations were cancelled. Workers were frozen on their jobs. Many were urged to work a ten-hour day. Every able-bodied person, male or female, was urged to join the labor force and take a job. During this period, millions of men were moved out of the labor force and into the armed forces. But even so, the value of the total output of the U.S. economy more than doubled between 1940 and 1945: from less than $100 billion in 1940, to more than $200 billion in 1945.

### The Soviet Union

In the Soviet Union, the planned production goals and constant push for more outputs have speeded things up greatly. Outputs have been expanded and more capital has been produced. This is the way so much economic growth has been generated in the U.S.S.R. over the past several decades.

### The Economic Plan Must Answer The Macro-Question

*In a completely planned economy* (where the political process is in full charge of all the economic choices) *somebody must decide on the rate at which the system will run.* In the governmentally planned "sectors" of every economy, these planning decisions must be made. For example, in the U.S. economy we face the question of whether or not our public schools should go on a 12-month year. We must decide how many class hours and how many students should be assigned to each teacher. These are the kinds of normative macro-questions which the political process (the "resource administrators") must decide.

### The Advanced "Mixed Economies" Could Produce More

The advanced, highly productive economies don't work as hard as they could. They don't seem to be interested in maximizing their rate of output. It seems that as an economy grows and as output per person increases more and more, there is a tendency for the society to let its "total desired" output fall farther and farther below its "maximum physical potential." People like to take more time off. Young people sometimes don't start doing anything productive until they're in their twenties. Many people retire early.

There's a great deal of slack in the American economy and in all the other modern economies of mixed socio-capitalism. Why? Mostly because most people don't seem to want to produce as fast and as much as they could. So if it's left up to the individual members of the society, the "total desired" output (the "normative macro-level") of the economy will be less than the amount the economy is *physically capable* of producing. Most free people in a free society will produce less than they would if they were forced to work as long and as hard as they could. Nothing surprising about that!

### The Positive Macro-Question In The Planned Economy

From what you've been reading about the *normative* macro-question, a good bit about the *positive* macro-question has already come to light. Suppose we're talking about a completely planned economy. You know that the positive macro-question (what actually influences or determines the speed of the economy) will be answered by government directive. The "resource administrators" give the orders and the people respond. Everyone does what he is told to do, so the economy runs at the planned speed.

Of course you know it isn't quite as simple as that. People don't always do exactly as they're supposed to. Administrative controls very often are inefficient. Nevertheless, the speed of the planned economy generally is controlled by the economic planners and resource administrators.

A good course in public administration or administrative management, or one on the Soviet economy, will clue you in to how the "planned economy" carries out its micro and macro decisions. It isn't difficult to see how economic controls and government resource management are supposed to work. But what about the market system? How does it work there?

How is the "automatic pure market mechanism" supposed to take care of the positive macro-question? What forces are supposed to determine the overall rate of employment and output and income in the "pure market" system? Then, in the real world where there aren't any *pure* systems (only "mixed economies"), how does it work? It's time to get into these questions, right now.

## HOW THE SPEED OF THE "PURE MARKET ECONOMY" ADJUSTS AUTOMATICALLY

In the "model market system" the speed of the economy just takes care of itself. It is automatically determined, just like everything else in the "model market system." The "normative question" just doesn't arise. In the "pure model," however fast the economy *does* run, that's how fast it *should* run. The speed of the economy responds directly to the wishes of the society as those wishes are expressed in "the market." The "natural laws of economics" take care of everything.

If everybody wants to work hard and long, then a lot of output will be produced. A lot of income will be earned by all those industrious people. There will be a lot of spending. If all the manufacturers and farmers and businessmen and workers and everyone are really producing as much as they can, pushing hard to make as much income as they can get, then employment and output and income in the society will be high. But if nobody feels stimulated to work very hard or produce very much, the society's output and income will be low. See how the normative and positive macro-questions are automatically answered in the pure market system?

### Say's Law Again

As people want more, they work harder and produce more. That way they get more income and can buy more things. The economy speeds up. But if people decide to work less, take long vacations, and if businessmen don't bother to try to keep producing and trying to make profits, then the economy slows down. The more the people want for themselves, the more they are stimulated to produce and sell to others. The more they produce and supply to the markets, the more income they get. The more income they get, the more they buy from the market. The more they produce the more they sell and the more they earn and the more they buy.

See how Say's Law works in the "model market economy"? People produce more, so they get more income, so they buy more. No one needs to decide how fast the economy *should* run. No one needs to plan and direct the people to produce at the right speed. It's all determined automatically by the natural market forces—by "the basic laws of positive economics."

### Prices Adjust To Prevent Depression

But suppose that sometime, for some unexplained reason, suddenly there is too much output for sale in the market. Businessmen and farmers are going broke because they can't sell all their products. Many workers lose their jobs. The economy goes into a depression. How does the market process solve this problem of depression?

In the "pure model," it's easy. Prices will adjust themselves until the surpluses are all bought up, and until the unemployed workers all have jobs again. Let's talk about how it's supposed to happen.

**Product Prices Go Down.** The producers who can't sell all their products start offering them for sale at lower and lower prices. As the prices of the surplus products go down, money becomes more valuable. Each person now can buy more products with his dollars. Also, people are more willing to spend, to take advantage of all those good bargains! So as the prices fall more and more, people (and businesses) buy more and more. Soon all the surpluses are cleared out of the market, and everything's okay again.

**Wages Go Down.** At the same time that product prices are falling, wages (labor prices) are also falling. The unemployed workers (the "surplus labor") will be seeking jobs, and they will be willing to work for lower wages. This will push wages down. As wages go down, each businessman's dollar will buy more labor than before, so the cost of producing his product goes down. Therefore his chance to make a profit gets better and better. So he hires more and more labor. Soon the unemployment problem is solved.

**Interest Rates Go Down.** Also, there's another "price" working to help get rid of the surpluses and unemployment: the interest rate. Businesses usually borrow much of the money they need to build their factories and to buy their machines and equipment and raw materials and to stock their shelves with inventories. If business is booming and everybody is trying to expand, then everyone will be trying to borrow money. The "money market" will be very "tight." And the interest rate? High, of course!

But suppose the economy is depressed. Not many businessmen are trying to borrow. The money market is very "easy." Interest rates are low. As the economy slows down, interest rates go down. With "easy money" available, the businessman sees the chance to expand his business at very low interest cost. This induces him to borrow more, to buy inputs, to expand. This helps to bring prosperity back again.

### Depression Is Self-Correcting, In The "Model Market System"

So, in the "model market system," depression is self-correcting. As soon as surpluses and unemployment begin to appear, automatic price adjustments come into play and begin to solve the problem. The price adjustments keep doing their thing until the problem is solved. It's so neat you just wouldn't believe it! So don't believe it. It doesn't work quite that way in the real world.

### Real-World Prices Aren't Very Flexible

In the real world, all these "market tendencies" really do exist. The tendencies really do try to work in the real world, just as the model says. When there are surpluses and widespread unemployment, it is a fact that prices and wages do *try* to go down. Some prices and wages actually *do* go down, some. But most prices and wages *can't* go down. People won't let them go down. Nobody wants the price of what he sells to go down. Those who sell their labor don't want their wages to go down, either.

In general, everyone would like for prices to go down on what he *buys*. But who wants the price of what he *sells* to go down? Nobody. Most often, it is the *seller* (not the buyer) who has the most control over the price. So guess which way the prices of most things are likely to move. Up? Yes. Down? No.

All modern nations have such things as minimum wage laws, agricultural price controls, and lots of other "price-fixing" laws. For most products the prices are set by the manufacturers or wholesalers or retailers. These businessmen are very reluctant to cut their prices. Usually they think they would make smaller profits (or greater losses) if they did cut their prices very much. Usually they are right.

It's easy to see how our "pure market model" economic system overcomes depression. It all works very neatly and automatically, just like everything else does in the "model pure market system." Very neat and beautiful. But how well does the model explain how the depression problem is solved in the *real world*? (Brace yourself!) It doesn't. It can't.

### If Prices Won't Move, The Market Model Won't Work

Once prices are frozen, the pure market model is of no help in explaining how to get out of a depression. All it can tell us is this: "With inflexible prices, if everything else stays the same the depression will continue indefinitely." That's no help to the hungry workers, the dispossessed farmers, the bankrupt businessmen, or to the harassed economist who's trying to advise the government what to do!

For the market system to maintain full employment—that is, to prevent depression—wages and prices must be flexible. All wages and prices must be able to move freely up or down to reflect changes in demand or supply. But are all prices flexible? Of course not!

Generally, prices can move upward without too much resistance. But downward? No. Not without a long, hard struggle. Many prices would never move down, no matter what!

### The "Gut Issues" Of Macroeconomics

If we can't depend on *price adjustments* to bring "total spending" into line with full employment, then *what can* we depend on? If price adjustments don't occur to bring about the right amount of total demand (total spending), then what does? Whatever does, can we depend on it to keep us out of bad depressions? Or bad inflations?

Now you have your finger on the real questions—on the real "gut issues" of macroeconomics. It will take you some time to gain a thorough understanding of the complex forces involved and to work your way to the answers to these questions. It isn't particularly difficult. It just takes time. All of Parts Three, Four, and Five of this book are concerned with these issues.

In the real world, price adjustments don't work the way they're supposed to, to keep the economy out of depression. So can we be sure that an economic system of "mixed capitalism" will stay out of depression? That's a good question. To get at the answer we have to back off, then re-approach this whole issue from a different angle—from a new and different direction. It will take several more chapters for you to *really understand,* but for now, let me give you a tentative answer.



We *can* depend on the U.S. economy as well as all the other economic systems of mixed socio-capitalism, to stay out of serious and prolonged depression. To the extent that the economy does it by itself, that's fine. But when the economy doesn't do it by itself, the government will enter the picture and do whatever it must to prevent a serious and prolonged depression. The government has the "tools" it needs to do the job. You may be quite certain that whenever it becomes necessary these tools will be used, and that serious depression will be averted. After a few more chapters you will understand how it all works. But for now, on with our story.

## HOW TO ESTIMATE THE SPEED OF THE ECONOMY

Suppose the economy is running too fast. Or too slow. How can you tell? What do you look at to find out?

You could go around asking businessmen: "How's business?" Or you might go down to the state employment office and see how many people are out of work and looking for jobs. If all the businessmen are complaining about slow sales, and if there are long lines of unemployed people at the employment office, then we can say the economy is running too slowly. But suppose the businessman says: "It would take me three years to fill all the orders I have, right now!" Suppose the employment office manager says: "Everybody's trying to hire more workers but nobody's looking for a job!" Then you know the economy is trying to run too fast.

There are lots of ways to see if the economy is booming or depressed. If you see lots of trucks on the highway, trains going by, factories running night shifts, crowds of people spending lots of money in the supermarkets and department stores and bars and everywhere, then you know the economy must be booming. If you see a lot of idle factories and glum people pinching pennies and looking for work, then you know the economy must be depressed. But isn't there some "more exact" way to observe and measure the speed of the economy? Can't we use some *numbers* to show the "overall level of economic activity"? You bet we can! And we do.

These numbers are very helpful in telling us what's going on in the economy—in telling us that everything is all right, or that it isn't. The numbers can tell us what's wrong, and how wrong, and where, and (to some extent) *why,* and what needs to be done about it. Magic little numbers! The most basic ones are found in what we call the national income and product accounts. During the depression of the 1930's we didn't have any such numbers to help us to see what was happening, and why. But we do now.

By the time you finish the next chapter you will understand all about the magic little numbers—the "national income and product accounts." But before you go ahead, be sure you have a good solid understanding of the circular flow idea, of the basic questions of macroeconomics, and of the way the macro-questions are automatically taken care of in the pure market model. These are all easy and simple things. But learn them well. They're basic. Then, on to the next chapter and another step deeper into macroeconomics!

---

## REVIEW EXERCISES

### MAJOR CONCEPTS, PRINCIPLES, TERMS (Try to write a paragraph explaining each of these.)

total spending equals total receipts
real output equals real income
the basic questions of macroeconomics
the normative macro-questions
the positive macro-questions
how the market process answers the macro-questions
the "gut issues" of macroeconomics

### OTHER CONCEPTS AND TERMS (Try to write a sentence or phrase explaining what each one means.)

the total spending flow
real output
real income
"value added'
normative
positive
national income and product accounts

### DIAGRAM (Try to draw and explain.)

The Circular Flow Between Households and Businesses.

### QUESTIONS (Try to write out answers, or jot down the highlights.)

1. Suppose, next spring, *everyone* decides to really be stingy with himself and save up for a nice long vacation trip, in July. Nobody buys any new cars or appliances or clothes, or goes out to restaurants or shows.

Everybody just saves. What do you think will happen? Will everyone succeed in his savings and vacation plans? Discuss.

2. It's interesting to think about how much the normative macro-questions are determined by tradition and custom and habit, even in the highly sophisticated, modern economies of the 1970's. Think about such things as the work-day, the work-week, school vacations—how many of these arrangements have been "passed down" to us by the needs of the agriculturally-based, daylight-restricted societies of the past? Do you think it would be a good idea to make some basic changes in some of these "customary" arrangements? Discuss.
3. Explain how depression is supposed to be automatically prevented, and self-correcting in the "pure market economy." Then explain why it doesn't (can't) work that way in the real world of the 1970's.

# 11 HOW TO MEASURE NATIONAL OUTPUT AND INCOME

*The national income and product accounts, and price indexes.*

How fast should the economy be running? And what determines how fast it will run? These are the questions you have been thinking about for the last two chapters. These are the most basic questions of macroeconomics. The next thing you need to know is how we *measure* the speed at which the economy is running. You're getting ready to find out about that, right now. After that you'll find out about price indexes.

## THE NATIONAL INCOME AND PRODUCT ACCOUNTS

There are several things we might look at and measure to find out the level of "total activity" in the economy. We could look at and measure total employment—which could mean "the total use of all the productive factors," or it could mean just "the total employment of labor." Or we could look at and measure "the total amount being produced"—total output. Or we could look at and measure "the total amounts being received as income by the owners of all the factors of production"—total income. Any one of these figures would give us a measure of "total economic activity," and would tell us about the speed at which the economy is running.

Suppose you look at the figures for "total employment" and "total output" and "total income" and see that all three figures are getting larger from month to month. You know that the rate of total economic activity is increasing. Right? But do you really need to look at all three figures? Of course not. All three figures will be moving along together, won't they? Sure. You know that total output is equal to total income.

And you could guess that the rate of employment is closely related to the rate of output and income—the more employment the more output and income.

Now you know what to look at. But where do you look? How can a person find out what these "employment" and "output" and "income" figures really are—for the American economy, or for any other economy?

Most nations of the world actually try to add up their "national output" and "national income" to see how much is being produced in the country. Everyone wants to know "how's business?" You are already familiar with the term gross national product (usually called "GNP"). Probably you have heard about national income, too. These, together with the "employment" and "unemployment" figures, are the most often used measures of "the overall level of economic activity," or "the rate at which the economy is running."

### How The Government Gathers The Statistics

How does the government get the "national product" and "national income" numbers? To get the "product" statistics, the government statistics gatherers ask each business to fill out forms telling how much they are producing. As Willie Wonka is operating his candy factory, one day he will get a questionnaire asking about the size of his output. They will ask him to send in a report several times a year indicating his "rate of production." They won't ask for his daily or weekly or monthly rate. They want his *annual rate* of output. So instead of telling them he is producing 10 truckloads a week, he tells them he is producing "at the annual rate of 520 truckloads per year" (10 truckloads, times 52 weeks = 520 truckloads). But he needs to tell them more than that.

They need the *dollar value* of his "annual rate of output." Why? Because they are going to add up his output of candy, together with everyone else's output of everything else. How can they possibly do that? Only by translating each output into its dollar value. Then they add up the dollar values. This gives them one number, a dollar figure, to indicate the "rate of flow of output" for the entire economic system. That number, the dollar figure, is called the *gross national product*—GNP.

So Willie Wonka tells them that during the first quarter of this year he was producing at the rate of 10 truckloads of candy a week, @ $10,000 worth of candy per truckload, for a total of $100,000 worth per week. So he's producing at the rate of $5,200,000 worth per year ($100,000 worth per week, times 52 weeks = $5,200,000 worth per year). When they get Willie Wonka's and everyone else's figures for the first quarter, they add them all up. Suppose they come up with a total figure of $1,000,000,000,000 (one trillion dollars). Then the U.S. Department of Commerce publishes a report saying that during the first quarter (January–March) gross national products (GNP) in the United States was being produced *at the annual rate* of one trillion dollars worth of goods and services per year.

## The GNP Rate Of Flow Changes During The Year

Suppose this rate of output continues for all twelve months of the year. Then the total GNP for the year will amount to one trillion dollars worth of output. But that isn't the way it's really going to happen. The "annual rate of output flow" isn't going to stay the same for all twelve months of the year. During some months people may buy more candy and Willie Wonka may put on an extra shift of workers to produce enough to meet the extra demand. During other months, demand may slack off, and he may cut back production. During July he may shut down the plant for two weeks while everybody goes on vacation. This will cut his July output almost in half. Even if nothing else happens, he won't produce as much candy in the month of February. Why? Because there aren't as many days for him to produce candy, or for his customers to buy candy, in February!

If this turns out to be a good year for business, then the annual rate of flow of the GNP will probably get larger from month to month. But if this turns out to be a bad year for business, the flow will get smaller. As the year progresses, we can watch these GNP figures and see how fast the economic machine is running and whether or not it is speeding up or slowing down. As the new GNP figures become available they are reported in the *Survey of Current Business* and the *Wall Street Journal* and the *New York Times* and *Business Week* and *Time* and *U.S. News* and in all the news media, including your own home town newspaper. It's easy enough to find GNP figures!

So much for the "product" figures. Now, how does the government get the "national income" figures? The same way they get the product figures. They ask the businesses how much they are paying out to the owners of the factors of production—for labor, land, and capital.

The government's "national income statisticians" try to find out the total amount of income being received as wages, rents, interest, and profits. Then they add up all the figures and come up with the "national income." You can see that these figures would not be too difficult to find. All the businesses and several government agencies keep records of all these payments anyway—for income tax purposes and for several other reasons. It's a simple matter for them to report the figures to the "national income statisticians."

## Why The GNP And National Income Figures Are Not Equal

Suppose all the figures of "income payments to the factors of production for the first quarter" have been reported. Then all the figures are adjusted to annual rates and added up. And suppose the total comes to $850,000,000,000 ($850 billion). The government issues a report saying "national income during the first quarter was running at the annual rate of $850 billion." But you know that's got to be wrong. A minute ago you found out that the GNP was running at the annual rate of one trillion.

You know that if *output* in the first quarter is being produced at the annual rate of one trillion, then the rate of *income* being received must also be one trillion. Output is *always* equal to income. Right?

You know that national output and national income *must* be equal. Every dollar paid for output must go to somebody as income. There's no other way! Still, the government *does* come out with a trillion dollar GNP figure, and a figure of only 850 billion for national income. How can this be? The answer is easy. The GNP figure isn't really the *true value* of the output. The GNP figure is partly fictitious. It includes some make-believe values, so it overstates the value of the output flow. But for the national income figure, the government statisticians stick to the true values. So the national income figure is always smaller. Let me use an example to explain.

Suppose you are talking to me about my backyard corn-producing business. I tell you that, using only my own labor, I produced and sold $1,000 worth of corn last year. I produced 100 bushels of corn and sold it all for $10 a bushel (100 bushels times $10 = $1,000). You congratulate me for making $1,000 of income from my garden. (You know that if I produced and sold $1,000 worth of product I must have received $1,000 of income.) But I say, "No. I really didn't earn a thousand dollars. I only made about $850. Let me explain why."

### Deduct Depreciation From GNP To Get NNP

"First of all, I really didn't produce a hundred more bushels of corn. I had five bushels of corn to begin with, which I planted as seed corn. So really, I didn't add a hundred bushels of corn (or $1,000 worth) to the society's output. All I added was 95 bushels, or $950 worth. See? My net product was only 95 bushels. That's all the 'value' I really added through my production process."

It's easy to see why the seed corn must be deducted to find the "net product." It's just as easy to see why *all* the capital used up in the productive process must be "replaced in its original condition" before we can start bragging about how much we have produced. If you wear out a thousand dollars worth of tractor parts producing a thousand dollars worth of corn, or tear up a thousand dollar fish net catching a thousand dollars worth of fish, or wear out a thousand dollars worth of canning machinery producing a thousand dollars worth of canned peas, you really haven't "produced" *anything*!

So the first "make-believe" part of the GNP figure is that it's too "gross." (That's why we call it the *gross* national product.) We must take out enough of this "gross product" figure to make up for what we used up, tore up, and wore out in the process of producing the product. We call this deduction the depreciation deduction. When we subtract the "depreciation deduction" (also called the "capital consumption allowance") from the GNP figure, we get a *net product* figure. We call it net national product (NNP). That's closer to a "true value of what really was produced" than is the *gross* GNP figure. But even the NNP figure includes some fictitious value. So we're going to have to subtract something else. Guess what? You'd never guess, so let's go on with the corn example. You'll find out.

### Deduct Indirect Business Taxes From NNP To Get NP = NI

The "seed corn replacement deduction" (depreciation) takes off $50 from the $1,000 I received when I sold my corn. My "net product" is $950. But my income from my corn garden was only about $850. Remember? So what happened to the other $100? Why did my $950 worth of net product only bring me about $850 in income? Here's why: The *true* value of the corn I sold really wasn't $10 a bushel. The corn was only worth $9 a bushel. The price was pushed up to $10 a bushel by "embodied taxes" (indirect taxes) of one dollar a bushel. I collected $1,000 from my customers, but the corn I sold them was really only worth $900. I only got to keep $900, not the whole $1,000. Let me explain.

Last year, the local county board levied their "dollar-a-bushel and pass-it-on-to-the-tourist" corn tax. All roadside-stand corn sellers must pay this tax. If it wasn't for the tax, corn would be selling for $9 a bushel. The corn is really only worth $9 a bushel. The tax pushes the price up to $10 a bushel. So the *real value* of the corn I produced last year was only $9 a bushel. Since I really only added a net product of 95 bushels, the true value of my total output was really only $855 (95 bushels @ $9 = $855). And that's exactly the amount of income I received from my corn-producing operation! (I had to pay the county $100 for the indirect taxes I collected, of course, and the true value of the seed corn I used up was really only $45: 5 bushels @ $9 = $45.)

I collected $100 in "indirect taxes" for the county by raising the price of my corn from $9 to $10 a bushel. Indirect taxes are always collected that way. Indirect taxes become embodied in the prices of the products people buy. The buyer pays the tax without even knowing it. But because of the tax, the price of the product is higher than the true value of the product. That's the reason why all indirect taxes must be deducted from the prices of things—that is, from the total value (price) of the net national product—in order to get to the "true value" of the output of the economy. After we deduct the fictitious value added in by all these indirect taxes, then we arrive at the *true value* of the national product—the value of national product which is equal to national income.

Let's summarize for a minute. The gross national product (GNP) is a "partly fictitious" figure. It includes two kinds of "fictional value" which must be taken away before we get down to the true value of the output of the economy. First we must deduct for all the capital being used up in the production process. We call this the *depreciation* deduction. *"Depreciation" simply means "the rate at which we are using up capital in the production process."* It doesn't make much sense to talk about the additional output we're producing unless we first replace the things we're using up in the process (like seed corn).

Second, we must get rid of the inflated prices which have resulted from indirect taxes. Indirect taxes always get embodied in the prices of things. The prices of almost everything bought—in the United States or Canada or Great Britain or Germany or anywhere else in the world—reflect some embodied indirect taxes. The prices of automobiles, tires, TV sets, furniture, and almost everything else, are pushed up quite a bit by indirect taxes. For some products—for instance cigarettes and whiskey—the embodied indirect taxes account for more than half of the price! We must get rid of this "embodied tax" if we want to get down to a real, honest value of the product. So that's why indirect taxes must be deducted from the value of net national product (NNP) in order to get to the true value of the national product.

### National Income Equals National Product

After we subtract depreciation and indirect taxes from the GNP figure, what do we have? We have the "real, honest value of the national product." Is this "real value of the national product" exactly equal to the "national income" received by the owners of the factors of production as wages, rents, interest, and profits? You bet it is!

The national product is exactly equal to the national income. The two *must be equal*. Every dollar paid for new output is a dollar received by someone as new income. The income is shared among the owners of the factors—labor, land, and capital—which helped to produce the product. After all the factor owners are paid their shares, any money left over is profit. So all the money received from the sale of the product is distributed: as wages, rents, interest, and profits. All of it goes as income to somebody.

The national product and income statisticians actually go through this very process we have been talking about. They add up all the payments to the factors of production. They also take the gross national product and subtract depreciation and then subtract indirect business taxes.

When they do both of these things—figure "national income" one way, and figure "the real value of the national product" the other way—do they come up with exactly the same figures, for "product" and for "income"? No, not exactly. But they would if all their figures were precisely accurate.

We know that the national product and the national income are equal. Any differences in the figures are obviously statistical errors. When the government statisticians do it both ways, the two figures usually come out very close.

Now you know what we look at, to see and measure the speed of the economic system. You know that the government statisticians collect the figures and publish the "national income and product accounts"—GNP, NNP, NI figures and all that. This work is done by the Office of Business Economics in the U.S. Department of Commerce. The Bureau of Labor Statistics in the U.S. Department of Labor publishes statistics on employment and unemployment. Several other agencies of the federal and state governments and several local governments and regional planning commissions and chambers of commerce and industry groups and labor groups and others are busy gathering statistics to see and measure the speed of economic activity in the various parts of the economy. But we don't have time to get into all that. However, we do need to go a little further with the "national income and product accounts."

### Finding Disposable Personal Income

The *income and product accounts* do more than show national product and national income. The accounts also show how much of the national income the people actually get to keep, and how much they actually spend for consumer goods.

So I really made $855 last year on my corn operation. That was the real value I added to the national product and that was my part of the national income. Did I get to spend it all? Did all of it flow to my "household"? You know it didn't. The government made me pay $170 in income taxes. All I actually got to spend out of my corn crop was $685 ($855 – $170 = $685). That's a lot less than the $1,000 I took in. But I guess it's a lot better than nothing.

You know that all the "national income" doesn't stay in the hands of the people who earned it—that is, it doesn't all become "disposable personal income." You just saw me pay $170 in taxes. That's one example of why all the national income doesn't become disposable personal income. People must pay taxes.

Anything which keeps you from being able to spend a part of your income, *reduces* your disposable personal income. Anything which gives you *more* money to spend, *increases* your disposable personal income. Let's talk about some of these adjustments.

**Personal Income.** First of all, some of your income never actually gets to you, "in person." Social security taxes for the OASDHI (old age, survivors, disability, and health insurance) program are taken right out of your paycheck. You never see it. All the social security taxes collected in the country must be deducted from national income before we get to "personal income." But that isn't all.

If you own stock in a corporation, the government takes about half of the corporation's income in taxes before you get to see any of it in your dividend checks. It's really your income, but you'll never see it! Even more than that, most corporations hold about half of their "after tax" income and use the money for growth. We call this "unpaidout" income "corporate retained earnings." Maybe the corporation wants to build a new plant, or install new equipment or something. "High growth" companies sometimes retain *all* their after-tax earnings and use the money for growth. They don't pay out any dividends at all! So the stockholders never get to see the money. Just like the social security taxes, it doesn't get into anybody's "personal income."

So how much of the total national income gets distributed to the people as personal income? Only what's left after the social security taxes and the corporate profits taxes and the "retained earnings" (undistributed profits of corporations) are deducted. All these dollars are pulled out before the people ever get their hands on them. So if you want to find out how much personal income the people in the nation are receiving, just deduct all these things from national income and you will have it.

But wait! What about the people who are on the receiving end of all those social security taxes? What about the retired people and the dependent children and all the other people who are receiving the government's retirement and welfare and disabled veterans and all the other money-sharing payments? All those people get money from the government as *personal income*, don't they? Sure they do. But that's an easy thing to handle in the national income and product accounts.

After the statisticians finish *subtracting* the social security taxes and the corporate profits taxes and the retained earnings of corporations, then they *add* the amount of money the governments are giving to people. These "government gifts of money" are called "transfer payments." Income is taken from the people who originally earned it and then "transferred" to other people.

**Disposable Income.** If we take the national income figure, then *subtract* social security taxes and corporate profits taxes and corporate retained earnings, then *add* government transfer payments to individuals — what figure do we wind up with? Personal income, of course! Then what do we do to get to the disposable personal income figure? Just subtract all the *other taxes* (the direct taxes) the people have to pay out of their incomes. Then you have disposable personal income. That's what the households *really* get to spend.

What taxes are deducted to get to disposable income? All of them. All the taxes you pay directly out of your income. The income taxes are the biggest for most people. But sales taxes and property taxes and all the others take some personal income and leave less disposable income.

You may wonder why the social security taxes are treated differently than the income taxes. Income taxes for most people are deducted from their paychecks. The wage earner never gets his hands on that money. It doesn't ever become a part of his *personal income*. So why the difference? It's just that the national income statisticians decided to do it this way many years ago, back when income taxes weren't being deducted from everybody's paychecks. It isn't precisely logical anymore, but as long as we all understand, it's okay. No harm done.

## Where To Find The National Income And Product Figures

Now you know all you need to know about the national income and product accounts. If you ever want to find out what these figures actually are for any quarter, or year, you can find the figures in a recent issue of the *Survey of Current Business*, published monthly by the Office of Business Economics, U.S. Department of Commerce. It's in most libraries.

There wouldn't be much point in your simply trying to memorize all the steps in the national income and product accounts. The important thing is that you understand the logic of the system. The national income and product accounts attempt to measure the economic flows—the size of the spending stream—the output and income flows in the economy. These accounts represent the real-world application of the output and income concepts you have been learning.

## National Income And Product Have Been Expanding Greatly

Over the past 40 years, in the advanced nations all over the world, national income and product have been expanding greatly. The income and product figures show how fast economic growth has been going on.

Just take a look at Table 11-1 and see the national income and product figures for the United States over the past several decades. Can you see the Great Depression of the 1930's? Sure. And the big boom of World War II? And then the economic expansion of the 1950's and the 1960's? Sure. But one word of warning: much of the increase you see in the table represents rising *prices*—not just rising *outputs.*

So how do we get rid of this "rising prices" illusion? How can we find out how much the output *really* increased over these years? We have to "deflate" the figures to get rid of the effect of rising prices. That's easy enough to do. You'll learn how to do it in the next section. But first, stop awhile and study Table 11-1. Then, on to the next section.

**Table 11-1: NATIONAL INCOME AND PRODUCT FIGURES FOR THE UNITED STATES, SELECTED YEARS* 1929–1972 (in Billions of Dollars)**

| | 1929 | 1933 | 1939 | 1944 | 1950 | 1960 | 1970 | 1972 |
|---|---|---|---|---|---|---|---|---|
| GNP | 103 | 56 | 90 | 210 | 285 | 504 | 976 | 1,152 |
| *- depreciation* | 8 | 7 | 7 | 11 | 18 | 44 | 84 | 104 |
| NNP | 95 | 49 | 83 | 199 | 267 | 460 | 892 | 1,048 |
| *- indirect taxes (and other minor adjustments)* | 8 | 9 | 10 | 16 | 26 | 45 | 92 | 113 |
| NI (=NP) | 87 | 40 | 73 | 183 | 241 | 415 | 800 | 935 |
| *- social security taxes* | a | a | 2 | 5 | 7 | 21 | 57 | 74 |
| *- corporate taxes* | 1 | a | 2 | 13 | 18 | 23 | 37 | 42 |
| *- retained earnings* | 3 | -3 | 1 | 7 | 11 | 13 | 14 | 26 |
| *+ payments, transfer (etc.)* | 4 | 3 | 5 | 7 | 23 | 43 | 109 | 143 |
| PERSONAL INCOME | 86 | 47 | 73 | 165 | 228 | 401 | 801 | 936 |
| *- direct taxes* | 3 | 2 | 3 | 19 | 21 | 51 | 116 | 141 |
| DISPOSABLE INCOME | 83 | 45 | 70 | 146 | 207 | 350 | 685 | 795 |

*Source: Economic Report of the President*, 1973. (a=less than one billion)

*Some of the figures presented in this table are not precisely accurate because of adjustments made by the author, to take care of statistical discrepancies and to make the table internally consistent without introducing confusing and irrelevant detail.

This table shows annual rates of flow of output, income, tax payments, etc. You can see a lot of interesting things by looking at these figures.

You can see that the rate of flow of GNP, NNP, and NI have all increased by about ten times since 1929. But taxes and government transfer payments have increased much faster. The rate at which direct taxes are flowing to the government has increased by more than 50 times, and transfer payments flowing from the government to the people have increased by more than 25 times.

Just look how much change there has been since 1960! Almost all the flows have more than doubled in size. Corporate retained earnings is the only big exception.

You might guess that one of the reasons for the big increases in all these numbers is *inflation*. That's right. You are getting ready, right now, to find out what to do about that.

## Comparing GNP From Year To Year

Now when a friend asks you "how's business?" you will know how to find the answer. You will be able to say: "During the first quarter of this year, GNP was flowing at an annual rate of more than one trillion dollars. This is an increase of about five percent over the rate of flow of GNP during the first quarter of last year. It looks like business this year is going just fine."

But that isn't quite all. You must also remember to say: "Of course some of this increase in the dollar GNP figure doesn't really represent an increase in real output in the economy. Yes, prices have increased. And yes, we are valuing each unit of output at a higher price. For example, I hear that Willie Wonka's candy output is up from an annual rate of $5,200,000 last year, to $5,720,000 this year. But guess what? He is still producing at exactly the same rate, 10 truckloads a week. What's the difference? It's just that this year the price is up from $10,000 to $11,000 a truckload. Prices went up and costs went up. If Willie's candy business is any indication, the economy really isn't producing any more this year than last."

So your friend wants to know whether or not Willie Wonka's case is typical. "Has there really been any increase in the rate of output of the economy between the first quarter of last year and the first quarter of this year? How can I find out?" You tell him that the way to find out is to "deflate" the GNP figures, using a price index.

Anyone who is going to compare "levels of economic activity" from one year to another needs to know something about price indexes. A price index is one of the world's simplest things. But lots of people don't understand it. In a few minutes, you will.

## HOW TO MAKE AND USE A PRICE INDEX

Let's suppose that just for fun you wanted to make your own price index for the cost of mailing first class letters. You know that in 1970 the cost was 6¢. Then in 1971 the cost went up to 8¢. Suppose you decide to use 1970 as your base year—that is, your reference point—the year to start from. Then you want to find out what the "first class postage price index number" is, for right now. How do you do it? Easy.

### Divide The Base-Year Cost Into The Present Cost To Get The Index Number

You know how much it cost to mail a letter in 1970 (the base year)—6¢. And you know it costs 8¢ now. Simply divide the base-year cost (6¢) into the present cost (8¢), and you will have your index number for the current month. (8 divided by the base-year cost of 6, comes out to 1.33, or 133 percent, or an "index number" of 133.) What could be easier than that?

The price index for the base year is always 100. Why? Because you are dividing a number into itself! You must come out with one, or 100 percent, or an "index number" of 100! In our postage case, you would be dividing 6¢ (cost in the base year) into 6¢ (cost in the base year). (6/6 = 1, or 100 percent, or an index number of 100.) In order to show that 1970 is your base year, you might say: "1970 = 100." That simply means that 1970 is the base year.

### Divide The Index Number Into Present Cost To Find "Base-Year Cost"

Now that you know how to get an index number, the next question is: "how do you use it?" For one thing, you can use your "first class postage price index number" to find out how much it *would have cost* to mail your letters, back in the base year. The index number is 133. This tells you that it's now costing you $1.33 to get as much "first class postal service" as you got for $1.00 in 1970.

Suppose that last month you spent $12 for first class postage. How much would this have cost you in 1970? Just divide $12 by the index number of 133, and you'll see! ($12 divided by an index number of 133 equals $9.) The value of your "first class postage dollar" has gone down, all right. In 1970, $9 would have bought as much "postal service" as $12 can buy, now!

### Divide The Index Number Into A Dollar To Compare "Present Value" Of The Dollar

Suppose you say to yourself: "As compared to 1970, what is the present value of my 'first class postage dollar'?" To find out, just divide the index number into a dollar. You come out with 75¢. ($1.00 divided by an index number of 133, equals about 75¢.) So for first class postage services, the dollar this year is worth only as much as 75¢ was worth in 1970. Sad, isn't it?

Suppose this little exercise intrigues you and you decide to build a first class postage price index on a 1960 base. In 1960 it only cost 3¢ to mail a first class letter. So with 1960 as your base year (1960 = 100), what is the first class postage "price index" for 1974? It is 8¢ divided by 3¢, or 2.67, or an index number of 267. So to get as much postal service as you are now getting for $12, how much would you have had to pay in 1960? You find this out by dividing your index number into $12. In 1960 it would have cost you only about $4.50 to mail as many letters as you are mailing these days for $12. ($12 divided by an index number of 267 equals about $4.50.)

There's been a big decline in the value of your postage dollar! In order to buy as much postal service in 1974 as you could have bought for $1 in 1960, you must now pay $2.67! So what is the value of your first class postage dollar now, as compared with 1960? It's only worth about 37¢! ($1.00 divided by 267 equals about 37¢). Now that you know what a price index is and how to use it, let's take another look at the problem of trying to compare GNP figures when our "measuring rod" (the value of the dollar) keeps shrinking from year to year.

### Divide The Index Number Into The GNP Figure, To See How Much The Output Really Changed

When the gross national product for the present quarter is added up, it will include a figure for first class postal services. The figure will be considerably larger now than it was for the same quarter in 1970. Why? A big flurry in first class correspondence? No. The higher figure comes partly from the higher price charged for postage services, and only partly from the higher volume of mail. How do you get rid of this "fictitious product value" which results from the inflated postage prices? Simply divide the present "value of first class postal services performed" figure, by the index number (index number 133; 1970 = 100). This will give you a figure which accurately reflects the *real* increase in "the output of first class mail services" between 1970 and today.

You know that postal services make up only a very small part of the GNP. But suppose you could go around and do, for *each* product and service, exactly what you just did for postal services. Then when you got the dollar value of *each* product and service "deflated," you could add up the deflated figures and get a new GNP figure. The new figure would show the present output, not measured by the *inflated* "current" dollar units, but measured by the *same size* "constant" dollar units that were used to measure the gross output back in 1970. The new (deflated) "constant dollar" GNP figure would show how much the output of the economy really has changed since 1970.

Can you see it? If prices go up and output value gets bigger, you don't know if more is *actually* being produced, or not. But if prices stay the same and output value still gets bigger, you know for sure that more is being produced. Right? Of course.

So, do the government statisticians make up a price index for postal services and one for tonsillectomies and one for hamburgers and gasoline and tires and heating oil and subway fares and beer and dorm rent and phone calls and textbooks and everything else, and then "deflate" the "present output value" of each, and then add up all the "deflated output values" to get a new (deflated) GNP figure? That's sort of the way they do it—only they don't need to go into such detail. They take some shortcuts.

The Office of Business Economics of the U.S. Department of Commerce uses the GNP price deflator. This is an index number which tries to reflect the "average change in all output prices." With this index number you can deflate the whole GNP figure in one fell swoop! How? Simply divide the current GNP figure by the index number! That brings the GNP figure down in size so that it's comparable to the base year GNP figure. Isn't that neat?

Remember Table 11-1, where you saw how GNP in the United States has changed so greatly, since 1929? How much of the change was caused by increased output? And how much of it was just caused by higher prices? You know how to find out, now. Just go back and revalue the output produced during each year. Only, instead of using the "current prices" which existed in each year, use the *same* prices for each year. Use "constant prices"—the prices which existed in the year you choose as your "base year."

How do you do this? How do you convert the "current dollar" GNP figures into "constant dollar" GNP figures? Simple. Just divide the "current dollar" GNP figure for each year by the "price index" (the "GNP price deflator") for that year. That's all there is to it! Table 11-2 shows that when you deflate the GNP figures it can make quite a lot of difference! Take a few minutes now and study Table 11-2. After that we'll get into some other kinds of, and uses for, price indexes.

### The Consumer Price Index (CPI)

There are all kinds of price indexes. One of the most often used is the "consumer price index," or "cost of living index." This index is put together by the Bureau of Labor Statistics, U.S. Department of Labor. What they do is make up a list of all the things that the "average consumer" would buy, and in what quantities, in the average year. This list will include postal services, medical services, tires and gasoline, meat and potatoes, clothing, and everything else bought by "the average family."

It isn't always easy to decide exactly how much of which things to put into this list. But the government statisticians make the best estimates they can, and go ahead. Once they get the list made up, they then assign a "base year" cost figure (say, 1967 = 100) to each item on the list. Then they add up all the costs of all the things. This gives them the "cost of living" of the "average family" in the base year. That's the first step.

The next step is to take the same list of things and assign the *present* cost to each item on the list, then add up the total. This shows the "cost of living" of "the average family" at the present time (present month, or week). Then the *base year cost* is divided into the *present cost* to get the consumer price (cost of living) index.

Suppose the list of things would cost $8,000 in 1967, and $9,300 in September of 1971. Then the index number for September, 1971 (on a 1967 base) would be 116. ($9,300 divided by $8,000 equals about 116.) This means that it would have cost the average family $1.16 to buy in September, 1971 what they could have bought for $1.00 in 1967. It

**Table 11-2: A COMPARISON OF CURRENT DOLLAR VALUES AND CONSTANT DOLLAR VALUES OF U.S. GNP, SELECTED YEARS, 1929–1972 (in Billions of Dollars) 1958 = 100**

| | 1929 | 1933 | 1939 | 1944 | 1950 | 1958 | 1960 | 1970 | 1972 |
|---|---|---|---|---|---|---|---|---|---|
| GNP *(current dollars)* | 103 | 56 | 90 | 210 | 285 | 447 | 504 | 976 | 1,152 |
| GNP *(constant dollars 1958 = 100)* | 204 | 142 | 209 | 361 | 355 | 447 | 488 | 722 | 790 |
| PRICE INDEX *(GNP Deflator)* | 51 | 39 | 43 | 58 | 80 | 100 | 103 | 135 | 146 |

Source: *Economic Report of the President*, 1973.

What a difference price changes can make!

The "current dollar" figures show a tenfold increase in GNP since 1929. The deflated (constant dollar) figures show that the real rate of flow of output has increased by only about four times. That's still a lot of increase, but it's a lot less than ten times!

The rapid price increases which have occurred since 1960 are easily visible in this table. See how the "current dollar" figure has more than doubled since 1960? But the "constant dollar" figure shows that the rate of flow of real output has increased by only a little more than 60 percent. That's still a lot of increase, but it's a lot less than 100 percent.

means that the "average family's consumer dollar," in September, 1971 was only worth as much as 86 cents would have been worth to the "average family" in 1967. ($1.00 divided by the index number of 116, equals $.86.)

By May of 1974, suppose the cost of the things on the list had gone up to $10,400. Then the consumer price index (CPI) would be 130. ($10,400 divided by $8,000 equals 130.) The May, 1974 dollar would buy only about as much as 77 cents would have bought in 1970. ($1.00 ÷ 130 = about 77¢.)

## Other Price Indexes

Price indexes are interesting, simple, and helpful numbers. The best known and most widely used index is the one you just heard about—

the consumer price index. Then there is the "wholesale price index" which compares prices of things that businesses are buying. And there is the "construction cost index," and lots of indexes for individual products, such as the housing cost index, the medical cost index, the food cost index, the transportation cost index, and so on.

Each of these indexes simply gives a quick and easy way of finding out how much the prices of these things have changed, on the average, since the "base year." You can look at the index and tell immediately how much it would cost you *now* (on the average) to buy what was "a dollar's worth" in the base year. If the index is 130, it now costs you $1.30 to buy what you could have bought for $1.00 in the base year.

If you want to find out how the "real purchasing power" of your present income compares with the real purchasing power of your income in the base year, simply divide your present income by the cost of living index (CPI). Just as the GNP figure can be "deflated" by dividing by an index, so can your own income.

If the price index is 200, it means that your income now must be twice as large as it was in the base year, for you to have as much "real income" now as you had then. Your income must have doubled just to keep you in the same place. Also, it means that the pension funds or insurance policies that you contracted for in the base year are now going to bring you only *half* as much real goods and services as you thought you would get when you made the contract. That, of course, is a major problem—and a great social injustice—of inflation.

### A Dollar's Value Depends On What It Buys

One small word of warning about price indexes. An index number always refers to one thing, or to an "average list" of things. If you are spending your dollars exactly the same way as represented by this "average list," then your dollars will change in value exactly as indicated by the index number. But no one spends his money exactly like the "average list." Let's take a wild example.

Suppose medical services, automobiles, gasoline and related products, and meats and bakery products are all items which have increased considerably in price since 1967. But suppose you use very few of these things. You have decided to sell the car and ride a bike. No one in your family has been sick. You and your family are vegetarians and you do all your baking at home. Further, you always go camping on your vacations and you buy winter clothes for the children at the local rummage sales. You make your own wine and brew your own beer in your basement. So what does the "consumer price index" have to do with the value of *your* dollar? Not very much!

*The actual value of each individual's dollar is determined, not by the consumer price index, but by how much value that individual gets when he spends his dollar.* One family may live better on $6,000 a year than another family would on $9,000. Why? Because the people in the first family spend their money carefully and maximize the value of their dollars! All this does not mean that the "cost of living index" is not a valuable tool. It really is. But recognize that it doesn't apply in a precise specific way to anyone. You can have considerable control over the value of *your* dollars—if you want to. A good course in consumer economics can help you learn how!

## SUMMARY

Now you know how to visualize and measure the speed of the economy. You know about the equalities between spending and receipts, between output and income, and between savings and investment. You know about the national income and product accounts, and you know how to use a price index to deflate the income and product figures.

But you don't yet know enough about the important role of *money*, in the macroeconomics process. You don't know how the money supply expands and contracts. It's time for you to get into that now. That's the subject of the next chapter.

---

## REVIEW EXERCISES

**MAJOR CONCEPTS, PRINCIPLES, TERMS (Try to write a paragraph explaining each of these.)**

the purpose of the NP and NI accounts
why depreciation is deducted
why a price index is needed
what determines the value of *your* dollar

**OTHER CONCEPTS AND TERMS (Try to write a sentence or phrase explaining what each one means.)**

gross national product
national income
depreciation deduction
net national product
indirect taxes
personal income
transfer payments
price index
base year
GNP price deflator
consumer price index (CPI)
cost of living index

**QUESTIONS (Try to write out answers, or jot down the highlights.)**

1. Suppose you produced macramé belts and sold them to people at the concession stand at Seashore State Park last summer,

and you took in $1,000. Do you think all of that thousand dollars would be included as a part of the "national income"? Or what deductions do you think would need to be made? Make up a list, and then explain why each deduction would be necessary.

2. I'm sure you don't spend your money exactly like the "average family." Think about the specific things *you* usually buy, and about how their prices have been changing in the past year or two. Do you think you could make up a sort of "roughly estimated" personal price index, and then figure out what has been happening to the value of *your* dollar? (Maybe you've found out how to buy things cheaper—wholesale, maybe—and the value of your dollar is going up!) Want to take a few minutes and try to make up your own personal price index? It might be interesting to see what you come up with!

# PART 4

WHEN YOU LOOK AT TOTAL SPENDING WHAT YOU SEE DEPENDS ON WHICH WAY YOU LOOK AT IT!

Money and spending and employment and output and income are all tied in and move up and down together, and there are several different ways to visualize and analyze this total "macroeconomic process."

# 12 HOW THE MONEY SUPPLY EXPANDS AND CONTRACTS

*As banks make loans the money supply expands automatically and people can spend more.*

You already know a lot more about money than most people do. You learned about it back on the island when the fishstick warehouse burned down. And remember when the northside chief took all the paper money out of the vault and burned it? Yes, you really do know quite a bit about money. But you need to know more.

Money plays a very important role in the modern economy. How money works may determine how the economy itself works—how well, or how poorly. That doesn't surprise you. I'm sure you are already quite impressed by the economic magic of money. What about the effect of money on total spending?

If you want to understand total spending, one of the most important things you must look at is money. How much money is available? Is money easy and cheap, or difficult and expensive, to borrow? If money is easy to borrow, at low interest, people will borrow and spend more. Spending is stimulated by easy credit. Tight money retards spending. Of course!

## The Money Supply Influences The Rate Of Total Spending

You can see right off that the government might be able to use some kinds of "money adjustments" to get people to spend more, or less. This might be a good way to overcome recessions or inflation. Yes, money *is* an important tool of economic policy. Monetary policy is being used constantly for economic stabilization. You need to know about "monetary policy."

Monetary policy, just like everything else in economics, is really very simple and easy to understand, so long as you move into it carefully, slowly, one step at a time. There are several steps to take. First, you need to know more about money itself — how it expands and contracts in the modern economy. Next, you need to see how its expansions and contractions influence total spending, total output, and total income. Then when you have a good feel for how it all works, the "monetary policy" approach to economic stabilization will be very simple and obvious. This chapter takes you step by step, deep into the "magic world of money."

### The Money Supply Expands And Contracts Automatically

In the last two chapters you have been hearing about increases and decreases in aggregate demand — that is, about changes in the rate of total spending in the economy. When total spending speeds up, doesn't the money supply need to expand? Or if spending slows down, not as much money will be needed. Right? Yes, that's true — at least to some extent it's true.

A speedup in total spending could possibly result from each person spending his money *faster*. If the money supply stayed the same size, but the velocity of circulation increased, then we could have more spending (each day or each week, or month) without having any increase in the money supply. But in the real world it isn't easy for the "velocity of circulation" to change very much, very fast. (More on that, later.) So, sure enough, if the economy is going to speed up very much, the "money supply" — the actual number of "spendable dollars" in existence — must somehow increase, to permit the increased spending to occur.

*Reminder:* The money supply consists of checking accounts — that is "demand deposits" — and currency. Most of the money is made up of demand deposits in the banks throughout the country.

When the economy is trying to speed up, where does the extra money come from? It's newly created. Nobody "creates" it. It just sort of naturally expands itself through the normal workings of the banking system. Surprising? Probably not. By now you know enough about the magic of money not to be too surprised by anything! So now that you know *what* happens, it's time for you to find out *how* it happens. But first, let me level with you.

This is no trick. It isn't some "economics professor doubletalk." It's the honest truth. The amount of money in existence really does expand and contract to meet the "money needs" of businesses and consumers. It really does happen automatically through the normal processes of the banking system. Most of the money in existence in the United States and in the other modern nations, has been created by the banks. More is being created every day. Here's a little story that will show you exactly how it happens.

## HOW ONE BANK CAN CREATE MONEY

Mr. Zimmer, the local banker, is sitting at his desk one morning musing about the ups and downs in the banking business. "The 'money market' is a strange 'market'," he says to himself. "Six months ago we had more money to lend than we knew what to do with. I was buying government bonds just so we could earn some interest on all the money we had on deposit. But today? It's hard to believe. We're all loaned up, and the demand for loans seems to just keep on increasing.

"Interest rates are going up too, but people keep on borrowing. I guess those 'monetary policy' people in Washington are doing something to make money hard to get. Anyway, it's sort of good to be all loaned up. No need to worry about having to keep our money busy by buying government bonds! On the other hand, it's tough to tell old customers, old friends, that you can't lend them any money because you're all loaned up. Ah well, such is life. . . ."

### Banks Can't Lend Unless They Have Some Money To Lend

While Mr. Zimmer is thinking all these thoughts, Mr. Baker walks into the bank. He interrupts Mr. Zimmer's thoughts with a shocking request. Mr. Baker had an economics course in college some years ago. He flunked it, but he remembers the part that said the banking system really does create money. So, full of desire to show his sophisticated knowledge about banking (only we inside boys know this—wink) he says: "How about creating $4,000 and lending it to me? I'd like to buy a camping trailer that's on sale."

Zimmer calms himself enough so that he says nothing worse than: "Don't be ridiculous! You think I can create money for you, right out of thin air? Preposterous! I can only lend you the money I have available to lend. And I don't have any money available to lend. Money is very tight these days."

Baker is confused. He's *sure* he remembers the part about banks creating money. That's the only question he got right on the final! He's thinking, "No money to lend? Is this guy pulling my leg?"

### A New Deposit Creates Excess Reserves

About that time one of Mr. Zimmer's good customers, Mr. Alber, comes striding into the bank, waving a government check, saying: "I knew the government would finally understand my case and refund all those unjustified taxes I have been paying over the past five years!" The man is carrying a treasury check for $5,000. The check is drawn against the U.S. Treasury's account in the Federal Reserve Bank of New York.

(The Treasury keeps accounts in all twelve of the Federal Reserve Banks. When the government spends money it just writes checks against those accounts. When it collects taxes or sells bonds, it deposits the money in those accounts.)

Mr. Zimmer's bank is not a very big bank. In fact, there's only Mr. Zimmer, one teller, and a secretary. Right now the teller is out to lunch, so Mr. Zimmer has to interrupt his conversation with Mr. Baker to take Mr. Alber's deposit. Baker doesn't mind the interruption. In fact, he isn't too happy about the way things are going, so he's getting ready to leave anyway. He heads for the door.

Zimmer calls him back: "Mr. Baker, wait! Now I will be able to lend you the money!"

### A Bank Can Lend Its Excess Reserves

Baker looks puzzled. He turns and slowly walks over to the other two men and says, "I'm afraid I don't understand. You just said . . ."

Zimmer interrupts: "I know what I just said. But that was before we received Mr. Alber's $5,000 deposit. Now we have money to lend! Let me introduce you to Mr. Alber. Why don't both of you have a seat here at my desk and let me explain to you something about banking procedures." So they do, and he does.

"We keep checking accounts for people. We call them demand deposits because they are payable 'on demand' to anyone the owner tells us to pay them to. Whenever Mr. Alber writes a check to somebody he's just telling us to pay that somebody some of the money out of his (Mr. Alber's) account. Right, Mr. Alber?"

Alber nods.

"Mr. Alber has had demand deposits in this bank for several years. He's one of our best customers. But we have lots of good customers. In fact would you believe that in our little bank we have a total of one million dollars in demand deposits? Don't ask to see the money in our vault. We only keep about $50,000 in the vault. There's no profit made on money in the vault! But on the books we really do have total deposits of $1,000,000. I'll bet that surprises you!"

Baker doesn't look too surprised, or too impressed either. In fact he's a little bit bored listening to Zimmer brag about how "big" his little bank really is. "Okay," he says. "So what's the point?"

"The point I was coming to is this: The only way we can make any money in this business is to lend out the money our customers deposit in their accounts. Mr. Alber just made a deposit. So now we can make you a loan."

"There, he said it again," Baker thinks. "He keeps trying to tell me that banks don't create money when I know they do! And I'm getting kind of tired of all this . . ."

Zimmer continues. "Mr. Alber just deposited $5,000 but we can't lend you all of that. The banking regulations of the Federal Reserve System require that we keep some money 'on reserve.' The more money we have in our demand deposit accounts, the more we must keep 'on reserve.' So how much must we keep? Right now the reserve requirement is something less than 20 percent, but let's say it's 20 percent, just so the numbers will come out even. Twenty percent of what? Twenty percent of the total amount of money our customers have in their demand deposits, that's what! So you can see that whenever our customers deposit more money in their demand deposit accounts, we also must keep more in our reserve. Understand?

Baker nods. "I think I see it, but I'm not sure."

### Federal Reserve Banks Hold Reserve Deposits

Baker is beginning to get interested. "You said you're holding a million dollars in demand deposits, and suppose the reserve requirement is 20 percent. That means you must be holding $200,000 on reserve. (20% of $1,000,000 is $200,000.) Right? And you just said you only keep about $50,000 in your vault. Where's the other $150,000? Or are you just pulling my leg again?"

"No, no leg-pulling. It's all very true. And you're right. There is another $150,000. It's in our bank account at the Federal Reserve Bank in this district."

"You mean your *bank* has a bank account? Like a demand deposit account? At the Federal Reserve Bank?"

"Right!" says Zimmer.

### Banks Pay Each Other Out Of Their Federal Reserve Accounts

Zimmer goes on. "My bank has an account in our Federal Reserve Bank, but we can't write checks to people or businesses on our *Federal Reserve Bank Account*. The account is only used for making payments to other banks. Whenever another bank has a 'money claim' against this bank—that is, whenever we owe money to another bank—the other bank gets paid out of our account at the Federal Reserve Bank. It's a neat system!"

"But wait!" Baker thinks something's a little bit fishy here. "If you are going to be using that account to pay off other banks, you aren't going to have enough left. You're supposed to keep that $150,000 there on deposit all the time to meet the 'reserve requirement' you told me about. If you pay some of the $150,000 to another bank, you won't have enough left to meet the 20 percent requirement."

Zimmer's eyes sparkle. "Aha! Now you see it! That's why I couldn't lend you the $4,000 you wanted to borrow awhile ago. You would sign a promissory note and I would open a checking account for you. Then what would happen? You would go right over to that sporting goods dealer in Tonawanda and spend the money. You would write him a check for $4,000. Then what do you think would happen after that?"

Baker says, "The sporting goods dealer would come right over here and cash the check and get the money out of my account."

### No One Ever Cashes A Large Check

"Come now, Mr. Baker," Mr. Zimmer says. "Who would ever cash a $4,000 check? No, the dealer will deposit the check in his own demand deposit account at the bank where he does business in Tonawanda. Then he will have the deposit in his account at the Tonawanda Bank. The bank will have the check. So what do you suppose the Tonawanda bank will do with the check?"

"Come over here and cash it and get their money out of my account?"

"There. You said it again. No. Nobody *ever* cashes a $4,000 check! The Tonawanda bank will send the check (along with a bunch of other checks) for deposit to its account at the Federal Reserve Bank of New York (or maybe to the Buffalo branch). The Federal Reserve Bank will then deposit your check—that is, add the $4,000—to the account of the Tonawanda bank. But the Federal Reserve Bank is not just going to add $4,000 to the Tonawanda bank's account without getting that money back from somewhere. Where does the $4,000 come from? Can you guess?"

"I'll bet they take that $4,000 out of your $150,000 deposit at the Federal Reserve Bank. Right?"

"Right!"

### Insufficient Reserves Must Be Made Up Somehow

Zimmer goes on. "If they take $4,000 out of my reserve account, I'm suddenly in an illegal position. I won't have enough left to meet the 20 percent reserve requirement, to 'back up' my demand deposits. I will have to do something quick to make up the difference. I might *borrow some money from the Federal Reserve Bank* to make up the difference. Or maybe I would *sell some of my government bonds to them, or to somebody.* Or when someone *comes in to renew his loan I might refuse to renew it and make him pay it off right away.* Then I could deposit that money in my Federal Reserve account.

"Anyway," Zimmer continued, "I'd rather not have to do any of these things. The Federal Reserve Bank is charging a high discount rate—that is, high interest on the loans it makes to banks—these days. And the bond market is depressed. Bond prices are so low that if I sell my bonds now, I'll have to take a loss. As for refusing to renew loans, I certainly don't want to do that to my good business customers. They would have to sell off their inventories to pay off their loans. They would never do business with me again!

### New Excess Reserves Permit New Loans

"Now you see why, before Mr. Alber's deposit, I couldn't lend you any money. I had no money to lend. I had only enough to cover my required reserve. No excess at all. But now, with Mr. Alber's $5,000 deposit I do have some excess reserves. I can lend you any amount you want, up to the amount of my excess reserves. Can you figure out how much excess reserves I have now? Can you tell me how much I can lend you?"

"Let's see. Mr. Alber deposited a government check for $5,000. You're going to send that check to the Federal Reserve Bank and they will deposit it to your account. Right?"

"Right!"

"Then you will have $155,000 in your account. You only need $150,000. So you can lend me $5,000. Right?"

"Almost right, but wrong. You forgot something."

### Part Of The New Deposit Becomes Required Reserves

"When Mr. Alber deposited his check, he added $5,000 to my bank's total demand deposits. So now I must . . ."

Baker interrupts. "Wait! I see it! Now you must have more reserves, because now you have more deposits! You must have enough new reserves to cover Mr. Alber's new deposit. If the required amount is 20 percent (one-fifth) then for $5,000 more in deposits you must have $1,000 more in reserves (20 percent of $5,000 equals $1,000). So you only have $4,000 in excess reserves. You can lend me any amount up to $4,000! Right?"

"Right. Now you understand. So how much do you want to borrow?"

"I'll take the whole thing! That's exactly how much I wanted to borrow anyway. Let's draw up the loan papers. I have about $6,000 worth of bluechip stocks here to leave with you as collateral. You can open a checking account for me. Then you can just deposit the $4,000 in my new checking account. Okay?"

"Fine. It will only take about ten minutes to get the papers ready for your signature."

While the loan papers are being prepared, Mr. Baker is re-thinking this whole banking process. "They only need a 20 percent reserve to back up their checking accounts. I'm borrowing $4,000 and it's going to be put into my checking account. Well then! All they *really* need to have on reserve to back up my account is 20 percent of $4,000—only $800!"

He thinks about this a few more minutes and then decides to confront Mr. Zimmer with this new bit of logic. "You really only need $800 behind this $4,000 deposit of mine! Why do you insist on having $4,000 in your Federal Reserve account to 'back up' my checking account balance of $4,000? You said you only needed 20 percent. If you're only holding $1,000 behind Mr. Alber's $5,000 deposit, how come you insist on holding $4,000 behind my $4,000 deposit?

"You're holding a 100 percent reserve behind my new checking account, but only 20 percent behind Mr. Alber's. How come? I don't want to borrow but $4,000 anyway, but I sure would like to know what's going on!"

### Loaned Money Soon Leaves

Mr. Zimmer seems pleased that Mr. Baker has gotten so interested. "That's an intelligent question! But I'm sure that if you thought about it awhile, you'd figure out the answer. Really, *your* account is very different from Mr. Alber's account. Both accounts look exactly the same on the bank's books, but still they're very different. Mr. Alber's account consists of *deposited* money. Yours consists of *borrowed* money. Let me explain the difference to you.

"You are going to go out and spend that $4,000, right away. When you spend it to buy the camping trailer the check is going to be deposited in the sporting-goods dealer's account in the Tonawanda bank. Then the Tonawanda bank is going to send the check to the Federal Reserve Bank. Remember?

"The Federal Reserve Bank will deposit the check to the Tonawanda bank's account. Then what? The Federal Reserve Bank will take the money away from my account. How much money? $4,000, of course! Then they will send the check back to me and I will take it off your checking account. Then, poof! Your money (account balance) will be gone!

"See what would have happened if I only had $800 in my reserve account to back up your $4,000 demand deposit? When your check clears, I am going to lose $4,000 from my reserve account at the Federal Reserve Bank. So I *must* have a "100 percent reserve," or $4,000 behind your account. Your account was created, not by a *deposit* but by a *loan*. That is what makes it different. Do you understand?"

### Deposited Money "Never Leaves"

Baker is still puzzled. "But then, why do you feel so confident about keeping only $1,000 behind Mr. Baker's $5,000 deposit? Suppose he goes out and writes a check tomorrow for $5,000? When the check clears, aren't you going to be in trouble? What about that?"

"You're right," Zimmer admits. "If that happened tomorrow, and if that's the *only thing* that happened tomorrow, we would be in trouble. But that's not going to happen. Or if it does, it's not going to be the only thing that will happen. I guess you would say that we are 'protected by the law of averages.' Let me explain it this way.

"We have many customers who come in and make deposits every day. These regular customers are also writing checks, spending their deposit balances every day. We know that on the average day in the average month, the number of people coming in and *depositing* money, will be almost exactly equal to the number of people who are going out and writing checks and *spending* their money. So the two balance each other. There's no 'net change' in our Federal Reserve account."

### In A Growing Bank, Deposits Keep Expanding

Zimmer continues. "Look at it this way. On November 1 of *last year* we had total demand deposits of about $990,000. On November 1 of *this year* we expect to have total demand deposits of slightly more than $1,000,000. *Next year* on the same day, we expect to have even more demand deposits on our books. Do we need to worry about the fact that any one individual is spending some of his demand deposit money, and that some of our reserves are going to other banks?

"So long as the money coming back and being deposited in our Federal Reserve account is more than enough to make up for the money leaving our Federal Reserve account, we have nothing to worry about. As long as we are a growing bank, our deposits will always be more than enough to offset the 'spending withdrawals.' Doesn't that make sense?"

Mr. Baker sort of understands. But he asks, "What about me? Why don't I fall into the same category?"

"Because you are borrowing money. Borrowers go out and spend money *immediately*. The amount you borrow will be leaving our reserve account right away. We can be certain of that. There won't be any 'average return flow of deposits' to offset your expenditure. So you don't fit into the normal pattern. Therefore we can't lend you any more money than we actually have free to lend—that is, we can't lend you any more than we have in excess reserves. We can only lend you as much as we can afford to lose from our reserve account. Why? Because we know that when we lend it to you, you are going to spend it and we are going to lose it. Right away."

Mr. Baker seems to understand. Anyway, now the secretary has prepared all the loan papers, ready for signatures. Mr. Baker signs the promissory note. Mr. Zimmer signs a receipt for the stock Mr. Baker is leaving as collateral—as "security" to guarantee that he will repay the loan.

The teller, who has returned in the meantime, gets Mr. Baker to sign a signature card. Then he gives Mr. Baker a new checkbook and a deposit receipt for $4,000. Mr. Baker writes down the balance in his checkbook ($4,000), waves goodbye to Mr. Zimmer and starts walking toward the door. Mr. Alber, who has been quietly listening, walks out at the same time.

Mr. Baker smiles and comments to Mr. Alber that this has been a very educational day. "I always thought banks created money. They taught me that in college. But today I found out how wrong that is. Banks don't create money at all. Banks don't lend money unless they already have that money to lend, dollar for dollar! How about that?"

**Lending The Excess Reserves Creates Money**

Mr. Alber, who happens to be taking college courses at night, working on his MBA, looks back and smiles. "I think you missed something," he says. "Banks really do create money. You just watched this bank create $4,000."

By this time they are in the parking lot. Baker stops for a second before entering his car and says, "No, I think you must have misunderstood."

"No, my friend," says Alber. "It is you who has misunderstood. Open your checkbook and look at your account balance. What do you see there?"

Baker says, "There's no need for me to look. I already know. But let me show you. Here's the balance, $4,000! That's just the amount of money the bank had free to lend. That's not bank-created money. It's real solid money that has existed all the time."

Then Alber says: "Yesterday I had to pay my fall semester tuition and buy books. This morning I was flat broke. My deposit balance was zero. Then I got the tax refund check in the mail. You saw me deposit it. Right?"

Baker nods. He has no idea what Alber is getting at.

Alber continues: "Here. Look in my checkbook and read the balance. See? $5,000. That's my money. Now look in your checkbook there. See? $4,000. That's your money. Between the two of us we have a total of $9,000. Right? Baker nods again.

Alber goes on. "When I walked in the bank I had $5,000. A government check. Now the two of us are leaving the bank and we have a total of $9,000 between us. An extra $4,000 came from someplace. Where do you suppose the extra money came from? Where could it have come from? Only one place. Our banker friend, Mr. Zimmer, created it."

Baker's jaw drops. He just stands there, looking at the balance in both checkbooks. There's $9,000 all right, and he knows it's true. But he only half believes his eyes. Sort of half-consciously he mutters, "Well, I'll be!"

## HOW THE BANKING SYSTEM CREATES MONEY

Yes, the banking system really does create money. Do you see how it happened in the Alber-Baker-Zimmer case? A man brought in some "new money" and deposited it in his checking account. This was a "primary deposit"—a deposit of new money, of some new reserves flowing into the banking system. The "new money" went into the bank's reserve account.

Only 20 percent of the newly deposited money had to stay in the reserve account to "back up" the new deposit. So 80 percent of the amount deposited is now in "free reserves" or "excess reserves." It can be lent to someone else. It is through this process of *lending the free reserves*, that money is created. Once the process gets started—once the new "primary deposit" enters the banking system—the money expansion can go on and on from bank to bank to bank. The potential expansion of the money supply is limited only by the size of the reserve requirement.

### The Lower The Required Reserve, The More The Money Supply Can Expand

Suppose the reserve requirement was 100 percent. Then how much money could be created by Mr. Zimmer's bank? None. All the money deposited would be held in reserves. Suppose the reserve requirement is 50 percent. How much can be re-lent? Half of it. So the lower the reserve requirement the more the money supply can expand. *The whole purpose of having a reserve requirement is to limit the extent to which the money supply can expand.*

We just watched Mr. Alber's $5,000 turn into $9,000. Alber still has $5,000 and Baker has $4,000. The money supply of the nation has been expanded. There's no question about that. But all you have seen so far is the first step in the expansion process. In terms of the nation's money supply, Mr. Alber's $5,000 government check has already grown into $9,000. But before it's all over with, *that check can grow into $25,000.*

How do we know $5,000 can grow into exactly $25,000? Because if the required reserve ratio is 20 percent (that is, 1/5), then $5,000 in new bank reserves can "back up" $25,000 in new demand deposit money. The supply of money can expand up to *five times* the amount of the new $5,000 in reserves. Suppose the reserve requirement is only 10 percent (that is, 1/10). Then $5,000 in new reserves can "back up" $50,000 in new demand deposit money. The money supply can expand up to *ten times* the amount of the new reserves. If the reserve requirement is 25 percent (1/4), the money supply can expand up to *four times* the amount of the new deposit.

As soon as the initial deposit (Mr. Alber's deposit) brings new reserves to the banking system, the money supply can start to expand. It can keep on expanding until all the new reserves are being used to "back up" new demand deposits. If the required ratio of reserves to deposits is 20 percent (*one* to *five*), then every one dollar of new reserves can "back up" five dollars of new demand deposit money. If the required ratio is 10 percent (*one* to *ten*), then every dollar of new reserves can "back up" ten dollars of new demand deposit money. That's all simple enough to see, isn't it? But how does all this expansion come about?

### The Money Expansion Process Goes From Bank To Bank

You know all about the first step in the expansion process. Mr. Alber, Mr. Baker and Mr. Zimmer have shown you that. But what happens next? Let's watch.

Mr. Baker drives over to the Tonawanda dealer to buy the camping trailer. He writes a check for $4,000 to the dealer, Mr. Culver. As soon as Baker has hooked up the trailer and pulled it away, Mr. Culver jumps in his car and speeds off to his bank in Tonawanda. He parks in the lot and dashes into the bank waving the $4,000 check in the air saying, "I finally got my money out of that old-model trailer!" He fills out a deposit slip and walks up to the teller's window to deposit the check. As he is making the deposit he notices a sad-faced young man walking slowly toward the door leading out to the parking lot. Then he hears Mr. Yeager, the loan officer, call out:

"Mr. Dover! Come back!" As Mr. Dover comes back and approaches Mr. Yeager's desk, Mr. Culver hears Mr. Yeager say: "Mr. Dover, I think we are going to be able to handle your request for a loan, after all. We have just received a new deposit for $4,000. With a required reserve of 20 percent we only need to keep as reserves, one-fifth of the amount ($800). We will be happy to lend you the other four-fifths. That's $3,200. Isn't that the amount you need to pay the contractor for the new plumbing in your house?"

So Mr. Dover gets his $3,200 loan. That's new money. Mr. Culver still has his $4,000 in his checking account, and Mr. Dover has $3,200 in his. And remember, Mr. Alber still has his $5,000 in his checking account. The original $5,000 has now grown to a total of $12,200 ($5,000 + $4,000 + $3,200 = $12,200).

Notice that Mr. Baker doesn't have his money any more. He spent it. The one who borrows it *always* spends it. His money went to Mr. Culver, so his bank's "excess reserves" went to Mr. Culver's bank in Tonawanda. That's why Mr. Culver's bank can now make the loan to Mr. Dover.

So what's Dover going to do? Spend his borrowed money right away to pay the plumbing contractor, Mr. Elder. And Mr. Elder will go into his bank waving the check in the air, then deposit it in his account. His bank must hold one-fifth of the amount in its reserves ($640), but it can lend the other four-fifths ($2,560) to Mr. Fuller who wants to build an addition and more shelf space so his shoe store can carry more inventory.

Mr. Fuller will write a check to pay Mr. Garner, the carpentry contractor. Mr. Garner will deposit the check in his bank. The bank will keep one-fifth ($512) as reserves and lend the other four-fifths ($2,048). The borrower will spend the money. And on and on it goes. Until when? Until there aren't any excess reserves left to lend!

The money supply can continue to expand until *all* of Mr. Alber's initial $5,000 of new reserves is being used as reserves to "back up" new demand deposits. That is, the expansion can continue until the total of new demand deposits amounts to $25,000. The $25,000 will include Mr. Alber's original demand deposit of $5,000, plus another $20,000 of demand deposit money created through bank lending. Figure 12-1 on page 252 shows a summary of this money expansion process. It would be a good idea for you to take a few minutes to study that figure, now.

### The Purpose Of The Reserve Is To Limit The Money Expansion

Alber's new deposit sets off a process of lending, spending, and relending which continues from bank to bank. As the excess reserves move from bank to bank, they leave a trail of newly created money in their wake. Every bank they pass through has an increase in its demand deposit money. But in each succeeding bank the addition gets smaller and smaller. Why? Because each bank keeps a part of the "moving excess" as it goes by.

If there was *no reserve requirement,* each bank could relend *all* of the excess. The "moving excess" could go on and on from bank to bank, leaving a new $5,000 demand deposit in *each* bank. The money supply could just keep right on expanding. A new $5,000 "primary deposit" could go zooming around, touching banks, creating $5,000 more at each stop. It could create millions and millions!

### Unlimited Expansion Would Destroy The Value Of Money

Can you see why a reserve requirement is necessary? If we didn't have the reserve requirement the money supply could continue to expand indefinitely! There would be no way to limit the size of the money supply. Each time a bank received a new deposit it could re-lend it all. Each time this happened the money supply would get larger. It would just be a matter of time before the money supply would increase so much that runaway inflation would occur. The monetary unit would be destroyed. We would have to abandon the "dollar" and establish some other kind of money.

In the "olden days" of gold-backed and silver-backed money, the limited availability of these metals limited the expansion of the money supply. But in modern times, when the account numbers on the books of

Fig. 12-1 **HOW MONEY EXPANDS AS BANKS MAKE LOANS**

*A Summary of the Steps in the Depositing-Lending-Spending-Redepositing-Relending-Respending Process*

| Deposited by | Amount of Deposit | Required Reserve (20%) | Excess Reserve (80%) | Lent to | Paid to |
|---|---|---|---|---|---|
| **Alber** → | **$5,000** → | **$1,000** → | **$4,000** → | **Baker** → | **Culver** |
| **Culver** → | **4,000** → | **800** → | **3,200** → | **Dover** → | **Elder** |
| **Elder** → | **3,200** → | **640** → | **2,560** → | **Fuller** → | **Garner** |
| **Garner** → | **2,560** → | **512** → | **2,048** → | **Horner** → | **Ilter** |
| **Ilter** → | **2,048** → | **410** → | **1,638** → | **Joker** → | **Keller** |
| **Keller** → | **1,638** → | **328** → | **1,310** → | **Lester** → | **Miller** |
| **Miller** → | **1,310** → | **262** → | **1,048** → | **Nader** → | **Olter** |
| **Olter** → | **1,048** → | **210** → | **838** → | **Palmer** → | **Quaver** |
| **Quaver** → | **838** → | **168** → | **670** → | **Richter** → | **Salter** |
| **Salter** → | **670** → | | | | |
| | | | | | **etc., etc., etc.** |
| **Ultimate Totals:** | **$25,000** (total demand deposit money) | **$5,000** (total required reserves) | **-0-** (total excess reserves) | No more loans are possible until new reserves come from somewhere | |

This table shows Mr. Alber's original deposit of $5,000 growing into a total of $25,000—that's twenty-five thousand real, spendable dollars of demand deposit money. Mr. Alber has $5,000 of it, Mr. Culver has $4,000 of it, Mr. Elder has $3,200 of it, Mr. Garner has $2,560 of it, and so on.

Only five thousand dollars existed in the beginning. Where did the extra twenty thousand dollars come from? It was created by the banks. The banks created it just by lending their excess reserves.

But don't forget: If times are bad, this expansion in the money supply may not occur. If Mr. Baker or Mr. Dover or Mr. Fuller or any of those other people get pessimistic and decide not to borrow, then the expansion will stop, right there!

banks serve as our money, the only effective limit on money expansion is the reserve requirement. The reserve requirement is vital!

For money to have value, its supply must be limited. In the United States and Canada and the other modern nations, the supply limitation rests on the legal reserve requirement. If you want to, you can go around telling people that *the money supply of the United States is backed by the "legal reserve requirement" of the Federal Reserve System!*

Back in Chapter 8 you learned that the total money supply in the United States in 1973 was about $250 billion, and that about $200 billion of it (about 80%) was made up of bank deposit (checking account) money. Where do you suppose all this demand deposit money came from? Would you believe it came from bank lending? That really is where most of it came from, just like in the illustration you just read about. Each time new reserves are introduced into the banking system, new money can be created. That's where most of our existing money came from.

## New Reserves Come From The Treasury And The Federal Reserve System

Where did all the new reserves come from, to permit the money supply to expand? Most of the new reserves came from the Treasury and the Federal Reserve System. You just saw $5,000 come from the Treasury and become new reserves in the system. Whenever the government runs a deficit—spends more money than it collects in taxes—it makes up the difference by selling bonds. If people and businesses and insurance companies and other private corporations buy the bonds, then no new reserves go into the banking system. But if the Federal Reserve Banks buy the bonds and create new demand deposits for the Treasury, this creates new money —new high-powered money.

Why "high-powered" money? Because the Treasury will give out government checks to pay the bridge builders and the civil service workers and the welfare recipients and all the others. Then the checks will go into the banks. The banks will send them to the Federal Reserve Banks. Then the Treasury's newly created "high-powered money" deposits will be transferred to the banks' reserve accounts. So what do we have? Excess reserves!

Not just Mr. Zimmer's bank, but *all* the banks get some new excess reserves. So now they can *all start making more loans.* The money supply can begin to expand, and it can keep on until a five-fold expansion has occurred. High powered money? You bet! A dollar will get you *five*! And is this where most of the U.S. money supply really came from? That's exactly right.

Did you ever wonder where the government would get the extra money if it wanted to run a deficit?—that is, to spend more than it collects in taxes? Now you see how the government can get all the money it wants. It simply creates the money by printing up bonds and "selling" the bonds to the Federal Reserve Banks in exchange for new Treasury deposits! Then it can write checks and spend these new deposits. The Federal government can spend as much as it wants to. It never needs to worry about not being able to get the money! But you can see how such a program would pump new reserves into the banking system. This would let the money supply expand. It could create some serious problems. We'll talk more about this, later.

### Excess Reserves Do Not Guarantee An Expansion

Putting new reserves into the banking system is not certain to result in a multiple expansion of the money supply. For example, suppose Mr. Baker is out of a job, and has already had to sell his stocks to pay the rent and buy groceries. And suppose no one else comes in to borrow the $4,000 of excess reserves created by Mr. Alber's deposit. What will happen? Nothing. New reserves *permit,* but don't *guarantee* an expansion of the money supply.

The money supply will not expand unless people want to borrow. People will not borrow unless they want to go into debt for the purpose of *spending.* If people don't want to spend these days, no amount of free reserves in the banks' reserve accounts will force an expansion. (Someone once said that trying to force a monetary expansion by pushing excess reserves into the banking system is like trying to push with a string!) Excess reserves *permit* people to borrow. That's all.

When excess reserves are high, banks want to lend. There is no profit in keeping money in excess reserves! If the banks have excess reserves and borrowers aren't borrowing, interest rates will go down. Money will be easy to borrow. This may stimulate borrowing. But it may not. It all depends on how bad things are. How much unemployment? How much overstocked inventories? How many shut-down plants? How gloomy and pessimistic is the outlook? But let's not get ahead of our story. We'll get into these issues later.

### Loan Repayments Destroy Money

Now that you know about how money expands through the banking system, it's easy to see how it contracts. When everyone is in debt and business is bad, people try to pay off their loans and get out of debt. Businesses try to sell off their excess inventories and pay off their loans. Each time a loan is repaid, unless someone else borrows an equal amount, the money supply gets smaller. Excess reserves start piling up in the banks.

Unless someone wants to borrow a bank's excess reserves, the bank can't lend. Obviously. And who wants to borrow and build inventories during a depression? Nonsense! So what's happening to all the money that used to seem so plentiful, back during boom times? It's being destroyed, more and more every day. The businessmen sell hard to get the consumers' dollars. Then they use the dollars to pay off loans. Poof! The dollars are gone! The businesses are gathering back up and destroying the dollars they were creating and paying out, before. The money supply is contracting and total spending is getting smaller. No wonder we're having hard times!

The excess reserves just keep piling up in the banks' Federal Reserve accounts. The banks can always use their excess reserves to buy government bonds. This will give them some interest income. But this doesn't expand the money supply. It doesn't help to increase business spending, or consumer spending. It doesn't help to speed up the economy.

## SUMMARY

The way money is created and destroyed through the banking system is a fascinating subject, and it's a very real and serious one. The real problems of inflation and unemployment which seem to be continuously—sometimes simultaneously—attacking the economies of today's world, can't be understood or treated without understanding about money. We'll be getting into all these issues before long. But not yet.

There are still a number of concepts and principles you need to understand before we dig into today's real-world macroeconomic problems. The next chapter will go deeper into the issue of how the level of spending is determined. But before you go on, take your time and be sure you understand how money is created and destroyed through the banking system. It's crucial!

---

## REVIEW EXERCISES

**MAJOR CONCEPTS, PRINCIPLES, TERMS (Try to write a paragraph explaining each of these.)**

how banks create money
why borrowed money always leaves
why deposited money "never leaves"
role of the reserve requirement
how the government creates money
how new reserves are generated
how loan repayments destroy money

## OTHER CONCEPTS AND TERMS (Try to write a sentence or phrase explaining what each one means.)

the money supply
velocity of circulation
demand deposits
reserve requirement
Federal Reserve Bank
Federal Reserve account
discount rate
excess reserves
primary deposit
"high powered" money

## QUESTIONS (Try to write out answers, or jot down the highlights.)

1. Try to describe in detail the step-by-step process through which money is created by the banking system. Start with a new "primary deposit" coming in from somewhere, and take it from there.
2. Suppose Zimmer's bank finds that its Federal Reserve account is too small to meet its reserve requirement. How might it get more money to deposit in its Federal Reserve account?
3. If the federal government can create all the money it wants just by printing up bonds and selling them to the Federal Reserve Banks, then why don't they do that? Why does the Federal government keep making people miserable by collecting taxes all the time? Why don't they just print up the bonds and get the Federal Reserve Banks to create new government deposits, as needed? What would happen to the money supply (and to the value of our money and to the economy) if they did that? Discuss.
4. There seems to be a sort of "natural tendency" for the money supply to adjust to the wishes of the consumers and businesses. When buying and spending are picking up, people and businesses are going into debt, and the money supply is expanding. When things are slowing down, loans are being repaid and the money supply is contracting. Can you visualize and explain just how this all happens? Try.

# 13 HOW THE LEVEL OF TOTAL SPENDING IS DETERMINED

*Consumers, investors, governments, and foreigners all add to the total spending-output-income flow.*

Total spending is the key to understanding macroeconomics. You already know that. And you understand about banks creating money. In fact you know a lot about macroeconomics. Let's take a minute and review the total "overview picture."

## An Overview Of Macroeconomics

It is total spending which supports the total level of employment and production—the total level of economic activity. Total spending pulls forth total output. As spending goes down, total output goes down. As spending increases, output increases—that is, output increases until the economy reaches full capacity. If spending continues to increase after full capacity is reached, there will be shortages and prices will rise. We will have inflation.

The total level of spending must be just right if we are going to have anything approaching "full employment and stable prices." If total spending is too low we will have unemployment and depression. If total spending is too high we will have shortages and inflation. That's it, in a nutshell. But what determines the total level of spending? What makes it just right? Or too low? Or too high? We need to get into that now.

## People Spend For Consumer Goods And Businesses Spend For Investment Goods

Who spends? Mostly people and businesses. Why do they spend? To buy things, of course! And why do they want to buy things? People buy things to have, to use, to consume, to enjoy. People buy "consumer goods." Businesses buy things to use in production, hoping to increase their profits. Businesses usually buy because they think it will be profitable to do so.

So there are two basic motives for spending. One is to consume and gain personal satisfaction. The other is to buy capital and become more productive. People spend for consumption purposes, and businesses spend for investment purposes. And that's all! What other purpose could there be for people and businesses to spend? There isn't any. Now let's look at some of the reasons why consumers spend. Later we will talk about why businesses spend.

## WHY CONSUMERS SPEND

What might make people spend more? Or less? Various things. Anything which gives people more money to spend usually causes them to spend more. No mystery about that! Suppose the economy is depressed and many people are unemployed. Then for some reason business picks up. Unemployed people are hired, the "unemployed" buildings are rented, idle factories start producing things. People start earning incomes again. What do you suppose will happen to the total level of consumer spending? Do you think people will buy more TV sets? Movie tickets? Frozen lobster tails? All sorts of things? Of course. *Consumer spending reflects people's incomes.* You know it does.

### Expectations Influence Consumer Spending

*Consumer spending also reflects people's expectations.* If people are concerned about their economic future they will usually spend less (save more). People will save more if they think they are going to lose their jobs. Or if they think the union is going to call a long strike. Or if they expect that all their teenagers will need some money to help them go to college. Or if they expect they will have to retire early or quit work because of poor health. Or if for *any* reason they expect their incomes to fall, or their expenditures to rise. And the opposite is also true.

If people expect to be earning more (or if they expect their spending obligations to go down) they will be willing to spend up their present incomes right away. They may even buy lots of things "on time." Why should they try to save if they think they are going to be earning more and more all the time?

Anything which changes the attitudes or the outlook of people will be likely to have some influence on how much they spend. The level of consumer spending even reflects such things as whether or not it rains on the Fourth of July weekend and whether or not it snows just before

Christmas. Consumer spending is influenced by radio, TV, and newspaper announcements of developments at home or elsewhere in the world. Anything which makes people more optimistic or more pessimistic is likely to influence their rate of consumer spending.

### Consumers Can Spend Past Income Or Future Income, And Defy Say's Law

It's important to remember that people don't just spend their current incomes. Most of the big items people buy are not bought with current income. When a person buys a new car he is spending either his past income (his savings) or his future income. (He buys on time.) So consumer spending is influenced by how much savings people have, and their willingness to spend these savings. Also, it is influenced by how much the consumers are already in debt, and their willingness to go further into debt.

Consumer spending is influenced by how easy it is to buy things on credit. You probably could go out and buy a $500 TV set today by spending less than $50 (the down payment). Maybe you could buy one with *no* money down. With easy credit, you don't have to have very much current income to be able to buy things. But then your future income is already spent before you get it. That will force you to cut down on your future spending.

You can see how consumer spending is influenced by how much savings and how much debt the "average consumer" already has, and by the availability of easy credit. You can see, too, why a really big boom in the sales of autos, appliances, and furniture *last* year, is likely to be followed by a slump in the sales of these consumer durables *this* year. If everybody bought new things last year, this year they'll all be paying off their debts!

See how the existence of *money* and *credit* keeps Say's Law from working? The amount a person "puts into the market" this year may have very little relation to the amount he "takes home" this year! For this reason it's very easy for shortages to develop (like in consumer durables *last* year, when everyone was trying to buy new things) or for surpluses to develop (like in consumer durables *this* year, when everyone is trying to pay off his debts).

### Taxes Reduce Consumer Spending

Taxes exert an important influence on consumer spending. When you pay taxes you don't have as much money left to buy things. Income taxes take hundreds of dollars from you. Social security taxes take some, too. This money is taken right out of your paycheck. You never even get to see it! This part of your income goes directly to the government. Then, in addition, you must pay property taxes and automobile taxes and gasoline taxes and utility taxes and alcohol and tobacco excise taxes and state and local sales taxes and lots of other taxes. Taxes put a real crimp in the amount you have to spend. Taxes reduce your *disposable personal income*. Remember?

The basic idea behind taxing is to keep you from spending—that is, to force you to save (not consume) and leave some goods in the market. Then the government can spend money to buy these "leftover" goods and use them to make highways and schools and things. Or it can give money to poor people who can then buy these "leftover" goods for themselves. The whole idea of taxing is to get all of us not to spend so much—to leave some things in the market so the government can buy them.

In capsule form, we can summarize the highlights of the "consumer spending decision" this way:

> The amount the consumers are going to spend will be determined by how much current income they have, as modified by how much taxes they have to pay, how much money they already have saved up, or how far they are in debt, how easy it is to borrow or to buy things on time, and what they expect to happen to their *future* incomes and expenditure needs. Anything which causes any of these factors to change is likely to cause consumer spending to change.

Think about it. Aren't these the things which influence the amount *you* spend for consumer goods?

### "Total Consumer Spending" Is Less Than "Total Consumer Income"

People receive incomes and spend for consumer goods. Some people spend all they earn and more. Others save some. In highly productive, high income economies like the United States, Canada and other advanced nations, a sizable share of the total income is saved—that is, is not spent for consumer goods. This means that on the average, people are receiving more income than they are spending for consumer goods. They are producing and pushing more goods into the market than they are pulling back out and consuming. If this is so, then why aren't there great piles of surpluses, piling up in the market? Why don't we have a depression?

### Investment Spending Offsets Consumer Saving

You already know that people's savings can be offset by investment spending. Businesses spend to buy capital goods. If investment spending is high enough to offset the savings, then there will be no surpluses and no unemployment. But investment spending must be high enough to pull out of the market all of the extra products and to employ all of the factors of production left in the market by the savers.

The savers just take their money and go home! So we had better hope that in all the advanced "free enterprise" nations, businesses will keep pushing enough spending into the market (investment spending) to make up for the amount the consumers are pulling out as savings. If not, surpluses will pile up, production will be cut back, and unemployment and recession will result.

As long as investors are injecting just enough spending to offset the withdrawals by savers, the economy will keep running along at the same level. If this level of spending is the "full employment" level, then that's fine. The "macroeconomic situation" of the nation is just fine. There might be some microeconomic problems. The savers might be deciding not to buy canned peas while the investors have no interest in spending for canned peas. They want gear grinders. But this is only a microeconomic problem. You know how this kind of a problem will be solved, right?

The production of canned peas is going to be cut back and the production of gear grinders is going to be increased. The factors of production will flow from the declining industry to the expanding industry. The output of the society will still be at the maximum—at a point *on* (not below) the production possibility curve. The only thing that happens is that canned peas will be "transformed" into gear grinders. Society is answering the question "What to produce?" by saying: "Produce more capital goods (gear grinders) and less consumer goods (canned peas)." Of course, you know that in the real world the adjustment will not come about as quickly and smoothly as we might wish. But it will come about. You may be sure of that!

## WHY BUSINESSES SPEND

Now let's ask the critical question. Do the businesses try to spend enough for investment to exactly offset the amount the people are saving? (that is, *not spending* for consumer goods?) Do investors want to spend exactly enough to offset the savings withdrawals? or not? You probably already know the answer to that. It's: "Sometimes they do, and sometimes they don't." Sometimes investors want to spend too little, and sometimes they want to spend too much. So sometimes we see problems of recession and sometimes problems of inflation. Why? Because the amount the businesses decide to spend for investment goods is not exactly equal to the amount consumers decide to "not-spend" for consumer goods. So what influences investment spending?

There's a lot that we don't know about the reasons why businessmen spend for capital goods. Surely there are many, many reasons why a businessman might decide to invest more or less in his business. But in general, it would be safe to say that businesses spend whenever they expect the investment to be a "good deal." If a businessman expects a high enough return (a big enough addition to revenue or a big enough reduction in cost) then he will make the investment. If not, then he won't. That's obvious enough, isn't it?

How high does the "expected return" need to be to induce the businessman to invest in a machine? High enough to "pay off" all the expected costs of the machine (including the interest cost) and leave a *reasonable profit* for the businessman. Anything which causes the "expected return" to increase (or anything which causes the "expected cost" to decrease) is likely to stimulate businessmen to increase their investment spending. Anything which lowers the expected return or raises the cost is likely to slow down investment spending. Let's take a closer look at these important things — these things which influence the rate of investment spending.

## Investment Spending Is Influenced By Expected Returns And Interest Rates

Why would anyone spend money for capital goods? Why would anyone want to buy a tractor to use to produce corn and tomatoes? Why might you buy a machine to expand the output of your macramé belt shop? How do you decide? You consider the "expected returns" you will get from the investment. And you consider the "interest cost" you will have to pay on the borrowed money you will use to buy the machine.

(If you plan to use your own money to buy the machine, the interest cost is still just as real. You must consider the *interest income* you will have to give up — the opportunity cost — when you withdraw your savings or sell your bonds in order to buy the machine.)

You already know that capital is *productive*. An extra piece of capital has a marginal "value product" just as does an extra worker. Whether or not you will buy a piece of capital depends on whether or not you expect the marginal value product of that piece of capital to be greater than the cost of the piece of capital to you. If you think the additional output will be worth more than the cost of the capital, then you will invest in it. But if you don't think the piece of capital will add enough value to your output to justify its cost, then you will not buy it. Nobody pays more for something than he expects it to be worth to him. Not ever!

## The Marginal Efficiency Of Capital

Everyone who is producing something has some "demand for capital." His demand for capital will be high or low depending on whether or not he thinks the capital will add a lot or a little to the value of his output (or perhaps subtract a lot or a little from his cost of production). Is your demand for the capital *high*? Then that means you expect a *high return* from the capital. Of course!

This idea of expected return as the thing which determines the demand for capital, was developed by John Maynard Keynes (pronounced KAYns) in his famous and somewhat revolutionary economics book, *The General Theory of Employment, Interest and Money* (1936). Keynes emphasized the importance of expectations as the thing which determines whether or not the businessman will invest in a piece of capital.

Keynes referred to the additional return the businessman expects to get from an additional piece of capital as the "marginal efficiency" of the capital. He said that if the capital is expected to be productive enough to more than cover its cost (including the interest cost), then the businessman will invest in the capital. If it is not, then he will not. Using the Keynesian approach, if we want to understand what determines the total rate of investment spending in the economy we need to understand what determines the expected returns of capital, and the interest rate. So let's think about both of these things for a little while.

### What Determines The Expected Return?

What causes a businessman to think some piece of capital—some machine—will bring a high additional return—that is, will have a high "marginal efficiency"? Actually, all kinds of things influence the businessman's expectations. But there are some fairly straightforward ways of getting at this issue. First, what kinds of "returns" can a capital investment offer? Two kinds. It can add to the output value, or it can reduce the production cost. Often the businessman expects the new piece of capital to do some of both.

If the local union gets a big wage increase, then the businessman's labor cost goes up. This increases the expected return of capital, to him. He will be more likely to invest in labor-saving machinery. If he thinks the demand for his product is going to go up, this also increases the expected return, the "marginal efficiency of capital," for him. He will be more likely to buy new machines and expand his plant.

When a businessman thinks good times are coming, he sees the marginal efficiency of capital as high. He is likely to want to spend a lot for new capital investments. But when he is pessimistic about future conditions in the economy—when he expects unemployment and recession—he sees the marginal efficiency of capital as being low. His investment spending is likely to be low. See how close all of this is to just "good common sense"?

### Some Examples Of The Investment Decision

If the businessman sees the expected return of a certain piece of capital as "high," he will have a "high demand" for that piece of capital.

But will he buy it? Or not? How does he decide? It depends on the price. Actually, it depends on two prices. It depends on the *cash* price the businessman must pay for the piece of capital. It also depends on the price he must pay for the *money* he will have to sink in the investment. That is, it depends on the *interest rate*. Let's look at some real-world type examples to see how all these pieces might fit together.

Suppose a small seafood packing house down on the Eastern Shore of Virginia is considering investing in an oyster-breading machine. This is a "latest technology" piece of equipment. It has a stainless steel breading mixer, then a small conveyor belt to carry the oysters from the mixer, through a roller to flatten them, and on into the final package for freezing. The minimum wage law has just pushed up the cost of labor and the seafood packing houses have raised their prices for frozen-breaded oysters. Several restaurants and hotel chains have complained about the high prices of frozen-breaded oysters. Some customers have shifted from Chesapeake Bay oysters to Gulf shrimp and Cape Cod flounder fillets. The oyster-dependent economy of the Chesapeake Bay area is hurting.

The oyster-breading machine sells for only $20,000. It is designed for a small operation. Remember your uncle who runs the Seaside Seafood Packing House in Chincoteague, Virginia? He remembers you! And he knows you are taking a course in economics so he writes to you and asks you whether or not he should pay $20,000 for this new stainless steel oyster-breading machine. You are going to have to answer the letter and say something. What are you going to say?

Before you get too upset about having to make such a tough decision, think of the plight of your roommate's father. He is a high-powered management consultant receiving $500 a day to help a major steel company decide whether or not to install a new iron ore pelletizing plant at Ishpeming, near the Lake Superior shore of the Upper Peninsula of Michigan. The steel company is already operating a pelletizing plant which is less than 10 years old. The present operation crushes the iron ore and tumbles it around and gets the good part—the iron part—to stick together in little pebble-type pellets. Each pellet comes out about 40 percent iron and 60 percent other stuff—"inert ingredients" you might say.

But suppose that a new process has been developed. It would cost $2.8 million to install the new equipment, but it can turn out pellets of 75 percent iron and only 25 percent other stuff. And it operates with only half as much labor. Also, it can handle twice as much iron ore per hour as the present equipment. A shipload of the 75 percent pellets will actually have in it about twice as much usable iron (and only half as much "inert ingredients") as a shipload of the 40 percent pellets. So in addition to all the other benefits, the cost of transportation will be cut almost in half. But remember? The present plant is less than 10 years old!

Should the steel company buy the new equipment and scrap its existing plant which is still working as good as new? Or should they keep using the old plant until it wears out, and then put in the new equipment? Your roommate's father is going to advise them. Tough task, huh? But these are the kinds of questions on which real world investment spending decisions are based. And you'll see that it's really not so difficult to figure out.

In both the oyster-breading case and the iron ore pelletizing case there is no question that there is a *demand* for the capital. That is to say, there is *some* price at which it would be "good business" to buy the capital. But is the price right? Would it be good business to pay $20,000 for the oyster-breading machine? Or $2.8 million for the new iron ore pelletizing plant?

### How Much Will Output Value Increase And Cost Decrease?

How does a businessman decide whether or not to buy new equipment in cases like these? How complicated is it? Not very. The first step is to estimate each way in which the piece of equipment is expected to reduce the cost. Next, estimate each way it is expected to increase the revenue. Then you put the two together, and you have it. What could be simpler than that?

When you make these estimates you would like to be as precise as you can. But don't try to overdo it. Make-believe accuracy in the face of real-life uncertainty is a luxury which only college professors can afford! (Businessmen know better.) Remember that over the life of this piece of equipment, *all* the cost and revenue conditions are going to change in several unforeseeable ways. Some of the changes may be completely unexpected. The situation in three or four years may not even resemble today's situation. So there is no need to really knock yourself out trying to achieve some kind of precise accuracy. That's a waste of energy.

How much cost and effort should a businessman spend in trying to estimate the "expected returns" of a piece of capital? He should follow the marginal principle. Each time he puts any more effort or money into trying to increase the accuracy of his estimates, he should expect the increased accuracy to be worth at least as much as the cost of the extra effort. It would be a waste to try to get more accuracy than that!

## HOW TO FIGURE THE EXPECTED RETURN AND FIND "CAPITALIZED VALUE"

You know that the new capital will bring savings in labor costs and perhaps in other input costs—maybe in transportation costs, power costs, and other costs. And you know that the value of the output probably will increase as a result of the new machinery. But how much decreased cost and increased revenue will there be? Enough to justify the investment? How can you figure this out?

Just take it step by step. It isn't difficult to work out an "expected return" for the first year. Then you project to the second year, third year, and so on for the estimated life of the capital. Take the oyster-breading machine for example. Suppose you figure that during the first year it would save about $2,000 in labor costs and other production costs, and it would probably increase your uncle's output ("value added") by about $1,000 worth. So you estimate that the "expected return" of the machine is about $3,000 in the first year.

Next you might try to guess what is going to happen to wages, to other costs of production, and to the prices of breaded oysters in the following years. Then you could make a different calculation for each year, depending on your cost and revenue projections. Or you might take the more practical approach—the easy way—and assume that the return for each year will be approximately the same—$3,000 each year.

Suppose you figure this machine will last about 10 years and its expected return will be about $3,000 each year. Then its total "expected return" over its life will add up to $30,000. Does this mean the machine is now worth $30,000 to your uncle? To get at the answer to this question, and to emphasize the importance of the interest rate, let's ask a different question.

### The Effect Of The Interest Rate

Would you give me $30,000 today if I agreed to give you back $3,000 each year for the next 10 years? You certainly wouldn't! Why not? Because you could put your money in a savings account or in government bonds or in AT&T bonds or in GM stocks or in any kind of investment you could think of and get back *more than* $3,000 each year for the next 10 years. If you could invest your $30,000 at an interest rate of 10 percent, you would get back $3,000 a year *forever*! And always own the $30,000 to boot! See what a bad deal it would be to invest $30,000 when you only expect to receive a return of $3,000 a year for 10 years?

So how much would you advise your uncle to pay for the oyster-breading machine. Certainly not $30,000. But how much? $25,000? $20,000? $15,000? $10,000? How do you figure it out? Let's go on with our example and soon you will see.

Suppose your uncle bought the machine. By the end of the first year it would have brought him an "expected return" of $3,000. Just to make it easy, let's assume that he receives all his "return" on the last day of the year. So, looking at it from the point of view of the *first* day of the year, how much is that $3,000 return (to be received on the *last* day of the year) worth? If he had the $3,000 on the *first* day of the year, he could invest it or lend it out, or just put it in a savings account, and by the *last* day of the year he would have the $3,000 *plus* one year's interest. At a 5 percent interest rate, he would have $3,000 plus $150. So: $3,000 to be received a year from now is worth about $150 less to you than the $3,000 would be worth if you had it right now. That is, if the interest rate is 5 percent, "the right to receive $3,000 a year from today" is worth only about $2,850, today ($3,000, minus the $150 you lose by *not* having the money now).

Sometimes people have trouble with this concept because it seems strange at first. But it's really quite simple. Suppose somebody offers you a choice. He will give you $3,000 now, or he will give you $3,000 one year from now. Which would you prefer? You would rather have it now, of course. How strong would be your preference to have the money now? That depends on "the going interest rate" for the kind of investment you are thinking about. If you are thinking about investing the money at an interest rate of 5 percent, then it "costs" you $150 to wait a year. If you are thinking about an 8 percent rate, then it "costs" you $240 to wait a year. (8 percent of $3,000 is $240.) That isn't hard to see, is it?

### Future Payments Must Be "Discounted"

Suppose someone comes to you and says: "Here's a piece of paper which says I will pay you $3,000 one year from today. How much will you pay me for this piece of paper (promissory note) right now?" If you're thinking of a 5 percent interest rate you will pay about $2,850. ($3,000 minus $150.) If you are thinking of an 8 percent rate, you will pay about $2,760. ($3,000 minus $240.)

What you are doing is buying a $3,000, one-year promissory note, at a discount. If the interest rate is 5 percent, you will "discount" it about $150; if the interest rate is 8 percent, you will "discount" it about $240. Suppose it is a *two*-year promissory note. At 5 percent per year the discount would be about $150 each year, or about $300, total. The present value would be about $2,700. At 8 percent per year the discount would be about $240 per year, or $480, total. The present value would be about $2,520. (You will notice that these figures are not exactly correct, because of the "compounding effect." But don't worry about that now).

See how the present value of an expected future return goes down more and more as we look farther and farther into the future? Also, notice how much difference it makes whether the interest rate is 5 percent or 8 percent! Maybe you have always thought of the interest rate as something small and insignificant. But as the years go by, the interest rate can become a very big thing. (Find out how much of the average car payment or house payment goes for interest. You may be surprised!)

### Discounting the "Expected Return"

Your uncle expects a return of $3,000 at the end of the first year. If the interest rate is 8 percent, the present value of that expected return is only about $2,760. At the end of the second year he expects to receive another $3,000. But the present value of that is only about $2,520. Actually it's a little less than $2,520, because of the "compounding effect"—he's losing the chance to earn interest on the interest!

You can see how the *present* value of the money to be received gets smaller and smaller as the "due date" gets farther and farther into the future. The longer you have to wait before you get your money, the more interest you lose. It's just the opposite of having your money in a savings account. The longer you *have it* in your account, the more you *gain*; the longer you *don't have it* in there, the more you lose. Get the idea? So the *present* value of the $3,000 your uncle is going to receive in the third year, the fourth year, the fifth year and so on, gets smaller and smaller and smaller.

So what do you advise your uncle to do? Should he pay $20,000 and buy the machine? How do you decide? Simply find the discounted value of the "expected return" for each of the years over the expected life of the machine. Then add them all up. How simple! This little exercise will give you the capitalized value of the machine. It will tell your uncle how much the machine is worth to him, on the basis of your estimates. *The "capitalized value" of any asset is the present value of the asset, as figured on the basis of how much money that asset is expected to bring to its owner over its lifetime.*

Suppose you spent a thousand dollars (or a million dollars for that matter) building a machine, and then found out that it would cost more to run the machine than the output would be worth. What would be the capitalized value of the machine? Zero, of course. The machine might have some scrap value or some "aesthetic value" or something, but capitalized value? No. Anything which won't make you any money doesn't have any capitalized value. The more money it will make for you, the more its capitalized value will be.

You can see that the higher the interest rate happens to be, the greater the discount which must be subtracted from the expected returns. So *as interest rates rise higher and higher, capitalized value gets lower and lower.* At higher interest rates, fewer capital investments appear profitable. High interest rates can have a very discouraging effect on the flow of investment spending. Now you can understand why.

How do you figure out the capitalized value of a machine, or of any other asset? First you figure out what you think the expected return is going to be in each year over the expected life of the asset. Then you "discount" the dollar figure for each year. This gives you the present value of each future-year's expected income. Then you add up all the *present values* of all the future-year incomes. And what do you have? You have the present capitalized value of the asset. Of course!

The easiest way to do the discounting is to look up the figures in a "present value" or "interest and discount" table. (Ask your friendly neighborhood banker to show you one.) I looked up the present value figures for your uncle's investment problem and found out that at an interest rate of 5%, the present capitalized value of the oyster-breading machine comes out to about $23,000. This assumes, of course, that the machine really will last exactly ten years, and that it really will bring your uncle a return of $3,000 a year. And of course it assumes that the interest cost for the money he invests in the machine really will be 5 percent. These are big "ifs"!

### Should The Businessman Allow Some "Margin For Error"?

So what do you advise your uncle to do? Should he buy a machine that costs $20,000, when it looks like the machine is going to be worth $23,000 to him? Probably so.

But suppose the interest rate is considered to be 8%. What then? Then the capitalized value of the machine comes out at about $20,000. Should your uncle buy it? Probably not. There certainly doesn't seem to be much margin for error! It's chancy, at best. Maybe you should just send your uncle all the facts and figures and let him decide. Do you see how uncertain, how indefinite the investment spending decision can be?

Your roommate's father goes through the same process. The only difference is that he deals with bigger figures, and he may use a computer to make it easier. Besides, the computer makes everything look so scientific and precise! If he comes up with a "capitalized value" for the iron ore pelletizing machinery *greater* than the $2.8 million it will cost (plus some satisfactory margin for error and risk) then he will advise his client to buy and install the equipment. If not, he won't.

You can see that both of these decisions will be highly sensitive to *changes in expectations* about future returns. Also, the decisions are sensitive to *changes in the rate of interest* in the "money market." So what determines the level of investment spending in the economy at any moment? Expected returns and interest rates have a lot to do with it. Higher expected returns are likely to stimulate investment spending; higher interest rates are likely to retard investment spending. Common sense? Sure.

## THE COMPONENTS OF THE TOTAL SPENDING STREAM

So far in this chapter we have been talking about the two "basic spending sources" in the economy—consumer spending and investment spending. You now have some understanding of how the level (rate) of each of these is determined. But you haven't yet looked at *all* the components of the spending stream and their relationships to each other. It's time to do that now.

## Consumer Spending (C) And Investment Spending (I)

You know that total spending by people for consumer goods (consumption), plus total spending by businesses for capital goods (investment) add up to total spending for the economy's output of consumer and capital goods. The total rate of spending in any day, week, month, or year for these goods, must be equal to the total value of the consumer goods and capital goods being bought; and the total amount of money received by the people producing all this output, is equal to the total amount being spent by all those buyers.

Expenditures for consumer goods (C), plus expenditures for investment goods (I), equals total spending for consumer goods and capital goods (C + I). It must be so. And it also equals the total value of the output of these goods, and it equals total income received by those who produced these goods. You've known about all this ever since Chapter 10. Right? Well, now that you understand it so well, it's time to admit that it doesn't really tell the whole story. So let's take another step to complete the picture.

## Government Spending (G)

Since all things produced are either produced to be consumed, or to be saved and reused later, all goods produced must be either consumer goods or capital goods. That's all there is! Still, expenditures by individuals for consumer goods and expenditures by businesses for investment goods do not really tell the whole story.

The governments—*all* levels of government—spend money. They buy goods which would have been consumer goods if individuals had bought them, and they buy goods which would have been capital goods if businesses had bought them. But since the government buys them, we call them "government goods." So now we add a third "spending source." We simply refer to this as "government spending" (G). The government spends for "government goods." "Government goods" include everything from school buildings and highways and teacher's services to Polaris submarines and the canned peas to be used to feed the crew of the aircraft carrier *Enterprise*.

So now you know that total spending in the economic system (the total spending stream) generates from at least three sources: spending by individuals for consumer goods (C), spending by businesses for investment goods (I), and spending by government for "government goods" (G). We might say that "C + I + G equals total spending, which is equal to the total value of the output, which equals total income in the economy." We could say that. But it wouldn't be precisely correct, as you will see in just a moment.

Why do we put *government spending* into a separate category? Couldn't we just include government under "investment spending" and only have two categories? Yes, we could do that. Why don't we? Because it's useful to look at government spending separately. The spending choices (how much to spend, in which ways, at what times) are made *differently* by the government than by businesses or by individuals.

The whole purpose of breaking down total spending is to identify and understand each "spending source." If we can understand why each "spending source" decides how much it will spend, then we will have a much better understanding of how *total spending* is determined. Then we will know much more about macroeconomics. Perhaps sometimes we can even predict what is going to happen. And perhaps then we can establish government policies to bring better macro-conditions in the economy. We hope so. More on this, later. But now, *another* (the last) spending source.

## Foreign Spending For Our Goods (Ex)

The fourth and final spending source we break out and identify separately is "spending by foreigners for our goods." Just as with the government, this "spending source" could be included in the "C + I" two-way breakdown. But you can see that it might be helpful to separate "spending by foreigners for our goods" into a category all its own. The term usually used to mean "spending by foreigners for our goods" is "exports." We can use "Ex" as the symbol for this.

Suppose a Frenchman decides he would like to have a case of Del Monte canned peas. He goes to the Bank of France and exchanges some of his deposit account francs in that bank for some deposit account dollars which the Bank of France has on deposit in an American bank. Now he can use those U.S. dollars to buy the Del Monte canned peas. So he does. This adds to total spending in the American economy. And suppose someone in Italy decides to buy an Oldsmobile Cutlass. He goes to the Bank of Italy and exchanges some of his lira deposits for some dollar deposits, and he sends the American dealer a check to pay for the Cutlass. Suppose somebody in Great Britain decides to buy an oyster-breading machine from the Baltimore manufacturer. He goes to the Bank of England and trades some pounds for dollars and then sends the manufacturer a check for the machine. Suppose a Japanese businessman decides to buy a shipload of American chemicals to be used in making plastic toys. He exchanges yen for dollars and sends a check to pay for the chemicals. (These "foreign exchange" checks may be called "bills" or "drafts" or "orders" or whatever. But they're really checks.)

All purchases by foreigners add spending into the spending flow, stimulate production, add to the value of output and add to incomes in this country. The effect on spending, output, and incomes is the same as if the canned peas had been bought by someone in Willimantic, Connecticut, and the oyster-breading machine by someone in Pascagoula, Mississippi, and the chemicals by someone in Rahway, New Jersey, and the Cutlass by someone in Snook, Texas! Can you see that from the point of view of the seller of the goods, or of the effect on the total spending stream, it really doesn't make any difference at all whether the buyer lives in Paris or Poughkeepsie? In Tokyo or Kokomo? Bangkok or Little Rock?

Some of the foreign buyers are buying consumer goods. Others are buying capital goods. But still we lump both together and treat them the same. Why do we do it this way? Only because it's convenient, and useful. This gives us our fourth and last spending source: "spending by foreigners for our goods." Exports (Ex).

### The Four Spending Sources

Now we have the entire four-way breakdown of the spending stream. Each spending source is identified in the most useful way. Now our "spending flow breakdown" becomes C + I + G + Ex. Now we have the whole thing. Now can we say that C + I + G + Ex = GNP? Can we say that? Or not? Think about it.

You know that anything which would cause any one of these "spending sources" to get smaller would pull down the size of total spending. That would cause some of the goods being produced not to be bought. This would cut down on the amount businesses would produce. Production would slow down and total output and total income would decline. Reduced spending means reduced output and reduced income. GNP would fall.

You also know that anything which might happen to cause spending by any one of the "spending sources" (C, I, G, or Ex) to increase, would push total spending upward. This would mean that the goods available in the market would be bought up more rapidly. Producers would expand their production and increase their outputs. GNP would rise.

Yes, C, I, G, and Ex are the only sources from which spending can flow into the total spending stream, to support total income and total output (GNP). So can we say that C + I + G + Ex must be equal to total spending and total output? to GNP? No. Almost, but not quite. There's another adjustment we have to make, because C + I + G + Ex doesn't *exactly* equal GNP. Why not? Because we must account for the part of the spending flow which is "leaking out" to foreign countries.

## THE FOREIGN TRADE BALANCE

We can't really say that C + I + G + Ex equals total spending, and equals GNP. It almost does, but not exactly. Why not? Because some of the spending for consumer goods (C) is *not* for consumer goods *produced in this country,* but for consumer goods produced in foreign countries and imported to this country. Some of the capital goods bought by the investment spending (I) are capital goods *produced in other countries* and imported to this country. Some of the things that the government is spending for (G) are things that are being *produced in other countries.* Do you begin to get the point?

### Some Of The Spending Flow Is "Leaking To Other Countries" (Im)

You know that this country's GNP (gross *national* product) is pulled forth by the flow of spending in *this* country, spent for *this* country's output. Right? So we must be careful not to include the part of our spending flow which is "leaking overseas"—that is, the part which is going to foreign sellers to pay for foreign-produced goods. Let me say it again, slightly differently.

You know that the "C + I + G flow of spending" is the spending which supports total output and total income. But the part of the "C + I + G flow" which flows to other countries *does not* support output and income in this country. Therefore, that part of the flow which goes to buy foreign-made goods must be subtracted. Then we will know just how much of the total spending stream is being used *to pay for goods produced in this country* (and to pay incomes to the producers and workers and everybody, in this country).

So how do we handle this adjustment? this subtraction for the part of the spending flow which is "leaking overseas"? We could subtract from "C," that part which is being spent for imported goods, then subtract from "I," that part which is being spent for imported goods, then subtract from "G" the part which is being spent for imported goods. That would take care of it. Right?

Yes, it would. But that would be the hard way. It's much easier to add up all the dollars "leaking overseas" (to pay for imports) and then subtract the whole thing, all at one time. That's just exactly what we do. We call the subtraction item "imports" (Im). What it means is "total spending by buyers in *this* country for goods produced in foreign countries."

You already knew that the spending flow in this country is always being *increased* by the amount foreigners are spending to buy our goods—that is, the amount they are spending for our exports (Ex). Now you know that the spending flow in this country is always being *decreased* by the amount that buyers in this country are spending for foreign goods—for our imports (Im). So wouldn't it be a simple thing to just subtract the "spending outflow" (Im) from the "spending inflow" (Ex)? Sure. That would take care of the problem. That's exactly what we do. It looks like this:

$$C + I + G + (Ex - Im) = GNP.$$

### We May Have A Positive Or Negative Foreign Trade Balance (F)

You can see that if the *spending outflow* for foreign goods (Im) is *greater* than the *spending inflow* for our goods (Ex), then "Ex – Im" will be a *negative* number. More spending is flowing out than is flowing back in. This holds down the size of output and incomes in this country. In this case, we would say that the foreign trade balance (F) is negative.

If the *spending inflow* from foreigners for our goods (Ex) is greater than our *spending outflow* to buy foreign goods (Im) then "Ex – Im" (F) will be a *positive*. There will be a net addition to the spending flow in this country. The positive "foreign trade balance" (F) will help to support our total employment, output, and income. If you want to, you can say it this way:

$$C + I + G + (Ex - Im) = GNP;$$

or you can say it this way:

$$C + I + G + F = GNP.$$

Either way, our equation is now exactly and precisely correct. You know that "F" can be either plus or minus, depending on whether foreigners are spending more for our goods, or we are spending more for theirs. If imports and exports should happen to be exactly the same, the foreign trade balance (F) will be zero.

### A Positive Foreign Trade Balance Increases GNP

You can see that a positive foreign trade balance (F) could be important in helping to support employment, output, and incomes in a country. You can see why government, during times of recession or depression, would like very much to see increases in their exports and decreases in their imports. But every country is in the same boat as far as the foreign trade balance is concerned.

If one country tries to force reductions in imports (by tariffs, or "quota restrictions" on foreign goods), and at the same time, tries to force increases in exports (through paying subsidies to exporting businesses, and helping them in other ways), that hurts the foreign trade balances of all the other countries. Or if one country decides to "devalue" their money (sell their money cheaper to foreigners so that domestic goods will be cheaper to foreigners, and foreign goods will be more expensive to domestic buyers), that hurts the foreign trade balances of all the other countries, too.

What do you suppose would happen if one country decided to restrict imports, subsidize exports, and devalue their money? Other countries would retaliate? Of course. History shows plenty of examples of times when countries have tried to restrict imports and stimulate exports. All the countries begin competing with each other and they all wind up with everybody getting hurt. In the end, none of them is as well-off as they would have been if the whole process had never been started. But would you believe that some of this sort of thing is going on, even today? Economic *intelligence* doesn't always control economic *policy*!

This little discussion of the foreign trade balance (F) and its importance in influencing the macro level of the economy doesn't take you very far into the subject. But it gives you a little bit of basic understanding. In the final chapters of the book, you will learn more about international trade and finance. For now, first review this chapter, then on into Chapter 14 for more macroeconomics!

---

## REVIEW EXERCISES

### MAJOR CONCEPTS, PRINCIPLES, TERMS (Try to write a paragraph explaining each of these.)

what influences consumer spending
what influences investment spending
how investment spending offsets consumer saving
finding the capitalized value
the components of the spending stream
the foreign trade balance
C + I + G + F = GNP

### OTHER CONCEPTS AND TERMS (Try to write a sentence or phrase explaining what each one means.)

consumer durables
consumer expectations
investor expectations
expected return
John Maynard Keynes (KAYns)
capitalized value
discounted value
C, I, G, Ex, Im, F

### QUESTIONS (Try to write out answers, or jot down the highlights.)

1. There are several ways in which *optimism* can be very stimulating to the economy. Can you explain several of the ways? Try.
2. What would be likely to happen to investment spending in the economy if the interest rate suddenly went up from 6% to 10%? Can you explain

exactly why, using the idea of "capitalized value" to illustrate your explanation?

3. If you were "the uncle in Chincoteague," would you pay $20,000 for the oyster-breading machine when the interest rate was 8% and the capitalized value was about $20,000? Or not? Explain why you would (or wouldn't).
4. If there's too much spending, and lots of inflationary buying in the U.S. economy, then it would be good for the economy if some people would stop buying Vegas and Pintos and Gremlins and Dusters, and start buying Volkswagens and Toyotas and Volvos instead. But if the economy is in a recession, it would help for more people to "buy American." Discuss.

# 14 HOW TO VISUALIZE AND ANALYZE THE TOTAL SPENDING STREAM

*A closer look at the component parts of the spending-income stream, at the spending injections and income withdrawals, and at macro-equilibrium.*

What determines the macroeconomic level of the economy? What determines the overall level of economic activity? Of production, employment, and income? What are the conditions of macro-equilibrium for an economic system? You know the answers. But let's summarize.

## THE CONCEPT OF MACRO-EQUILIBRIUM: AN OVERVIEW

Total spending supports (pulls forth) total output. At the same time, total spending generates the *incomes* flowing to the owners of the factors of production (as wages, rents, interest, and profits). The total income received by all of these factor owners must be *totally spent* if the *total output* is going to be all bought.

The total income, remember, is just exactly big enough to buy the total output. So unless all the income is spent, some of the output will be left, unbought, in the market. The producers of the unbought output will cut back production. They will fire workers and buy less raw materials. The rate of production will slow down. Incomes will drop. The macro-level of the economy will go down.

If all the *output* is going to be bought, that means all the *income* must be spent, to buy it. But we know that the people who receive the income are *not* going to spend it all for output. Some of the income will be saved. So will we have unemployment? Will a depression come? That depends on whether or not somebody else—the investors—will buy up the goods left in the market by the savers.

### Investment Spending Must "Clear The Market"

To keep the economy in macro-equilibrium, investment spending must be great enough to "clear the market." The investors must buy up all the goods left by the savers. If the businesses try to buy *more* output than the savers are leaving, then there will be shortages in the market. These shortages will stimulate producers to produce more, up to the point where all of them are producing all they can. Output will expand until the economy reaches full capacity. We could say that the economy will "move up to a higher macro-equilibrium."

Suppose the amount of savings is too great. Suppose the rate of savings is not offset by the rate of investment spending. Some output will be left in the market, unbought. The unbought surpluses will result in reduced production, and unemployment. The economy will "move down toward a new macro-equilibrium." Perhaps the "new macro-equilibrium" will be one of widespread unemployment and depression.

You can see that sometimes the issues of macroeconomics could give us some things to worry about. Surpluses, unemployment, and depression; or shortages and inflation—not very pleasant to contemplate. Sometimes these problems can get really serious! And that's not all. Consider this: whenever macroeconomic conditions start getting bad, there's a tendency for them to get even worse. Unfortunately, macro-equilibrium in the economy is not a stable, dependable condition.

### Macro-Equilibrium Is Unstable, Like A "Perched Boulder"

Macro-equilibrium is always sort of shaky. The "spending injections" and the "withdrawals from the spending stream" determine the macro-level of the economy. All these injections and withdrawals are sort of unstable. Macro-equilibrium is like a round boulder resting at the very top of a gently sloping mound, or hill. Anything which pushes the boulder off balance is likely to cause it to roll quite a distance before it once again comes to rest—that is, before it finds a new equilibrium. That's the way it is with total spending. So that's the way it is with macro-equilibrium.

This chapter gets into more of the details about macroeconomics—about what determines total spending and what causes it to change, and how. You will get a chance to see how unstable this macro-equilibrium really is. Ultimately we will talk about what can be done to stabilize this "perched boulder"—this macro-equilibrium—at the right place. But that's still a few chapters away. First you need to go farther into the basic concepts of macroeconomics.

## WAYS TO BREAK DOWN THE TOTAL SPENDING-INCOME STREAM

Everybody who spends money for a part of the nation's output is adding to the total spending-income stream. When you buy a box of cough drops, that adds to the total spending-income stream. When your mother buys a can of peas, that adds; when your father buys a new pair of shoes, that adds; when your college buys another box of chalk, that adds. What a mass of little additions to the total spending-income stream!

### We Can Look At Big Flows, Or Little Flows

We can look at total spending as one big flow, going on all the time; or we can look at it as thousands and thousands of little flows, going on all the time. Back in Chapter 10, we looked at the circular flow between households and businesses. There we were looking at the spending flow as one big flow. When we look at "national product" or "GNP" we are looking at one big flow. The same is true if we look at "national income." These are ways of looking at the spending flow (and the output flow) as "one big flow."

In Chapter 13, you learned how to look at the spending flow as a "four sector" flow, each sector defined according to the *source* of the flow: consumer spending, investment spending, government spending, or foreign spending for our goods. That's a useful way to look at the spending flow. But there are other ways.

Each one of these "four sectors" could be broken down into much smaller parts. The consumer spending flow (C) might be broken into expenditures for: consumer durable goods (automobiles, appliances, furniture, etc.), non-durables (gasoline, canned peas, paper towels, toothpaste, etc.), and services (auto repair, attorney's fees, doctor bills, airline tickets, motel bills, etc.). The investment spending flow (I) can be broken down into fixed capital (buildings, machines and equipment, etc.), and circulating capital (raw materials, gasoline for the tractor, seed corn, etc.).

Government spending and foreign spending can be broken down in many ways, too. And each of the smaller spending categories can be broken down even more. For example, expenditures for non-durables can be broken down into food, and other things. Expenditures for food can be broken down into meats, canned goods, etc., etc.

### Why Should We Break Down The Spending Flow?

Why would anyone ever want to break down the total spending flow into all these little component parts? We want to try to understand what is going on, and why. Sometimes it is helpful to know *what kinds* of consumer spending are increasing or decreasing, or *what kinds* of investment spending are increasing or decreasing. By knowing these things, sometimes we can better understand what is going on. Then we may be able to figure out and get ready for what the future holds in store for us.

How far do economists go in breaking down the total spending flow? As far as necessary to do what they want to do, to find out what they want to find out. Sometimes the basic "C + I + G + F" breakdown is sufficient. At least, that's usually a good way to start. But often it's necessary to go farther. If something is "going wrong" with consumer spending, a more detailed breakdown of the consumer spending flow (C) may help us to see what's going wrong, and why.

### Sources Of "Spending Flow" Statistics

Do government statisticians actually break down the spending flows into these smaller parts? You bet they do! Where do you think you could find the figures? In the *Survey of Current Business* from the Office of Business Economics, U.S. Department of Commerce? Right! Or in the *Economic Report of the President* (annual), or the *Federal Reserve Bulletin* (monthly), or on page two of *Business Week* magazine (weekly), or from time to time in the *Wall Street Journal* (daily), the *New York Times*, or in almost any news magazine or big city newspaper, or in any publication or article entitled: "Indicators of Economic Activity." It might be fun for you to look up some of these figures, sometime, just to see what's going on in the economy. (Why don't you go to the library and do it right now?)

All the ways we have been looking at and breaking down the spending stream have been more or less the same. We have been looking at the spending stream as made up of various components, with each component identified by the *source* of the spending: the consumer, the business, the government, or the foreign buyer. But that isn't the only way of looking at the spending flow. There is an entirely different way of looking at it. You can think of the total spending flow as "the quantity of money" (number of dollars) in existence, and the *speed* at which the money (each dollar) is circulating. Let's talk about this new approach.

### Another View: The Money Supply And The Velocity Of Circulation

If the total money supply consisted of $250 billion, and if each dollar changed hands (was spent for new output) four times each year, then the total size of the spending flow in the country would be $1,000 billion (or one trillion dollars) a year. ($250 billion in existence, each dollar spent four times a year: $250 billion × 4 equals $1 trillion.) The total amount of output bought in the country would amount to a trillion dollars worth. Total spending equals total output. Right?

Can you see that this is another perfectly good way to look at the spending flow? The quantity of money in existence, times the velocity at which each dollar is flowing, equals total spending. If we wanted to understand what was going on, using this approach, we would try to find out (a) what might cause the total money supply to get larger or smaller, and (b) what might cause the velocity, or speed at which each dollar is turning over, to get faster or slower. This "quantity of money" approach to understanding total spending has been used by economists for many years. You will learn more about this approach later.

### Other Breakdowns Can Be Useful

There are many other ways of breaking down the total spending-income flow. We might look to see which parts of the country do the most of which kinds of spending (geographic breakdown). Or we might look to see how much and which kinds of spending are being done by young people or older people. Or we might find out how much spending is done by cash, or on credit. There are lots of breakdowns we could use.

Several of these ways of breaking down the spending stream actually are used. When? Whenever someone thinks such a breakdown will help him to find out something he wants to know, of course! The basic "four-sector" breakdown by spending source (C + I + G + F) and the basic breakdown which focuses on the *quantity* and *velocity* of money, are generally the most useful. Those are the two we will use.

Right now we need to go farther with the four-sector breakdown (C + I + G + F). After that, we'll go deeper into the "monetary" approach and you will learn more about "the quantity of money (M) times the velocity at which it is circulating (V)." But that must wait until the next chapter.

## INJECTIONS INTO AND WITHDRAWALS FROM THE SPENDING STREAM

People stand on the receiving end of the total spending stream. They get checks in payment for the use of their labor, land, and capital. Each person receives his little share of this total spending stream as his income. When he receives it, what does he do with it? What are his choices?

### Income Must Be Used In One Of Four Ways

When a person receives his income he has four choices about what to do with it. He can either spend it for consumer goods (C), or he can save it (S). You already know that. But there are two additional ways he can dispose of his income. He can pay taxes (T), or he can buy goods from foreign producers—that is, he can spend his money for imports (Im). That completes the list. Those are his only choices. That is, the total income received by each person *must* be disposed of in one of these four ways: for domestically produced consumer goods (C), saving (S), paying taxes (T), or for imported goods (Im). These are the only choices anyone has. (Of course he could spend it for capital goods, for business purposes (I), but then we would say he *saved.* Then he invested.)

The spending-receiving-spending cycle is going on all the time, every moment of every day. The consumers, investors, government and foreign buyers (C + I + G + Ex) are pushing money into the spending stream. The people who are receiving the money are pushing some of it back into the stream, to buy consumer goods (C). They are pulling the rest of it out as savings, to pay taxes, and to buy imports (S + T + Im).

## Each "Withdrawal" Has An Offsetting "Injection"

Notice that when the income receivers spend their money for domestically produced consumer goods (C), that money goes right back into the income stream. But the parts of their income they use as savings (S), taxes (T), and for imports (Im), do *not* go right back into the stream. So, does the stream get smaller? Not necessarily.

You have already noticed that for each kind of withdrawal from the income stream there is a corresponding reinjection into the stream. Savings are offset by investment (S = I). Taxes are offset by government spending (T = G). Imports are offset by exports (Im = Ex). Since each withdrawal is offset by a corresponding reinjection, does it all come out even? Maybe so. The problem is that it doesn't always work out that way. When it doesn't, that can create quite a problem! A macroeconomic problem. A problem of people trying to spend too little, or too much.

If the system is working along smoothly and you look at it at any moment, what you will see is this: Consumer spending is flowing at a certain continuous level, around and around. Taxes (T) are being pulled out of the stream and flowing away to the governments. Savings (S) are being pulled out of the stream and going into idle pools where they may be available for investment. Payments for imports (Im) are being pulled out of the stream and flowing into the bank accounts of foreigners.

Also, at the same moment, you will see government spending (G) flowing into the stream to offset the tax flow. You will see business investors pulling money out of the idle pools of savings and from the banks (creating money) and adding spending (I) back into the stream. Foreigners will get some dollars and push them into our spending stream to pay us for our exports. (Ex). All these things are going on simultaneously, every moment, all the time. If you were to look at the overall macroeconomic picture at any moment, that's exactly what you would see.

It's obvious what would happen if at any moment the *reinjection* flows (I, G, Ex) should either fall below or rise above the *withdrawal* flows. The size of the spending stream would change—would get smaller, or larger, wouldn't it? Sure.

## The Expenditure And Income Equation

There are four ways, then, in which the people dispose of their incomes: spend for domestic consumer goods (C), save (S), pay taxes (T), or spend to buy foreign goods (Im). And there are four ways money can get pushed into the spending stream: by people spending for consumer goods (C), by businesses spending for investment goods (I), by government spending for goods and services (G), and by foreigners spending to buy our goods (Ex). *The amounts the spenders are spending* (putting into the income stream) *must be equal to the amounts the receivers are receiving* (taking out of the stream). You recognized this right off as an equality, an "equation," didn't you? The equation looks like this:

$$C + I + G + Ex = C + S + T + Im.$$

Figure 14-1 on the following page explains this equation in detail. You should study Figure 14-1 now.

## An Overview Of The Injections And Withdrawals

Income (spending) is flowing in a circle all the time. Consumer spending stays in the circle and goes around and around. At each instant, some of the money in the income stream is being pulled out, but simultaneously new spending is being added in. The stream has a stable level only when the amounts being drained off are equal to the amounts being added in. The stream will be the same size if each withdrawal is exactly "neutralized" or offset by a corresponding injection into the stream.

If the amount the investors are adding to the stream is exactly equal to the amount the savers are pulling out, then S = I, and the effect of both is neutralized. But if the investment injection gets *larger* than the savings withdrawal, then the stream will get larger. Or, if investment spending drops *below* the savings level, savings will be pulling out more than investment is putting back in. The stream will get smaller. The same thing can be said about G = T, and about Ex = Im.

If the amount people are spending for consumer goods continues constant, and if all savings are offset by investment, all government expenditures are offset by taxes, and all imports are offset by exports, then the spending stream will be stable. The size of the stream cannot change until one or more of these equalities is broken. So long as C = C, S = I, T = G, and Im = Ex, the size of the spending stream (income stream)

Fig. 14-1 **THE EXPENDITURE AND INCOME EQUATION**

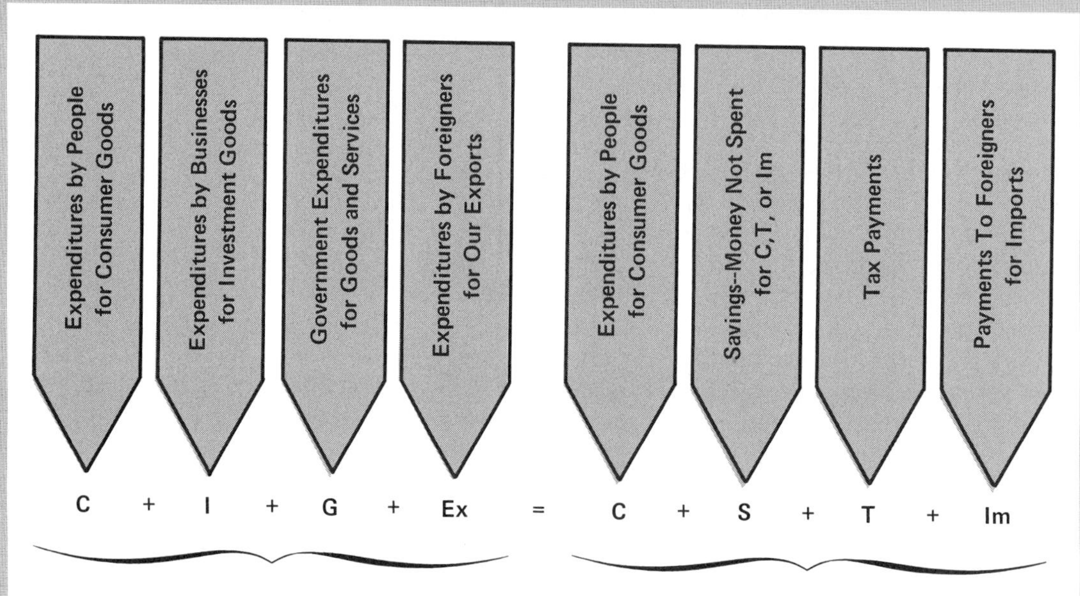


The Expenditure And Income Equation Shows Where The Income Comes From, And Then How It Is Used By Those Who Receive It.

You receive income because consumers, businesses, governments, and foreigners are spending to buy your output. You dispose of your income by spending for consumer goods, or by saving, paying taxes, or buying imports.

cannot change. But any time these withdrawals and the offsetting reinjections are *not* equal, the size of the spending stream *must* change.

Of course, any one of these injections (I, G, or Ex) could offset any one of the withdrawals (S, T, or Im). For example, if investment spending was not great enough to offset savings, the government could spend more. That might offset some of the savings and keep the income and spending flow from going down. The important thing is that S + T + Im (total withdrawals) must be exactly offset by I + G + Ex (total reinjections). Otherwise, the economy will be speeding up or slowing down. No doubt about it. Now, let's look at it another way.

## A "Fire Hydrants And Drain Pipes" View Of The Economy

Try to visualize this total spending stream as a stream of water flowing around in a big circle. There are three big "drainpipes" (savings, taxing, and imports) draining off the water as it flows around. There are three big "fire hydrants" (investment spending, government spending, and exports) gushing water into the stream as it flows around. It's easy to see that if the hydrants keep gushing exactly as much water into the stream as the drainpipes are draining out, then the total size of the stream will stay the same.

The "basic flow" in the stream is consumer spending. That just keeps going around and around. The savings drainpipe drains off some of this basic flow. So do the taxes drainpipe and the import drainpipe. But if the outflow through the savings drainpipe is just offset by the inflow from the investment hydrant, and the taxes drainpipe outflow is just offset by the inflow from the government spending hydrant, and the import drainpipe outflow is just offset by the export hydrant inflow, then the stream will go on flowing around and around. The size, or level of the stream will not change. A lot of water will be gushing in from the hydrants at the top and a lot will be gurgling out through the drains at the bottom. But the size of the stream will stay the same. Can you picture it?

## There Are Millions Of Little Faucets And Drainpipes

Now that you have the basic picture, let's complicate it a little. You know that in the real world all the savings and taxes and import spending don't really drain off like in one big pipe. All the investment spending and government spending and export payments don't gush in at one place, either. Actually, it's more like millions of little faucets and millions of little drainpipes. Each person is in charge of regulating the withdrawal flow through his own tiny little "savings drainpipe." Each businessman is in charge of regulating the reinjection flow through his own little "investment faucet."

With all these people involved and with each one doing his own thing, it seems that only a miracle would make the savings withdrawals and the investment reinjections come out even! The same is true for foreign buyers of our goods, and for domestic buyers of foreign goods. Each person decides on his own whether to buy a Saab or a Pinto, or a Datsun or a Vega. What he decides determines whether or not he will be draining some of his income out of the U.S. spending stream. The government's decisions on taxing and spending are not quite as scattered as this. But when you think about all the state and local governments, you'll see that there's a lot of scatter! Governments usually try to adjust the "tax drain" to be about the same size as the "expenditure injection"—but not always. Even when they *try* to "balance the budget," they aren't always successful.

### The Federal Government Has A Big Hydrant And A Big Drainpipe

There is only one spending unit which has control of a really big "spending hydrant" and "drain pipe." Which spending unit? The Federal Government, of course! Even the largest business corporations are not big enough to have much influence on the total spending flow. But the Federal Government has control over a big enough injection hydrant (government spending) and a big enough drainpipe (taxes) to have a really big effect on the size of the spending flow if it wants to.

We will talk more about these issues later. For now, just be sure you understand how it all fits together. And remember that all these withdrawals and injections must offset each other to keep the economy in "macro-equilibrium." If the total amounts people are trying to withdraw (drain off) are not exactly equal to the total amounts the people are trying to reinject, then the level of total spending will be changing—will be getting larger or smaller. The economy may be heading for inflation or recession.

Figure 14-2 shows a diagram illustrating the injections into and the withdrawals from the total spending stream. It would be a good idea for you to study that diagram now.

## THE QUESTION OF EQUALITY BETWEEN INJECTIONS AND WITHDRAWALS: DOES S = I?

Just about now there's something that may be beginning to bother you a little bit. Something may be beginning to seem illogical about this "spending, receiving, spending" flow. Let's talk about it.

**Fig. 14-2 A DIAGRAM OF THE TOTAL SPENDING-INCOME STREAM, SHOWING INJECTIONS AND WITHDRAWALS**

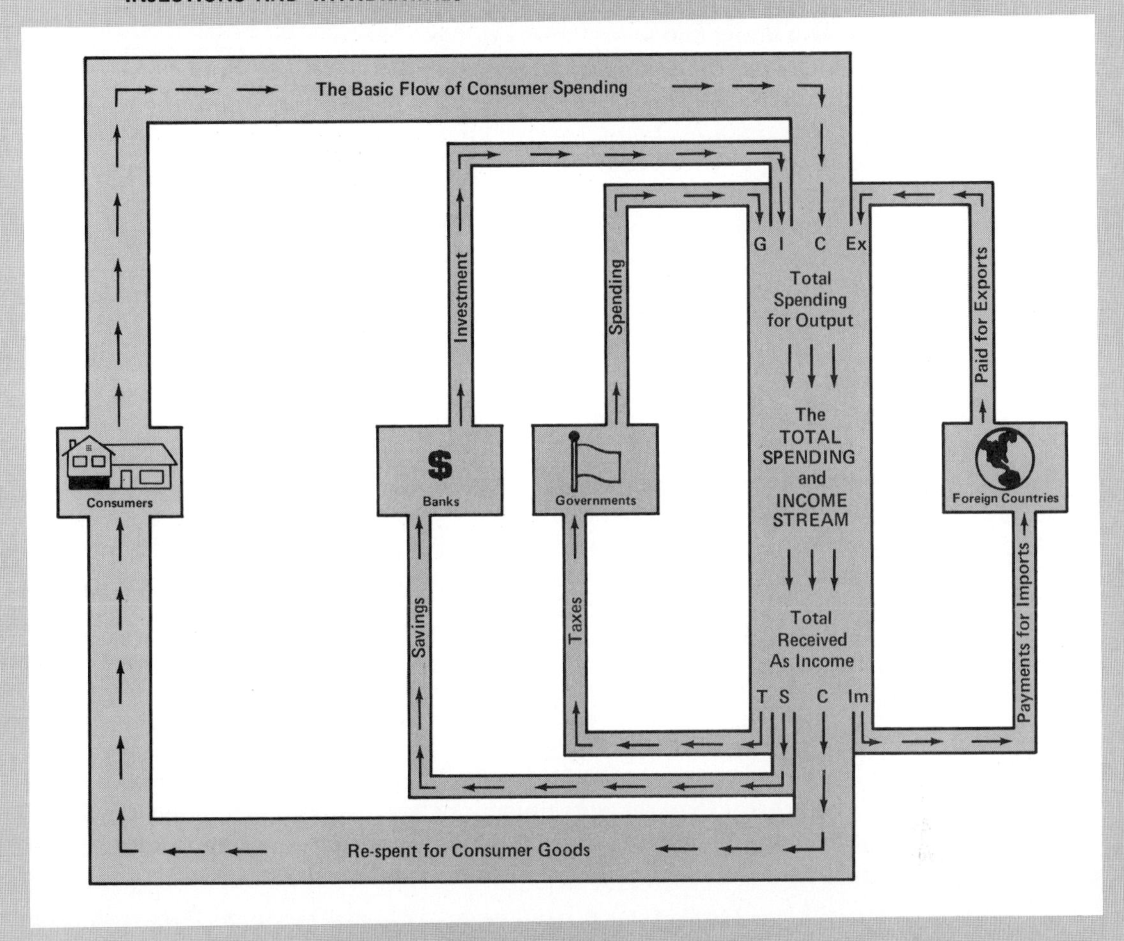


The Spending-Income Diagram Is Just A Picture Of The Expenditure And Income Equation.

The total spending stream is generated by the basic flow of consumer spending, plus spending injections from the investors, the government, and the foreign buyers. All these "spending sectors" are spending to buy outputs of goods and services.

The total spending stream flows as incomes (as wages, rent, interest and profits) to the owners of the factors (labor, land, and capital) and then is "disposed of" either as consumer spending, or for savings, taxes, or imports.

From this diagram you can see what would happen if any of the spending injections got larger, or smaller. The spending flow would get larger, or smaller. Right? Of course!

## Does "Total Spending" Equal "Total Receipts"? Or Not?

You know that every time a dollar is spent, somebody is receiving it. So total spending equals total receipts. Right? No doubt about it. If people are receiving dollars, somebody must be spending them. Therefore, the amounts being received must be exactly equal to the amounts being spent. That's so obvious that there could be no question about it. But now, the dilemma.

You also know that if all the income received is respent—that is, if *spending* is as great as *receipts* (and no greater)—then the total spending flow will stay the same size—no larger, or smaller. The macro-equilibrium of the economy will be maintained.

The only time the macro-equilibrium (the rate of spending, employment, output, and income) can possibly change is when the rate of spending changes. And the only way the rate of spending can possibly change is for total injections (I + G + Ex) to be smaller or greater than total withdrawals (S + T + Im).

Do you see what all this means? It means that for us to have an increase in spending (and output and income), total *spending* must be *greater* than total *receipts*! And you know that spending and receipts are (must be!) equal. What a dilemma!

It takes no genius to see that there's something wrong with the logic of this statement:

> Since every dollar *spent* must also be *received*, spending (C + I + G + Ex) must always equal receipts (C + S + T + Im). But whenever the national income flow is *increasing*, spending must be *greater* than receipts (injections greater than withdrawals); whenever the national income flow is decreasing, spending must be *less* than receipts (injections less than withdrawals).

This is a dilemma all right! What's the answer? The key to the answer is this: the words "spending" and "receipts" don't always mean the same thing. It's sort of a question of *which* spending equals *which* receipts! Let me explain.

## Which "Spending" Equals Which "Receipts"?

As I am standing here receiving income, *somebody* must be spending it. No question about that! People only receive what other people spend. So spending equals income. The amount of income I am receiving is equal to the amount of spending somebody else is doing. Of course. But then, as I am receiving the money, am I going to be *spending* the same amount I am *receiving*? Not necessarily!

The total income everybody is receiving right now obviously is equal to the total amount being spent (to generate income) right now. But the amount the people are receiving right now as income, is not necessarily the amount that they are going to turn right around and respend! So money being spent now, is money being received now. But money being received now doesn't necessarily equal money that's going to be respent right away!

The total amount of money people are receiving today, is equal to the total amount of money people are spending, today. No question about it. But look at it this way: We *can't* say that the total amount of money the people *receive today* is necessarily equal to the total amount of money they are going to *spend tomorrow*! Today's spending equals today's receipts. But today's receipts don't necessarily equal tomorrow's spending! That's the key to the dilemma. The amount of money people are spending at any moment is equal to the amount of money people are receiving at that moment. But the amount people received yesterday isn't necessarily the amount that they are going to respend today!

If the amount being spent today is equal to the amount being received today, that means that savings (and other withdrawals) today must be equal to investment (and other injections), today. It's obvious that C = C, today. (It's the same thing!) So then it's obvious, too, that all the other spending injections (I + G + Ex) must be exactly offset by withdrawals (S + T + Im). At any moment, spending must equal receipts, and since C = C, withdrawals must equal injections.

How is it ever possible for withdrawals and injections to become unequal? Only when we compare today's injections with tomorrow's withdrawals, or today's withdrawals with tomorrow's injections. That's when injections and withdrawals can be unequal, and that's the inequality that explains increases and decreases in the size of the national income.

### Each Economic Unit Is Locked Into The Macro-System

Let's go back to something we were talking about before, back in Chapter 9. Remember how the economic system is sort of a "total mechanism," and that all the "little wheels" are locked in, and turning together? Yet each little wheel has its own engine. Each "little wheel," by itself, can try to speed up or slow down. But just the same, all the little wheels are locked in together so they *must* all be turning together! Suppose most of them are *trying* to speed up. What then? The total mechanism will be speeding up. Right? Or if most of them are *trying* to slow down, then the whole mechanism will be slowing down. Do you begin to see what I'm getting at?

Let's forget about the government sector (G and T) and the foreign trade balance (Ex and Im) for a few minutes. Pretend that the only withdrawals are for savings (S), and the only injections are for investments (I). *Whenever investors are "trying to" spend less than the savers are trying to save, the economy will be slowing down. Whenever investors are "trying to" spend more than the savers are trying to save, the economy will be speeding up.* Nevertheless, at any moment, the *actual amount* of the investment injection is going to turn out to be exactly the same size as the *actual amount* of the savings withdrawal. This will be true whether the investors and the savers like it or not!

### Increased Investment Can Force An Increase In Savings

If the investors are trying to invest *more* than the savers are trying to save, the increased investment buying will *force* people to cut back on their consumer buying—that is, to save more. At any moment, if the investors are going to be buying more of the output, how is that possible, unless the consumers are buying *less* of the output? To say that people are *buying less consumer goods,* is just another way of saying that people are *saving more.*

As it actually turns out, S = I. But look. *Because* the investors are *trying* to spend more for investment than the savers want to save, there are shortages. Producers expand output. The total spending flow begins to increase. There is more employment, more output, more income.

As long as the investors are trying to spend more than the savers want to save (want to "leave in the market for the investors"), there will be some "forced saving." The people can't spend as much for consumer goods as they want to, so they are *forced* to save. Shortages exist, and output, employment, and incomes expand. The pace of the economy quickens.

### Output Cannot Expand Beyond Full Capacity

As long as the economy has "excess capacity" (unemployed workers and machines and things) the output can continue to expand. But when all the people and factories and machines are busy "producing at full capacity" (producing as much as they want to produce), then if the *desired rate* of spending by the investors ("planned I") continues to be greater than the desired rate of saving by the consumers ("planned S")—that is, if the investors are still trying to buy more than the savers want to leave in the markets for them to buy—then the shortages can't be overcome. Output cannot be increased. The economy is producing all it can produce. So what happens?

The consumer-goods buyers really want what they want. They offer to pay more for consumer goods. They try to "bid things back" from the investor-buyers. But the investor-buyers want what they want, too. They offer to pay more for the investment goods they want, to "bid things back" from the consumer-buyers. If there aren't enough goods for everybody, then somebody is going to come up short. The ones who will pay the most will get the goods. What's happening? Inflation? You bet!

Why are we having all these shortages? and inflation? Because the economy is producing as much as it can, and still, "planned I" is greater than "planned S." The investors are trying to buy more goods than the savers want to leave in the markets. The investors are trying to spend more than the savers want to save—trying to buy more of the output than the consumers want to release. Total "spending pressure" is too great. Prices go up.

### Increased Savings Can Force An Increase In Investment

Now that you know what happens when "planned I" is greater than "planned S," you can easily figure out what happens when "planned S" is greater than "planned I." But let's go through it anyway.

Suppose the people are trying to save more than the investors want to invest. Does this force the investors to invest more than they had planned to invest? It really does! Watch how it happens.

Suppose the consumers start leaving more things in the market than the investors want to buy. This means that some of the businesses can't sell all the output they have produced. So what happens to the unsold output? The businesses are forced to keep it. They have no choice. And is that an *investment*? You bet it is! It's "forced investment." "Unplanned investment." Here's an example.

### An Example Of Unplanned Investment

Suppose Chrysler Corporation discovers, one day, that its cars are piling up in the dealers' lots. The dealers are saying: "Don't send any more cars! We have already invested to the limit in our new car inventory! People just aren't spending enough for cars these days."

It seems that people are *saving* too much. So Chrysler Corporation finds itself with a lot of cars it can't sell. It has unwillingly *invested* a lot more than it had planned to, in its "stock of new car inventories."

See how, if the consumers decide to save more—that is, to leave more in the markets than businesses want to buy up—that *forces* businesses to increase their investments? Sure. And you know exactly what is going to happen next. The economy is going to slow down.

Chrysler Corporation isn't going to keep on producing more and more cars and pushing them into its new-car inventory. Of course not! It has invested more in its new-car inventory than it had planned to, already. So what does Chrysler Corporation do? It cuts back production. It puts some of its plants on a four-day work week. It may even close down some of its plants altogether. See what's happening? Output and employment and incomes are getting smaller. Total spending is going down. Why? Because people are trying to save more than the businesses want to invest; "planned S" is greater than "planned I."

All this time, while Chrysler Corporation is having its problems, what's going on at GM? and Ford? and American Motors? The same thing: unplanned investment in new car inventories. What's happening at RCA and Magnavox and GE and Westinghouse and Whirlpool and Maytag? The same thing. Unplanned investments in inventories of new TV's and refrigerators and stoves and washing machines and things. Why? Because consumers are saving too much for the level of planned investment; "planned S" is greater than "planned I."

It's obvious what's going to happen. All these producers are going to cut back production. The economy (total spending and production and output and employment and income) will slow down. When? Immediately!

### The Economy Begins To Adjust Immediately

As soon as the unplanned investments (in new product inventories) begin, the producers begin cutting back. When consumers save more (spend less), it doesn't take long for the companies to see that their sales are slipping!

The moment the consumers begin to save more, that's the moment the unplanned inventory investments begin. It doesn't take long after that for the companies to start cutting back their production. The economy slows down.

The response is just as fast when the opposite occurs—when the investors are trying to spend more than the savers want to save—when "planned I" is greater than "planned S." As soon as shortages begin to show up in the markets (that is, as soon as the businesses see their new-product inventories being completely sold out) they quickly order more. The producers begin to produce more. The economy can start speeding up very quickly whenever "planned I" is greater than "planned S"; it can start slowing down very quickly whenever the savers are trying to save more than the investors want to invest (whenever "planned S" is greater than "planned I").

### Planned Savings And Planned Investment: An Overview

At any moment, the amount the savers are *trying* to (want to) pull out of the spending stream may be *greater* or *less* than the amount the investors are trying to (want to) pour back in. So "planned S" may be *greater* or *less* than "planned I."

If "planned S" (withdrawal) is greater than "planned I" (injections), surplus products will be left in the market. Businesses will cut back. Total spending will get smaller. The economy will slow down. But if "planned I" (injections) is greater than "planned S" (withdrawals), total spending will get larger. There will be shortages of things. The economy will speed up. If the economy is already working at "full capacity," there will be inflation.

Does "S" always equal "I"? Sure. It always comes out that way. But if by "S" you mean "the amount people are *trying* to (want to) save" and if by "I" you mean "the amount businesses are *trying* to (want to) invest," then S doesn't equal I, unless the economy is in macro-equilibrium.

Whenever "planned I" is *greater* than "planned S," the spending flow will be increasing and the economy will be speeding up. Whenever "planned I" is less than "planned S," the spending flow will be decreasing and the economy will be slowing down. If anybody ever asks you if savings are equal to investment, tell them: "Yes. Or maybe no. It all depends on what you mean by 'savings,' and 'investment'."

### Planned Withdrawals And Planned Injections: An Overview

Now, just to be sure it's all tied together, let's bring all three kinds of "withdrawals" and "injections" back into the picture. You can be sure that at any moment the total flow of injections (I + G + Ex) into the income stream is exactly equal to the total flow of withdrawals (S + T + Im) out of the income stream. You can be *perfectly sure* of that. If it wasn't true, it would mean that someone was spending money and no one was receiving it—or that someone was receiving money and no one was spending it. And that, obviously, is too ridiculous to contemplate!

However, as people, businesses, governments, and foreigners are *receiving* their money, they are *planning* the amounts they are going to save and to spend, and the ways they are going to spend it. In this *planning process* there's no assurance at all that the planned injections (I + G + Ex) will be equal to the planned withdrawals (S + T + Im). If the planned injections are greater than the planned withdrawals, the size of the total spending flow will be increasing. The economy will be speeding up. But if the planned injections are less than the planned withdrawals, the size of the total spending flow will be decreasing. The economy will be slowing down.

### Macro-Equilibrium Depends On What The Spenders Are Trying To Do

Just think of all those little wheels, all locked in and turning together like the wheels inside a watch. They're all moving along together.

For one to speed up, all must speed up. There's no doubt about that. But if several of them are trying to go a little faster (*planned* injections greater than *planned* withdrawals), the whole mechanism will speed up. If several of them are trying to slow down (*planned* injections less than *planned* withdrawals), the whole mechanism will slow down. Macro-equilibrium works like that. It depends on what the various economic units—consumers and investors and the government and the foreigners—are *trying* (planning) to do. If they are trying to withdraw more than they are trying to put back into the spending stream, then the economy is going to slow down. But if they are trying to inject more, then the economy is going to speed up. Get it?

It is the *equality* or *inequality* between what the "injectors" (I, G, and Ex) and the "withdrawers" (S, T, and Im) are *trying* to do, that determines whether the economy will be speeding up or slowing down. Macro-equilibrium requires that the injection plans are exactly large enough to offset the withdrawal plans. Really, that's all there is to it. Now, how would you like to see it on a graph?

## THE KEYNESIAN "NATIONAL SPENDING AND INCOME" GRAPH

As you might have guessed, economists have a graph to show the relationship between spending and income. We call this a "Keynesian" graph after John Maynard Keynes, who once described and analyzed these spending and income flows in his famous book, *The General Theory of Employment, Interest and Money* (1936).

The Keynesian national spending and income graph shows total spending on the vertical axis (the y axis) and total income on the horizontal axis (the x axis). Think about that: spending on one axis; income on the other. What kind of a curve will that make on the graph? As spending goes up, what happens to income? As income goes up, what happens to spending? Both go up together? Sure! When one increases, the other increases, and *by exactly the same amount*. Of course!

The curve is going to be *positive* (sloping upward) and linear (a straight line). And more than that. If both the vertical (y) axis and the horizontal (x) axis are marked off on the same scale, the straight line will be rising at a 45-degree angle, exactly. You can see why, can't you? (When you increase by one unit on the spending (y) axis you must increase one unit on the income (x) axis. Get it?)

There are two of these graphs shown on the next two pages. You should study and practice drawing them now. Then, when we start to really use them in the next chapter, you will be ready. Now look at Figure 14-3 and you will see that it looks exactly the way you expected it to.

The Keynesian national spending and income graph doesn't tell you very much yet. It isn't supposed to. But you will soon see that it provides a very useful tool for looking at and analyzing of the total spending flow.

**Fig. 14-3 THE KEYNESIAN NATIONAL SPENDING AND INCOME GRAPH: THE TOTAL SPENT FOR OUTPUT ALWAYS EQUALS THE TOTAL RECEIVED AS INCOME**

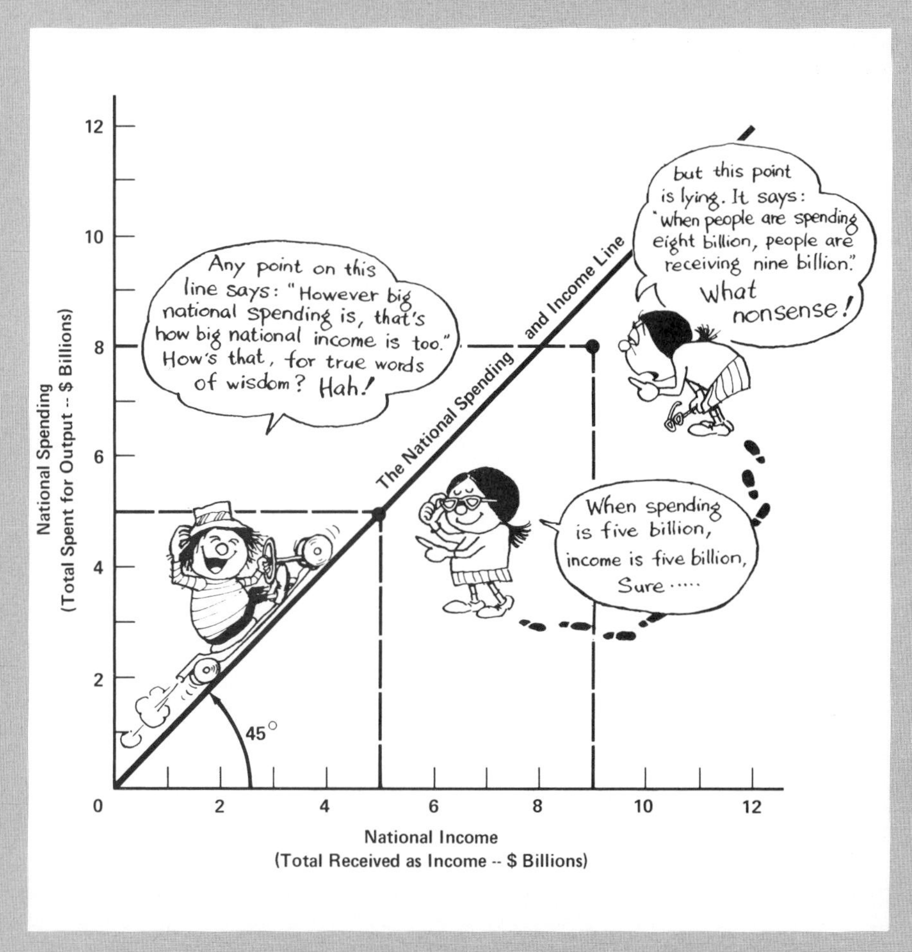


This Is Just Another Way To Show That Spending Equals Income.

This graph is a simple illustration of something you already know so well— that the total amount being spent for output is always exactly equal to the total amount being received as income. Of course! It's the same flow. (But you know that.)

This is such a simple graph and it illustrates such simple and obvious things that you may wonder why I would ask you to learn it. The reason soon will become obvious. Before long we are going to start breaking down the total spending flow into its component parts (C, I, G, and F). Then you will see just how helpful this graph can be.

**Fig. 14-4 THE KEYNESIAN NATIONAL SPENDING AND INCOME GRAPH: A VIEW OF THE EQUALITIES**

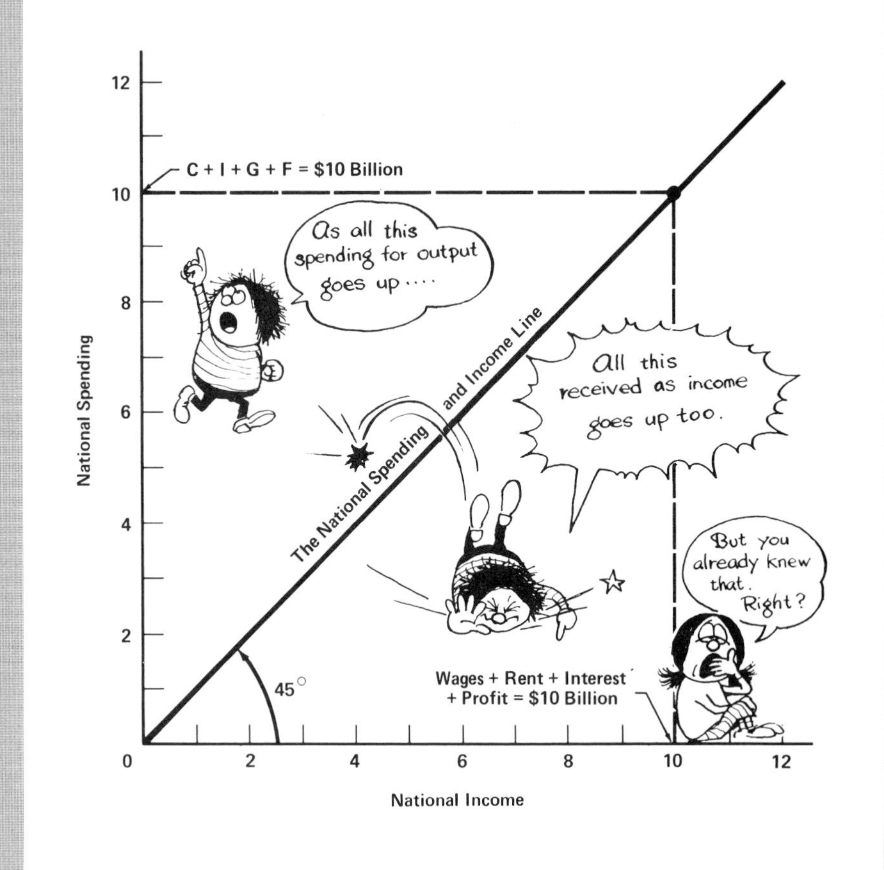


This Graph Ignores The Adjustments For Depreciation And Indirect Taxes.

At "Point A," National Spending (C + I + G + F) Is $10 Billion, And National Income Is $10 Billion. That's All You Can Read From This Graph.

If all spenders (C, I, G, and F) are *trying* to spend more than $10 billion, then spending and income will be moving up along the line. If all the spenders are *trying* to withdraw more (spend less), then spending and income will be moving down along the line. But there's no way you can look at this graph and tell which is happening.

Soon we will draw in more curves. Then you *can* answer these questions. But no matter what lines we draw in, you may be sure that national spending and national income will always be shown by a point somewhere along the "national spending and income" line. Obviously!

The Keynesian national spending and income graph can be a very helpful tool for "national income analysis." You will find out about that in the next chapter.

## SUMMARY

Now that you have completed another chapter on macroeconomics, you should begin to feel sort of comfortable with the idea of "spending flows and income flows." Much of what we have been doing in this chapter is just good common sense. Now that you think about it, isn't it sort of obvious that as people are out spending money, they are adding to total spending and to the total incomes of other people? Sure. And if all of us suddenly decide to take our paychecks and go home and hide them in the bureau drawer, pretty soon the economy is going to be very depressed. Right? All of us might wind up losing our jobs!

The equalities between spending and receipts, and between injections and withdrawals are all very logical. They really do work that way, too. I hope you have a "comfortable feel" about all these *spending flow* concepts and principles, now. That's important, because unless you do, the next chapter may be confusing. If you know this chapter well, the next one will be a real breeze! See you there.

---

## REVIEW EXERCISES

**MAJOR CONCEPTS, PRINCIPLES, TERMS (Try to write a paragraph explaining each of these.)**

the concept of macro-equilibrium
the "four sector" breakdown
the "quantity and velocity of money" breakdown
the Expenditure and Income Equation
how increased investment can force increased saving
how increased saving can force increased investment

**OTHER CONCEPTS AND TERMS (Try to write a sentence or phrase explaining what each one means.)**

velocity of circulation
full capacity
planned investment
planned savings
planned injections
planned withdrawals
unplanned investment

## DIAGRAMS AND GRAPHS (Try to draw, label, and explain each.)

A Diagram of the Total Spending-Income Stream, Showing Injections and Withdrawals
The Keynesian National Spending and Income Graph

## QUESTIONS (Try to write out answers, or jot down the highlights.)

1. What does it mean to say that macro-equilibrium is sort of unstable, like a "perched boulder"?
2. If a person could really understand and predict the *planned injections into* and the *planned withdrawals from* the spending-income stream, then he would be able to tell with certainty whether the economy was going to speed up or slow down. Explain.
3. Total spending must equal total receipts because every dollar spent must have a receiver. But total receipts are not necessarily equal to total spending. The only time national spending and income can change is when the two are *not* equal. Can you explain this apparently illogical statement?
4. Suppose "consumer spending" at your macramé belt concession stand at Seashore State Park dropped off, and you got caught with a lot of money tied up in belt-making materials and in finished belts you couldn't sell. Is that "unplanned investment"? Do you think you would slow down your future investment spending? maybe all the way to zero? If this sort of thing was happening all over the economy, what do you suppose the result would be? Explain.

# 15 NATIONAL INCOME ANALYSIS: THE KEYNESIAN AND MONETARY APPROACHES

*How to use Keynesian and monetary techniques to analyze the spending and income flows.*

You already know what this chapter is going to be about. You're going to learn how to use the "Keynesian national spending and income graph" to find the macro-equilibrium of the economy. I told you in the last chapter that we would do that in this chapter. And there's another thing this chapter is going to do. Remember about the other important way of looking at the total spending flow? Remember about "the size of the money supply" and "the velocity at which the dollars are being spent"? In this chapter you will learn more about that, too.

First we will analyze the spending flow in a new way, using the Keynesian national spending and income graph. This approach will emphasize the four "spending sectors": C + I + G + F. Next, we will look at the spending and income flow the other way: as the quantity of money (M) times the velocity at which each dollar is being spent (V). Now, the Keynesian analysis.

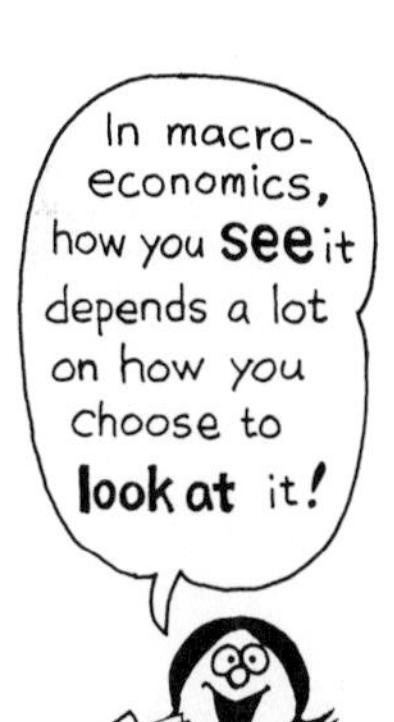


## KEYNESIAN NATIONAL INCOME ANALYSIS

What determines the total size of the national income? Total spending, of course! What determines total spending? Lots of things. Consumers spend for one set of reasons, investors for another, and governments for another.

If we could just know for sure why the spenders in each sector behave as they do, we could understand the whole thing. But we don't. Too bad. Still, we do know quite a lot about it. Let's start with the *basic flow* in the spending stream, and build up from there. What's the basic flow? You already know. It's "spending for consumer goods" (C).

## The Propensity To Consume

What determines the rate of total consumer spending (C) in the nation? It's influenced by several things. One of the things is the level of national income.

You know that if people are receiving high incomes, they are likely to be spending more for consumer goods — more than if their incomes were low. That's just plain common sense. But also, there are statistics to prove it. As people's incomes get larger they really do spend more for consumer goods. As their incomes get smaller they really do spend less for consumer goods. Usually people with more income buy more things. I'm sure that doesn't surprise anybody!

We know something else about how consumer spending relates to income, too. We know that as a person's income gets higher, he is likely to save a larger share — a larger percentage of it. People with very low incomes usually don't save anything. They spend all the money they earn and maybe even go into debt to pay their bills.

How about you? If you are like most college students, you're spending more than you're receiving in income. How is that possible? Well, maybe you worked awhile and saved some money and now you are living off your savings. Or maybe your parents are helping you out. Or maybe you borrowed money so you can make it while you're going to college. But one way or another, almost all college students are consuming more than they are receiving as income.

If the income of the whole nation was very low, most people probably would be trying to spend more for consumer goods than they were receiving as income. That makes sense, doesn't it? Then if the national income moved higher and higher, most people would spend more and more for consumer goods. Right? Of course!

Suppose your income (disposable personal income) was only $5 a week. Would you spend it all for consumer goods? and save nothing? I'll bet so. I'll bet you'd be trying to borrow, too, so you could spend more. Most people would.

## People With High Incomes Usually Save More

Suppose your income was $1,000 a week. (That might be a little difficult for you to imagine, but try to imagine it anyway, just for a minute.) What would you be doing with all that money? Most people could live pretty high on $1,000 a week! Most people would save some of it. But pretend that somehow you managed to spend your whole income ($1,000 a week) for consumption. No savings.

Then suppose your income suddenly *doubled*, to $2,000 a week. What then? Would you double your rate of consumer spending? Would you spend the whole $2,000 for consumer goods and services? I doubt it. But suppose you did. Then suppose your income doubled again. Do you suppose your spending for consumer goods would double *again*? And suppose it happened again? and again? How far can this thing go?

Do you get the point? The idea is that the higher a person's income is, the more likely it is that he will save some. That isn't too hard to believe, is it? As a rule, a person with a large income will save a larger proportion (percentage) of his income than will a person with a low income.

What happens when a person gets an *increase* in income? What proportion (fraction) of the increase will he spend to buy consumer goods? Suppose he's almost starving. He may spend it all! Suppose he's very wealthy. He may save it all.

## The Propensity To Consume Is An "If . . . Then" Concept

What we are talking about is something economists call the propensity to consume. "Propensity to consume" is a sort of "what would you do if . . ." kind of thing. It talks about: "If your income was such and such, what fraction of it would you be spending for consumer goods?"

Keynes suggested that if the national income was low, then the "propensity to consume" (the proportion or fraction of it that people would spend for consumer goods) would be high. If the national income was high, then the propensity to consume (the fraction of it that people would spend for consumer goods) would not be so high. See how "the propensity to consume" is a kind of "if . . . then" concept? It doesn't tell us what fraction of the national income *is* being spent for consumer goods. It tells us what fraction *would be* spent for consumer goods, at all the *various levels* of national income that *might* (sometime) exist.

For each level of national income you might choose, there would be an average propensity to consume—that is, some fraction (percentage) of the total income would be spent for consumer goods. For example, if national income was $10 billion, and total consumer spending was $8 billion, that would mean that out of every ten dollars being received as income, eight dollars were being respent for consumer goods. Then the average propensity to consume (APC) would be 8/10ths (or 4/5ths). The average propensity to save (APS) would be 2/10ths (or 1/5th). Out of every five dollars of income being received, one dollar would be saved—that is, not spent for consumer goods.

Let's take another example. If the national income was $8 billion, and consumer spending was $7 billion, then the APC would be 7/8ths. The average propensity to save would be 1/8th. Get it?

## The APC Determines The Size Of The Basic Spending Flow

What does all this have to do with total spending and total output and macro-equilibrium, and all that? Quite a lot! Just think. Consumer spending is the basic flow in the income stream. The average propensity to consume (APC) tells us, for any level of national income, just how big that *basic flow* would be!

Let's take an example. Suppose, at a national income (NI) of $10 billion, consumer spending would be $8 billion (APC would be 8/10ths). Then the only way the NI could *ever* get up to $10 billion would be for $2 billion of spending to come from someplace to *add* to the "basic consumer spending flow" of $8 billion. See how important the APC is?

Suppose I told you that if NI was $5 billion, then APC would be 5/5ths, or "100 percent." That means that if NI was $5 billion, consumer spending would be $5 billion. Consumer spending would support the entire national output and national income, right by itself!

With APC of 100%, the average propensity to save (APS) would be zero. There would be no "savings withdrawals." So if the APC is 100 percent when the national income is $5 billion, then NI can *never* drop below $5 billion! How about that? (That is, it can't unless taxes, or a negative foreign trade balance are draining off some of the basic consumer spending flow. But we don't have to get into that. So let's don't.)

See the importance of the average propensity to consume? It has a lot to do with setting the macro-equilibrium level of spending for the economy. At any level of national income you might choose, the APC tells how much of the income received will be *automatically* returned to the total spending flow (as consumer spending). So it tells you (implicitly) how much "spending injection" (investment) would be required to bring national spending (and income) up to any size (level) you might choose!

These relationships can be shown very clearly on the Keynesian national spending and income graph—the one you learned in the last chapter. We'll get into that in just a minute. But first we need to talk briefly about one other concept: The marginal propensity to consume.

## The Marginal Propensity To Consume

Suppose national income went up, just a little. Do you suppose the basic flow of spending for consumer goods would go up, too? Of course.

The "marginal propensity to consume" (MPC) is concerned with the question of how much consumer spending (C) would go up if national income (NI) went up. *The MPC tells you what fraction of an increase in income, would be respent for consumer goods.*

Suppose NI increases by $4 billion, and C increases by $3 billion. Then MPC would be 3/4ths. So you know, implicitly, that the marginal propensity to save (MPS) would be 1/4th. Right? If 3/4ths of the increase is respent for consumer goods, then the other 1/4th *must* be saved.

Now, to get some practice with this new idea, look at the marginal propensity to consumer (MPC) from the point of view of an individual. You. Suppose you were receiving an income of $1,000 a week. Not bad!

And suppose your average propensity to consume (APC) was 8/10ths. That means you would be spending $800 out of your $1,000 income for consumer goods, and saving $200. Then suppose your income increases by $100, to $1,100 a week. How much of the *extra* $100 will you spend for consumer goods? Maybe none of it. Maybe $800 a week is all you want to spend for consumer goods. See how MPC is likely to be low, if your income is high?

But let's guess that you would spend an extra $50 a week for consumer goods. That means you're spending half of the extra $100, and saving the other half. So your marginal propensity to consume (MPC) is 1/2, and your marginal propensity to save (MPS) is 1/2. See how simple this "marginal propensity" concept is?

### The MPC Causes A Spending Increase To Multiply

What does your MPC mean from the point of view of the total spending-income stream? Quite a lot! If your marginal propensity to consume is high, then when you receive an increase in income, you will respend a lot of the increase to buy consumer goods. That way, you will pour most of your new income right back into the basic spending-income flow. If your MPC is high, it means that when you receive an increase in income, you "pass it along." The basic spending-income flow expands. If everyone has a high MPC, any increase in income will have a high "multiplying effect," so the initial increase will just keep on pushing up total spending, more and more.

Suppose that when you receive an increase in income, you decide to save most of it. That means your marginal propensity to consume is low. You won't pass along much of it. You don't put much of it back into the basic spending flow. Instead, you withhold it as savings. Can you see the difference it could make if everyone had a high MPC? or a low MPC?

A person with a low income is likely to have a high MPC. At high levels of income, marginal propensity to consume (MPC) is likely to be low. *The higher your income, the higher your marginal propensity to save (MPS) is likely to be.*

So far we have covered a lot of ground in this chapter. Maybe we should stop right now and see what all this looks like on the Keynesian national spending and income graph. There are four of these graphs coming up, all four showing the same picture in slightly different ways. The graphs show exactly the same things we've been talking about all along in this chapter.

You probably could understand all of this from only one graph, but I think it's better to show it in small, comfortable steps. So what's coming up now is really only one graph, looked at four times in slightly different ways. Take your time and study each one and learn it well.

Fig. 15-1 **THE CONSUMPTION FUNCTION: HOW LARGE WE EXPECT THE BASIC FLOW OF CONSUMER SPENDING TO BE, AT DIFFERENT (ASSUMED) LEVELS OF NATIONAL INCOME**

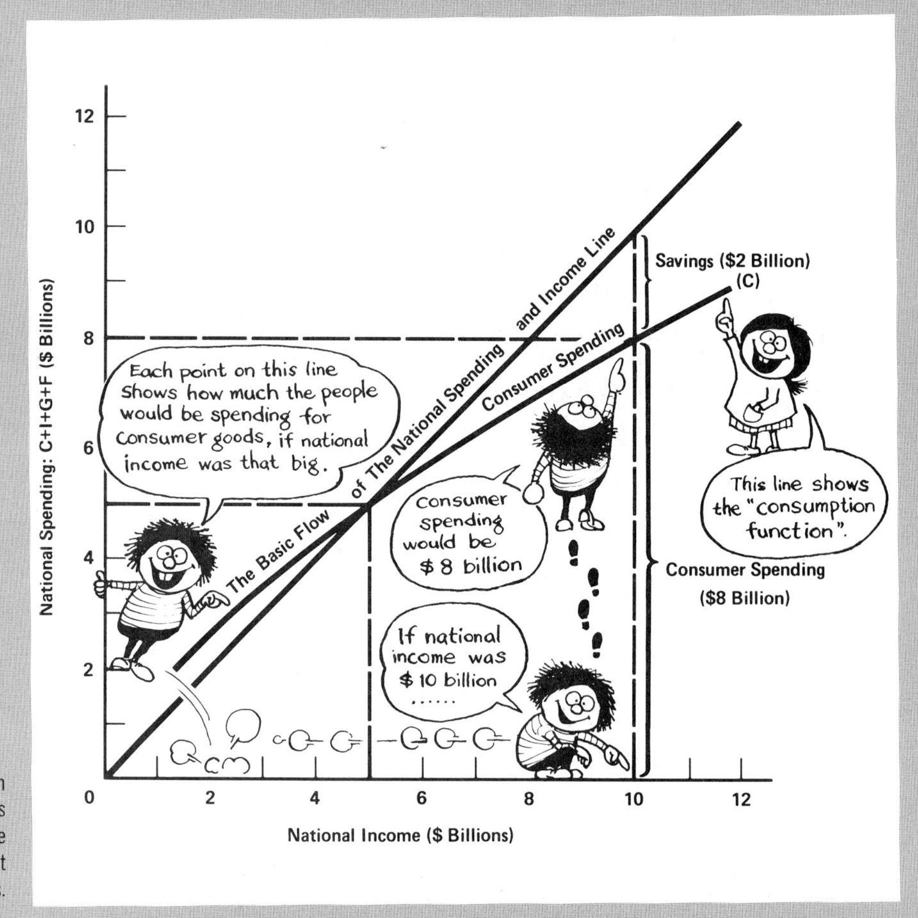


The "Consumption Function" Line Shows What Fraction Of The Income Will Be Spent For Consumer Goods.

The basic flow of consumer spending would be larger at the higher levels of national income. But also, notice how much more the people would save if the national income was high!

The higher the national income, the smaller the proportion (fraction) that would be spent for consumer goods. At very low levels of national income, the people would be trying to spend more than they were receiving. At high levels of national income, the people would spend less than they were receiving. They would save some.

Just look how high the savings withdrawal would be, if the National Income was $12 billion!

**Fig. 15-2 THE CONSUMPTION FUNCTION: THE AVERAGE PROPENSITY TO CONSUME (APC) WOULD BE SMALLER AT HIGHER LEVELS OF NATIONAL INCOME**

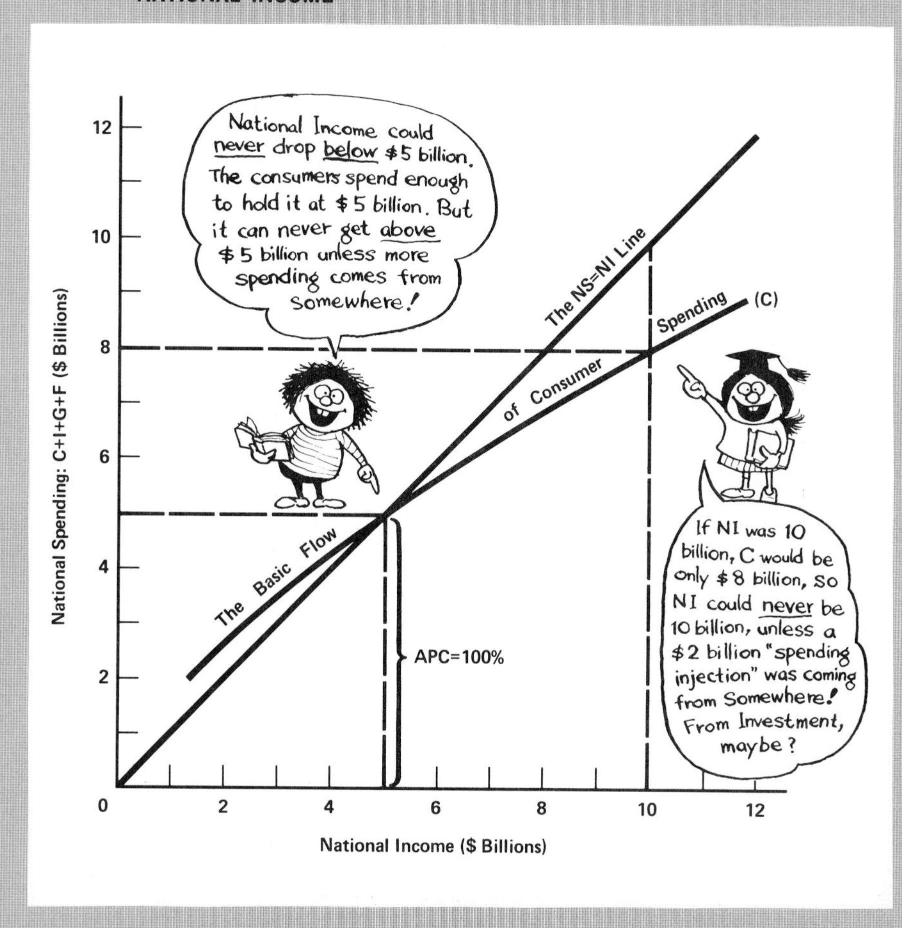


The Higher The National Income, The Smaller The Fraction Spent On Consumer Goods.

You can see that as long as the "consumption function" line stays where it is, the only way national income can ever get above $5 billion is for some *other* source of spending to be added to the income stream.

If the consumers are spending only $5 billion, and if they are the only ones spending, then total spending will be $5 billion. And that's all. So national income will be $5 billion. And that's all!

Fig. 15-3 **THE CONSUMPTION FUNCTION AND THE SAVINGS FUNCTION: TWO DIFFERENT WAYS OF LOOKING AT THE SAME THING**

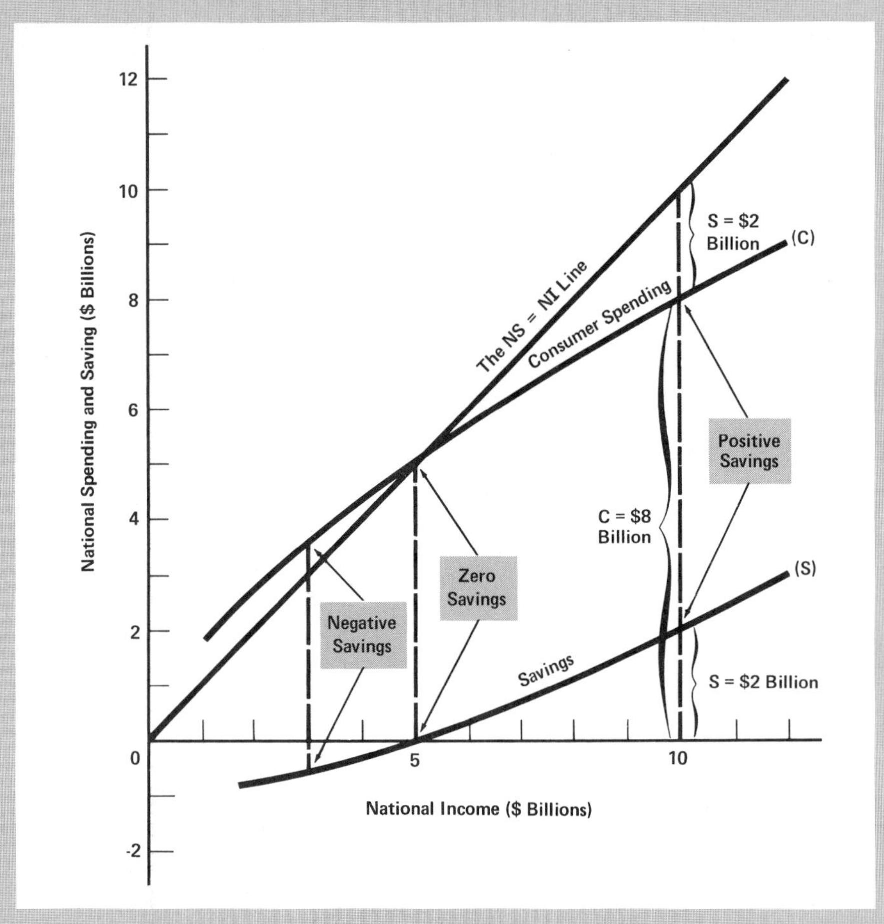


As Income Gets Higher, The "Savings Gap" Gets Larger.

The relationship between the "savings function" (S) and the "consumption function" (C) is obvious. The two together must always add up to "100 percent" of national income. For example, if C = 80% (8/10ths), then S = 20% (2/10ths). If C = 100% (10/10ths), then S = 0.

What about "negative savings"? If S = – 10% (people are borrowing, and spending more than they are earning) then C must equal 110% of national income—which means: "Total *consumer* spending is 10% bigger than total spending"—which you know has got to be nonsense!

*Some* people can spend more than their total incomes, but *all* people can't. If they tried, NI would be forced up, by the amount of the increased spending. Of course!

**Fig. 15-4 THE MARGINAL PROPENSITY TO CONSUME AND THE AVERAGE PROPENSITY TO CONSUME**

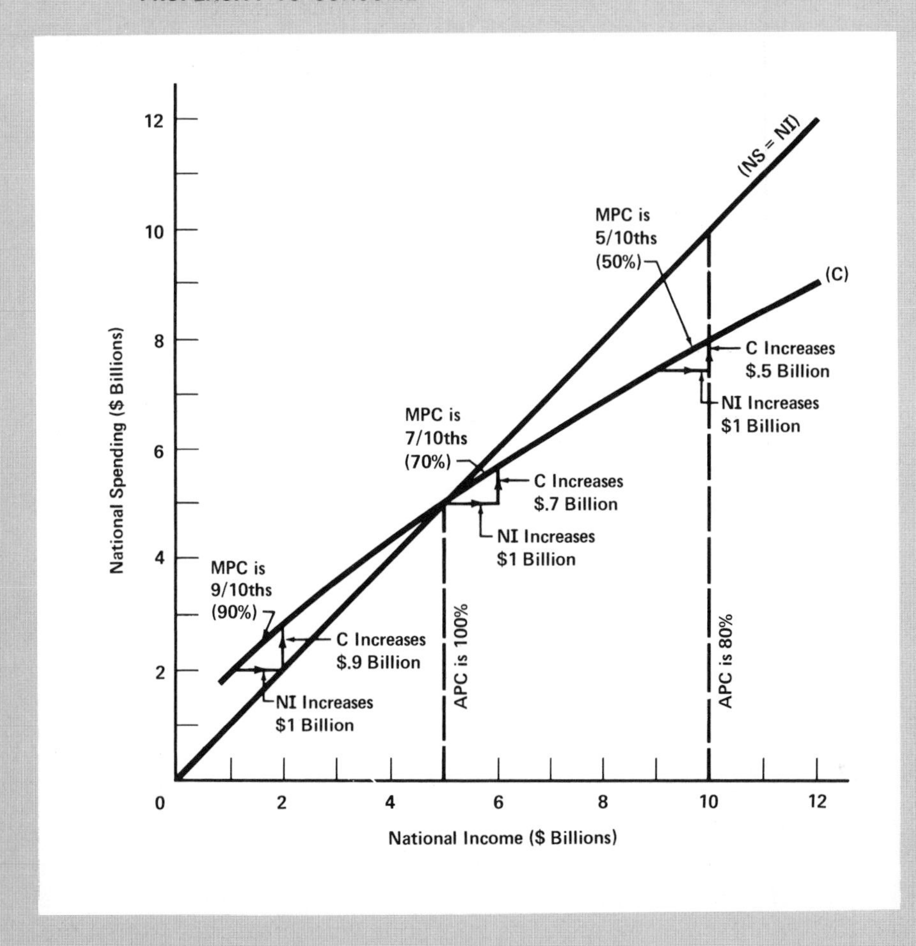


A Prosperous Nation Requires A Lot Of Investment Spending.

For any level of national income, the *marginal* propensity to consume is not as high as the *average* propensity to consume. So no matter what the existing consumption rate (APC) might be, an *increase* in NI is going to bring a lower consumption rate (a higher savings rate) than before.

Therefore: The higher the national income, the higher the *savings* rate; and therefore, the higher will be the *investment* rate, required to sustain the high level of national income.

So therefore: a very prosperous nation must depend on investment spending (reinjections to offset the savings withdrawals) to sustain its prosperity. How about that!

Now that you have had time to study and learn the graphs, let's take some of these things a little farther. We will still be talking about the same basic situation we have been talking about all along. It's the same set of conditions pictured in the four graphs you just learned.

### The "Full Employment" Level Of National Income

Suppose, in this little make-believe nation we're talking about, we need to have a national income of $10 billion. Let's say that a $10 billion NI would bring us full employment, an acceptable rate of economic growth, and no problems of shortages or inflation. But if NI was $10 billion, the average propensity to consume (APC) would be only 8/10ths ($8 billion). Remember?

If NI was $10 billion, the people who received the $10 billion would only be spending $8 billion of it for consumer goods. If the consumers are the only ones doing any spending, the national income can't be $10 billion. That's obvious. Unless $2 billion of spending is coming from somewhere else—from investment spending, or government spending, or from a positive foreign trade balance—NI can *never* get up to $10 billion.

Now, for the next step, there are two questions which need to be answered:

> *Question one:* In the little make-believe nation we're talking about, if the only people doing any spending were the consumers, then the national income would be just $5 billion. Right?
>
> If NI was $5 billion, APC would be 100% ($5 billion). All the national income received would be poured back into the basic spending flow. National spending and national income would be in equilibrium at $5 billion. That would continue to be the macro-equilibrium level of NI until some new spending injections were introduced from somewhere, or until something happened to cause the consumption function (the APC) to change. That's all very obvious. Nothing to worry about yet. So, on to the next question.
>
> *Question two:* Suppose our little nation was dragging along with a national income of only $5 billion, and then the businesses decided to start spending at the rate of $2 billion a year. That would push national spending and national income up to $7 billion. Right? No. Wrong. Let me explain why.

### Investment Of $2 Billion Increases NI By $5 Billion

Remember how much investment spending it would take to reach the desired level of NI ($10 billion)? Only $2 billion, right? But $5 billion (the

NI with *no* investment injection), plus $2 billion (the investment injection), only adds up to a total spending flow of $7 billion. Where does the other $3 billion of spending come from?

[$5 billion (C) + $2 billion (I) + $3 billion (?) = $10 billion (NI)] I think you know the answer already. The extra $3 billion comes from *more consumer spending*. It comes from the *marginal propensity to consume* (MPC). Remember? There is one level of NI (and only one level) at which consumers would spend exactly $5 billion. What level of NI? NI = $5 billion. Look at the graphs again, and you'll see.

At any level of NI *higher* than $5 billion, consumer spending will be *higher* than $5 billion. As incomes rise, people spend more for consumer goods. Of course! So it's nonsense to talk of NI = $7 billion and consumer spending of only $5 billion. That just couldn't be.

If the NI was $7 billion, the people would *not* be respending *all* of their incomes on consumer goods (as they would be if the NI was $5 billion), but they certainly would be spending more than $5 billion! Let's say they would be respending $6.5 billion and saving $.5 billion. (That's what it looks like on the graphs.) So, can we say that the new level of national spending and income will be:

[$6.5 billion (C) + $2 billion (I) = $8.5 billion (NI)]?

No, we can't say that, either. Why not? Because if NI was up to $8.5 billion, spending for consumer goods would be up to *more* than $6.5 billion. See how NI and C seem to be working back and forth on each other, pushing each other up, higher and higher? If NI was $8.5 billion, C would be up to about $7.4 billion. The graphs show that. So can we say that the new equilibrium level of national spending and income will be:

[$7.4 billion (C) + $2 billion (I) = $9.4 billion (NI)]?

By now, you already know the answer. No, we can't say this either. It just can't be true! If NI was $9.4 billion, then C would not be $7.4 billion any more. Therefore $9.4 billion can't be the new level of national spending and national income. Get it? Got it!

When the rate of investment spending is $2 billion a year, what will be—what *must* be—the macro-equilibrium level of national spending and national income? It *must* be that level which will induce the people to want to *save* $2 billion. With *investment spending* of $2 billion, the national income automatically will move up to the level where the *savings withdrawals* become $2 billion. The national income must expand to that rate of flow at which the savings withdrawals are equal to the investment injections. Remember? Of course! So here it is:

[$8 billion (C) + $2 billion (I) = $10 billion (NI)].

How do we know that this is the equilibrium level? Because we know (that is, we assumed in the beginning) that APC would be 8/10ths ($8 billion) when NI was $10 billion.

(*Note:* We "assumed" (decided to use) these round numbers so that it would all come out nice and even. Whenever you're using numbers to illustrate things, on graphs or anywhere, be careful to "assume" some round numbers that are easy to work with. It's so much easier, that way. But more important, it lets you concentrate on the important things—on the *principles and concepts*, instead of the numbers.)

### The Macro-Equilibrium Level Of National Income

It's all as simple as 1, 2, 3.

1. If we know what the average propensity to consume would be at each "might exist" level of national income, and
2. if we know how much "injection spending" (investment) there is going to be, then
3. we can see right off what the macro-equilibrium level of national income must be—where withdrawals (savings) are equal to injections (investments).

Isn't that neat? You can see it on the Keynesian national spending and income graph, too.

In the Keynesian graphs you have already studied (Figures 15-1, 2, 3, and 4) the only equilibrium level of National Spending and Income you can see is NI = $5 billion. If consumer spending is the *only* kind of spending in this nation, then $5 billion is just where NI is going to be. That's where it *must* be! If the consumers are the only ones spending, and if $5 billion is all they're spending, then $5 billion is all there is.

We all know that the consumers really aren't the only ones spending. So we know that the macro-equilibrium for our little nation will come out at a higher national income than just $5 billion. But how much more? That depends on the size of the *injections*. Let's introduce some investment spending into our graphs and see what happens to national income. Figures 15-5 and 15-6 tell the story. Study them now.

Thinking of the Keynesian graph, how would some other kinds of spending injections influence the macro-equilibrium? And how would it look on the graph? Just exactly like the investment injection looks, in Figures 15-5 and 15-6. Exactly. Government spending (G), or a positive foreign trade balance (F) would shift the curve upward, just as the "investment injection" did. Then NI would expand to a new level of macro-equilibrium.

What about withdrawals? The savings withdrawal is already shown in the graph. But how would taxes, or a negative foreign trade balance influence the picture? Just as you would guess. These withdrawals would pull down the basic consumer spending flow. The "consumption function" curve would shift downward. The withdrawals would cause the macro-equilibrium level of NI to be smaller. That's obvious isn't it?

**Fig. 15-5 THE EFFECT OF AN INJECTION OF INVESTMENT SPENDING (THE MULTIPLIER EFFECT)**

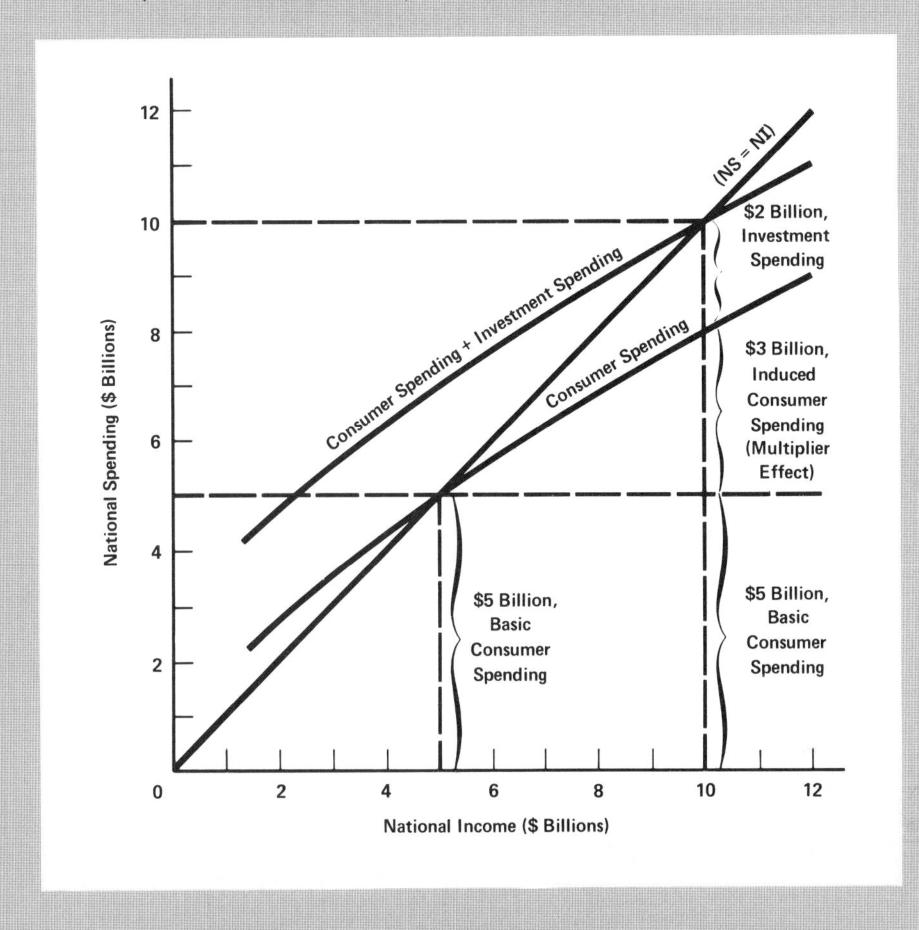


Investment Spending Has A "Multiplier Effect" On National Income.

Investment spending of $2 billion brings an increase in national spending and income of $5 billion. Why? Because of the income multiplier!

National spending and income are increased by 1 1/2 times the amount of the investment spending. How? By causing (inducing) consumer spending to increase by $3 billion.

This multiple effect of an increase in investment spending (caused by the induced consumer spending) is called "the multiplier." In this case the multiplier is 2.5. [$2 billion (I), times 2.5 (the multiplier) equals $5 billion. This $5 billion is the *total* increase in spending, resulting from the initial increase of only $2 billion. The initial increase, times the multiplier, always gives the total increase in spending and income.]

Can you see that the size of the multiplier depends on the size of the marginal propensity to consume (MPC)? The MPC is "the respending effect" of an increase in income. The more respent, the steeper would be the "consumption function" line in the graph, and the higher would be the multiplier.

**Fig. 15-6 MORE INVESTMENT SPENDING INDUCES MORE SAVINGS; MACRO-EQUILIBRIUM IS WHERE S = I**

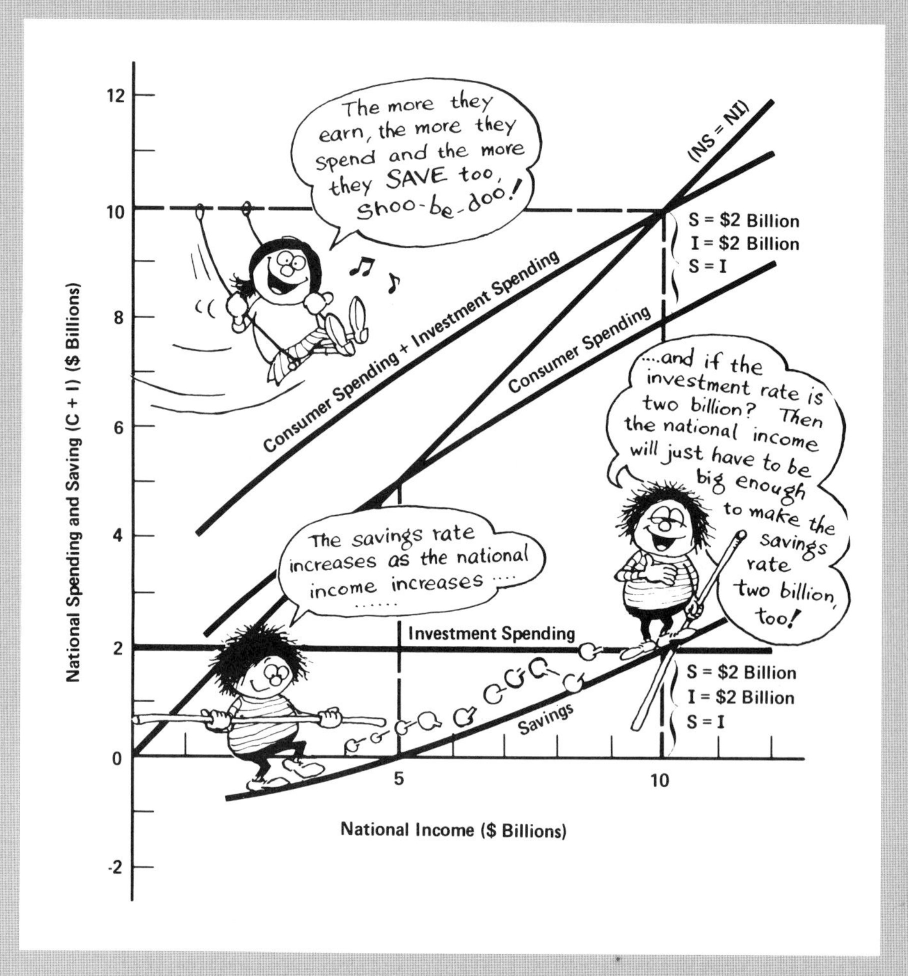


More Investment Spending Causes National Income To Increase, Which Causes Savings To Increase.

National income must continue to increase until the size of the flow of the savings withdrawals is great enough to equal and offset the size of the flow of the investment injections.

When investment spending is zero, then the savings rate also must be zero. That means APC must be 100%. All income received is respent for consumer goods. The economy is in macro-equilibrium at a very low level of employment, output, and income (NI = $5 billion).

With the injection of $2 billion of investment spending, NI must rise, not just by $2 billion, but by enough to induce the income receivers to withdraw from the consumer spending flow (that is, to *save*) $2 billion (to offset the investment re-injection). National income must rise to that level at which the income receivers will be saving $2 billion. (NI = $10 billion.)

What about the "consumption function" curve itself? Is that a stable, dependable thing that just stays where it is until some injections or withdrawals push it up, or pull it down? Not really. But it doesn't usually jump around too much.

During times of high expectations and optimism, we would expect the consumption function curve to shift upward. There would be more consumer spending at any level of NI you might choose. But during gloomy times, or when everyone is deep in debt and trying to get things paid off, we would expect the average propensity to consume (APC) to get smaller. The curve would shift downward.

### The Keynesian Graph Provides A Useful Approach

It would be good, just for practice, if you would stop for a few minutes now and draw a few of the Keynesian graphs, and move the consumption function curve around. If you do, you will notice that anytime you move it, there will be a multiplier effect. (The "induced respending effect" will see to that!)

By now, you're getting a real feel for this Keynesian national spending and income graph, right? See what a helpful framework it gives for illustrating and analyzing the effects of all these different kinds of withdrawals and injections? As soon as you're sure you have it down pat, move on into the next section and learn about the quantity and velocity of money — but first, be sure!

## A MONETARY APPROACH TO MACROECONOMICS

All this time we have been looking at total spending as the sum of its parts. We have defined four basic parts, or sectors, or spending sources: consumers (C), businesses (I), government (G), and foreigners (F). Now we are going to do something entirely different. We are going to look at total spending as "a big mass of dollars flowing around and around."

### The Spending Flow Is A Mass Of Moving Dollars

Total spending can be looked at as "the total amount of money in existence in the nation" and "the speed at which it is flowing." You can see that if the total money supply is a billion dollars, and if each dollar is spent to buy some output once each month, then the total size of the spending flow will be one billion dollars each month. Right? Or $12 billion a year. Nothing very mysterious about that is there?

Economists use the symbol "M" to mean "the supply of money in existence" (the money stock), and the symbol "V" to mean "the velocity of circulation of money" (the number of times each dollar is being spent). It is obvious that the quantity of money in existence (M) times the velocity at which "the average dollar" is being spent (V) equals total spending. So we can say that M times V (or just "MV") equals total spending.

If each dollar is turning over one time each year, then total spending during the year is equal to the size of the money supply. The velocity of circulation is one. If each dollar is spent three times during the year, then the velocity is three, and the total spending flow is three times as large as the total money stock. All this is very easy to see, isn't it?

Remember that in this discussion we are not going to count all "spending velocity." We are only going to count the velocity of spending for *new* output—that is, only the "output velocity," or "income velocity" will be counted. We will ignore spending and respending for the same *old* goods. Economists frequently use the "MV" approach and consider *all* transactions in figuring "velocity of circulation." But for our purposes it's better to count only the "output and income" velocity.

Why should anyone ever look at the total spending flow as a big mass of dollars flowing around? What is the purpose of this "MV" way of looking at the spending-income stream? It's just that when we look at it this way we can see things that otherwise we couldn't see. The "MV" breakdown lets us concentrate on the *size* of the money supply (the money stock) to see how the *quantity of money* (and changes in the quantity of money) might influence (and be influenced by) the rate of spending in the economy.

### What Determines The Spending Velocity Of Money?

What determines "V"? What influences the number of times each dollar will be spent during a week, month, or year? Several things. Some of the things are sort of built into the system. If people get paid once a week, then each dollar will turn over faster than if people get paid once a month. If everyone in the economy was on a weekly pay period, the money stock would support a higher rate of total spending than if everyone was on a monthly pay period. Each dollar would be changing hands more often, "doing more work"—buying more things.

If a person earns $100 a week, but only gets paid every four weeks, then it takes four-hundred dollars of "money stock" to pay him for his four weeks of work. But if he earns the same wage and is paid at the end of *each* week, then only $100 of money stock is needed to pay him. When he gets paid $100 at the end of the first week, he spends it. Then he could get paid the same $100 at the end of the second week, then spend it, then get it back again, spend it again, and so on.

Of course he wouldn't get back the *same* dollars, week after week. But I think you get the idea. It's simply that the more frequently people get paid, the more work each dollar will do. If people get their money once a week, they usually spend it by the end of the week. If they get it once a month, they usually spend it by the end of the month. On a monthly payments system, some of the dollars sit idle all month long, until the last day of the month.

If the economy was on a *daily* payments system, each person would get paid and then spend his day's wages, every day. Then the velocity of circulation would really be high! A small money supply could support a great amount of spending.

### Velocity Is Fairly Stable

At any moment the velocity of circulation is fairly stable, not likely to change very much. It is sort of like the consumption function we were talking about awhile ago. It has a tendency to stay where it is, but there are many things which can cause it to change—to move up or down, in response to the people's mood or expectations. Still, it isn't unrealistic to think of "V" (the rate at which dollars are changing hands) as being fairly constant and stable.

### The Money Stock Is Closely Related To Total Spending

If we can think of the velocity of money (V) as being fairly stable, then we can say that the size of the money stock is very important when we are trying to understand the size of the total spending flow. If "V" stays the same, then if the money stock is increasing, that means that total spending also must be increasing; if the size of the money supply is decreasing, that means total spending must be decreasing. How about that! Would you believe this really is what happens? It really does.

Since the velocity of circulation of money is fairly stable, it's obvious that *an increase in the size of the total spending-income flow requires an increase in the size of the money supply!* Also, *if spending and income are slowing down, the money supply must be contracting.* And that's just what happens. It happens automatically through the banking system. Remember?

In order for the economy to speed up, bank loans (and therefore, the money supply) must be expanded. But as the economy slows down, bank loans are repaid and the money supply contracts. See how important the money supply can be? And see how it can be useful to look at the total spending flow as "a mass of moving money"?

Since the spending velocity stays fairly constant, if the money supply also stays fairly constant then the total spending flow will stay fairly constant. But if "M" increases, spending increases; if "M" decreases, spending decreases. So, if we want to, we can say that *all* increases or decreases in spending can be "explained" by increases or decreases in the money supply! How about that?

### As "M" Increases, Total Spending Increases, And Vice Versa

We can say that *no matter what the consumers and investors and the government spenders and the foreign buyers are doing, as long as "M" is gradually increasing, total spending and output and income will be gradually increasing, and everything will be just fine!* So why don't we just forget about C + I + G + F, and concentrate our attention on the money supply? Wouldn't that be easier and neater? Sure. But it may not be wise. Perhaps we should talk a little bit about the question of "cause and effect."

The whole purpose of our macroeconomic analysis—taking the spending flow apart—is to try to find out what *causes* changes in the size of the spending-income flow. So we know that the size of the money supply and the size of the spending flow move very closely together. That's important. But here's the real question: Is it changes in the size of the money stock (M) which *cause* the spending flow to expand or contract?

Do you see the problem? Spending doesn't increase unless "M" increases. But "M" doesn't increase unless spending increases! What causes what? Do you suppose that each could have some causal influence on the other? Sure. That's why it's a good idea not to ignore either "MV," or "C + I + G + F" as ways of looking at the total spending-income flow.

### Economists Disagree About The Causal Role Of "M"

Many economists these days disagree about the best way to look at and analyze the macroeconomic forces at work in the economy. The "monetarist school of thought," which includes Professors Milton Friedman, Paul McCracken (former chairman of President Nixon's Council of Economic Advisers) and others (the "Chicago School of thought") emphasizes the necessity of keeping our eye and our finger on the size (and changes in size) of the money supply. The monetarists emphasize the importance of "money stock" as the key to understanding economic conditions and prices in the economy. *"Keep your eye on the money supply!"*

On the other side of this argument stand the followers of Keynesian economics (the "new economics"). These economists focus their attention directly on the spending sources which contribute to the total spending flow. This group includes Professors Paul Samuelson (Nobel prize winner in economics in 1970), Walter Heller (Chairman of the Council of Economic Advisers under President Johnson), and many others. They believe that the supply of money is primarily a responsive (dependent) rather than a causal (independent) force. They believe that in order to understand and/or influence macroeconomic conditions it is necessary to focus on the "Keynesian components" of the total income-spending flow: C + I + G + F.

Who is right? The monetarists? The Keynesians? Until a few years ago, the Keynesian approach was clearly dominant in macroeconomics.

But the monetarist philosophy has gained headway in the last few years. The monetarist philosophy had a dominant influence on economic policy during the Nixon administration—especially during the first two years (1969 and 1970). Perhaps we are about to see the development of a "new" new economics, integrating the best from both the Keynesian and monetarist schools. Or maybe something else, better than either or both, is just over the horizon. Who knows? (Maybe YOU would like to become an economist and help us to get this thing figured out. If so: Welcome! But you needn't decide yet. You'll be getting a lot more involved in these issues in the next few chapters.)

### Another Way To Look At "Output"

Now you know two ways to look at the total spending flow: by "spending sector" (C + I + G + F) and by "money supply times velocity of spending" (MV). So now it's time to look at *output* a new way: as "the number of units produced" times "the average price per unit."

Let's take an example. Suppose we are producing a thousand units of output per day. Suppose the average price (per unit) is one dollar. Then the total value of the output flow will be one thousand dollars per day. Right? Then suppose the total output flow increases to *two* thousand units, and the average price is still one dollar. Then the total value of the output flow will be two thousand dollars per day. Obviously.

Let's say it this way: total output value is equal to the total quantity (number of units) produced (Q) times the average price per unit (P). You can see that "Q" times "P" is really "the value of the national product." It's gross national product (GNP) stated in a slightly different way. Now that we have the output flow defined in this way (Q times P), we can put together what the economists call the "equation of exchange."

### The Equation Of Exchange

The equation of exchange is another way of showing that the total spending flow is equal to the total output flow. We say it this way:

$$MV = PQ.$$

This only tells us that the total spending flow (looked at as the total money supply times the average velocity at which each dollar is being spent for output) is equal to the total value of the output (looked at as the total quantity of units produced times the average price per unit). The statement is obviously true. Isn't this an interesting way to look at the total spending-output flow of the economy? Interesting, yes. And different. But what can we do with it? Just watch.

The velocity of circulation (V) is fairly constant and stable. We talked about that. Remember? But what about the quantity of output (Q)? Sometimes that can decrease and increase quite a lot, when the economy is unstable. But as a rule, the economy may be expected to run along on a fairly stable course most of the time. If so, the quantity of output (Q) is not likely to change greatly in any short period of time. Now, do you see where this leaves us?

If "V" is fairly constant and "Q" is fairly constant, this gives us a direct tie between "M" (the money supply) and "P" (the average level of prices). How about that? It tells us that if the money stock (M) does not increase, then "prices in general" absolutely cannot rise. Inflation is impossible! Now do you understand the monetarists' position on fighting inflation? "Don't let 'M' increase any more than 'Q' increases; then there can be no inflation!"

This was the position which guided the Nixon administration's economic policy during its first two years. Monetary expansion was restricted. Total spending was restricted, but prices continued to rise. It was *output* (Q) that slowed down, instead! Unemployment increased while inflation continued. Why? what does all this mean? where does this leave us as far as "economic stabilization" policy is concerned?

The following chapters will go further into these interesting issues of how to fight inflation and unemployment. But for now, let's stop and be sure all this is tied up right.

## THE COMPONENTS OF THE SPENDING FLOW ARE ALL LOCKED TOGETHER

You can think of the spending flow as made up of the four spending components (C + I + G + F), or you can think of it as a total quantity of money, circulating at a certain rate (MV). Either way you look at it, you are looking at the same thing. If any one of the "spending components" slows down, (unless something else speeds up) spending and income will get smaller. If this is happening, it means that either M or V (or both) are getting smaller. These things must move together, because they are exactly the same thing, just looked at in different ways.

You can look at the output flow as the quantity of units of output (Q) times the average price per unit (P). Or you can look at it as "GNP." GNP, really, is figured up by adding the output of things, times the price of each. Remember? So in both cases, you are looking at the same thing: output times prices. Obviously the two must be equal. They're the same thing!

If anything happens to cause the total spending flow to increase, then the value of what the spending flow is buying (the output flow) must also increase. There can be no other way. Either the size of the output must increase, or prices must rise, but one or the other (or some combination of both) must increase. See how all this is "locked-in" together? Everything moves together, so it's impossible simply to look at what is happening and *decide what is causing it* to happen. Let's take an example.

## What's Causing What?

Perhaps I look and see an increase in investment spending. The investors borrow money from the bank (which expands the money supply) and spend the money to buy more goods. But if the economy is already fully employed, output cannot increase. So prices are forced up. What I see is: increased investment spending, bringing inflation."

Someone else looks at the same situation and says: "the expansion in the money supply resulted in inflation." Was it the increase in investment spending? Or was it the increase in the money supply which brought the inflation? The answer is obvious. It was both! The increased investment spending could not have occurred without the increase in the money supply; the increase in the money supply could not have occurred without the increased borrowing and spending by the investors!

If it happens that next spring, consumers start going into debt and are spending more for appliances and automobiles and things, the money supply will increase, the economy will boom, and prices may rise. What's responsible? The increase in the money supply? Or the "upward shift in the consumption function"? See the problem? It's difficult to separate inseparable things!

## Changes Are Likely To Be Cumulative

Another problem is that these macroeconomic variables are so closely interrelated that any change is likely to set off a chain reaction. If total spending begins to increase for any reason, this may bring a wave of increased spending. If consumers spend more, investors spend more. So incomes rise and consumers spend even more. The money supply expands. Employment and output increase. Probably, prices rise.

It works the other way, too. If businesses start cutting back spending and paying off loans, the money supply gets smaller. Workers receive less income. Some lose their jobs. Consumer spending drops. So businesses cut back even more. The money supply shrinks farther. More jobs are lost. Things get worse and worse.

See how macro-equilibrium is sort of like a "perched boulder"? You'll find out even more about that in Part Five. But we have been covering a lot of solid ground. Maybe you'd better not go ahead into Part Five until after you spend some time reviewing and thinking about the concepts and principles explained in Part Four. This Part (Four) and the previous one have spelled out the basic macroeconomic concepts and principles—money, spending, macro-equilibrium and all that. Part Five will show you the kinds of bad times (depression, and inflation) which sometimes develop, and then explain what can be done to overcome these macroeconomic problems. Review first. Then, on to Part Five!

## REVIEW EXERCISES

### MAJOR CONCEPTS, PRINCIPLES, TERMS (Try to write a paragraph explaining each of these.)

the propensity to consume
the multiplier effect
why money times velocity (MV) equals total spending
why output times prices (PQ) equals GNP
why MV = PQ
the problem of "what causes what"

### OTHER CONCEPTS AND TERMS (Try to write a sentence or phrase explaining what each one means.)

average propensity to consume (APC)
average propensity to save (APS)
marginal propensity to consume (MPC)
consumption function
full employment
the money supply (M)
the velocity of money (V)
"income (or output) velocity"
Milton Friedman
Paul McCracken
the "Monetarists"
the "Chicago school of thought"
the Keynesians
Paul Samuelson
Walter Heller
the "new economics"

### GRAPHS AND CURVES (Try to draw, label, and explain each.)

The Consumption Function
The Consumption Function and the Savings Function
The Effect of an Injection of Investment Spending
Macro-equilibrium, Where S = I

### QUESTIONS (Try to write out answers, or jot down the highlights.)

1. Can you think of anything that might cause *your* "consumption function" to shift up or down (either temporarily or permanently)— that is things that might cause you to spend more, or less for consumer goods, even though your income didn't change? Can you think of things that might cause the consumption function to increase or decrease for the entire nation? Explain.
2. What do you think of the "relative usefulness" of the Keynesian and monetary approaches to understanding macroeconomics? For which purposes would you prefer to use the Keynesian approach? the monetary approach? Discuss.
3. Using the "MV = PQ" approach (and thinking back to things you learned in the previous chapters) it becomes pretty obvious why it's necessary to prevent unlimited expansion of the money supply, and why the government must be very careful about how much it finances its spending programs by creating new money (high powered money) by selling bonds to the Federal Reserve banks. Right? Explain.

# PART 5

IT'S FEAST OR FAMINE, DROUGHT OR FLOOD;
WHENEVER IT RAINS, IT POURS!

When the economy speeds up it speeds up too much and when it slows down it slows down too much, so the government tries to stabilize things and nudge the economy into a "good macro-equilibrium."

# 16 BOOM AND BUST: THE UPS AND DOWNS OF THE ECONOMY

*The process of economic fluctuations and the role of the Income Multiplier and the Acceleration Principle*

By now you know a lot of the most basic theory of macroeconomics. In this chapter you're going to see some of these things working in "real world-type" examples. By the time you finish this chapter, you'll have a lot better feel for some of these things.

## The Process Of Macroeconomic Change

What makes the economy speed up and slow down? What's the *process* of economic expansion and contraction? What is it that makes total spending increase so much sometimes? And decrease so much at some other times?

You already understand how an increase or decrease in spending by anybody has a tendency to result in further increases or decreases by others. You saw that in the Keynesian national spending and income graph. Also, remember (back in Chapter 9) when the eastside fishermen reduced their spending to save for Christmas? Soon everybody was forced to spend less. Yes, all these things really are locked in. They really do move together. Here's a "real-world type" example.

## A Spending Increase Must Come From Somewhere

Suppose there are lots of students in your area who would like to have part-time jobs. They all want some income. Now suppose a national toy manufacturer hears about the great supply of low cost part-time labor available in your area. A "plant location team" visits your area, and decides not to build an automated plant, but to design a toy factory requiring a lot of cheap unskilled labor, and build it in your area. So they build the plant halfway between the college and the shopping center.

Soon they start running ads in the college paper saying that students who want to work can come in and put in a few hours any day they want to. All you have to do when you work there is sit on a swivel-stool between two moving "assembly line" belts. You pick up the pieces you need to assemble a toy from the moving belt in front of you. You put the pieces together, then put the finished toy on the belt behind you. It's really simple to do. Not much variety and challenge, but it's a job, and it pays money. To those who need money, that's important!

Many students go down and work in the plant between classes and in the afternoons. Would you say that this new plant is resulting in increased economic activity in your area? Is total output increasing? Is total income increasing? The answer, quite obviously, is "yes." An increase in spending by the toy business brings an increase in total spending in your area.

It results in an increase in output, and an increase in income. That's very easy to see. But this is just the beginning. Now the spending increase is going to *multiply*.

## THE INCOME MULTIPLIER IS THE "RESPENDING" EFFECT

Everyone seems to be glad about the new toy factory in the area except perhaps the few who used to like to go and sit and study under the trees in the field where the toy factory now is. Now the trees are gone, but such is sometimes the cost of economic growth. Many students have more income than before. They are happy.

What happens now? Is that the end of it? Or is this new toy plant going to have additional effects on the local economy? What do you think? Are all of the students going to keep all that extra money they're making? Save it? Put it into savings accounts or stocks and bonds or real estate or some such? I doubt it. They will probably spend it at the snack bar, at the student union, at the college bookstore (ugh!), and at the shopping center and downtown. Right?

### Employed People Spend More

You students are spending more for consumer goods these days. Why? Because the toy business decided to spend more in your area, that's why. See how an increase in spending is cumulative? The initial spending increase by the toy manufacturer seems to multiply throughout the community.

First, the toy manufacturer's rate of spending in the area increases. He pays wages to the student-workers. Then the student-workers' rate of spending in the area increases as they spend more each week at the shopping center and downtown. The rate of total spending is multiplying, all right! That's why economists call this "the income multiplier," or just "the multiplier." In the last chapter you saw how this looks on the Keynesian spending and income graph. Now you're seeing what it looks like in the real world.

### New Spending Starts The Ball Rolling

Now let's back off and really look at the big picture. Let's suppose that the toy manufacturers are hiring because they expect a big boom this year. They think the bicycle boom is going to continue to expand. And they think all kinds of toys for "children-children" and for "adult-children" are going to be in strong demand, this Christmas. So all around the country new toy factories are springing up.

In many places the manufacturers follow the same pattern started in your town. Instead of automated plants, they design their plants to use lots of low cost, part-time labor. So we see all these new toy factories and all these students working there, producing things, making income. We know the additional income they receive will let them spend more for consumer goods. Some of them will save some of the money they earn. But it's a pretty good bet that most of them will spend most of it.

You can see what's happening. The boom in your hometown is being repeated in lots of places. All these new jobs and new wages and new outputs are not just in your town. They're everywhere! What does the multiplier tell us is going to happen? It tells us that all this new spending by the toy producers is going to cause more money to be spent for consumer goods. There will be an increased rate of consumer spending in response to the increased rate of spending by businesses.

### The Extent Of The Respending Effect

Wonder how much the rate of consumer spending throughout the country will increase as a result of the increased spending by all these toy companies? We really don't know. But we know what it depends on. It depends on how much each individual respends out of his new income. That's the marginal propensity to consume (MPC). Remember?

If each student increases his weekly consumer spending by as much as his weekly income goes up, that means his MPC is 100 percent. When he respends all of his new income, he turns it into new income for other people. Then if all those other people respend all of it, they turn it into just that much more new income for still other people. Then if those people spend it all, and the next people spend it all, and if this continues over and over, the consumer spending stream will keep getting bigger and bigger and bigger! The rate of spending will keep going up and up.

If the MPC is 100 percent, then as more people get their incomes pushed up, they turn right around and respend and push up the incomes of others. The multiplier just keeps going on and on. The "investment hydrants" are pouring in more money, but the "savings drainpipes" don't open up to drain any of it off. What happens? The income flow keeps getting bigger and bigger!

### Income Expands Until New Withdrawals Offset The New Injections

Could the MPC be 100 percent? Under some circumstances, for some people, sure. But for the total economy, with incomes increasing rapidly? No. As the national income increases, people will start to save some. The more the income increases, the more the people will save. So how far will the national income increase? Just to the level where the amount the people want to withdraw (to save) is equal to the amount the businesses want to inject (to invest).

Now let's talk about the opposite example. Suppose each student who received new income from the toy factory was so proud to get his money that he dashed down to the bank, cashed the check (traded it for paper currency) then went home and hid the currency in a shoebox on the closet shelf. What's the multiplier now? It's one. Why? Because the new spending created new income only *one time*—just for the student workers. Then the money *immediately* leaked out of the spending stream and turned into savings. The "savings drainpipes" were immediately opened wide enough to offset the effect of the new flow from the "investment hydrants."

From these two extreme examples, you can easily figure out about the multiplier. If people save all the extra money they receive from the toy business, the multiplier will be one. That means it doesn't multiply at all. If people spend all the extra money and everybody who receives it after that also turns around and spends it all again, the multiplier will be infinite. The spending stream will keep getting bigger and bigger and bigger. But in the real world, neither of these two extremes would be likely to happen. The real situation is going to fall somewhere between the two extremes. (You could have guessed it.)

### At Higher Incomes People Save More

With the increased spending by the toy companies, the income stream will get larger and larger. But as incomes get larger, people will start to save more. Eventually the "savings withdrawal rate" will increase enough to exactly equal (and exactly offset) the new "spending injection rate." You saw this in the Keynesian national spending and income graph in the last chapter. Remember?

Suppose you work at the toy factory and receive $10 a day. That means ten dollars a day of new income is being generated by the plant. Then each day you take the $10 out and spend it for gasoline, clothes, records, food, movies, and all sorts of things. Then the people who produce and sell the gas and other things, will have an additional $10 a day in incomes. If all those people go out and spend $10 more each day, then some other people will have an additional $10 in income. The $10 just keeps going on and on, pushing up people's daily rates of income more and more.

If each person who gets a $10 increase in income immediately increases his spending by $10, you can see that the "pass it along" effect will keep going on forever! But usually when a person starts receiving an extra $10 a day as income, he doesn't run out and spend it all. He usually spends part and saves part. When this happens it means that each person withdraws a little of the increase (opens up his "savings drain valve" a little) before he "passes it along." As each person does this, the "moving $10" gets smaller and smaller, until finally, there is nothing left to pass on—nothing left to multiply. That means that the "savings withdrawal rate" has increased by $10. The total income-spending stream will expand no further.

Suppose you're receiving the extra $10 a day and you decide to save $2 of it and stick it in your closet shoebox. Then you spend $8 more each day for gas and picnic supplies and things. Suppose the gas people and the picnic supply people and all the ones who helped to make all those things—suppose they start getting this extra $8 a day in income and suppose they decide to increase their daily rate of savings too, and spend the rest.

Each time the multiplier effect (of the $10 a day increase) pushes someone's income up, it pushes up their spending rate. But usually it also pushes up their savings. Eventually the savings rate is increased enough to exactly offset the $10 a day increase in spending by the toy company. Then the income stream is again stable, but at a higher level. Ten dollars a day higher? No. More like 20 or 30 or 40 or 50 dollars a day higher. Why so high? Because of the *income multiplier,* of course.

You already know that economists have figured out a way to assign a number to the multiplier. We call the number "the multiplier." What it tells you is how many times total income will increase as a result of the initial increase in spending. If everybody saves all his increased income, the multiplier will be one. The increased spending increases the income only one time. Then it's all saved. But suppose everybody saved half of the daily increase, and spent the other half. Then the multiplier would be two. But if the marginal propensity to save is one-tenth, then the marginal propensity to consume is 9/10ths and the multiplier is 10. The higher the MPC, the higher the multiplier. Obviously. It's the "consumer respending effect" (the MPC) which *is* the multiplier!

**A Summary Overview Of The Multiplier**

The concept of the multiplier simply says that when businesses (or governments or foreigners — or consumers, for that matter) increase their spending, this immediately increases people's incomes by the amount of the increase in spending. But ultimately the incomes of the people will be increased by *more* than the amount of the initial increase in spending. The multiplier concept is simply a way of looking at the "respending effect." If the people respend almost all of the increases in their incomes, then "the multiplier" will be high because the marginal propensity to consume is high. (The marginal propensity to save is low.) But if people don't respend much — that is, if they save most of the increase — the multiplier will be low.

The "income-increasing effect" of an initial increase in spending will continue for how long? Until the "daily rate of savings withdrawals" increases enough to offset the "increase in the daily rate of spending injections." Once the "savings withdrawal rate" increases enough to exactly offset the new "spending injection rate," the spending flow will be stable again — at a higher level than before.

Let's look at the big picture again, going back to the beginning. Suppose the economy is moving along with a lot of unemployed student labor — people who really would like part-time jobs. The economy is in "macroequilibrium" — that is, it isn't expanding or contracting. But we have productive capacity that is not being used. So what happens?

The toy manufacturers invest in capital and hire the unemployed students. This increases total output and total income of the society. But that's not the end of it. The increased output increases the total product going into the markets of the society. The increased income increases the amount of money the students are receiving. As the students' incomes increase from working in the toy factories, their rates of spending also increase. They add more dollars to the spending stream.

The toy factories keep right on paying the higher incomes. The students keep on buying more things. Those who sell the things to the students keep on receiving more income, and spending more, themselves. So the income stream is larger by the amount of the initial spending injection, plus the amount of the new "multiplier-induced" spending.

New consumer spending is induced by the new (higher) incomes. That's the multiplier. The multiplier carries the total income of the society to a higher level, where it stabilizes. The new stable level must be the level at which the new (induced) rate of planned withdrawals (savings) is equal to the new rate of planned injections (investments).

You can see why a $10 increase in the daily rate of spending is going to keep on pushing up income until it induces a $10 increase in the daily rate of savings. If the toy factory in your area pays out a total of $100,000 a year to part-time student workers, then incomes in your area will immediately increase by $100,000. Then as the student-workers spend that money, people's incomes in your area will go up more and more. The only way the additional $100,000 spending injection can be offset is for the people to want to increase their savings withdrawals by $100,000 also. Until they do, the multiplier expansion will keep pushing up the level of spending. People's incomes will continue to rise.

Anybody who spends more money to buy more output, *automatically* generates more output, and more income. The people who receive this "more income" will spend a part of this "marginal increase," for consumer goods. If they spend *much* of the increase for consumer goods, the multiplier will be *large*. If they spend *little*, the multiplier will be *small*.

Now you really understand the multiplier. That's good, because this is one of the things which really does work in the real world. It isn't always as neat and predictable as I have described it, but it really does work. The multiplier can be very powerful. But the multiplier is only half of the "induced spending" picture. There is another half. We call the other half "the acceleration principle."

## THE ACCELERATION PRINCIPLE IS THE "INDUCED INVESTMENT" EFFECT

One afternoon you go by the new toy shop and work a few hours to earn some income. Then you go on down to the shopping center to spend your earnings. You go into the shoe store to buy a new pair of sandals. All the clerks are busy, so the store manager, Mr. McCoy, waits on you. He seems to be in a very good mood. He tells you that his business is better this month than ever before. Can you guess why? You know very well why. It's the multiplier effect of the new income generated by the toy factory!

### The "Multiplier" Can Set Off The "Accelerator"

Students now have more money to spend, so they're spending more for shoes. That shouldn't be a surprise. You're spending money at the shoe store, today. That's what the multiplier says you will be doing. When you spend your money for sandals, that's going to increase the incomes of a lot of people, each a little bit. Mr. McCoy will get to keep some of it for his own income, but not much of it.

Some of the increased spending goes to the shoe store employees. Some goes to the wholesaler who supplies the shoes. The wholesaler gets to keep a little of it, and some goes to his employees. Then some goes to the shoe manufacturer. The manufacturer keeps a little bit of the income and some goes to his workers. Then some goes to the leather company. The leather company gets to keep a little bit and then they pay their workers. Some goes to the bank as interest on loans. The banks keep a little and pay the employees a little. Some goes to the electric company, and to their suppliers and employees, and on and on it goes.

Ultimately, all of the increased spending for output must go as new income to somebody. Each person who receives some of it, respends some. All this new income is being generated by the multiplier. *The multiplier is the respending effect,* don't forget. But something else is getting ready to happen. The multiplier is going to set off the acceleration principle. Watch.

## As Sales Increase, Inventories Increase More

While you are in the shoe store trying on sandals, you hear a lot of hammering going on in the back. You ask Mr. McCoy what's going on. He says he is expanding his shelf space. He is going to expand his inventory of shoes, especially sandals. He explains that when sales are faster, he likes to have a larger inventory. That way he can be sure he won't run out of particular styles or sizes just when sales are going best.

"Each time you run out of someone's style, or size, you lose a customer," he says. "In the past, we have been selling at the rate of about $1,000 worth of shoes per week and we have been keeping about a $10,000 inventory of shoes on hand. But now our sales are up to almost $2,000 per week, and we think we should keep a much larger inventory, maybe even as much as $20,000 worth." That seems reasonable to you. You commend Mr. McCoy for his quick response to the increased demand for shoes. Then you pay for your sandals and go on your way.

As you are leaving, you happen to think about the shoe *manufacturers.* How happy they must be, to get Mr. McCoy's order! He has been ordering about $1,000 worth of shoes a week through the wholesalers, from the factory. Now, suddenly Mr. McCoy is going to place an order for $12,000 worth of shoes, all at one time! What an order! He needs $2,000 worth to cover his (larger) weekly sales volume, and another $10,000 worth to build up his inventory.

You say to yourself, "Just think. His sales of shoes increased from $1,000 to $2,000 a week. That's an increase of 100 percent. But his order for more shoes from the manufacturer is going to *more* than double, or triple, or quadruple, or even quintuple. His order is going to increase by *1200 percent*! He is going to order *12 times as many shoes* as he did before. Wow! I'll bet the economists have some word—some strange sounding term—to describe a situation like this!"

You bet we do. What you have just seen happening is what we call the acceleration principle. If you think about it for a minute you can guess exactly what's going to happen next. Shortages? Right! Let's back off again and look at the big picture.

### A Tidal Wave Of Derived Demand

Throughout the economy, students are working at toy factories earning extra money. Then they are going to the shoe stores and buying twice as many shoes as before. So shoe stores all over the country are sending in big orders to the shoe manufacturers—orders suddenly *12 times as great* as their previous orders. Now I ask you, how in the world are those shoe companies going to fill all those orders? They aren't. They just can't! There's just *no way* they can produce all those shoes. So what do they do? They hire all the extra workers they can find. They start working three shifts a day. They try to lease more buildings and buy new shoe-making machines.

The shoe machinery people are next to get hit by this tidal wave of derived demand. When a Wilmington manufacturer calls in a rush order for more machines, the shoe machinery people tell him: "You have been ordering only two machines a year from us, just to replace your old machines as they wear out. Now, suddenly you want to buy twenty new machines so you can set up a new factory. And you want immediate delivery? Would you believe that every shoe manufacturer in the country has called us and tried to buy 20 new machines? There's just *no way* we can produce all those machines.

"We will go on overtime, hire all the extra labor we can get, and expand our plant as quickly as we can. We will do everything we can to produce the machines as fast as we can. But we should warn you that we can't ship you more than one new machine every three months, until we get tooled up to expand our output. That may take more than a year. Please understand that *we can't get the machines we need* to expand our plant, either!"

See how the "acceleration principle" blows up an increase in consumer demand, all out of proportion? Of course this example is a little bit exaggerated—I hope you will forgive me for that. Used with discretion, exaggeration sometimes can serve a useful purpose as "the microscope of the social scientist." (But *always* with discretion!)

### Shortages Appear In Almost All Industries

Now look back to the day you bought your sandals. Suppose you had gone into the local appliance store or hardware store or sporting goods store or music store or automotive accessories store or almost anywhere else. You would have seen the same thing going on. Fast sales. Expanding orders for inventories. Then, shortages. It's happening everywhere!

Here's the pattern: Consumer demand increases some. Why? Usually because of the multiplier. Increased consumer demand is what triggers the acceleration principle. Retailers will increase their orders *much more* than consumer demand increases. Then wholesalers and manufacturers "accelerate" the demand even more. *Total demand* goes right out of sight!

If only shoes were involved, resources could be shifted out of other things and into shoe production. But when *everything* is in short supply, there's nowhere to get the needed resources. There's just no way all this inventory buildup can happen in a week, or a month, or even in a year! So there will be tight markets, waiting lists, and people offering extra money to get on the top of the lists. And inflationary pressures? You bet.

What started all this? The increased spending by the toy companies, which set off the multiplier. Remember? Then the increased consumer spending (multiplier effect) set off the accelerator. Consumer demand has gone up. *Derived demand* has gone out of sight! But that's not all. There's more.

## THE ACCELERATOR AND MULTIPLIER WORK TOGETHER

What kind of "spending" do we call it when the shoe store spends for inventory? "Investment spending," of course! People buy shoes as *consumer goods*, but retail shoe stores buy shoes as *investment goods*, for investment purposes. So when the shoe store increases its spending for shoes, this is an increase in investment spending. Also when the shoe manufacturing company leases and renovates the building next door and buys new machines, what kind of "spending" is all this? "Investment spending," of course!

Remember what happens to incomes and to consumer spending, when investment spending increases? If the initial increase in spending by the toy companies caused all this havoc, what do you suppose will happen when all these retailers and wholesalers and manufacturers and everybody else start increasing their spending by leaps and bounds? Talk about an uptight economy!

What's going to happen? How long can these high demand, shortage, inflationary conditions last? You could guess. *Until total production has caught up with and satisfied the pent-up demand* for factories and machines and inventories. But this could take quite a long time.

Each increase in consumer demand works up through the retailer and wholesaler and manufacturer, and further accelerates the demand for inventories and factories and machines.

At the same time, each increase in investment spending for inventories and factories and machines, works down through the multiplier to further increase consumer demand. Things are really tight. They're going to stay tight for a long time.

### When Shortages End The Boom Ends

The boom doesn't keep on going this way forever. One day Mr. McCoy heaves a big sigh of relief and says: "Finally I have my inventory built up to the right size. I have been trying to do that for almost two years, and finally, I've made it. My sales of shoes are now up to $5,000 per week, and my inventory is up to $35,000 worth of shoes. All my shelves are full. My sales seem to be stable at $5,000, and I think I will just let my inventory stabilize at the $35,000 level."

Mr. McCoy notifies the wholesaler and the manufacturer that in the future he will be satisfied with $5,000 worth of shoes per week. No longer will he write nasty letters saying that he must have another $15,000 worth. So the pressure is off, at least from Mr. McCoy's store.

What's the situation throughout the rest of the economy? The same. The level of sales is much higher than before and inventories are much higher than before. Everywhere you look, outputs and incomes are much higher than before. Spending is high. People are buying shoes and everything else like crazy. Prices of everything are higher, too.

But one thing is different now. There is no longer a need for the shoe store to build its inventory. So it cuts back its demand for shoes. What's going to happen when every shoe store and clothing store and appliance store and music store and furniture store and paint store and all the others stop building inventory? The manufacturers are going to be in for a real jolt!

### Manufacturers Are The First to Feel The Slump

It took the manufacturers a long time to build up enough productive capacity to meet the "accelerated" demand. But now things are different. The wholesalers and retailers don't want to buy but a trickle, anymore! The shoe manufacturers start closing down some of their plants. Other manufacturers do, too. Soon it happens to several manufacturing firms. See what kind of a problem is sneaking up on us?

What do the manufacturing workers do when they get laid off? They buy less shoes and things. Is this the multiplier going the other way? It sure is. The manufacturers are reducing their spending for labor and other inputs. Down go incomes. So down goes consumer demand. How about the acceleration principle? Does it work in reverse, too? I'm afraid so. It looks like the boom is all over. The economy is on the brink of a recession.

Here's how it happens. First a few manufacturing workers are laid off. They reduce their spending for consumer goods. With less sales, retailers don't need as much inventory. So they stop ordering from the manufacturers for awhile. They just sell out of inventory. When this happens, the manufacturers' sales drop to *zero*!

What do the manufacturers do? They close down their plants and lay off their workers for a few weeks. The laid-off workers don't have any income, so they don't spend much. That means retail sales drop even more. So the retailers decide to reduce inventories even more. No sense in carrying the expense of high inventories if it isn't necessary! More manufacturers close down.

How far does total spending (aggregate demand) go down? Is this a really bad recession? a depression? Who knows? We could exaggerate it enough to show a time when *all* retailers would be trying to sell out of inventory and *nobody* would be ordering anything from the manufacturers. Then nobody would be making any money working in the factories, so nobody would be buying very much from the retail stores.

If nobody is buying much of anything, how long is it going to take the retailers and wholesalers and manufacturers and everybody to get rid of all their overstocked inventories? A very long time. Such a depression would last a very long time. This is the kind of situation that existed in the United States in the 1930's. The inventory problem wasn't the only thing that caused the Great Depression of the 1930's to be so bad and to last so long. But it was one of the important things. I'm sure you can understand that, now.

## The Multiplier, The Accelerator, And The "Business Cycle"

It's easy to see how the multiplier and the accelerator reinforce each other. Any increase in spending by businesses (or by anybody) triggers the multiplier. Consumer spending increases. The increase in consumer spending triggers the accelerator and pushes investment spending up. The two work together, first to speed things up too much, then to pull the rug out from under us. The boom keeps going until the plant and inventory expansions have been completed. Then everything turns around. The investment spending cut-back brings consumer spending cut-backs which bring more investment spending cut-backs and down, down, down we go.

This "boom and bust" we are talking about is sometimes called the business cycle. The multiplier and accelerator reinforce each other and create a sort of "natural tendency" for the economy to overexpand, and then overcontract—to go from boom, to recession. Sometimes this tendency is referred to as "the problem of the instability of capitalism." It really can be a problem.

This "business cycle" of "boom and bust" really isn't desirable from anybody's point of view. The times of declining demand, recession, unemployment and low incomes are usually thought of as the worst times.

But the times of shortages and upward pressures on prices really aren't satisfactory either. I don't need to tell you that depressions and inflations are undesirable. Everybody knows that. But maybe it won't hurt, just for a few minutes, to talk about some of the undesirable effects of depression and inflation.

## THE GREAT SOCIAL COST OF DEPRESSION AND INFLATION

Think for a minute. What's the purpose of "the economy"? What is the "economic system" supposed to do, anyway? Why should all these people be working and burning up energy and using up resources and making things and all that? The purpose is to feed and clothe us and to try to improve our lives—to let us live better. Right? We produce for the good of ourselves. We do and make what "we the members of the society" think will benefit us—will bring us better lives, or a better world. Anyway, that's what we're *trying* to do, I suppose.

### The Tragedy Of Unemployment And Depression

When we say "there is widespread unemployment," exactly what does that mean? It means that there are people who want to be productive, to produce something of value for the society. But because of some hang-up, some snag, some flaw in the way the economic system is working, these people don't have the opportunity to produce things for the society. When there's a lot of unemployment, there's a lot of waste. Labor is being wasted. And not just labor. Factories are sitting idle, getting rusty. Trucks, railroad cars, machines, all sitting around, doing nothing. Becoming obsolete. While there's so much that needs to be done!

When unemployment exists, it means that the economic system somehow is failing in its job. The things society could be doing for its own benefit are going undone, while the society's productive resources are wasting away. Really tragic.

If we decided to produce less—maybe to conserve natural resources, or to go on a 32-hour work week, or to slow down the economy for some other reason, that's okay. But that's different! When we are talking about unemployment and depression, we are talking about people *wanting* to be productive and society *wanting* and *needing* the output, but nothing happens. The pieces just aren't fitting together.

People are walking the streets, looking for jobs, suffering the shame of having to admit failure. People having to accept handouts, losing their self respect. Sometimes just giving up. Sometimes turning against the society which has done this to them.

How would you measure the "social cost" of depression? We can't, really. But everyone knows that the cost is great. Not just *economic* costs, but *many kinds* of social cost. The breakdown in family harmony. The rise in the crime rate. Delinquent taxes. State and local governments facing fiscal crises. The quality of education, health, hospital, social welfare, other services, going down. The list goes on and on. The social costs of widespread unemployment are appalling. Intolerable.

### Persistent Depression Is No Longer Tolerable

You can understand why people refuse to put up with depression. If any modern nation began to experience serious depression, the government would be forced to do something about it. If it was necessary to change the economic system to solve the problem, then you may be sure that the economic system would be changed. An economic system which would permit persistent depression and widespread unemployment is no longer acceptable in today's world.

The American economic system has been greatly changed during the past four decades. Much of the change has resulted from the people's feelings of economic insecurity. People really fear the threat of unemployment. They demand that the economic system be set up to do something positive to prevent it, whenever necessary.

The purpose of an economic system is to serve the needs and objectives of the society, both in the short and in the long run. A system which permits the great waste and economic insecurity of recurrent periods of widespread unemployment is no longer acceptable among the people of the modern world. If "mixed and modified" socio-capitalism is to survive, it must be immune to serious depression. The following chapters explain what the government can do. But first, what about inflation? What's so bad about inflation?

### Inflation Tears The Society Apart

What's inflation? Prices of things are going up. That's all. Is that so bad? Yes. It's very bad. Much worse than you think. It hurts the society. It brings really serious injustices to people. It actually *robs* people of the economic assets the society owes them. It forces people who, by all rights, deserve a comfortable retirement, to live their retirement years in poverty. But that isn't all. Not by a long shot.

Inflation really does tear the society apart. It forces everyone in the society to become belligerently active in fighting to see that the value of *his income* and the value of *his assets* keep going up at least as fast as prices are going up on everything else. Inflation forces workers to fight for "exorbitant" wage increases. It forces businesses to announce "exorbitant" price increases. It forces the state and local governments to pass "exorbitant" tax increases, while costs of government services surge forward even faster. "Austerity budgets" are forced on the public schools, colleges, health and welfare services, highway departments. College students are forced to pay higher tuition and fees. People feel more and more cheated, and get more and more angry.

What does inflation do? You can see how it tears the society apart! Everyone must fight as hard as he can just to stay in the same place—to keep from being left behind. It breeds unreasonableness, dissension, mistrust, unhappiness, and even hate among otherwise reasonable people.

If inflation continues, people lose interest in saving money. Nobody wants to hold money. Spend it! Hurry, before prices go up any more! Buy a new car. Buy land. Or stocks. Anything. But don't hold on to any money.

Once the people get afraid to hold money, they spend fast. Then prices really skyrocket! The next step is the complete collapse of the markets of the economy. People just *refuse* to accept money. Suddenly, there is no medium of exchange. The markets collapse. When the markets collapse, production stops. The economy collapses. Depression!

## Gouge Thy Neighbor

Inflation instills in all the people—children and adults, workers, investors, businessmen, policemen, teachers, college professors, everyone—the philosophy: "Gouge thy neighbor because thy neighbor is certainly out to gouge thee!" We are all forced to get involved in bargaining and pushing and trying to get our own incomes up enough so that (if we are lucky) we at least stay in the same place.

We might all be surprised if we knew just how many of today's economic and social problems are born of the distortions and injustices of inflation. Inflation distorts the economic relationships among all the people of the society. Businesses which sell in markets where they can raise their prices can come out all right. Workers who work where they can keep their wages rising fast, can make out all right, too. But many people have incomes which are not easy to adjust upward. They are seriously hurt. Usually the burden falls most heavily on the ones who are least able to carry it.

It isn't "just a few people" who are hurt by inflation. Most of the "ordinary people"—those of us who make up most of the society—are depending on some savings, some insurance policies, perhaps some government bonds, and a retirement program of some kind to bring us the things we need in our later years. For many years, most people produce and contribute more to the society than they consume. They save. The extra value they produce for the society (but don't consume) is supposed to flow back to them (as goods and services) so they can help their children through college, and so that in later years they can live decently. But what happens?

Inflation robs them of what the society owes them. If the inflation rate is high, the people simply aren't going to be getting the things they've saved up for. The things that rightfully belong to them will be going to the businesses and workers and governments and all the others who have the opportunity and the power to "increase their take" enough to come out on top. Everyone else gets left behind.

Do you begin to understand what inflation does to a society? If so, then you can understand why it is essential that inflation be held in check. If it isn't, soon it begins to run away. The further it goes, the more the people push to get their prices and wages up. So the worse it gets. Ultimately it can destroy the economy, the government, the entire society.

The problem of inflation is very real. It is much more serious than most people realize—potentially as dangerous as depression. Neither inflation nor depression can be ignored. But what can be done?

### The Necessity Of Economic Stabilization

In the following chapters we will be talking about "monetary and fiscal policy"—techniques used by the government to stabilize the economy. While we're talking about what these policies are and how they work, try not to lose sight of the *vital purpose* of economic stabilization. Modern society somehow must prevent the intolerable social costs of both depression and inflation.

Now, before we go on into the issues of preventing and overcoming depression and inflation, there's one more thing we need to talk about. We need to ask this question: When the multiplier and the accelerator get going and everything is speeding up, where does all the *money* come from, to finance such a boom?

## THE MONEY SUPPLY MUST EXPAND TO FINANCE A BOOM

What role does money play in this whole "boom and bust" cycle of economic instability? The money supply must expand, or else the boom couldn't occur! Remember the toy manufacturers? Where do you suppose they got the money to pay the students for all that part-time work? Where do you suppose Mr. McCoy got the money to build up the value of his shoe inventory from $10,000 to $35,000? What about the money to lease and renovate buildings, and to buy all those shoe machines, and to pay for all the labor and other things? The multiplier couldn't get going very well without some new (bank-created) money to finance all those payments. Right?

Now think of the consumer goods boom. Where do you suppose all those people got the money to buy all those TV sets and new automobiles and furniture and all those things? All this buying, building, and stocking up of things just couldn't be going on unless a lot more money was coming from somewhere. Where did it come from? It was being created through the banking system.

Could the money have come from savings that consumers are making out of their current incomes? Obviously not. The consumers are spending up their incomes, and going into debt! Maybe the consumers are spending up the money they have been saving for a rainy day? Not a chance. Who is paying cash? Nobody. Well, almost nobody. Everyone's buying on time. How about the money the businesses have saved up? Or the money they're taking in from current business profits? No chance. The businesses are all going into debt, too. And selling things on credit. So *where could* the money be coming from? It's newly created. Of course.

The money to finance the boom is newly created, just for the purpose of financing the boom. It's created by the banking system, to meet the loan demands of businesses and consumers. You know exactly how it happens. You learned all about it back in Chapter 12, from Mr. Alber and Mr. Baker and Mr. Zimmer. Remember?

But what about the new excess reserves needed to let the money supply expand? Where did they come from? Do you suppose the government or the Federal Reserve System permitted bank reserves to expand? Of course. What if they hadn't? Suppose they didn't want this boom to go so far, so fast. Could they have held down the money expansion and slowed down the boom and relieved some of the shortages? You bet they could! And they would, too.

The "monetary control tools" can be used to work against both inflation and depression. In the next chapter you will find out what these tools are, and how they work. The next chapter is all about "monetary policy." As soon as you're sure you understand all about the multiplier and the acceleration principle and the other concepts in this chapter, you'll be ready to go on into "monetary policy."

---

## REVIEW EXERCISES

**MAJOR CONCEPTS, PRINCIPLES, TERMS (Try to write a paragraph explaining each of these.)**

the multiplier
the acceleration principle
the business cycle
the tragedy of depression
the injustice of inflation
a boom requires a monetary expansion

## OTHER CONCEPTS AND TERMS (Try to write a sentence or phrase explaining what each one means.)

macroeconomic change
inflation
depression
derived demand
the "consumer respending effect"
the "induced investment effect"
economic stabilization

## QUESTIONS (Try to write out answers, or jot down the highlights.)

1. Can you explain the relationship between the marginal propensity to save and the multiplier? Try.
2. Why is it that when the *shortages* end, the boom ends? Explain.
3. The multiplier and the accelerator always tend to work together and to reinforce each other, yet each one is a *distinctly different thing* than the other. Can you explain the distinct difference between the two principles?
4. When prices continue to rise year after year (as they have been in recent years in the United States and in most of the other countries of the world), the continuing inflation affects the lives of everyone in the society. What are some of the ways inflation has touched *you, personally*? Explain.
5. If the business cycle really does try to work the way it's explained in this chapter, and if recurrent depression and inflation really are as bad as this chapter says, then it's pretty obvious that we can't let the economy just go its own way, and "let nature take its course." It's pretty obvious that the government must play an active macroeconomic role—that it must carry out some policies of "economic stabilization." The next two chapters will be talking about that. But already, right now, you should be able to think of some of the things the government might do. Can you?

# 17 USING MONETARY POLICY TO STABILIZE THE ECONOMY

*Policies of easy money or tight money are used to try to prevent or overcome depression or inflation.*

How would you like to be President of the United States? Maybe just for awhile? I wouldn't either. But would you be willing to make believe? Do you have enough imagination to think of yourself as the President?

This chapter is concerned with "economic stabilization." We'll be talking about the government's efforts to influence total spending, income, output, and employment in the economy. If you could visualize yourself as the President—as "the guy who's really in the hot-seat," you really might be able to get into it. You might get a feel for the techniques and tools, and for the difficulties and frustrations involved in economic stabilization. So will you try? Then okay, Mr. or Ms. President, let's have a go at it!

Now that you're the President you realize what a tough job it is, trying to run the government. But you've become sort of attached to the White House, and you really would like to be elected to another four-year term. You know that if everyone is happy about everything that's going on in the nation, you probably will be voted in again next November. But if the people are unhappy they will blame you and out you will go! So is there any problem? Yes. The economy is sort of depressed.

The average unemployment level throughout the country is about 6 percent. That means that 6 out of every 100 people in the labor force are out of work and looking for jobs. In some sections of the country the unemployment percentage is more than 10 percent. For some segments of the labor force—young, unskilled, minorities—the percentage is over 20 percent in some areas. You know that these people aren't going to be very happy with your administration. Neither are their families or their friends or the grocers who sell them groceries or any of the others who are hurt by all this unemployment. Unless something can be done to improve conditions in the economy before election time, you may find yourself out of a job!

Mr. or Ms. President, you remember from the Econ 101 course you took in college many years ago, that total spending supports total employment, total output, total income, and all that. So you know that if you could get the people to go out and buy more refrigerators and vacuum cleaners and automobiles and clothes and shoes and household furnishings and fishing rods and things, everything would soon be all right. Or if you could get some of the businesses to build new plants or order new equipment or start building up inventories, all this new investment spending would solve the problem. But how do you get the people and businesses to start spending more? There are several things you might try. Perhaps monetary policy would be a good thing to try.

## WHAT IS MONETARY POLICY?

The first thing you think about is: "Maybe we could stimulate the economy by using monetary policy." You think: "If we had 'easy money' (all the banks with lots of excess reserves, and low interest rates for borrowers, and low down payments and easy credit terms for autos and appliances and all), then maybe people and businesses would borrow and spend. Then the economy would speed up, unemployed people would be able to get jobs, and the spending-output-income flow would be all right again." (If all that happens, you will be reelected in November!)

You call together your top economic advisers: Chairman of the Council of Economic Advisers, Chairman of the Board of Governors of the Federal Reserve System, the Secretary of the Treasury, and the Director of the Office of Management and Budget. You ask them, "What can we do with monetary policy to stimulate the economy, *quickly*?"

### Depressed Conditions Call For Easy Money

Everyone agrees that there is only one thing you can do with monetary policy to stimulate the economy: make it easier and cheaper for people to borrow money and to buy things on credit. That is, see to it that the banks have plenty of excess reserves, and see to it that the interest rates are kept down low. Your advisers agree that there will be some danger of inflation if you follow this policy. But they also agree that a policy of "easy money" at this time would probably stimulate both business spending and consumer spending. Then this initial spending increase would increase employment, output, and incomes, and perhaps would set the economy on the path to prosperity.

That sounds good to you, Mr. or Ms. President, so you urge your four top economic advisers to do whatever they can to bring about "easy money." You try to impress on them the need for haste. The political future of a great national leader is at stake!

When the meeting breaks up, your four economic advisers go their separate ways. The Chairman of the Council of Economic Advisers goes back to his Council and starts preparing some reports which he thinks will make bankers more willing to lend. But he recognizes that there really isn't very much he can do to help. There isn't very much the Director of the Office of Management and Budget can do, either. He can say the right things and hope to help to get the people in the right mood. But he really doesn't have any "monetary policy tools" to work with.

The Secretary of the Treasury has some direct dealings in the money market. The Treasury "markets" the government debt, so it is constantly dealing in the "money market." But it is the Chairman of the Board of Governors of the Federal Reserve System who really is in the driver's seat when it comes to monetary policy. First, let's talk a little bit about what the Treasury does, then we will be ready to really dig in on the Federal Reserve and its role in "monetary policy."

### The Treasury Deals In The Bond And Money Markets

The Treasury is dealing in the bond and money markets every day. It must do this to "manage the government debt." The U.S. government debt is in the neighborhood of $500 billion. It is made up of government securities: bonds (long-term), and Treasury "notes" and "bills" (short-term). The people, banks, insurance companies, business corporations, local governments, Federal Reserve banks, and others hold these government securities. The securities are sold by the Treasury in the "open market"—that is, to anyone who wants to buy them.

Where does the Treasury get the billions of dollars to pay off the securities (bonds, notes, and bills) as they come due? By selling more securities, of course. Where does it sell these new bonds, notes, and all? In the bond and money markets, of course. Where else? And where are these bond and money markets? Scattered around the country in cities and towns, wherever there are banks and securities brokers. Securities brokerage companies and banks deal in government securities. The "money market headquarters" (if there is such a thing) is in New York City.

So what can the U.S. Treasury do to bring easy money? It might try to pay off more of its bonds. It could create some new money by selling bonds to the Federal Reserve banks, and then use this newly created money to pay off bonds held by individuals and businesses. That would increase the money supply. Each time an individual cashes a bond he receives a government check and goes into his bank, waving it in the air shouting: "Look! A new primary deposit! Some new excess reserves for the banking system! High-powered, five-fold, multiple expansion money! Happy days are here again!" Well, maybe he doesn't do all this. But you get the point.

What else can the Treasury do to bring easy money? It might start issuing bonds that pay lower interest rates. This might help to push down interest rates in the bond and money markets of the country. The Treasury could sell these bonds to the Federal Reserve banks and create more money. But wait. See how the Treasury must keep leaning on the Federal Reserve if it wants to do anything to bring easier money? Suppose the Federal Reserve (the "Fed") doesn't want to buy all these new securities? What could the Treasury do then? Not much, I'm afraid. So you see, it's really the "Fed" which holds the key to monetary policy in the United States.

## The Federal Reserve Exercises Monetary Policy

There really isn't much the Treasury can do to bring easy money unless the Fed is willing to cooperate. The Fed really does hold the monetary control power. So before we go on we need to talk about the Federal Reserve System.

The Federal Reserve System was created in 1913. It is the "central banking system" of the United States. It is the system of "banks' banks." You already know that there are 12 Federal Reserve Districts with one Federal Reserve bank (and usually one or more branch banks) in each district. The Federal Reserve bank in each district is actually owned by the "member banks" in that district. Each "member bank" holds some "Federal Reserve bank stock." But the "stockholder banks" (the member banks) don't have control of their Federal Reserve bank.

The Federal Reserve System is controlled by a seven-man "Board of Governors." Each member of this Board is appointed for a 14-year term by the President of the United States. The major function of this board is to decide on and carry out monetary policy for the nation.

## The Federal Reserve System Has Much Independence

The Federal Reserve System is a sort of "independent part of the government." There is no question that it is a "government instrument."

It does governmental things. Regulating the money supply certainly is a governmental function. Yet the Federal Reserve System isn't a part of either one of the "three branches of government." The system was set up as "an independent watchdog agency" to keep an eye on the nation's money. The Board does not "report" to anyone—it isn't under anyone's direction and control. If it wants to, it can be as free and independent as the Supreme Court! (Well, almost.)

Maybe the Federal Reserve won't go along with the Treasury. It doesn't have to buy all those bonds if it doesn't want to. Maybe the Board of Governors of the Federal Reserve System won't even go along with your policy, Mr. or Ms. President! They may decide that the advantages of easy money look too small, and that the dangers of inflation look too great. If that's what they decide, then unless you can persuade them or pressure them into changing their minds, there isn't much you can do to bring easy money. Oh, the frustration of it all!

Maybe you can persuade Congress to pass a law to change the Federal Reserve System and give you more control. Some people have said that that might be a good idea, but some others strongly disagree. But no matter. It would take forever to get such controversial legislation through Congress, anyway. If it ever did pass, by then it would be too late to do *your* program any good. So, Mr. or Ms. President, the success of your proposed "easy money" policy—and of your economic recovery plan, and perhaps of your bid for reelection—seems to hinge on the decision of the Board of Governors of the Federal Reserve System. If you want things to go your way, now would be a very good time for some of your very best aides to start to do some gentle but persuasive arm-twisting!

### The Fed Usually Goes Along With The President

As it turns out, Mr. or Ms. President, you didn't have to be concerned about the Fed's decision after all. Following your meeting, the Chairman of the Board of Governors called a meeting with the other six members of the Board. After some arguments, they agreed to go along with your suggested "easy money" policy. They decided that they will take the necessary action to make money readily available to borrowers, and at low interest rates.

Can the Fed really do it? Can they bring easy money if they want to? Yes, they really can. And luckily, you know enough about money, to be able to understand the entire process with no trouble at all.

How do they do it? They have several techniques—several "tools of monetary policy"—which they can use to make money easy (or tight). Most of the rest of this chapter will be talking about what those tools are and how (and how well) they work.

## HOW "CHANGING THE DISCOUNT RATE" AFFECTS THE MONEY SUPPLY

Perhaps the first thing the Fed would do is to announce that they are going to lower the discount rate. What does that mean? Think back. Remember when Baker was first trying to talk Zimmer into lending him $4,000? Zimmer's bank didn't have any excess reserves. One way to get excess reserves is to borrow the extra money from the Federal Reserve bank. The "discount rate" is the "interest rate" the Federal Reserve bank would charge Zimmer's bank for "borrowed reserves." That's all.

### Lowering The Discount Rate Brings Easier Money

You can see how a lower "discount rate" will make it easier and cheaper for people and businesses to borrow money from the banks. This "easy money" stimulates more spending, output, and employment.

Suppose your uncle is still thinking about buying that oyster-breading machine. He has been to see his banker three different times to talk about borrowing the money. But he doesn't think the investment would be feasible unless he can borrow the money at 7 percent, or less. His banker keeps telling him that the lowest available rate for this kind of a loan is 8 percent. So your uncle has not borrowed the money and he hasn't bought the machine. Many people agree with your uncle on this. Nobody is buying oyster-breading machines. The breading machine plant in Baltimore closed down last month.

### If Banks Can Borrow At Lower Rates, They Can Lend At Lower Rates

When the Fed decides to lower the discount rate, the picture changes. Suppose the discount rate was 5 3/4 percent and the Fed lowers it to 4 percent. Now your uncle's bank calls and offers your uncle a 7 percent loan. Your uncle borrows the money and orders the machine. Other seafood packing houses do the same. Soon the Baltimore plant may be working three shifts. There may be waiting lists, and delayed deliveries for oyster-breading machines. When the economy "turns around," sometimes it turns around very fast!

Look at what's going on in upper Michigan, and in upper Wisconsin and Minnesota. Several iron and steel companies have borrowed money and are building new pelletizing plants. The lower discount rate has eased money, all right! Many businesses are taking advantage of the lower rate of interest. They're buying and building the capital they have been thinking about for the past several months, or maybe years. They were poised and ready to go! The multiplier effect of the new investment spending is beginning to set off a consumer spending boom. Mr. or Ms. President, soon you may be worrying about the opposite problem. Shortages! Inflation!

It's easy to see how lowering the discount rate makes money easier to borrow and lowers interest rates. The lower rate at the Fed's "discount window" pushes down the interest rates in every nook and cranny of the economic system. Even the corporations which are borrowing money by selling their own bonds don't have to pay such high interest to get people to buy their bonds. This makes expansion cheaper, so they may sell more bonds and expand more. When the Treasury issues new securities and sells them to get money to pay off maturing securities, the interest rate on the new issues will be lower. The lower rates will be reflected throughout the entire money market, and the effect of the money expansion and increased spending will be felt throughout the entire economy.

When the Fed offers to make loans to banks at a lower discount rate, this amounts to an increase in "the supply of money available for loans." Money becomes easier to borrow, and at lower cost. In the case described here, this "monetary policy tool" (adjusting the discount rate) has been very effective. There doesn't seem to be any need to use any other "monetary policy tools." But it doesn't usually work out quite this way. Let's look at an opposite example.

### When Times Are Really Bad, No One Will Borrow

Suppose your uncle just closed down his seafood packing house last week. Sales had been on a downtrend for more than a year. Labor costs, packaging costs, transportation costs, electric power costs, repair costs, taxes, and all other costs had been rising constantly. And the federal, state and local health and sanitation regulations and antipollution regulations and working hours regulations and minimum wage laws and bookkeeping requirements and all the other regulations and restrictions were getting more costly and time consuming and annoying all the time. In the letter you got yesterday, your uncle said ". . . and I dumped every penny of my savings into trying to keep the thing going. It will be *one cold day in July* before I ever put another penny into that money-leeching white elephant! I swear to that!" Things are bad in Chincoteague. And all over.

Your roommate just got a letter from his father. Remember the steel company he was consulting for, about the pelletizing plant? Well, that company is now in receivership (bankruptcy). All the steel companies are facing serious financial problems—sort of like your uncle, only on a much larger scale. The same is true for the railroads and the automobile producers and for textiles and petrochemicals and building supplies and food processing and wholesale trade and retail trade and automotive services and everything else. Talk about depression! This really is a bad one.

How much will it help to lower the discount rate when things are really this bad and everyone is really pessimistic about the future? How low must the interest rate go before your uncle will borrow $20,000 and buy the oyster-breading machine? Or before the (bankrupt) steel company will borrow the $2.8 million and install the new pelletizing equipment? Or before all the other businesses will start buying new equipment and expanding?

What banker would be so stupid as to lend money to all these "broke and going broker" businesses, anyway? Mr. Zimmer, the banker, figures that anybody who tries to borrow and expand under such frightening circumstances has got to be some kind of a nut! Certainly such a person would not be a good credit risk. Mr. or Ms. President, your "easy money" tool of lowering the discount rate is not going to work. Better bring on another tool.

## HOW "OPEN MARKET OPERATIONS" AFFECT THE MONEY SUPPLY

A second and even more important technique the Federal Reserve uses to bring "easy money" is called "open market operations in government securities," or just "open market operations." This is a neat and simple little trick for pushing more money into (or pulling money out of) the economy, and into (or out of) the reserve accounts of the banks. You can see that this could be a very powerful tool of monetary policy. If it lets the Fed put money into or take money out of a bank's reserve account (whether the bank likes it or not!) then it's a very powerful tool of monetary policy. Right? Yes. And that's what it does, so it is. It's *the most useful tool* the Fed has for controlling the money supply.

"Open market operations" means planned, purposeful, manipulation of the bond and money markets, by the Fed, through the buying and selling of government securities. That is, the Fed changes the supply of or the demand for (and therefore, the prices of and interest rates on) government securities. Then the effect spreads all throughout the money market and makes money easy, or tight, and interest rates high, or low. Does this surprise you? It may, because you don't understand it yet. But soon you will. It's really neat how it works.

You already know that the Federal Reserve banks hold billions of dollars worth of government bonds. The Federal Reserve banks can make interest rates move up or down and can make money "easier" or "tighter" just by buying or selling these bonds "in the open market."

The bonds are bought and sold through "normal investment channels," including banks, stock brokers, and other securities dealers. If the Federal Reserve is selling government securities (bonds), then anybody who wants to, can buy them. If it's buying, it buys from anyone who wants to sell. That's why they call it "open market operations." The Fed is buying or selling bonds in the "open market"—in the regular, normal "bond and money markets" of the economy. But how does all this make money easy or tight? and interest rates low, or high? Watch.

## The Fed Buys Bonds, Pushes Bond Prices Up And Increases The Money Supply

The economy is in pretty sad shape. You, Mr. or Ms. President, want to follow a monetary policy of "easy money." You think that with low interest rates and easy credit, more people will borrow and spend. That will stimulate the economy and touch off a rapid recovery. So what should the Fed do with its open market operations at a time like this? Buy bonds! What good will that do? Let's take an example.

Remember Mr. Alber? The businessman who's going to night school working for his MBA degree? Well, he owns five, $1,000 marketable government bonds. Each bond pays 6 percent interest on its face value. That is, each is a 6 percent, $1,000 bond, so it pays its owner $60 a year in income. (6% of $1,000 equals $60.) Mr. Alber thinks these bonds are a pretty good investment. That's why he bought them in the first place. But now the Fed is buying bonds in the open market. Will Mr. Alber sell his bonds? That depends. What do you think it depends on? On how much the Fed is offering to *pay* for his bonds, of course!

Suppose the Fed offers to pay $1,200 for each of Mr. Alber's 6 percent, $1,000 bonds? Do you think he will sell? Perhaps. Let's suppose he does. So Mr. Alber turns his five bonds over to his banker or to a local "stocks and bonds" broker. Then the banker or broker sends the bonds to the Federal Reserve bank and the Federal Reserve bank sends Mr. Alber a check for $6,000 (for 5 bonds @ $1,200 each). (The broker gets to keep a little bit of the money as his commission, but we won't worry about that.)

## The Fed's Payments For Bonds Become New Bank Reserves

What does Mr. Alber do with the $6,000 check he just received? What *can* a person do with a $6,000 check? Deposit it in his bank account, of course. When Mr. Alber does that, what does the banker, Mr. Zimmer, do with the $6,000 check? What *can* a banker do with a $6,000 check drawn on the Federal Reserve bank? He sends it to the Federal Reserve bank, for deposit to his reserve account, of course.

See what has happened? Just automatically the amount of money in that bank has gone up by $6,000 (Mr. Alber's account, which he can now spend). The money supply is now $6,000 bigger than before! But that's not all. The bank now finds itself suddenly having about $4,800 in excess reserves in its account at the Fed!

Now Mr. Zimmer has to worry about what to do with all that excess money—how to use it to bring in some income. He would like to lend it to someone. You can see why. See how the Fed's "open market operations" are bringing easy money?

Mr. Zimmer is going to try to expand his loans so he can start earning interest on the $4,800 he has in excess reserves. He hopes Mr. Baker will come in and borrow the $4,800. Then Mr. Baker will go out and spend it and the increased spending will help to stimulate the economy. Then Mr. Culver, who receives the $4,800 check from Baker, will deposit it in his bank. Then Culver's bank will have excess reserves. You know how this process can continue until the money supply expands to five times the amount of Mr. Alber's $6,000 deposit.

Yes, the money supply can expand by as much as $30,000, or maybe more, depending on the exact size of the reserve requirement. But will it? Maybe not. If conditions in the economy are really bad, maybe nobody will want to borrow and spend. It may take some very enterprising bankers to find people to lend the money to.

It's easy to see how the Fed, by buying bonds, places new money in the hands of the people who are selling the bonds. Also it's easy to see that this new money is "high-powered money." It flows directly into the reserve accounts of the banks.

If the Fed is buying bonds, it is pushing new excess reserves into the accounts of banks all over the country. Does this bring easy money? You bet it does! But wait. You don't know the whole story yet. To see the full effect of open market operations on "the bond and money markets" (on interest rates and bond prices and easy credit and all that), you first must know about the interesting relationship between *interest rates* (in the money markets of the economy), and *bond prices* (in the bond markets of the economy). (Both "markets" are the same market, of course!)

## BOND PRICES AND INTEREST RATES

Open market operations provide a very neat way to adjust the availability of credit, and to push interest rates up or down. A banker with excess reserves would be likely to offer loans at lower interest rates. Excess reserves push interest rates down. But there is another, more direct way that buying bonds in the open market pushes interest rates down. To understand this, you only need to understand this fact (rather obvious once you think about it): *When the market price of a bond increases, the "interest yield" you will get if you buy that bond, decreases. So as bond prices rise, interest rates fall. It must be so!* It's really simple, but sometimes it seems confusing. Let's talk about it.

### Market Prices Of All Kinds Of Bonds Rise And Fall

First of all, we aren't talking about the "savings bonds" (E and H bonds) that most individuals are familiar with. The E bonds and H bonds are special, *non-marketable* bonds. They don't fit into this discussion.

When the Fed is trying to ease money, it buys "marketable" government bonds. It pushes up the prices of these bonds. This entices people, banks, insurance companies and others to sell their bonds. As it pushes up the open market prices of government bonds, the prices of *all* bonds react the same way.

All the money and bond markets are very closely tied together. When the prices of government bonds are being forced up, this pushes up the prices of AT & T bonds, U.S. Steel bonds, General Motors bonds, Amoco bonds, Tenneco bonds, and all the other marketable bonds in the country (there are thousands). As the bond prices in the market go up, what happens to the actual interest a person receives on his money when he buys one of the higher priced bonds? It goes down.

Lots of people are buying and selling bonds every day—government bonds, corporation bonds, New Jersey Turnpike bonds, Mackinac Bridge bonds, all kinds of bonds. These aren't newly issued bonds we're talking about. These bonds have been around for years. But people buy and sell them every day. The prices go up and down in response to supply and demand.

Every bond has a stated face value, and a stated interest rate. The usual "face value" is $1,000. The "interest rate" stated on the face of the bond may range from about 3 percent or less up to about 9 percent or more. How much a person *actually pays* to buy the bond "in the open market" depends on its *price*, that day. The higher cost of the bond, the lower the "interest return" he will receive on his money.

### Bonds Paying The Highest Interest Sell For The Highest Prices

The face value of a bond may not be very close to its market price. For example, a 20-year, $1,000, 3% bond would not bring as much in the open market as a 20-year, $1,000, 9% bond. That's just obvious, isn't it? When the 3 percent bond was issued, interest rates were low. But as interest rates (interest rates on new bonds) got higher and higher, the market value of the 3 percent bond got lower and lower. That's the way it always is. *If interest rates are rising, bond prices are falling,* and vice versa. Or we can say it the other way and it's just as true: *if bond prices are falling, interest rates must be rising,* and vice versa.

Suppose we are talking about a $1,000 bond which pays 6% per year. The owner of that bond will receive $60 a year. Now suppose you paid $1,200 when you bought the bond. How much income would it pay you per year? Only $60, of course. That's what the bond says. It says it will pay 6 percent on $1,000 to its owner, whoever he may be! Even if you paid $1,200 when you bought the bond, you still aren't going to receive but $60 annual return on your $1,200 investment. That happens to be a return of only 5 percent. (5% of $1,200 is $60.) The bond is still paying 6 percent on its face value ($1,000). But *you* paid $1,200, and since you're only getting a return of $60 a year on your $1,200 investment, you are actually getting a 5 percent return on your money.

Why would anyone ever pay $1,200 for a $1,000 bond? Because they think it's the best investment for them available in the bond market, that's why! Suppose the new bonds being issued these days carry an interest rate of only 4 percent. Would you be willing to pay $1,200 for a long-term $1,000 bond that pays 6 percent? I would!

### Very-Long-Term Bond Prices Reflect Only The Interest Return

Just to make the point, let's suppose we are talking about 500 year bonds. The only reason a person would want a 500 year bond is for the interest return, right? He certainly wouldn't plan to hold it to maturity! So suppose you are considering buying some 500-year, $1,000 bonds. You can get 3 percent ones for $500 each, or 6 percent ones for $1,000 each, or 9 percent ones for $1,500 each. They're *all* $1,000 bonds, but they are selling at such different prices! Which ones would you buy? It really wouldn't make any difference, would it?

You are going to earn 6 percent on your money, no matter which bonds you buy, right? Guess what the going rate of interest is on new bonds being issued these days? It's 6 percent, of course! The 3 percent bonds were issued sometime back when money market interest rates were low. So what happened to the value of the 3 percent bonds when the money market interest rate went up from 3 percent to 6 percent? The bond values dropped from $1,000 to about $500.

The 9 percent bonds were issued sometime back when money market interest rates were high. So what happened when the money market rate went down from 9 to 6 percent? The bond values went up from $1,000 to $1,500. If you hold marketable bonds while the money market rates of interest are *rising,* chances are that the open market value of your bonds will be *falling*. But if money market rates of interest are *falling*, the market value of your bonds will be *rising*.*

Now you can see that whenever the open market values of existing bonds are being forced up, interest rates are automatically being forced down. Whenever interest rates are moving down, open market values of existing bonds are moving up. When interest rates are *falling*, people want to buy the high-interest, *old* bonds and they will pay a *premium price* to get them. When interest rates are *rising*, people want the high-interest *new* bonds. Anyone who wants to sell his low-interest, *old* bonds must sell them at a lower price, below "par." Otherwise no one would buy them. Would you? Of course not.

### People Hold Bonds To Get The Interest Income

This is an essential concept, and so often misunderstood. Let me say it just one more time, one more way. People buy bonds because they get income from owning bonds. How much you will pay for a bond depends on how much income you will get from owning that bond.

How much income do you insist on getting when you invest your money in bonds? The *going rate*, of course! Suppose the "going rate of interest" is getting lower. If you want to buy one of the old, high-interest bonds, you expect to pay more for it. Right? If the *new, low-interest bonds* are selling at *face value*, then the *old, high-interest bonds* will be selling at a *premium*. Why? Because they pay more interest, that's why!

Suppose the "going rate of interest" is getting higher. If someone wants to sell you one of his old, low-interest bonds, you expect him to offer it to you at less than face value. Right? Otherwise, would you buy it? Of course not. You would buy one of the new, high-interest bonds instead. As market rates of interest change, market values of existing bonds also change, but always in the opposite direction. It must be true, and now you understand why.

### Open Market Operations Directly Influence Bond Prices And Interest Rates

From all this discussion of the relationship between bond values and interest rates, you can see what happens when the Fed goes into the market and pushes up the value of government securities—like when it

---

*In real world markets, at any moment in time, this relationship may not be *precisely* true. Why? Because not only *current* conditions but also *expected future* conditions affect open market bond values and interest rates. But don't worry about this—not unless you're planning to speculate in the bond markets—in which case, please get some professional advice!

paid $1,200 each for Mr. Alber's $1,000 bonds. Mr. Alber's bonds have a face value of $1,000 each, and they are 6% bonds, so each bond pays its owner $60 a year (6% of $1,000 is $60). Those bonds will *always* pay $60 a year to their owner.

Now suppose Mr. Turner pays $1,200 for one of Mr. Alber's $1,000 bonds. Mr. Turner still only gets $60 a year. So how much interest is he making on his $1,200 investment? Only 5 percent! ($60 is 5% of $1,200.) So when the Fed goes into the open market and pushes the price of $1,000, 6% bonds up to $1,200, it automatically pushes "the effective rate of interest on government bonds" down to 5 percent. But that isn't all. When the effective rate of interest on government bonds is pushed down, the rate of interest on everything else goes down. All interest rates move together (more or less).

Now you can see that as the Fed forces up bond values, interest rates throughout the economy automatically are forced down. At the same time, high-powered money is being pushed into the reserves of the banks. You, Mr. or Ms. President, are hoping that the lower interest rates and the greater availability of loanable funds in the banks will induce more businesses and consumers to borrow and spend. This could set off the multiplier and the accelerator. Then soon the economy would be booming again.

Will the open market operation of "buying securities" really work? Nobody knows for sure. If conditions are really bad, then, no. It won't work. It might help a little, but it will not be likely to bring a quick economic recovery. But if conditions aren't really *too* bad the open market operations may work very well. It's never possible to be sure just how much effect a given amount of "open market operations" will have.

One of the best things about the "open market operations" approach to monetary control, is that the policy can be adjusted gently and quickly as circumstances change. The Fed has an "open market committee" which meets frequently to consider what the Fed's "open market policy" should be. This flexibility adds to the effectiveness of this *most important monetary control tool*—open market operations.

## OTHER MONETARY POLICY ACTIONS

Now you understand the two most frequently used tools of monetary policy—"discount rate changes," and "open market operations in government securities." A third "tool" consists of changing the reserve requirement itself. Suppose the amount required in reserves to "back up" the banks' demand deposits, is 20 percent. Remember Mr. Zimmer and his bank? He had $1,000,000 in checking accounts and $200,000 in reserves. If the reserve requirement was 20 percent, Zimmer's bank had just enough reserves to be "legal." What would happen if the reserve requirement was changed?

## Changing The Reserve Requirement

Suppose the Fed wants to ease money, so it decides to lower the reserve requirement to 19 percent. Suddenly Mr. Zimmer's bank has $10,000 in excess reserves, and so do other banks all over the country. Every dollar of these excess reserves can support an expansion of about five dollars in the nation's money supply. Billions of dollars of lending power have been created at one fell swoop! Talk about a meat cleaver effect! Reserve requirements can't be changed or adjusted as easily and sensitively as the other monetary policy "tools." For this reason, this "tool of monetary policy" is used only infrequently and with caution.

## For Tight Money, Use The Same Tools The Other Way

We have been talking about using monetary policy to bring "easy money" to overcome unemployment, recession and depression. Now let's look at the opposite kind of problem. What kind of monetary policy should we use to slow down a runaway boom? Suppose the people and businesses are trying to buy more than the economy can produce. Shortages and waiting lists are everywhere. Inflationary pressures are serious. Prices threaten to break loose and skyrocket.

What would be the proper monetary policy to slow down total spending? That's easy to figure out. Tight money! Make money difficult to borrow, and push interest rates up. How? Raise the discount rate. Sell bonds in the open market. We might even consider raising the reserve requirement. That would be a shock!

Raising the discount rate will discourage banks from lending and it will cause them to charge higher interest rates on their loans. Selling bonds will push down open market bond prices. That will push up interest rates. You know exactly how that works, now. Right? Furthermore, the checks people write to the Fed to buy the bonds, will be taken right out of the banks' reserve accounts at the Fed! This will pull down bank reserves. Banks will have to restrict their lending.

When loans are paid off, the banks will have to deposit this money (the repayment money) in their reserve accounts at the Fed. They can't use the money to make more loans. So the money supply will get smaller and smaller. People will have less money to spend, and it will be much more difficult to borrow, or to buy things on time. Will total spending slow down? You bet it will!

### Open Market Operations Can Force Down Lending And Spending

Let's look at the process of tightening money, using open market operations. Suppose our friend Mr. Alber has some excess cash in his bank account. He has been thinking about buying some government securities, but he just hasn't gotten around to it yet. Today he hears that $1,000, 6-percent government bonds are available for a price of $800 each. This seems like a good deal so he decides to buy some.

He buys five of the bonds—total par value of $5,000—for $4,000. He writes a $4,000 check. The check goes from the bond broker to the Federal Reserve bank. There, $4,000 is deducted from the reserve account of Mr. Zimmer's bank. Then the check is sent to Mr. Zimmer's bank, where $4,000 is deducted from Mr. Alber's account. The bank has lost $4,000 in deposits and also $4,000 in reserves. Is it in an illegal position? Yes! The bank needs less reserves because it has less deposits. But *how much* less reserves does it need? Only $800 less. But it has $4,000 less!

The bank must get another $3,200 from somewhere, to make up the "illegal deficit" in its Federal Reserve account. Where can it get the money? It could borrow the money from the Fed. But remember? As a part of its tight money policy, the Fed has just raised the discount rate. So borrowing from the Fed is expensive!

Zimmer's bank might sell some of their securities to get more money to deposit in their reserve account. But that's expensive, too. Bond prices are depressed now. Remember? What the bank is more likely to do is to reduce its lending. As old loans are repaid the bank just won't relend the money. They'll put the money in their reserve account, instead. That will get their reserve account built up again. Meanwhile, they will temporarily borrow from the Fed to avoid being in an illegal reserve position.

As the bank cuts back on new loans and refuses to renew old loans, businessmen will not be able to expand so much. Some may have to sell some inventories, reduce their expansion plans and pull in money to pay off their loans. You can see that this will tighten business spending.

The higher interest rates will reduce the demand for machines and equipment and materials, and for inventories. At the higher interest rates it becomes more expensive to carry inventories, and it's more costly to buy new machinery and equipment and things. Another thing. The capitalized value of everything goes down. You learned about high interest pushing down capitalized value, from your uncle in Chincoteague. Remember? It sounds like this tight money policy is going to succeed. Will it? Maybe. That's what we'll be talking about in the next section.

## HOW EFFECTIVE IS MONETARY POLICY?

Considering all the approaches, all the tools, how effective can monetary policy be? The Fed has several decades of experience.

How well have they done? Sometimes, apparently all right. Sometimes not so well. We can't really be sure. Everything the Fed has done with monetary policy has been criticized by someone.

The Fed has been accused of moving too slowly and doing too little. It has been accused of doing too much, of doing the wrong things, and of doing things at the wrong times. Even looking back, with all the wisdom of hindsight, it's hard to be sure which times the Fed did the right things, and which times it didn't. So it's difficult to be sure just how effective the tools of monetary policy have been.

### Our Picture Of The Economy Is Always Late

One of the frustrations in trying to use monetary policy (or any kind of stabilization policy, for that matter) is that we really don't know what's going on in the economy at any moment. We can't see what's going on today. We only see what was going on some weeks ago. Trying to prescribe stabilization policies in the real world is sort of like a doctor's trying to prescribe treatment for a patient he hasn't seen for several weeks.

Try to picture this situation: You don't feel very well, so you go to "a council of doctors" for tests, in mid-June. Then you wait. In late July the doctors figure out what was wrong with you, back in June. But the opinion is not unanimous. Several eminent doctors disagree with the diagnosis. But anyway, medicine is ordered. By the time the medicine starts, you may be well again. Or you may be dead! But once the medicine starts coming, you must keep taking it until the council of doctors agree that it should be stopped.

It may be several months before anyone can tell for sure if you're getting better or worse. Can you see why monetary policy—and all economic stabilization policy, really—is a sort of chancy thing? Only when conditions in the economy *really get serious* do the economists begin (more or less) to agree on what needs to be done, and how.

### People Will Spend The Easy Money Only If They're Ready

In general, we can say this: If businesses and consumers are about ready and have been waiting for some little additional nudge to shake them loose, then "easy money" is likely to have a quick effect. But if businesses are very pessimistic about the future and if the people are out of jobs and deeply in debt, then no amount of "easy money" is likely to induce very much increase in spending. No businessman is going to buy a new machine unless he expects to be able to sell his output! I'll bet that even if someone offered to lend your uncle $20,000, *interest free*, he still wouldn't buy that oyster breader!

If things are really bad, the consumers won't start borrowing and buying things. Who would lend money to an unemployed person who's already facing overdue debts, anyway? See how difficult it is to try to figure out what monetary policy to recommend—or which tool(s) to use? And just think how unresponsive the economy might be when things get really bad!

What about the opposite kind of macroeconomic problem—the problem of overexpansion, and inflation? How effective can monetary policy be in tightening down on an "overheated economy"? We know for sure that if money gets tight enough, people and businesses will spend less. Very tight money can definitely slow down spending in the economy. There is absolutely no question about that. But what we don't know is *when* to tighten, and *how much*. And still more serious, we don't know whether the induced *slowdown in spending* will have most of its effect in *reducing inflation*, or in *reducing production and employment*.

### Tight Money Could Result In "Over-Kill" And Recession

How much "tight money" do we need to "just do the job and no more"? A little bit of "over-kill" when we are trying to curb the boom could result in a serious downturn in the economy. Then the multiplier and accelerator, working downward, could bring widespread unemployment and depression. We just don't know how sensitive the economy will be, when we start to tighten money.

If a slight increase in the discount rate happens to shatter business optimism, the economy could move into an immediate recession. On the other hand, the business community may completely ignore a small change in the discount rate. It's impossible to know what the effect will be. If the business community overreacts it may be impossible for the monetary policy to be changed back in time to avoid a recession. With your knowledge of the accelerator and the multiplier, you know how hard it is to stop a cumulative expansion or contraction once it gets going!

Economic stabilization is always sort of chancy. But the modern world economies of mixed socio-capitalism require that the government be concerned with this issue. It can't be ignored. Sometimes problems arise which *require* action. At those times you may be sure that some action will be taken. If things don't soon improve, more (and maybe different) actions will be taken. I think you can be sure of that.

The tools of monetary policy are being used all the time. That's right. Every day, even! You will be hearing about these things all your life. When you buy a house or a car, or borrow money or use your "revolving credit account," or when you put money in your savings account you will be feeling the effects of monetary policy.

Monetary policy is very important in every modern economy. But it isn't the only approach to economic stabilization. Fiscal policy is another approach. You'll learn about that in the next chapter. But first Mr. or Ms. President, if I may venture a suggestion: Before you go on, maybe you should find a quiet spot out by the White House rose garden and spend some time reviewing and thinking about monetary policy.

---

## REVIEW EXERCISES

**MAJOR CONCEPTS, PRINCIPLES, TERMS (Try to write a paragraph explaining each of these.)**

monetary policy
changing the discount rate
open market operations
bond prices and interest rates

**OTHER CONCEPTS AND TERMS (Try to write a sentence or phrase explaining what each one means.)**

easy money
tight money
government securities
the "money market"
the "Fed"
the discount rate
the "over-kill" problem

**QUESTIONS (Try to write out answers, or jot down the highlights.)**

1. For many years there has been a continuing controversy among some economists over the question of the "independence" of the Fed. Some say independence of the Fed is essential to insulate the nation's monetary policies from the whims of politicians. Others say the president is supposed to be responsible for the health of the economy, but because the Fed is independent he is denied control over the monetary policy tools he needs to do the job. What position do you take on this issue, Mr. or Ms. President? Why?
2. Under what kinds of economic conditions in the nation would you expect monetary policy to be most effective? or ineffective? Explain.
3. Can you explain how open market operations work, *both* (1) through the direct effect on the money supply and bank reserves *and* (2) through

the direct effect on bond values and interest rates? It's important that you be able to do that.

4. After all is said and done, how effective do you think monetary policy really is in stabilizing the economy? Do you think it's more effective in overcoming depression? or inflation? Explain.

# 18 USING FISCAL POLICY TO STABILIZE THE ECONOMY

*Changes in government spending or taxes can be used to speed up or slow down the economy.*

How do you like pretending you are the President of the United States? A little credibility gap? Well, I guess so. But now that you are experienced at it, please stick with it a little longer. Okay?

Let's go back to where we were at the beginning of the last chapter. The economy is depressed. You, Mr. or Ms. President, are looking at the high unemployment figures and thinking, "Something must be done quickly!" You think: "The depressed economic conditions somehow must be overcome. We must get the recovery going and get those unemployed people back to work, at least before election day!"

You have already had a few meetings with your Council of Economic Advisers. Also, you have discussed this matter at length with the Chairman of the Council of Economic Advisers, the Secretary of the Treasury, the Director of the Office of Management and Budget, the Chairman of the Board of Governors of the Federal Reserve System, with some bankers and businessmen, and with a few economics professors. You are quite sure that the Fed is going to do something to bring "easy money" to the economy. But you aren't quite satisfied with that. You are still worried.

Remember, at your last meeting with your advisers, you said: "Monetary policy may be just fine. But I don't trust it. All we are doing is pushing money into the banks and into the hands of the people and making it easy for businessmen and consumers to borrow and buy things. *Suppose nobody wants to borrow and buy?* Suppose all anybody wants to do is put his money into a savings account and just let it sit there? Then what good will all this monetary policy do? No good! So that's why I'm worried."

## USING FISCAL POLICY TO INDUCE PROSPERITY

All the economists agree that monetary policy is a sort of indirect approach to the problem. It depends on what the people do. Are they going to borrow, and spend? Mr. Alber's new bank deposit certainly isn't going to increase the money supply any further unless Mr. Baker or somebody comes in and borrows the new excess reserves. If everyone is as pessimistic as your uncle in Chincoteague, there isn't going to be much borrowing and spending, easy money, or not!

### Fiscal Policy Can Be More Direct Than Monetary Policy

You are impatient, Mr. or Ms. President. You say, "Isn't there some way we can take *more direct* action to solve this problem? Can't we just take the bull by the horns and do something that doesn't depend on what a lot of consumers and businessmen decide to do?" Yes, there is something. Your economic advisers agree that you could *increase government spending*. That would be *sure* to put more spending into the total spending flow of the economy.

How obvious! Instead of trying to do something to induce the consumers and businessmen (C + I) to increase their spending, why not just come in and increase government spending (G) directly? Once you think of it, how obvious it is! The surest way to increase total spending is to increase it yourself! So you decide to do that. You decide to go to Congress and persuade them to pass some new spending bills—some new "appropriations legislation."

### Increase Government Spending Or Cut Taxes

You might try to persuade Congress to appropriate enough money so the government could *hire all the unemployed people*. That certainly would be the most direct way to solve the unemployment problem! Or you might get the Congress to appropriate money to build new school buildings and post offices and highways and recreation areas and military bases. That would create more jobs, more incomes, more spending. Maybe the multiplier and the accelerator would catch on and take it from there. Do you try it? Maybe not.

Maybe you decide that this is not a good time to try to get Congress to increase government spending. Maybe you decide to try to get them to cut taxes instead. If the government takes less taxes out of a fellow's paycheck, he gets to keep more money and, likely, he will spend more. If enough people spend enough more, that will solve the problem. Or maybe the government will cut taxes on business profits. Then businesses will be likely to spend more for investment, hire more people and produce more. That might do the trick.

So which will you do, Mr. or Ms. President? Why not try some of both? Increase spending some and cut taxes some. Use both "tools" of compensatory fiscal policy. When you adjust taxes or spending to "compensate" for too little (or too much) spending by consumers and businesses this is called "compensatory fiscal policy." Suppose you decide to try it. First you plan to cut personal income taxes and leave more money in everybody's paycheck. This should push total consumer spending up and help to get things going again. So you get some of your friends in Congress and some lawyers to start drafting a tax-cut bill.

### The Program Must Meet Political Realities

While the tax-cut bill is being prepared, you begin working with a group of Congressional leaders, drafting a program of expenditures for federal "public works" projects—new highways and streets, water and sewer systems, public buildings, things like that. Each senator wants several projects in his home state. Each Congressman wants one or more projects in his home district. All the "party faithful" governors and mayors throughout the country expect to be rewarded for their party loyalties.

This could get out of hand! Everybody can't have everything he wants. Too much government spending would create much more factor demand than there are factors available. *Surplus* would suddenly become *shortage*. Total spending, or "aggregate demand" would be too great for the productive capacity of the economy. Prices would break loose and go up. You don't want that to happen. You want to do enough, but *not too much!*

Finally, after many precious weeks have been lost in haggling, your legislative program for "compensatory fiscal policy" is ready to be offered to the Congress. You go to Congress and give an impressive speech in support of the tax cut and the "public works" spending program. You urge Congress to take speedy action on these high priority measures. But then what? The weeks drag by and nothing much seems to happen. Several leaders of Congress express concern about your program. "Spend more, and tax less? Blatant fiscal irresponsibility!" Some of your own party, but mostly the opposition, keep making noises about the "government deficit" which your proposed program is going to bring about.

Other Congressmen (and their economic advisers) say the program is one of over-kill. They say: "The economy is about ready to start booming anyway. This big push by government will bring shortages and inflation for sure!"

Still others say the opposite: "We are really on the verge of a serious depression. Inventories are high and general confidence is low. Just look at the recent stock market slide, for example! The President's proposed program is too little and too late. It's only a drop in the bucket. Let's build a public works program big enough to do the job!"

### How Much Fiscal Action Is Needed?

Who is right? Is your program the right size? With the right emphasis? Or not? Can't your economic advisers with their economist-statisticians (econometricians), tell you the answer? No, unfortunately they can't. They can make very scientific-looking guesses, but there's just no way they can tell what the people and businesses are going to do, once your program gets started. They can *estimate*, but they can't tell for sure.

Suppose inventories are down to low levels, and businesses are looking for some little excuse to begin re-stocking their shelves. Suppose consumers only need some little indication that things are going to get better. They are just waiting for some excuse to run out and start buying new vacuum cleaners and refrigerators and stoves and furniture and TV sets and automobiles and flared hip-hugger jeans and all. The multiplier and the accelerator are lying just under the surface, poised for action. In these conditions not much government spending, government hiring, or tax cutting will be needed to speed up the economy. But under different circumstances, things could be quite different.

### The More The Economy Is Overstocked, The Slower The Recovery

Suppose most businesses—manufacturers, wholesalers, retailers—are overstocked with (and are deeply in debt for) excess inventories. And suppose most consumers have just recently been on a buying binge. Everybody has a new car, new refrigerator, new furniture, new everything, and is up to his ears in debt. How much must the government cut taxes and increase spending to get the economy to speed up? Quite a lot! When taxes are cut and government spending increased, people and businesses will just take the extra money and pay off their debts. Nobody will start buying new cars, or TV sets, or things. Businesses won't hire anybody. The toy factory is so overstocked with toys that it would be able to meet the Christmas boom without hiring anybody to make another toy!

The economy is overchoked with inventory, with excess manufacturing capacity, and with debt. The multiplier and accelerator are buried deep under all this mass of inventory and excess plant capacity and debt. Until some of the inventories and debts are cleared away the "multiple expansion process" will work only very sluggishly, if at all. That's what happens in a big-big boom. We build big inventories and big debts. Now you can see why a very big boom is likely to lead to a very bad depression.

Is there any way that you, Mr. or Ms. President, can find out whether or not the economy is about ready to surge forward? Or if it is buried in excess inventories and debt? Luckily, there is. During the past few decades great progress has been made in keeping tabs on what's going on in the economy. Reports from the Departments of Commerce, Labor, Agriculture, and others will tell you how much the consumers are in debt, how old the average automobile is, how much inventory is being carried in each industry, what the businessmen and consumers are buying these days, and all sorts of things. These statistics help a lot, but they don't tell you everything.

### Reactions Of Consumers And Businesses Are Hard To Predict

The statistics can't tell for sure how businesses and consumers will react to changing circumstances. It's much easier to tell how old a person's car is than to tell whether or not he will buy another car this year if his taxes are cut by 8 percent. It's much easier to tell the level of debt in an industry than to tell how much sales must increase to set off an inventory expansion.

Individuals make their economic choices on the basis of their outlook —their expectations. They seem to all move at the same time. Now you know how a cowboy feels when he's trying to figure out how much noise it will take to move the cattle, without stampeding the herd! It's the same kind of question. Sometimes the cattle are spooky and poised for a quick move. How much noise? Just a little! At other times the cattle are feeling overfed and lazy. It takes quite a lot to make them move.

We take all kinds of surveys and opinion polls to find out what people and businesses *think* they are going to do. This information helps. But sometimes people don't really know what they're going to do until the time comes. People and businesses have a tendency to do whatever the people and businesses down the block are doing. This makes "economic forecasting" very hazardous.

So what now, Mr. or Ms. President? Want to change your mind and withdraw from the race? You really don't know how much, and what kinds of "fiscal policy" action to take. Cut taxes? How much? Increase spending? Spend for what? How much? See the problem? You face the real possibility of "ineffectiveness," or of "over-kill." The very best program you and your advisers could design might hit far from the mark. But no matter. The program you design isn't going to get through Congress anyway.

### Congress Doesn't Always Cooperate

Tax cuts and spending projects are of *great* interest to Congress. So no matter what you do, Congress is going to have the last word on this program. Some Congressmen will be fighting for what they see as "the good of the nation." Others will be fighting for reelection or to pay back a political favor, or for some other reason. Some of them may even be fighting for your job! But all of them will be fighting.

Which taxes will be cut, and by how much, and what projects will be undertaken, and where and when—these matters will ultimately be decided by Congress. Political considerations are likely to outweigh the economic considerations in *every one* of these taxing and spending decisions. Even though you and your advisers had known with absolute certainty what *should* be done, it is an absolute certainty that your "perfect program" couldn't have gotten through Congress anyway!

You, Mr. or Ms. President, are facing a tough problem. I suppose the best thing you can do is to get yourself some good economic advisers, then try to figure out what kind of program you can persuade Congress to pass, then design your program, cross your fingers and go ahead. If you do everything carefully and if you're lucky, maybe you will get to have a second term in the White House.

What we are talking about is "compensatory fiscal policy," or just fiscal policy. It's a powerful tool if you can get it to work right. You can adjust taxes and spending to overcome depression, or to hold down a boom and relieve inflation. Now, a look at how fiscal policy is used to fight inflation.

## USING FISCAL POLICY TO CURB INFLATION

Suppose almost everyone is trying to buy a new TV set and a new car and a new washing machine and new furniture and new hip-hugger jeans and lots of other things. And suppose every manufacturer of automobiles and home appliances and furniture is trying to double his plant capacity, and every wholesaler and retailer is trying to expand his inventory. Can all these demands be met? Obviously not. There's just no way!

Everybody is working overtime, making lots of money and trying to buy things. But everywhere you look there are shortages, waiting lists, people standing in line, delayed deliveries for everything. Some people have the attitude: "I want what I want and I want it now!" People are bidding against each other, trying to get the available things. Prices are being forced up.

Automobile dealers are making big profits. Many dealers start paying extra to try to get more cars than their normal allotments. The same thing is going on in the TV markets, in furniture, appliances, and in everything else. The manufacturers are twisting every arm they can twist, paying extra to try to get more machinery, more steel, more delivery trucks, more coal, more labor, more everything.

### The Threat Of Inflation

Do you see the picture? It's a booming economy, all right. An economy booming at this pace is headed almost certainly for two disastrous results. The first is the rapid price inflation which will eat up a lot of the purchasing power which is now making Mr. Average Consumer feel so wealthy. He's going to find out that he won't really be able to buy the car and the appliances and all the other things that he thought he was going to buy with all of his money. He's going to pay higher prices than he thought, and run out of money sooner (and/or be in debt deeper) than he expected.

What's the situation with the business firms? All businesses are trying to expand as fast as they can. They're ordering more inventories, machines, factories, everything. The railroads have ordered thousands of new railroad cars, but deliveries are slow. The steel companies have ordered several new iron ore pelletizing plants to be installed in upper Michigan, Wisconsin and Minnesota, but the installations are coming very slowly. The seafood producers and packers—shrimp on the Gulf, oysters in the Chesapeake Bay area, and flounder fillets and lobsters in New England—all have ordered the most modern labor-saving machinery and equipment. But deliveries are very slow.

These are "good times" from the point of view of employment. Everybody who wants a job gets a job. Income is good. Producers are begging workers to work overtime. Anybody who wants to work can get a job. That's great. But shortages are everywhere. Prices and wages are under serious pressure to break loose and start leap-frogging each other until they go out of sight over the horizon!

### The Threat Of Overbuilding, And Collapse

In addition to the inflation threat, another disaster is threatening. Everybody is ordering things, stocking up. Businesses are buying machines and equipment and inventories and going into debt. Consumers are stocking up on cars and refrigerators and other consumer durables and going into debt. Manufacturers are expanding their plants and installing new machines and going into debt. This rate of expansion can't possibly be sustained for very long.

If this big boom continues, soon there will come a day (in the next year or two) when everybody will have everything he wants and be head over heels in debt. Consumers will have new cars, furniture, appliances, well-stocked freezers and be head over heels in debt. Businesses will have all the inventory they could possibly want and be head over heels in debt. Manufacturers will have all the plant capacity they could possibly want, and be deep in debt. Then what will happen? Recession, of course!

Eventually the inventory buildup will cease. "Normal" levels of demand will return. The factories which have been producing to meet the boom-time demand for the build-up, will no longer be able to stay busy. They will cut back production. Overtime work will cease. Some unemployment will develop and you know what happens next. The unemployed people stop buying as much. So retail stores stop ordering so much. Soon the wholesalers and retailers will stop ordering and start selling out of inventory. Spending slows down even more. Unemployment spreads. Businesses and consumers stop spending and start trying to pay off their debts. Here comes our depression again! Scary, isn't it?

See why we want to hold down total spending in this supercharged economy? We want to keep from getting hit by the one-two punch of runaway inflation, followed by economic collapse and depression. How do we cool down this excessive boom before it's too late? You already know about the monetary policy tools of raising the discount rate and selling bonds in the open market. But what about the *fiscal policy* tools?

### Cut Government Spending Or Raise Taxes To Hold Down The Boom

One way to curb the boom with fiscal policy is to increase taxes on businesses and consumers. This will force them to cut back their spending. The other way is to cut back on government spending. Delay all new government projects and slow down or stop ongoing projects. This reduces the demand for labor and the other factors of production. Reduced government spending will release factors of production so they can shift to (and relieve shortages in) the "private sector" of the economy.

Would this "counter-inflationary fiscal policy" (increasing taxes and cutting back government spending) really work? Yes, it really would. Except that you will have all the problems you had before. Here are some questions to be considered:

> How much to raise taxes? Which taxes? How much to cut back spending? Which projects to stop? Slow down? Delay? Which government employees to lay off? Which private business contracts to cancel? How do I get the answers? Then how do I get the Congress to go along?

### Cutting Spending And Raising Taxes May Mean Political Suicide

Suppose you could get the "right" answers to all these questions (which you can't). Would you then prepare a message and go to Congress urging the tax increases and the spending cuts? Do you want to go down in history (or into the next election) known as the tax-raising, project-cancelling boom-killer? Suppose you're willing to take that chance. How many members of Congress do you think will go along with you?

Later, when you're stumping the country, campaigning for the next election, you will explain to the people why it was necessary to raise taxes and cut back their favorite projects. You will explain that the tax increase didn't take nearly as much of their purchasing power away as the *inflation* would have taken, so they're really better off, paying more taxes. You will explain that the reason you cancelled all their government projects—the new hospital and the pollution-free sewer system and the manpower training program and the promised bridge across the river—is that "they, the people" were spending so much for new cars and appliances and things that there weren't enough factors of production left over to keep the government projects going.

You will tell them all about opportunity costs—that "You can't have your cake and eat it too. The economy can't produce any more than it can produce." You will teach them all sorts of good lessons in sound economics. Guess what they'll tell you! (I'm sure you can guess.)

Even if they know you're telling the truth, this isn't the kind of truth people like to hear. They might ask why you didn't cancel the foreign aid programs instead of the domestic programs, and why you didn't cancel any of the programs in *your own* home state. Can you begin to see the difficulty of raising taxes and cutting government spending, to fight inflation? It can be made to work, yes. But it sure isn't easy.

Even if you didn't have to worry about the political repercussions, and even if you could get Congress to go along with your program, counter-inflationary fiscal policy is tricky business. How much should you increase taxes or reduce spending? That's hard to say. You always face the problem of choosing the right "dose"—between ineffectiveness, and over-kill.

### How Effective Is Compensatory Fiscal Policy?

So what can we say about compensatory fiscal policy? Is it effective? Maybe. Sometimes. If we are willing to take enough fiscal policy action during depression—cutting taxes, increasing expenditures, hiring people on government payrolls, buying up surplus outputs and so on—we can definitely overcome the depression. If the government adds enough spending to bring the economy up to full employment, then we will have full employment. Obviously!

We can also be sure that if we are willing to take enough fiscal policy action during times of shortages and inflationary pressures, we can cut back the inflationary pressure. But in both these cases we run the risk of "over-kill." Each time we move to overcome depressed conditions, we run the risk of driving the economy into inflation. Each time we move to hold down inflationary spending, we run the risk of generating unemployment and recession.

Fiscal policy is not a "fine-tuning knob" for economic stabilization. We can't hope to use it to maintain "just the right level" of spending, employment, output, and income. When we try to solve our macroeconomic problems with fiscal policy, we go after them, not with a surgeon's scalpel but with a meat cleaver. Any time "meat cleaver action" is justified, fiscal policy will work. It may work too much or it may work too little, but it will work.

Whenever serious depression threatens, you may be sure that Congress will take some fiscal policy action. And you may be sure that *it will work*. For this reason, you may be sure there will never be another disastrous depression in the American economy. The government will not let it happen.

We might overdo it and run into inflation. But no doubt we will take that chance. We might have to establish wage and price controls—maybe government rationing, even. But it seems clear that the people would prefer widespread economic controls to widespread unemployment. The point is this: There are, today, *known alternatives* to serious depression. It seems to be a safe bet that fiscal policy will be used to whatever extent necessary to prevent serious and prolonged depression.

Yes, the government will use fiscal policy sometimes—and it will work. Not perfectly, but well enough. If it doesn't work well enough at first, then more and different things will be done *until it does work* at least "well enough."

The government will unbalance the budget and run deficits and add to the debt as much as necessary to get things going again. But what will all these deficits and debts do to the economy? Let's talk about that.

## FISCAL POLICY, THE UNBALANCED BUDGET, AND THE NATIONAL DEBT

Maybe you haven't thought about it yet, but fiscal policy really is a policy of *unbalancing the federal budget, on purpose!* The idea is that if total spending is too small, let's increase government spending (that will *directly* increase the size of the spending flow in the economy) and let's reduce taxes (that will *indirectly* increase the size of the spending flow in the economy). But increasing spending while cutting taxes obviously means "unbalancing the budget." It means "deficit financing." It means "increasing the national debt." What about that? Isn't that bad?

### An Unbalanced Budget Increases The Debt

When the budget is unbalanced, that means the tax revenues coming in don't provide enough money to support the government's spending. So how does the government make up the difference? It sells bonds, of course! If it sells the bonds to individuals and businesses and banks and insurance companies and all, this pulls money out of the economic system. Then when the government spends the money it puts the money back into the economic system. When the government does this, the federal debt gets larger. The government owes the debt to the people who are holding the bonds.

Suppose the government did not sell the bonds to individuals and businesses and banks and insurance companies and all, but sold them to the Federal Reserve banks instead. What difference would that make? Quite a lot! A "Federal Reserve financed government deficit" would be much more stimulating to the economy. Why? Because then the government would be *creating new money* and putting it into the economy. New *high-powered* money!

When the Treasury borrows from the Fed, the Fed simply creates the money. It creates a new demand deposit (treasury deposit) for the government (just like Zimmer's bank created a new demand deposit for Baker). With the new deposit, the Treasury can write checks to hire people, or can buy up surplus things, or can start new programs, or whatever.

### Borrowing From The Fed Creates High-Powered Money

When the Treasury borrows from the Fed, it works like this: The Federal Reserve banks receive newly printed government bonds (just as Zimmer's bank received Baker's newly written promissory note). Then the Federal Reserve banks add the amount to the Treasury's deposit. It's as simple as that! The Treasury now has more money — a bigger deposit balance — in each Federal Reserve bank. The national debt is bigger, and the Treasury has that much more money to spend. See how easy it is for the government to get all the money it wants?

Soon the Treasury will start spending the new money. Checks will be paid to people and businesses throughout the country. What do you suppose all those people and businesses are going to do with their government checks? Deposit them, of course! Then when each check clears, see what happens? The new money is pulled out of the Treasury's deposit at the Fed and goes into each bank's reserve account at the Fed. The new money soon becomes *new reserves for banks* all over the country!

See how a government deficit, financed by borrowing from the Federal Reserve, creates new money? The new money is paid to people and businesses by government checks. Then the checks are deposited in the banks and the money becomes new reserves for the banking system. Then a multiple expansion of the money supply can result.

When the government uses deficit financing and "covers the deficit" by selling bonds to the Fed, that can create a lot of new money, all right! It can have a very stimulating effect on the economy, too. The increase in government spending pays money to people and businesses, so the total spending flow starts to increase. Then if the people have a MPC greater than zero (that is, if they respend *any* of their new income), there will be a multiplier effect. Spending will increase even more.

This kind of fiscal policy has a strong "easy money" monetary policy locked into it. When the government runs a deficit and creates more debt, that creates money? That's right! But suppose the government sells the bonds to individuals and businesses and banks and insurance companies and savings and loan companies and all. What about that? Does that create money? No. It just gathers up money from these people and organizations—money they wanted to invest in government bonds. Then when the government spends the money, that just pushes the money back into circulation again. No money is created.

### Is The Government Debt Good? Or Bad?

From what you know about the government debt, what do you think of it? Do you think it would be good if we could just "wish it away"? Would we be better off? To get at this question, first let's think about debt, in general. Surely I'd like to "wish away" the mortgage on my house! If you owe anybody money, I'll bet you'd like to "wish away" that debt, too!

But suppose somebody owes you money. Do you want that debt "wished away"? Do you have a savings account? Or a checking account? Then the bank, or savings and loan *owes you* your money. Right? How would you like it if their debt to you was suddenly "wished away"? Not so good, huh? If the mortgage debt on my house was "wished away" then your savings account in the savings and loan that lent me the money would have to be "wished away" too. So when we look at it from both sides, maybe debt is not such a bad thing.

Is the debt on my house a good thing? I must think so, or else I wouldn't have agreed to create it. I would not have borrowed. The savings and loan company must think it's a good thing too, or else they wouldn't have agreed to create it. They wouldn't have lent me the money. The people who sold the house to me thought the debt was a good thing, because it enabled me to buy the house. I wouldn't have been able to buy the house unless I could borrow the money and create the debt. That's for sure! So, all things considered, I guess the mortgage debt on my house must be a good thing.

### All Of Our Money Is Debt

What do we use as money in this country? Debt? That's exactly right. Nothing but debt. *Absolutely* nothing but debt! The money supply is made up of demand deposits and currency. The demand deposits are the debts of the banks. If you have a checking account in a bank, that bank *owes you* your money. (Economists sometimes define *money* as different from other kinds of debt—but no need to get into that issue here.)

What about the currency in your pocket? Currency is issued by the Federal Reserve banks. Each one-dollar bill or ten-dollar bill or fifty-dollar bill is a debt, a "liability" of one of the twelve Federal Reserve banks. It's their debt!

What do the banks use to "back up" their demand deposit money? You remember: their Federal Reserve accounts. More debt? You bet. Where did all those Federal Reserve deposits come from, anyway? Where did we get all that "high-powered money" to back up all these billions of dollars of demand deposit money? Can you guess? From the federal debt, maybe? Of course.

The government printed up bonds and "sold" them to the Federal Reserve banks. That gave the Treasury new deposits, which they spent. The money went to businesses and people, and then into the banks and became new "bank reserve deposits" at the Federal Reserve banks. That's exactly what happened.

### The Government Debt Supports Money Supply Expansions

The growth of the money supply in the United States (and in all other modern countries) has come about through the creation of new government debt. The government debt becomes "monetized." It becomes money! The newly created money moves into the banking system and creates new reserves. Then the new reserves support the further expansion of the money supply.

So is government debt good? Or bad? Would you like to "wish away" that part of the government debt which is supporting our money supply? You wouldn't want that any more than I would. Furthermore, as the economy grows, national income and national product grow. More people and more businesses are doing and making and buying and selling more things. So do we need a larger money supply? Of course. How do we get it? From the expansion of loan-created, demand deposit money? Sure. And the new reserves are created by the expansion of "monetized government debt."

So some of the government debt helps to support (and through the Fed's open market operations, helps to regulate) the money supply. That's okay. But what about the part of the federal debt which wasn't sold to the Fed? What about the "non-monetized" part of the debt—that is, the bonds and notes which were sold to the people and banks and insurance companies and all? What effect does that have? Would it be good if all this part of the debt suddenly could be "wished away"?

## Non-Monetized Government Debt Serves A Purpose, Too

One thing is sure. The two government bonds I own are not going to be "wished away"! (Not if I can help it.) And my bank holds millions of dollars worth of government bonds. If that part of the debt is "wished away" my bank is going to go broke and all my money (deposits) will be wiped out. I'm not in favor of that! The insurance company that I am depending on to help me in my retirement years, or to pay my survivors if something happens to me — that company holds millions of dollars worth of government securities, too. If that part of the debt is "wished away" the company is going to go broke and my retirement years will be lean and hungry!

I suppose the government could collect more money in taxes, and then use the money to pay off (buy back) all those privately-held bonds. But that would take a lot of tax collections. A lot of money would have to be pulled out of the spending-income stream. I don't think we want to do that. You can guess what the economic (and political!) consequences of that would be!

The people and companies holding the bonds don't want to get rid of the bonds. They own government securities because they've decided that's the best thing for them to do with their money. That's the best kind of "asset" for them to invest in. If they wanted to, they could get cash for their bonds, any day. Since they don't, it's pretty obvious that they don't want to.

So what do you think about the government debt? Is it good? Or bad? By now I think you can see that it is more good than bad. It's certainly better to have it than it would be to get rid of it! By far!

As time goes on, the federal debt is going to get bigger, rather than smaller. I think you can be sure of that. But there's no reason to get excited about that. When people get excited about the size of the government debt, it's usually because they don't understand it.

## Federal Debt, GNP, And Private Debt: A Comparison

If the size of the federal debt is alarming to you, perhaps it will be comforting for you to look at these comparisons:

1. At the end of World War II (1945) the federal debt was about the same size as the gross national product. It totaled about $200 billion. At that time, total private debt in this country was about $150 billion.
2. By 1960, the federal debt had increased by 25 percent (to about $250 billion), the gross national product had increased by 150 percent (to $500 billion), and total private debt in the country had almost tripled (to almost $600 billion).

See what's happening? The federal government debt is getting larger, but not nearly so fast as the rates of expansion of either the gross national product, or the total private debt in the economy.

3. From 1960 to the early 1970's, the federal debt increased by more than 60 percent (to more than $400 billion), gross national product increased by more than 100 percent (to more than $1,100 billion), and private debt *almost tripled* again (to about $1,500 billion).

Just look at the growth in that private debt! Should we be worried about that? My home mortgage is a part of that. Should we be worried about that part? Suppose the total size of the private debt started coming down? Would that be good? What would that mean? It would mean that people were paying off their home mortgages and that fewer people were buying houses. Or it would mean that businessmen were paying off their loans and were not re-borrowing to expand and produce more. It would mean that the economy was slowing down. Unemployment would be high. Hard times would be here.

Private debt and government debt are very different, in many ways. But both are functioning parts of our economic and monetary systems. It just doesn't make sense to look at the absolute size of either and, on that basis, to be alarmed.

## The "Burdens" Of Debt

Aren't there any burdens of debt? Of course there are. Debt can be very oppressive to some people, and at some times. If a person is trying to consume at a much higher rate than he is producing, he goes deeper and deeper into debt. If he keeps it up, someday his past will catch up with him. He will have a miserable time trying to pay off all those debts! But government debt, and the great mass of private debt in this country and in all the other modern countries, isn't that kind of thing at all. Debt is an essential part of the market system. It is a part of the exchange mechanism. As the economic system grows, debts will grow.

If anyone ever asks you if debt is bad, ask them these questions: Is money bad? Are government bonds bad? Are savings accounts bad? Are mortgages bad? Debt isn't bad. Back in the early chapters of this book you saw how efficiently debt can provide for the monetary needs of the society. You saw the wise old northside chief create all that money, those checking accounts. Those "bank debts." Remember?

To be sure, debt can create problems. The "borrowing privilege" can (and sometimes is) abused. This is as true for the federal government as it is for an individual. But the consequences of "debt abuse" by the federal government are different—are in fact *completely unrelated*—

to the consequences of debt abuse by an individual or a family or a business. That's one reason why there's always so much misunderstanding about the federal debt.

### The Consequences Of "Debt Abuse" By The Federal Government

There are several kinds of problems associated with the federal debt. But the greatest danger results from the *opportunity to be irresponsible* —to go to extremes. A rapid increase in the debt can generate too rapid an increase in the money supply. This can generate inflation and do serious harm to the economy.

Some members of Congress may try to use government debt to "make political hay." Congress can run a large deficit, create money, and undertake many projects to "woo votes" from the people back home. When a Congressman comes up for re-election (every two years) if he can show his constituents lots of government projects, and increased jobs and incomes—with no increase in taxes—he is likely to be re-elected. You can see how great the temptation might be!

It isn't the *existing size* of the federal debt that creates the problem. It's the rapid expansion in the money supply, and in total spending, which results from the *rapid expansion* in the size of the debt. Of course, if the economy is depressed, the increase in total spending will help the economy to get going again. And that brings us back to the idea of unbalancing the federal budget *on purpose,* to overcome depression or inflation.

### More Spending May Bring More Jobs, And More Tax Revenues

"Deficit financing" (spending more than is collected in taxes) can support public works projects and re-employ people. This can increase incomes and spending, and may generate a lot more tax revenues for the government. If the "deficit financing fiscal policy" is completely successful, the economy will regain its prosperity. People will have good incomes and will pay a lot more taxes. All those extra tax revenues may bring the government budget back into balance—or even generate a surplus!

When people pay taxes, the money goes into the Treasury accounts at the Federal Reserve banks. If the government runs a surplus, the extra money can be used to "buy back" the bonds from the Fed. So the new money, initially created by the deficit, can be pulled back out of the economy as tax revenues, and destroyed. Isn't that neat? Here's what happened:

> The deficit spending brought on an economic expansion, and higher incomes. The higher incomes produced a surplus of tax revenues.

The surplus was used to retire some of the debt. So the money which was *created* in the first place, was then *destroyed*. (It doesn't often happen this way, but sometimes it does.)

## THE "CHEER UP" APPROACH TO OVERCOMING DEPRESSION

You know about the importance of expectations in influencing the economy. Expectations influence the amount businesses will spend for factories, equipment, inventories, everything. Expectations play a big role in influencing consumer spending for new cars, TV sets and other consumer durables. If everyone expects high employment and good times, the consumer spending rate is likely to be high. But anyone who expects unemployment (and surpluses and special sales and price-cuts) tomorrow, will not go out and spend today. He will wait until the surpluses and price cutting begin. He may also be waiting to see if he still has a job!

### Optimistic Expectations Bring Good Times

See how important expectations are? If everyone thinks there is going to be high employment, a great demand for things, and rising prices, then that is exactly what will happen. Everyone goes out and buys what he wants before the shortages occur, before the prices rise, and this big spending increase *creates* the shortages and rising prices the people are expecting.

We can say it this way: If people expect rising prices and shortages and high levels of employment, that is exactly what they will get. But if people expect unemployment, surpluses, and widespread price cutting, then everyone waits. Because of the waiting and the drop in spending, the people get exactly what they expected! Many lose their jobs. The economy goes into recession. It's sort of a "self-fulfilling prophesy." Whatever the people *expect* to happen, that's what *will* happen.

If you're still playing the role of President, and trying to overcome a depression, maybe this discussion of the importance of "expectations" will give you some ideas. Perhaps if some very important person (maybe the President of the United States) makes optimistic announcements, maybe the people will start spending more. Maybe you should arrange for some prime time and go on radio and TV and try to convince all the people that a big boom is about to begin. Tell them there will be shortages, and lots of jobs and overtime for everybody, so if they want to buy anything they should go out and do it *tomorrow*. Maybe everyone will believe you and go out and borrow and spend. If they do, sure enough, it will turn out that you were right. But if they *don't* believe you, it will turn out that you were wrong.

## Hoover, Roosevelt, And Nixon Used "Cheer Up" Speeches

Back in the very early part of the depression of the 1930's, President Herbert Hoover announced to the nation: "Prosperity is just around the corner." If all the people had really believed him, he would have been right. But they didn't, so he wasn't. Then, in 1933, President Franklin D. Roosevelt in his "fireside chat" radio broadcasts, told the people: "All we have to fear is fear itself." That was about right. But we had plenty of fear to fear. Until we got rid of some of the fear, we weren't going to have much prosperity.

Then in the 1969–70 recession and stock-market crash, President Richard Nixon repeatedly assured the people that economic expansion was almost ready to begin, and that stocks were excellent buys at such low prices. But the people and businesses didn't believe him, so unemployment continued high, the stock market was in no hurry to regain its huge losses, and the vexing problem of inflation continued unabated.

All these Presidents tried to "psych" the people into spending more. Why shouldn't you? Really, the most powerful force for economic downturn and depression, or for economic upturn and prosperity, is the psychology of the people. If the people are convinced that the economy really is getting ready to go soft, then it will. If they are convinced that it is getting ready to improve, then it will.

## The People Need Confidence In The Government's Policies

Suppose the people are really confident that the government has the necessary tools, and can guarantee prosperity. Suppose everyone is sure the government knows how to take care of any serious economic problem. Then this confidence is likely to make it unnecessary for the government to use the tools.

If you, Mr. or Ms. President, have been able to inspire confidence in all the people to such an extent that they are *sure* that you can and will handle any serious economic problem which may arise, then the people will not be spooky, or gun-shy. They will all just keep going along, doing things in the normal way. No big "rush for the hills" for security, and no big "dash for the valley" to make a killing. If people expect stability and long-term growth in the economy, then they will act as though stability and long-term growth were the things which were going to occur. As long as they will *act* that way, stability and long-term growth is *exactly what will occur.* But it may take quite a bit of doing to convince all the people that you really do know what you are doing and that they really can trust and depend on you.

If the people are confident that the government can prevent, or quickly overcome depression or inflation, then *the confidence of the people* will take care of the situation. But if there's serious doubt in the minds of the people, the stabilizing tools are not going to work very well. It's almost like this: If the people believe you can handle the situation, then you can, quite easily. If they think you can't, then you can't, no matter how hard you try!

It isn't likely that we ever are going to be able to second-guess all the spending decisions of the masses of consumers and businessmen in the economy. We can't use fiscal and monetary policy to smooth out all the little ups and downs in the total spending stream. Anyone who thinks we can "fine tune" the economy on a perfectly stable course—either with fiscal or monetary policy or with any other approach—simply doesn't understand.

## A WORD OF SUMMARY AND FAREWELL TO MACROECONOMICS

How do you feel now, about macroeconomics? Do you understand what keeps the economy running? Sure. Total spending. So, what makes the economy speed up and slow down? Anything that influences total spending. Of course.

If you've been carefully working your way through all these macroeconomics chapters, I think you really understand it now. That's good. Macroeconomic issues are going to be staring you in the face, off and on, all your life. It's good to be able to understand what's going on.

We're leaving the subject of macroeconomics for awhile, now. We're moving on into a new part of the book—into microeconomics again. Remember about the natural forces of "the market"? About how these forces automatically get the people and resources to go to the right places and do the right things, for the good of society? Sure. It's time to get back into that now.

So now we leave macroeconomics—but not for good. Near the end of the book, in Part Eight, we'll be dealing with current economic problems. There, we'll get back into macroeconomics again when we're talking about the stabilization dilemma—the current problems of inflation and unemployment—and about *wage and price controls*. That's a very broad and perplexing issue; it involves both macro and micro concepts. It would be better for you to know more about microeconomics before we get into wage and price controls.

We'll wait for Part Eight to get involved in the pressing *current problems* of inflation and employment and direct controls. There will be time enough for that, then. Now, as soon as you're sure you and this chapter and this Part have really done your bit for each other, on to Part Six, and to new and interesting things about MICROECONOMICS!

## REVIEW EXERCISES

### MAJOR CONCEPTS, PRINCIPLES, TERMS (Try to write a paragraph explaining each of these.)

fiscal policy
how expenditure adjustments work
how tax adjustments work
the "burden" of the federal debt
the "cheer up" approach

### OTHER CONCEPTS AND TERMS (Try to write a sentence or phrase explaining what each one means.)

the "excess inventories" problem
the "excess debt" problem
the "excess manufacturing capacity" problem
"public works" programs
government deficit financing
unbalancing the budget
"monetized" debt
self-fulfilling prophesy

### QUESTIONS (Try to write the answers, or jot down the highlights.)

1. Suppose you were a Congressman and the President was trying to get you to vote to cut off the money for a planned (and badly needed) interstate highway in your home district. Also, the President wants you to vote to increase taxes. (You're up for re-election, of course.) Do you think you will vote for the President's program? In view of such problems, do you think it's politically feasible to use fiscal policy to fight an "over-boom," or inflation? Discuss.
2. The government can get all the money it wants just by printing up bonds and "selling them" to the Federal Reserve banks, but this is a dangerous procedure because a multiple expansion of the money supply is likely to result. Explain what this means and how it happens.
3. Write an essay on "the government debt"—what's good about it, and what ought to be done about it, if anything.
4. What do you suppose President Franklin D. Roosevelt meant (in 1933) when he told the people "The only thing we have to fear, is fear itself"? Was it true? Do you think the same thing could be said of *all* periods of economic recession and depression? Or are there some *basic economic causes* to blame (totally, or partially) for recession and depression? Discuss.
5. After all is said and done, how effective do you think fiscal policy really is in stabilizing the economy? Do you think it's more effective in overcoming depression? or inflation? Explain.

# PART 6

THE PRICE MECHANISM CAN TAKE CARE OF EVERYTHING, JUST LIKE MAGIC!

In the make-believe world of the "model pure market system", in the long run every price is just the right price, everybody earns just what he should earn, everybody gets just exactly what he pays for, the society optimizes the use of all its resources, and "social welfare" is maximized.

# 19 DEMAND, SUPPLY, AND PRICE

*The theory of how consumer demand activates the price mechanism and directs the economic choices in the "Model Market System."*

Welcome back to microeconomics! I suppose it would be good to review just for a minute, to let you think about some of the things you already know about microeconomics. What is microeconomics? It's the study of scarcity, and of choosing, and of how the market process automatically "senses" and then carries out the choices of the society.

**The Market Process, Again**

A person does something that "the society" wants done. So he gets income. The more, better, more valuable work he does, the more income he gets. Then he tries to "optimize" as he spends his income. He adjusts his spending pattern to try to maximize his progress toward his objectives. I'm sure you remember about all this.

The businessman tries to adjust his outputs and inputs and come out ahead, if he can. He tries to get enough revenues to cover his costs, with something left over as profits. He produces fewer of the things which are not in much demand and more of the things which are in strong demand. The things in strong demand are the things that offer him profits, so those are the things he produces.

The businessman always produces as efficiently as he can. He tries to use *fewer* of the *more expensive* input factors and *more* of the *less expensive* factors. If his efforts are successful, he will be adding a lot of value to the society's "total wealth" by producing the most valued (most scarce) outputs and using up the least valued (least scarce) inputs. He will be rewarded by big profits. The big profits will give him a big share of the society's output.

As all the individuals and businesses are making their choices, the production and distribution choices for the society are being made. Automatically. If people are demanding more cartridge tape players, the businesses which produce things like that will expand their outputs of tape players. More factors of production will move into the production of tape players.

The businesses and people who do the best job of producing the most tape players will receive the highest incomes. So they will get to buy the most output, and will have the most influence over the uses of society's resources. (The people and the businesses with the money are the ones who get to make the choices. Remember?)

The "market process" really is a very neat and simple system, in overview. But, as with most "neat and simple" things, when you look at it more closely, it isn't quite as neat and simple anymore. It's time, now, for us to look at it more closely. You'll be surprised how much there is to it—and how much of it you've been observing and participating in all your life, but never knew it.

### The Market Process Is Supply And Demand At Work

You probably have heard somewhere about "supply and demand." Actually, you already know quite a bit about supply and demand. Most of the basic concepts of economics are tied into and can be expressed in the simple phrase "supply and demand." The study of the market process is essentially the study of supply and demand. Demand reflects people's choices; supply reflects scarcity. Demand works through the market process to pull forth the output (the supply). The people who produce (who supply) the demanded output, get rewarded with income. See how the "supply and demand" idea sort of ties it all together?

Economists have a great body of economic theory dealing with supply and demand. And curves? Yes. But only two basic curves: the demand curve, and the supply curve. These two curves are quite simple. Really, they're sort of fun to play with, and they help a lot to see how the market process works—how all the pieces fit together. So now, after waiting so long and getting so ready, let's plunge into supply and demand!

## THE CONCEPT OF DEMAND

A person buys something because he wants to. He would rather have the "something" than to have the money it costs. To say it another way: He would rather have that "something" than to continue to have the opportunity to spend the money for some *other* "something." The more a person *wants* something, the more he is likely to *pay* to get it. The more he would pay for it, the higher is his *demand* for it.

If you are ready to give up a lot of money for something, then your demand for it is high. If you aren't ready to give up anything (any money) for something, then you have no demand for it. You may have some *desire* for it. But if you aren't ready to pay a price to get it, you do not *demand* it.

If you want something but still wouldn't pay to get it, there must be some reason. Perhaps you want to keep your money for some other purpose. Or maybe you just don't have any money. Either way, it makes no difference in the operation of the market process. Unless you are *ready to pay a price* for something, you do not demand it, no matter how much you may *want* it. Desire? Yes. Demand? No. You aren't going to influence any of society's choices through "the market process" unless you are able and willing to spend some money to direct some of society's resources. Your *wants* alone won't have any effect.

### Demand Is "The Propensity To Buy"

Demand, as we economists use the term, is another one of our "if . . . then" concepts (sort of like the consumption function you were working with a few chapters ago). The consumption function doesn't tell how big the basic flow of consumer spending *is*. It only tells "how big it would be if. . . ." for example, we might say: "If national income (NI) happened to be $5 billion, then consumer spending (C) would be $5 billion; if NI was $10 billion, then C would be $8 billion." This is the same way we use the term "demand."

The "demand" for something doesn't mean "how much of it people are buying." It only means "how much of it people *would be buying if . . .*" If what? If the price happened to be $5. or $10. or $30. or maybe 50 cents. So "demand," in economics, doesn't mean what it normally means to most people. In economics, it really means "propensity to buy." It means "how much I would buy, if. . . ." Let's make a comparison.

The "consumption function" (Chapter 15) relates "propensity to buy" (for consumer goods in general) to the different possible sizes of the national income. Remember? The "demand function" relates "propensity to buy" (for some specific good) to the different possible *prices* which might be charged for that good.

If you have a "demand" for something (a "propensity to buy"), whether or not you *actually will buy* depends on how high the price is. You may be willing to pay to get something, but you may *not* be willing to pay a price which you think is too high. If you have been buying something regularly (say, a dozen eggs a week) and the price goes up, that changes the picture.

If the price goes up you must give up more money to buy your usual weekly dozen eggs. Maybe you decide you would rather buy something else instead. Perhaps breakfast cereal looks like a better deal than eggs, now. And here's something else. Your "consumption possibility" gets smaller if you keep on buying the high-priced eggs.

### At Higher Prices, Your "Consumption Possibility" Is Smaller

When you spend your money for the high-priced eggs, you don't get as much for your money as you did before. Each dollar you spend for eggs buys less than it bought before. Suppose the high price turns you off, so you just quit buying eggs altogether. Suppose you decide to buy only things which haven't gone up in price. Then your money buys just as much as before! Your consumption possibility is just as large as it was before. (The value of *your* dollar depends on what *you* buy with it. Remember?)

There are always some people who are sort of undecided about which product to buy. They are "right at the margin." If the price of one thing goes up even a little bit, they will stop buying that thing and start buying other things instead.

You could say that demand is the "functional relationship" between the various prices which might exist, and the various quantities people would buy. The quantity people would actually buy is "a function of price." That is, the quantity bought depends on how high (or low) the price happens to be. Once you *demand* something, whether or not you will actually go out and try to buy it (and how much per week you will try to buy) depends on the price. I wonder how many eggs you would buy if the price went up to $10 a dozen! (You would still have a demand for eggs—a "propensity to buy"—but would you *actually* buy any? I doubt it.)

### The "Substitution Effect" Of A Price Change

When the price of something goes up, people buy less of it. There are two reasons why people buy less. The first reason is that they decide to stop buying the high-priced good and to spend their money for other things instead—that is, they *shift* their spending to other goods. Maybe they buy breakfast cereal instead of eggs. They *substitute* cereal, for eggs. Whenever the price of something rises, the people who are right at the margin—sort of undecided—will stop buying the higher-priced good and substitute something else. Economists call this the substitution effect of the higher price.

The "substitution effect" of a price change also works when the price goes down. If eggs get cheaper, people will buy fewer breakfast steaks and corn flakes and pancakes, and start buying more eggs. They "substitute" the (cheaper) eggs for the pancakes and things.

### The "Income Effect" Of A Price Change

The second reason people buy less of something when it gets more expensive, is this: When goods are higher priced your "money income" can't buy as much "real income" as it could before. Suppose you are buying eggs and the price of eggs goes up. If you keep on buying eggs, the higher price actually reduces your "total purchasing power." Your real income (the things you get to have and enjoy each week) gets smaller. Here's an example.

If you have only two dollars to spend for breakfast foods this month and all you really like in the morning is scrambled eggs, then if the price of eggs goes up from 50¢ to $2 a dozen, you are in trouble. Before the price increase, your $2 bought four dozen eggs. Now it will only buy one dozen eggs.

Your purchasing power for eggs has dropped from four dozen to one dozen. You might still want four dozen, and you might be willing to buy them even at a price of $2, but you just don't have the money. So when the price goes up from 50¢ to $2, the quantity you buy drops from four dozen to one dozen—not because you "substitute" other goods for eggs, but because your purchasing power has been reduced by the price increase. Your "real income" has been reduced.

Whenever the price of something goes up, the people who keep buying the good at the higher price actually experience a reduction in "real income." This causes them (really, *forces* them) to buy less. Economists call this the income effect of the price increase. The "income effect" also works when the price goes down. People who are careful to buy the goods they want only when the goods are on sale, can enjoy more "real income." (Of course this doesn't mean that you can save money by buying lots of things you don't *usually* buy, just because the things happen to be on sale!)

We can summarize this discussion this way: There are two reasons why people will buy more of something as its price goes down (and less of it as its price goes up): (1) the *substitution effect*, and (2) the *income effect*. Think about it. Don't both of these "effects" influence your buying choices? (They do mine.)

### Consumer Demand Depends On Wants, And Income

If you have a demand for something, that means that two requirements are being met: (1) you want it, and (2) you are willing and able to spend some money to get it. If these two conditions are met, you have a "demand." You have a "propensity to buy." Then, the higher the price, the *less* of it you will *actually buy*; the lower the price, the *more* of it you will *actually buy*. This little capsule statement explains about all there is to the basic concept of "consumer demand."

### The Law Of Demand

Economists make up "demand curves" to show the relationship between the various prices that might exist and the quantities people would be trying to buy at each of those various prices. Think about it. Will the demand curve show larger quantities bought at high prices? Or larger quantities bought at low prices? See? You already understand the demand curve. And you already know the *law of demand.*

The law of demand says that people would buy more of something at a lower price than they would at a higher price.

You know this is true, because of the substitution and income effects. You know it is true from your own personal experience, too. Right? The demand curve (Figure 19-1) illustrates the law of demand. Take time now to study it for a few minutes.

## ELASTICITY OF DEMAND

Figure 19-1 tells you something about elasticity of demand. But you need to understand a little more about it than is explained there. You already know that when the price goes down, people will buy more. That's the "law of demand." The elasticity of demand answers the question: "How much more?" You already know from Figure 19-1 that if it is "a lot more," that means the demand has a *lot* of elasticity. If it is "only a *little* more," then that means the demand has only a little elasticity.

If the price goes down a little and a great deal more is bought, then we say that the demand is "highly elastic." If the price goes down a lot and only a very little bit more is bought, then we say the demand is "highly inelastic" (meaning "very little elasticity"). What we are talking about is the price elasticity of demand—the responsiveness of buyers to *price* changes.

Most economists like to be very precise about things in economics (and about everything else!). It makes us nervous to talk about "a lot more" and "a little more"—or to say that the quantity people demand is "very responsive" or "not very responsive" to price changes. We get nervous about things like that, because we don't know what you have in mind when you say "a lot more" or "a little more." So we use a more precise way of talking about elasticity of demand.

Fig. 19-1 **THE DEMAND CURVE ILLUSTRATES THE PROPENSITY TO BUY. THE LOWER THE PRICE OF EGGS, THE MORE PEOPLE WILL TRY TO BUY.**

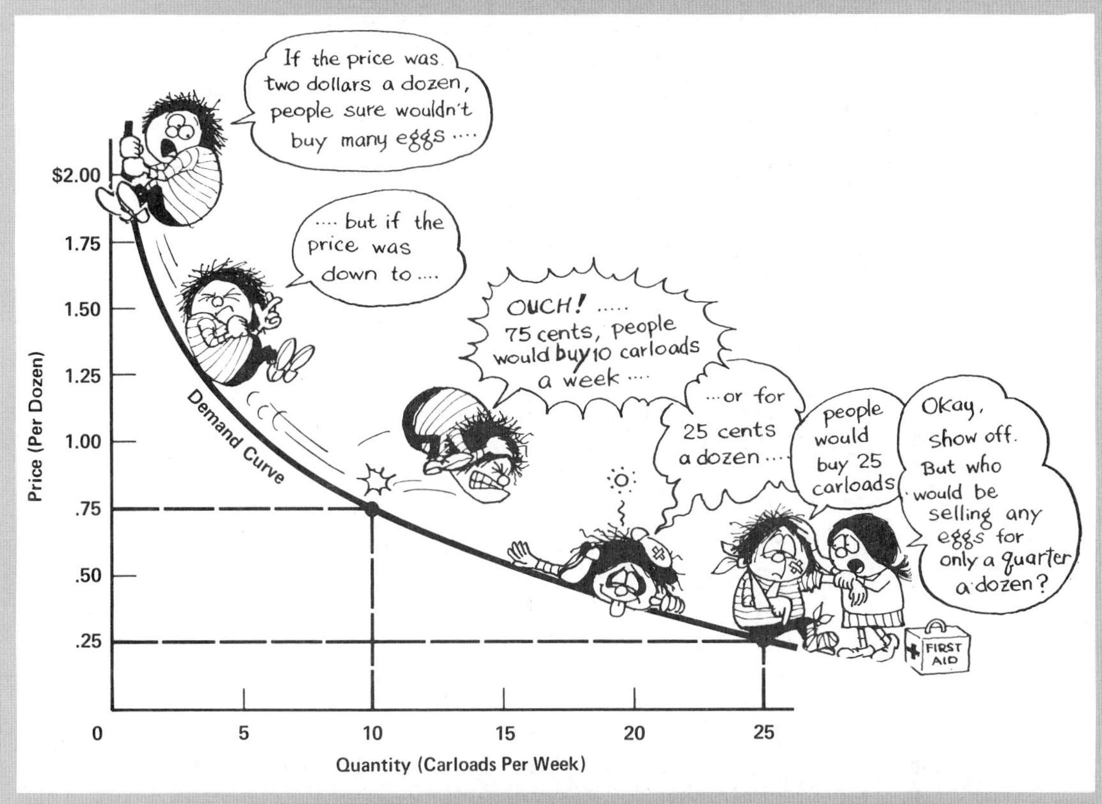


Think Of This As The Weekly Demand For Eggs In Your Local Metropolitan Area.

The demand curve illustrates the "law of demand." It shows that at lower prices people would buy more. For example, if the price moves down from $1.00 to 75 cents per dozen, the quantity people will buy increases from 6 carloads to 10 carloads per week. This shows that the quantity people buy is very responsive (stretches a lot) when the price goes down. The economist would say that the demand for eggs in this example is *elastic*. Elasticity means "responsiveness." When the price changes, if the quantity bought does not respond very much we say the demand is "inelastic." The responsiveness is low.

The concept of elasticity, simple though it is, is very helpful in understanding how the market process works. You will see it again, very soon.

## Elasticity Depends On Percentage Changes

We say it this way: If a change in price (say, a 10% change) would cause a *larger* percentage change (say, a 25% change) in the quantity people would buy, the demand is relatively elastic. "Relatively elastic" demand means the *percentage change in the quantity bought* would be greater than the *percentage change in the price.* It means the quantity bought is highly responsive to price changes—that the quantity people would buy "stretches a lot" when the price goes down and "shrinks a lot" when the price goes up. That's all it means.

If the change in price (say, a 10% change) would cause a *smaller* percentage change (say, a 5% change) in the quantity bought, we say that the demand is relatively inelastic. "Relatively inelastic" demand means the *quantity* changes by a *smaller percentage* than the *price* changes. The quantity people would buy isn't very responsive to price changes—that is, it doesn't "stretch" (or "shrink") very much if the price goes down (or up).

Suppose it happens that the change in price (say, a 10% change) is *exactly equal* to the change (say, a 10% change) in the quantity bought? Then we say: "the elasticity of demand is unitary." Why unitary? Because if you *divide the percentage change in price, into the percentage change in quantity,* what do you get? You get *one,* which is *unity,* of course! (10% ÷ 10% = 1.) If you got a number *larger* than one you would know the demand was relatively elastic. The "quantity response" was *greater* than the price change. If you got a number smaller than one you would know the demand was relatively inelastic. The "quantity response" was *less* than the price change. Why *relatively* inelastic? Relative to what? Relative to *unitary elasticity*! Of course! Here's an example.

Suppose American Airlines decides to offer a 25 percent reduction in air fares for students. Following the announcement of the reduced fare, suppose twice as many students start buying tickets and flying American. The demand turned out to be relatively elastic—highly elastic, in fact. A 25 percent price cut brought a 100 percent increase in "quantity bought." The elasticity of demand would be 4. The 100% quantity increase, divided by the 25% price decrease, equals 4 (100% ÷ 25% = 4). What if there had been only a 15 percent increase in ticket sales to students? Then we would say the demand was relatively inelastic (elasticity of less than one). The elasticity of demand would be .6 (15% ÷ 25% = .6). What if the student ticket sales increased by exactly 25 percent? Well, unitary elasticity, of course! (25% ÷ 25% = 1.)

The concept of elasticity of demand is very helpful. It's a way of indicating the responsiveness of buyers to a price change. If the "percentage change" idea is helpful to you, good. If not, don't let it worry you. Just understand and hang onto the elasticity *concept*—the idea of *responsiveness of buyers to price changes.* Now, here's another way to look at it.

## Elasticity Affects The Amount Spent

Another way of looking at "elasticity of demand" is to see whether or not people spend more money on something after the price goes down (or up). Suppose you really like bananas. When the price is 15 cents a pound you usually buy two pounds a week. You spend 30 cents a week on bananas (2 pounds @ 15¢ = 30¢).

Now suppose the price of bananas comes down to 10 cents a pound. What will you do? If you *spend more* than 30 cents a week for bananas (buy *more than* 3 pounds @ 10¢), your "banana demand" is "relatively elastic." If you *spend the same amount* (buy 3 pounds @ 10¢) your "banana demand" has "unitary elasticity." If you *spend less* than 30 cents (buy *less than* 3 pounds @ 10¢) then your demand for bananas is "relatively inelastic."

We can summarize it this way. When the price of something goes down, people will buy more. But will they spend more money? Or not? The lower price *saves* them some money, but the greater quantity *costs* them some money. Which one will win out? The price change? Or the quantity change? It all depends on *how much the quantity changes* in response to the price change. Right? That's just another way of saying "it all depends on the elasticity of demand."

Suppose the quantity doesn't change at all. That means the demand is absolutely inelastic. Then, if the price went down, total spending for that good would go down for sure! If the price went up, total spending for that good would go up too. Why? Because the quantity didn't change, to offset the "expenditure effect" of the price change. So now, the general principle:

1. If the quantity does not change enough to offset the "expenditure effect" of the price change, then the total amount spent for the good will move the same direction as the price moves. If price goes up, total amount spent goes up. If price goes down, total amount spent goes down. Demand is *inelastic*. (Quantity doesn't "stretch" enough to offset the "expenditure effect" of the price change.)
2. If the quantity changes by enough to *more than offset* the "expenditure effect" of the price change, then the total amount spent for the good will move the *opposite* way from the price: If price goes up, total amount spent goes down. If price goes down, total amount spent goes up. Demand is *elastic*. (Quantity "stretches" enough to *more than* offset the expenditure effect of the price change.)
3. If the quantity changes just enough to exactly offset the expenditure effect of the price change, the total amount spent for the good will stay the same. The elasticity of demand is *unitary*.

Now that you know all this about elasticity of demand, what's the purpose? Its just that this is one of the most significant concepts in economics, that's what! The concept of elasticity is always lying there under the surface, influencing the behavior, the decisions, the choices of all individuals and businesses and the society as a whole. The concept of elasticity of demand will be relevant to much of what goes on in this and in the next several chapters. If you aren't sure you understand it, please review this section and be sure you understand before you go on.

## HOW THE PRICE MECHANISM WORKS: THE FUNCTIONS OF PRICE

The demand curve shows how the buyer responds to price changes. But don't be misled. It isn't the market which directs and controls the economic choices of the people. It's the other way around. The people, as consumers, direct and control the society's choices through the market. But the market does let the consumer know how scarce each thing is, so that the consumer will know which of society's goods need to be most carefully conserved and economized.

### Prices Conserve Scarce Things: The Rationing Function

If the price of something goes higher, people don't buy as much of it. The substitution effect and the income effect see to that! We can say, then, that one thing which the *price* of something does, is to convince the people (the consumer-buyers) to conserve, to limit the use of scarce things. We call this role—the role which price plays to discourage the use of scarce things—the rationing function of price.

The more scarce something is, the higher the price will be. The higher the price of something, the more it will be conserved and economized. That's what is meant by "the rationing function of price."

The demand curve is a *graphic picture of the rationing function of price*. Remember the demand curve for eggs (Figure 19-1)? It shows that at a price of $2 a dozen, only a few eggs would be bought—less than one carload per week. Only those people who really like eggs (or people with lots of money to spend) will continue to eat two eggs for breakfast every morning when the price is $2 a dozen! See how much more carefully eggs will be conserved at $2 a dozen, than at 50¢ a dozen?

If the price of eggs would suddenly go up to $2, this would tell us that eggs must have suddenly become very scarce for some reason. People who are really crazy about eggs will probably go around scowling and muttering all the time about the high price of eggs. But they really should be very thankful that the price went up to $2. At a price of $2, every Tom, Dick and Harry will not be eating up all the very scarce eggs!

A price of $2 a dozen saves the eggs for the true egg-lovers. If for some reason the price of eggs had not been allowed to go up—if some "consumer protection committee" had convinced the legislature to pass a law holding the price of eggs down to 50¢—the eggs soon would all be used up and *nobody* would get any more eggs for awhile. You can see how the high price performs a use-limiting, or "rationing" function. If the price does not go up, the eggs soon will be all used up. There will be a shortage. The high price prevents this from happening. As the price moves up, it notifies the people who don't really care very much about eggs that they should stop buying eggs and leave the eggs in the market, for those who really go for eggs!

Free goods have no prices. They are used freely. No one tries to conserve, or economize, or limit, or ration the use of free goods. But once a good is scarce it needs to have a price. It needs to be *owned*—to *belong to* someone. Then the use of the good will be automatically rationed. If the buffalo on the plains had belonged to someone, and if a price had been charged for killing them, we would have saved many roaming buffalo herds!

The use of something is always limited to those purposes which are important enough to be worth the price. The higher the price of something, the more its use will be restricted. At 50¢ a dozen, eggs are used only by those who think having and eating a dozen eggs is worth 50¢. At a price of $2, eggs will be used only by those who think a dozen eggs is worth $2 to them. What nut would pay more for something than he thinks it's going to be worth to him? Nobody would be that stupid (I hope).

The "rationing function of price" limits the use of scarce things. Only those people who want something badly enough to pay the price, will get any. Things will be used *only* for those purposes which someone considers *important enough* to be worth the price. Are you beginning to be impressed by this thing we call "price"? Be impressed. It's truly the key to understanding the market process.

### Prices Direct Resources, And Induce Output: The Production-Motivating Function

While the high price is conserving the use of eggs, it is also doing something else. The high price is *pulling in factors of production*—more labor, land, and capital are pulled into the production of eggs. At a price of $2 a dozen, big profits can be made in the egg business. Of course! So anyone who can get some laying hens and start producing eggs can make himself a bundle!

It's easy to see why high prices motivate production. What we are talking about is called the production-motivating function of price. When we're talking about "motivating production," then we're running into the other half of our discussion of "supply and demand." We're getting into the concept of *supply*.

## THE CONCEPT OF SUPPLY

If, in the precise world of economics, demand means "propensity to buy," what do you suppose supply means? Maybe "propensity to sell"? Exactly!

We have been talking about people buying things—you buying bananas, people buying eggs, and all that. Goods are flowing out of the market to the buyers. But where are the goods flowing into the market from? From the sellers, of course.

### Supply Is "The Propensity To Sell"

Is it the amount flowing across the market from the sellers to the buyers? Is that the supply? Nope. The supply is an "if . . . then" concept, just like demand. The supply tells how much the sellers would be trying to sell if the price was $1, or $5, or $10, or $30, or whatever. It is "the *propensity* to sell."

The word "supply," like the word "demand," has a very precise meaning in economics. You could say that supply is the "functional relationship" between the various *prices* which might exist, and the various quantities the producers would be trying to sell. Once the "supply" (the "propensity to sell") exists, then the *actual quantity* each producer will offer for sale, will depend on the *price*. At higher prices, producers will offer more for sale; at lower prices they will offer less.

Now we're about ready to look at the supply curve. Do you know what it's going to look like? Of course. Positive slope. At higher prices greater quantities would be offered for sale. Does the supply curve illustrate the production-motivating function of price? Sure! Look at Figure 19-2 and that's exactly what you'll see.

### What Price Is The "Right" Price?

If the price of something goes down, people will try to buy more. The "law of demand" tells us that. Remember? But what about the sellers? They don't want to sell as much if the price is low. Only at higher prices would the sellers like to sell more. But at higher prices the buyers won't buy as much. So how does it all work out? There must be some "just right price"—a price just high enough and just low enough so that the buyers want to buy just exactly as much as the sellers want to sell. And so there is. We call this "just-right price" the equilibrium market price.

Think of supply as "a potential daily flow of the product into the market." Think of demand as "a potential daily flow of the product out of the market." Producers are producing so much per day in response to the price. Buyers are buying so much per day in response to the price.

**Fig. 19-2 THE SUPPLY CURVE ILLUSTRATES THE PROPENSITY TO SELL THE HIGHER THE PRICE, THE MORE EGGS PEOPLE WILL TRY TO SELL**

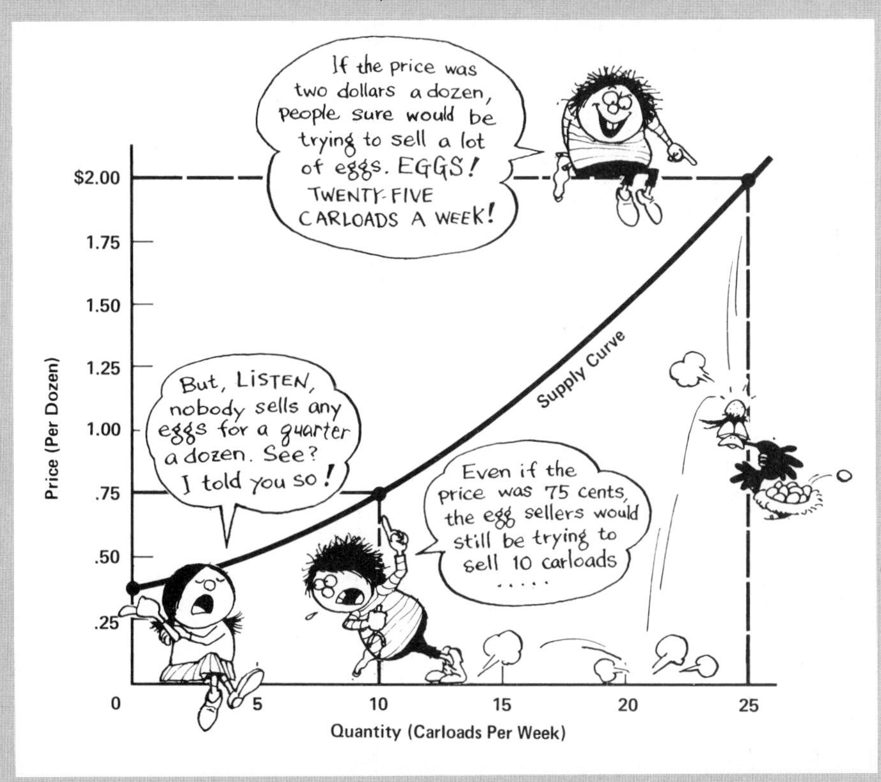


Weekly Supply Of Eggs In The Local Metropolitan Area.

The supply curve shows that at lower prices, smaller quantities of eggs will be offered for sale. How much the quantity offered for sale expands as the price rises (or how much it declines when the price falls) depends upon the elasticity of supply.

This supply curve only makes sense in the "short-run." It could not apply to the industry in the long run, because in the long run more producers can enter the industry, or some of the present producers could leave. If more farmers entered the egg-producing business, or if some left, the curve would shift. More on this later.

Can you see how the elasticity of this short-run supply curve depends on the law of diminishing returns? As output expands, cost goes up. As price goes higher and higher, businesses will produce more and more, because the higher price covers the higher cost and makes the higher output profitable. More on this later, too.

The goods are flowing across the market, from the sellers to the buyers. If price is just right, the flow being pushed into the market by the sellers will be just equal to the flow being pulled out of the market by the buyers.

Now that you know what the supply of eggs looks like, you know that it doesn't make any difference how many eggs people *would be* trying to buy at a price of 25¢ a dozen. Such a low price has got to be a make-believe price, anyway. The demand curve in Figure 19-1 shows us that the buyers would like to buy 25 carloads a day at 25¢ a dozen. But that could never happen. Look at Figure 19-2 now and you will see that the supply curve shows that at such a low price (25¢ a dozen) the suppliers are not willing to sell *any* eggs. None at all!

At a price of 25¢ a dozen, the egg producers would put the eggs in cold storage and wait for better times. If better times didn't come, a lot of egg-laying chickens would soon become stewing chickens and a lot of poultry farms would soon become potato farms or hunting preserves or something else. That's the way it's supposed to happen when the society isn't willing to pay the cost of keeping the factors of production working in the "egg industry." So eventually the egg industry will disappear.

What about a price of $2 a dozen? The demand curve shows us that people would buy less than one carload a week. But the supply curve tells us that at a price of $2, the egg suppliers would ship in 25 carloads a week. Soon we would all be hip-deep in eggs! So $2 can't be the right price, either.

The demand curve from Figure 19-1 and the supply curve from Figure 19-2 are shown together in Figure 19-3. This "supply and demand graph" shows that there is a price (but only *one* price) where the weekly flow of eggs being pushed into the market by the sellers is exactly equal to the weekly flow of eggs being pulled out of the market by the buyers. At what price are the supply flow and the demand flow equal? At the "equilibrium market price" of 75¢ a dozen. Figure 19-3 shows that the equilibrium market price *must be* 75¢ a dozen. At any other price there's either too much *supplied* (a surplus) or too much *demanded* (a shortage). You should spend a few minutes studying the "supply and demand" graph on page 397 now.

### The Equilibrium Market Price

Only one price can exist for long in our hypothetical egg market, or in any market. It is the *equilibrium market price*. It is the price where the quantity supplied equals the quantity demanded. The quantity flowing into the market from the sellers is equal to the quantity being pulled out of the market by the buyers. Each seller is selling all he wants to sell at that price; each buyer is buying all he wants to buy at that price.

Fig. 19-3 **THE SUPPLY AND DEMAND FOR EGGS**

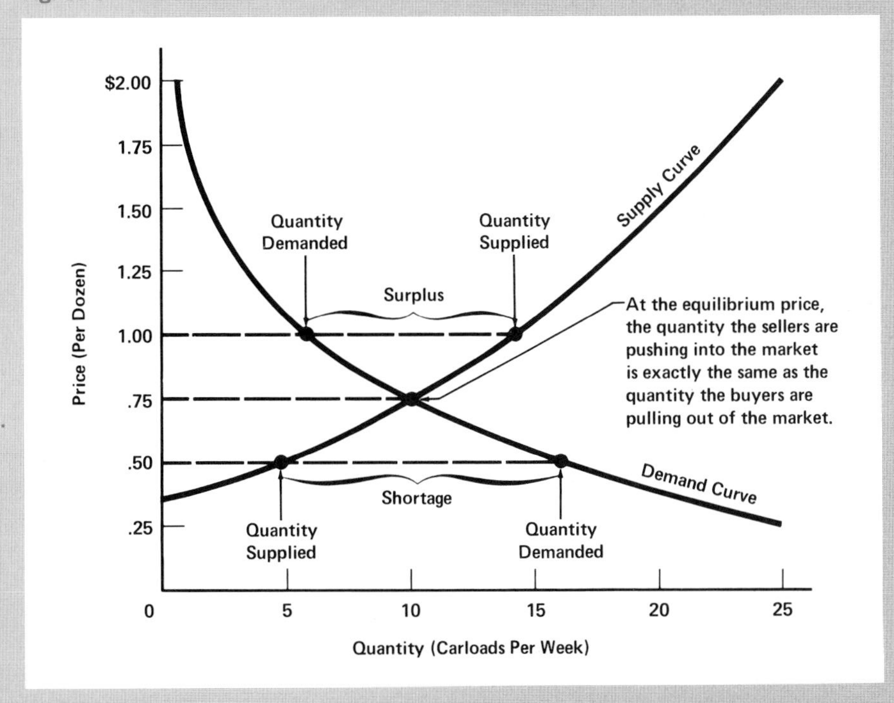


Only When "Quantity Supplied" Equals "Quantity Demanded," Will The Market Be "Cleared."

It is obvious from the graph that only the price of 75¢ a dozen can be the real price. Any lower price would leave demanders trying to buy more eggs than would be available for sale. Any higher price would have suppliers shipping in more eggs than people would buy.

If the price was too low (say 50¢ a dozen) buyers would be trying to get more eggs than the suppliers would offer. There would be a *shortage*. Some of the buyers would not be able to get any eggs from anywhere! Soon, these unhappy buyers trying to get some of the very scarce eggs would start offering more for eggs. They would push the price up.

If the price was too high (say $1 a dozen) sellers would be offering more eggs than people would buy. There would be a *surplus*. Some of the sellers would not be able to get rid of all their eggs. These sellers trying to get rid of their overly-plentiful eggs would start offering them at lower prices. This would push the price down.

Everyone who wants to sell at the equilibrium price can find a buyer. Everyone who wants to buy at that price can find a seller. That's what is meant by "equilibrium market price."

No one is *completely happy* with the equilibrium price. The buyers would rather get eggs cheaper. The sellers would rather sell for a higher price. But at this "equilibrium market price" the flow of eggs across the market is just the right size. There is no shortage and no surplus.

As long as the supply and the demand remain the same, the "equilibrium market price" (75¢ a dozen) and the "equilibrium quantity flow" (10 carloads a week) will continue indefinitely. But something will always happen to change either the demand or the supply, or both. Soon we will talk about the kinds of things which might happen to change the picture. Then you will see how the market system responds to changing conditions. You will see how the market process works — how price and quantity move toward a new equilibrium, and how each new equilibrium brings with it a new set of "production and distribution choices" for the society.

### The General Equilibrium Situation

Suppose the market process is working perfectly. Society is making all its choices through the *price mechanism*. All the prices of everything are "in equilibrium." Just the right amounts of all goods are flowing across all the markets from the sellers to the buyers. There are no surpluses of anything and no shortages of anything. Can you picture that, in a sort of "model situation"?

In the "general equilibrium" situation, all the people are carefully conserving the most scarce, most expensive, highest-priced things because of "the rationing function of price." Producers are working hard to produce more of the most scarce, most wanted, highest-priced things because of the "production motivating function of price." Everyone is free to do his own thing, and each person gets rewarded (receives income) according to how highly the market values what he "sells." Each person's "reward" depends on the price he gets for the product he makes, or for the services of his labor, his land, or his capital.

This view of the market system "in general equilibrium" shows our neat "model system" with all the choices being made automatically. But suppose something happens to disturb the equilibrium. What happens then? Suppose the people decide they want a larger "output flow" of some kind of good. Maybe they want to eat more eggs each week. What happens?

## HOW DEMAND DIRECTS THE ECONOMY

Suppose the egg market is in equilibrium, as shown in Figure 19-3. The price is 75¢ a dozen and 10 carloads a week are flowing across the market from the sellers to the buyers. Then suppose that a respected scientist makes a startling announcement: "Eggs, if eaten regularly, will maintain your youthful appearance and your vim and vigor, and will add ten years to your life!"

### Changing Wants Change The Demand And Upset The Equilibrium

This idea of "eggs for perpetual youth" causes many people to want to eat more eggs. The "propensity to buy" increases. People who always hated eggs, suddenly become egg-eaters. With all these people trying to buy more eggs, the weekly supply of 10 carloads gets sold out in a hurry. An egg shortage develops.

Everybody's trying to buy eggs but there are no eggs to be found. The disappointed buyers go around offering to pay extra to get some eggs. This pushes up the price. As the price goes up, the rationing function of price will squeeze some of the "would-be buyers" out of the market. The production motivating function of price will pull more eggs into the market. As the price goes up, these two "functions of price" work together to force the quantity supplied and the quantity demanded into a new equilibrium.

How high will the price go? It will continue to go up until the market moves into a new equilibrium. This situation is shown in the supply and demand graph in Figure 19-4 on the following page. Take a few minutes and study that graph now.

### Increased Demand Pushes The Price Up

The supply and demand graph shows that the price will have to go up to $1 a dozen. At that price, a lot of buyers are squeezed out of the market by the rationing function, a lot more eggs are pulled into the market by the production motivating function, and a new market equilibrium exists. Fourteen carloads of eggs will be sold by the sellers and bought by the buyers, each week. But this new equilibrium price of $1 a dozen can't last for long. Can you guess why? Competition, of course!

Remember I said that everything was in stable equilibrium back when the price of eggs was 75¢ a dozen. So a price of 75¢ must be high enough to convince the egg producers to keep on producing eggs. This means that a price of 75¢ was high enough to cover the production costs, and also to pay enough profits to the producers to keep them satisfied to stay in the egg business. If that's so, then think what a good deal $1 a dozen must be! Big profits in the egg business? Right!

Fig. 19-4 INCREASED PROPENSITY TO BUY: THE DEMAND CURVE SHIFTS

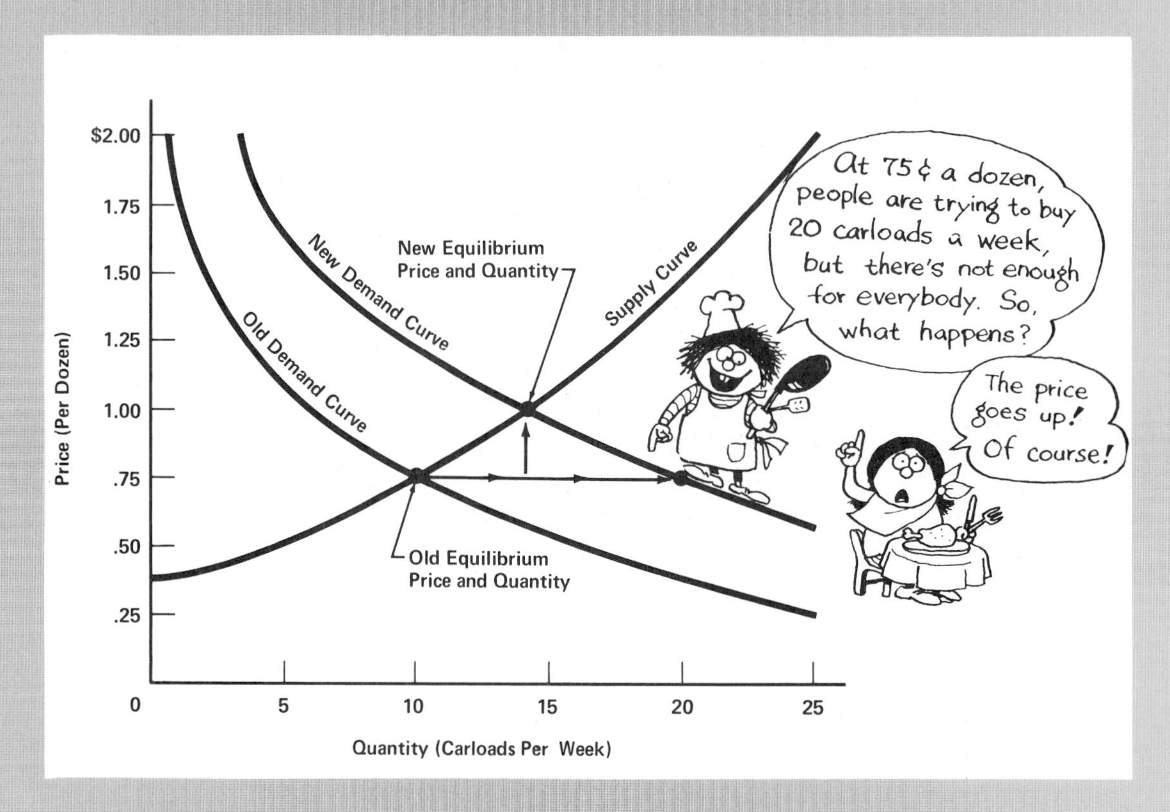


When Demand Increases It Creates A Shortage, Forces The Price Up, And The High Price Pulls In More Eggs

When people think eggs will bring "perpetual youth," the demand for eggs increases. At any price you choose, the quantity people would buy is greater. The demand curve must shift to the right to show that a larger quantity would be bought at each price. The "propensity to buy" has increased!

Seventy-five cents can no longer be the equilibrium price. At that price people now want to buy 20 carloads a week, but the sellers only ship in 10 carloads. A serious shortage! The buyers, trying to get eggs, push up the price.

The new "equilibrium market price" is $1 a dozen. This higher price (a) squeezes some of the buyers out of the market (the quantity demanded goes back down from 20 to 14 carloads) and (b) pulls forth more output from the sellers (the quantity supplied is expanded from 10 to 14 carloads).

### A High Price Brings Big Profits

Suppose you are producing frying chickens. You're doing okay, just making "normal profits." Then one day you see your egg-producing neighbor drive by in his new Cadillac. He's on his way to the airport to fly to the Virgin Islands for a three-day holiday. You suddenly realize that you should be producing eggs instead of fryers! So what do you do? As soon as you can you'll shift into the egg business where the big money is. Your neighbor doesn't mind. He knows that one more little egg producer won't hurt the egg market at all.

### Big Profits Attract More Producers

But what's going to happen? You have already guessed it. You aren't the only poultryman or cattleman or farmer who likes to make big profits. It isn't long before everyone who can shift his factors into the egg business, starts to do just that. Everyone wants to get some of the big profits from those dollar a dozen eggs!

This is competition in action. Soon, turkey producers, cotton farmers, beef ranchers, vegetable growers and others start producing eggs. Every day more people become egg producers. Each time a new producer enters the egg industry and starts shipping his output to the market, the total quantity of eggs going into the market increases a little more. Can you see what's happening? And what's going to happen?

### More Producers Bring More Supply

It isn't many months before the quantity offered for sale at a price of $1 a dozen increases from 14 carloads to 15 carloads a week. But at $1 a dozen, buyers will only buy 14 carloads. The demand curve in Figure 19-4 already told us that. At a price of $1 a dozen, with 15 carloads a week coming into the market, we have a surplus of one carload of eggs a week.

Some of the egg sellers can't sell all their eggs. So what do they do? They start cutting prices to get people to buy up all the eggs. See it? The "production motivating function of price" is still working! The high price is causing the egg industry to expand. Each time a new producer enters the industry, the supply increases. In terms of the supply and demand graph, the *supply curve* is now shifting to the right—a delayed response to the *initial* shift to the right, of the *demand curve*!

### More Supply Pushes The Price Back Down

How long will new producers keep on entering the egg business? That's easy. As long as it is more profitable for them to produce eggs than it would be for them to produce turkeys or broilers or beef or whatever else they might produce *instead* of eggs. This means that the egg industry will continue to expand and the supply of eggs will continue to increase (the curve will keep shifting to the right on the graph) until the price of eggs is pushed *back down to "normal."*

### The Normal Price Brings Normal Profits

You remember that in the beginning the normal price for eggs was 75¢ a dozen. The "normal price" is the price that brings the producers normal profits—that is, enough profits to keep them from leaving the industry, but not enough to entice other firms to stop producing other things and shift into this industry. At the "normal price," with "normal profits" being made, an industry will be stable. At prices and profits higher than "normal" the industry will be expanding. At prices and profits lower than "normal" the industry will be contracting. Businessmen are always moving into or out of one industry or another, trying to make more profits, or to eliminate their losses.*

As the egg industry expands, this may cause egg production costs to go up. If so, the new "normal price" will be higher than before (maybe up to 80¢) because of the higher costs. But let's just suppose that the "normal price" stays at 75¢. How long will the egg industry continue to expand? Until the price is pushed all the way back to 75¢ a dozen. Obviously? Sure.

### The New "Long-Run Equilibrium"

There's a graph coming up that shows the new "normal" or "long-run" equilibrium situation in the egg market in your metropolitan area. The price is back down to 75¢ a dozen. But now, instead of the original quantity of ten carloads, there are *twenty carloads* a week flowing across the market from the sellers to the buyers. Where did all those extra eggs come from? and why? It all happened through the automatic adjustments of the market process. The price mechanism took care of it for us.

First, people heard that eggs bring perpetual youth, so right away they started trying to buy more. The demand increased. So there was a

---

*Sometimes economists explain that it might be better not to think of "normal profit" as being "profit" at all. Since the "normal profit" is necessary to keep the firms from leaving the industry, "normal profit" is really a part of the cost of production—a "necessary cost" required to keep the product flowing to the market. It's only when the profit gets higher than "normal" that profit becomes a "surplus," and serves the purpose of attracting new firms into the industry. What's the point? Just this: Be sure to recognize the distinct difference between "normal profit" and "excess" or "surplus" profit. Really, they're different kinds of things, and they perform different functions in the operation of the market process. You'll see more on this later.

temporary shortage of eggs and the disappointed egg-seeking buyers went around offering more for eggs and pushed the price up. The higher price (a) conserved the use of the very scarce eggs, (b) induced the egg producers to increase their outputs as much as possible (to go out and stroke the hens to get them to lay more eggs), and (c) made the egg business very profitable and attracted many new producers into the business.

As each new producer came in and added his output to the market, the supply increased a little. A little surplus developed. The egg-sellers had to cut prices a little, to get rid of all their eggs. But more and more producers kept on going into the egg business and pushing the price down. For how long? Until the price finally got back down to "normal" (back down to 75¢ a dozen.)

Once the price gets back to normal, that's as far as it goes. Until there is some new change in demand or supply, 75¢ a dozen will continue to be the price and 20 carloads a week will continue to be the quantity exchanged in the egg market in your city. Figure 19-5 shows each step in the process: (1) the increased demand, (2) the higher price, (3) the increased supply, and (4) the new long-run equilibrium price and quantity. Be sure to spend enough time with Figure 19-5 to learn to show each of the steps. This graph is very important. It shows the "market process" and the "price mechanism" in action. It shows, in a "model system," how the market process directs society's productive resources into the desired uses.

See how the people, through their demands, make the choices about which things will be produced, and how much? An increase in demand pushes up the price, discourages consumption, and brings forth more output in the short run. Then in the long run, after there has been enough time for the industry to expand, more output comes forth and the price goes back down to its "normal" level. Then people are no longer severely discouraged from eating eggs. Everyone who is willing to pay the "normal price" (covering the cost of production plus a normal profit) can buy all the eggs he wants.

Now you know how "the market system" directs more of the society's scarce resources into egg production. You know how the market automatically carries out the decisions of the people. That's the way it works in the "pure model" of the "market system."

## Consumer Sovereignty And The Invisible Hand

In the "model market process" who holds the choice-making power? The consumer, of course. The consumer is *sovereign*. He is the holder of the ultimate power. He has the final control in deciding what will be done with society's scarce resources. His decisions about how to spend his money, ultimately determine all the economic choices for the society—all the production choices and all the distribution choices. The consumer has the ultimate control. That's what "consumer sovereignty" means.

Fig. 19-5 IN THE LONG RUN, THE SUPPLY INCREASES

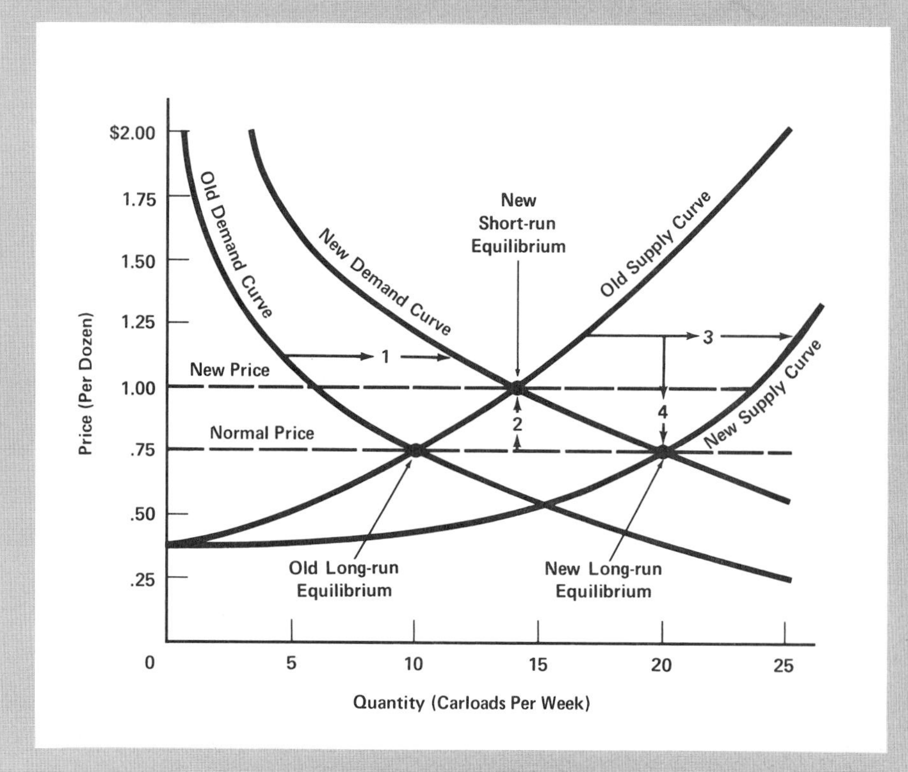


The High Price Pulls More Producers Into The Industry—A Delayed Response To The Demand Increase.

(1) The demand increases. At the price of 75¢ a dozen, the quantity the buyers want to buy increases from 10 to 20 carloads a week. This creates a shortage.

(2) The buyers, bidding against each other for the limited eggs, push the price up to $1. At the higher price the new quantity demanded moves back from 20 to 14 carloads per week, and the quantity supplied increases from 10 to 14 carloads. At the price of $1, we have a new short-run equilibrium.

At the price of $1, big profits are made in the egg business. New producers come in and start producing and selling eggs. Each time one more producer adds his output to the market this increases the total market supply a little bit. As the supply increases bit-by-bit, the price inches downward.

(3) The supply continues to increase and the price drops until the "normal price" is reached. Then (4) the industry is again in stable long-run equilibrium. The price is again 75¢, but the quantity crossing the market has increased to 20 carloads per week.

Anyone who has money and spends it, will influence "what is produced." Anyone who produces (or who owns productive factors which produce), will receive money to spend and can influence "what is produced." Of course we never have all the necessary conditions for a "model market system" to work exactly right. But if we did, then the consumer really would be "sovereign." He would have complete control over the economic choices of the society.

You have heard about Adam Smith, who, back in 1776, wrote a big and famous economics book called *Wealth of Nations*. Adam Smith understood and explained the market process, and about specialization and the gains from trade. He was strongly against the governmental controls of "mercantilism." He said that the nation would be most wealthy and everyone would be better off if the government would get rid of its economic controls and just let each businessman and each person go his own way.

One of Adam Smith's most famous statements was that if you leave people alone to follow their own personal interests, each will be automatically guided "as though by an invisible hand" to do the best things for the society. The idea of "the invisible hand" is the idea of the market process at work. It's the idea that society's economic choices will be made automatically, in response to the wishes of the people. The "sovereign consumers," through the market process, will cause the labor, land, and capital to do the best things for the society, as though all were guided by "an invisible hand."

## Consumer Spending Really Does Influence The Economic Choices

You know that in the real world no country has an economy just like the "pure market model" of consumer sovereignty and the invisible hand. Still, in the economic systems of most of the countries throughout the world, the consumer's choice (about what to buy with his money) really does exert a major influence. The consumer really does have an important say in deciding what will be produced, which resources will be used for which purposes, and how much each person will get. "Consumer sovereignty" in the real world is far from pure. On the other hand, it is far from dead.

How the real consumer in the real world decides to spend his money *does* have *some* effect on the economic choices in every country—in every society in the world today. The more a nation allows for and protects open and free competitive markets and private property rights, the more "sovereign" the consumer will be—the more "ultimate power" the consumer will exercise over the society's economic choices. I'm sure you realize that the "ultimate power of the consumer's choice" is much greater in the economy of a country like the United States or Canada or West Germany or Great Britain or Japan, than it is in the Soviet

Union or Communist China. But even in the communist countries the consumer's choices cannot be ignored completely.

## SUMMARY OF DEMAND, SUPPLY, AND PRICE

This chapter has carried you a long way forward in your understanding of the precise way in which the "pure market process" works. You know how a change in demand can move the price around. And you know how the price rations (conserves the use of) existing goods and stimulates the production of more goods. The higher goes the demand, the higher goes the price, and the more the good is rationed and the more its production is motivated.

You know, also, that a high price (higher than "normal") will cause an industry to expand. Producers enter the industry seeking some of the big profits. As the industry expands, more supply is pushed into the market. Soon the greater supply begins to create some surpluses.

Some of the sellers can't sell all their goods, so they start cutting prices. More sellers keep entering the industry, expanding the output, and cutting prices until the price is pushed all the way back down to "normal." It's really neat! But there's something about all this that you really don't quite understand, yet.

You can't really understand supply (the propensity to sell) until you understand more about the cost of production. You know that the supply of something depends on (reflects) the cost of producing it. If the cost is "too high," none will be produced. That's obvious. But you need to know more about it than that.

The next chapter is all about "cost." First, be sure you've really learned about supply and demand and the functions of price, and how to draw and explain the curves. Then you'll be ready to go on into the next chapter and start finding out about supply—about "the cost of production."

---

## REVIEW EXERCISES

**MAJOR CONCEPTS, PRINCIPLES, TERMS (Try to write a paragraph explaining each of these.)**

demand
elasticity of demand
the two "effects" of a price change
the two functions of price
supply
demand, supply, and price

## OTHER CONCEPTS AND TERMS (Try to write a sentence or phrase explaining what each one means.)

consumption possibility
substitution effect
income effect
law of demand
price elasticity of demand
relatively elastic demand
relatively inelastic demand
unitary elasticity of demand
absolutely inelastic demand
rationing function of price
production motivation function of price
equilibrium market price
change in demand
change in supply
normal price
normal profits
price mechanism
consumer sovereignty
the "invisible hand"

## CURVES AND GRAPHS (Try to draw, label, and explain each.)

The Demand Curve Illustrates the Propensity to Buy
The Supply Curve Illustrates the Propensity to Sell
The Supply and Demand for Eggs
Increased Propensity to Buy: The Demand Curve Shifts
In the Long Run the Supply Increases

## QUESTIONS (Try to write the answers, or jot down the highlights.)

1. Explain how supply, demand, and price work together in the "pure model market system." Use supply and demand curves to show how *first, price* and then, *supply* respond to changes in consumer demand.
2. When you're trying to explain how supply and demand work, it's not easy to always keep it straight whether you're talking about
   a. a change in the *demand*—that is, a change in your overall propensity to buy (a shift of the demand curve), or
   b. a change in the *quantity bought*, which results from a change in the price (that is, a movement up or down along the demand curve, to a higher or lower price and a smaller or greater quantity bought).

   Can you see what would happen to your explanation of how the market mechanism works if you let these two things get mixed up? Think about it and try to explain.
3. Think about your own "elasticity of demand" for some of the things you buy. List some things for which you think you would spend *more* if the price went *up* (inelastic demand). Then list some things for which you think you would spend *more* if the price went *down* (elastic demand).
4. The only price that really could exist in a free market would be the equilibrium market price. In the supply and demand graph, all prices above or below the equilibrium price, and all quantities greater or

less than the equilibrium quantity are really irrelevant. There's only one price (the equilibrium price) and one quantity (the equilibrium quantity) which would exist in that market at any moment. Can you explain why this is true? What forces guarantee that it must be true?

# 20 HOW COST DETERMINES SUPPLY

*As the rate of output changes, the cost per unit changes because of the Law of Diminishing Returns.*

Now you know a lot about supply and demand. You know, too, that the more you understand about supply and demand, the better you will understand how the market process works. No question about that. But what is there to understand about supply and demand? Quite a lot.

**Supply And Demand Analysis Only Provides A Framework**

"Supply and demand analysis" gives us a very convenient and useful framework for looking at the market system. It lets us see how the force of demand, pulling against the restraint of supply, activates the price mechanism; then prices direct the resources and products into the right places, and to the right people. You already know how that works. But there are many more interesting things to know about supply and demand!

Anything that influences supply or demand, automatically influences the microeconomic choices of the society. So the next question is, "what influences supply? or demand?" You know the "simple overview" answers to these questions. But let's have a quick review.

**Supply, Again**

Supply (the propensity to sell) is determined by the cost of production. Producers are likely to be willing to sell a lot more of something if they can produce it at low cost than if the cost is high. If the *production cost* is *higher* than the *selling price,* they won't want to produce and sell *any.* What if the selling price is much higher than the production cost? Then the producers will want to produce and sell a lot.

It's easy to see how cost determines "the propensity to sell." But the next step—really understanding a business firm's cost situation and seeing exactly *how* supply is determined by the cost—gets a little more involved. We're going to spend this whole chapter talking about that.

## Demand, Again

What about demand? Demand (the propensity to buy) depends on people's desires for things and the amounts of money they have. The more a person wants something and the more money he has to spend, the greater will be his "propensity to buy" it.

Back in Chapter 2, we were talking about "how individuals choose." You read about how each person would spend his money trying to maximize his progress toward his goals. You read about the "marginal utility" and "marginal rate of substitution" theories of consumer demand. Then about the important influences of income and savings and expectations and all. The chapter you just finished talked about demand. In addition to all that, don't forget that you have been a consumer all your life! You know a lot about the kinds of things that influence your desire and willingness to buy one thing or another.

What about the other part of demand—"non-consumer" demand? You know quite a bit about that, too. Remember Chapter 2, the section on "How Businesses Choose"? You found out that it's profitable for a business to buy inputs (labor, land, capital), as long as the marginal "input cost" is less than the value of the extra output (the marginal "value product"). Then in Chapter 9, we talked about "Why Businesses Spend." You remember about the importance of profit expectations. Then in the last chapter we talked about supply, and how much the businesses would produce at various selling prices. The more a business produces, the greater will be its *demand for factors of production*. Right? But there are still many unanswered questions.

This book isn't going to attempt to dig up and expose all the things which influence the supply or demand for anything. But there are a few other things you need to know about what determines supply and demand and how the "pure market system" fits it all together.

This chapter and the next one explain what economists call the "theory of the firm." You will find out how the businessman makes his decisions on the basis of his costs and revenues. This chapter explains the costs. The next chapter will integrate the revenues into the cost picture. Chapters 22 and 23 will talk about the "factor markets," explaining how supply and demand are determined and how they work in the markets for the factors of production (the labor, land, and capital markets). But right now it's time to take the first step in understanding the "theory of the firm." It's time to talk about the cost of production.

## TOTAL COST IS MADE UP OF FIXED COST PLUS VARIABLE COST

Have you ever thought of yourself as a producer? We have been talking about your role as a buyer, a consumer. But now, for a while, try to think of yourself as a businessman who owns and runs a small manufacturing plant. Your plant is located up in the Appalachian Mountains, at Cullowhee, North Carolina. You produce hip-hugger jeans. Let's talk about your cost of production, and about how the cost increases as your daily rate of output increases.

### Fixed Cost Doesn't Change As Output Changes

Suppose your daily output last week was zero. You didn't produce anything. What do you suppose your total cost was, last week. Zero? No. More than zero. Even if you produced nothing you still have to pay your fixed cost. You have to pay the rent, and the interest you owe, and insurance, and the night watchman, and the maintenance man who cuts the grass around the plant and lubricates the machines to keep them from getting rusty. And taxes. Don't forget about taxes.

Yes, even if you don't produce anything, there are always some "fixed costs" you must pay. "Fixed cost" is made up of all the costs you must pay even if you don't produce anything. If you produce nothing, or if you produce a thousand units a day, it makes no difference to your fixed cost. The fixed cost is the same, regardless of your rate of output.

Does the fixed cost tell you anything about how much output you should produce? No. But you know that the greater your rate of output the lower your "per unit fixed cost" will be. For example, suppose your fixed cost is $150 a day. If you produce only one unit of output per day, what will be your "per unit fixed cost"? The whole thing: $150. But if you produce 10 units per day, "average," or the "per unit" fixed cost will be only $15. ($150 total fixed cost divided by 10 units of output, equals $15 per unit fixed cost). Suppose you were producing 150 units per day. The fixed cost per unit would be only $1!

As far as the fixed cost is concerned, you would like to produce as much output per day as possible. The more you produce, the more you "spread the fixed cost." At a high rate of output the per unit fixed cost is very low.

This concept we're talking about is what businessmen call "spreading the overhead." Every businessman knows that if he has a high volume of output, his overhead (fixed cost) will not be anything to worry about. But when output slows down, high overhead can bring heavy losses. Many small (and some large) businesses fail because they don't give enough thought to their fixed cost.

## Variable Cost Increases As The Rate Of Output Increases

If you're going to produce any output you're going to incur some *additional* costs. To make jeans, you are going to have to buy some denim and some thread and some zippers (and maybe some brass buttons and rivets to "jazz 'em up a bit") and some other things. You're going to have to buy some cardboard shipping cartons, and gasoline for the delivery trucks, and electricity to run the machines. What about labor? You are going to have to hire some people to run the sewing machines, and some cutters, and some supervisors and inspectors and all, and some salesmen to sell your jeans to the department stores and specialty stores over in Asheville and Knoxville and Rock Hill and Texarkana.

All these *additional* costs—which increase as your rate of output increases—are called variable costs. When output is zero, the variable cost is zero. But as soon as you produce something you must pay some variable cost. The greater your daily output, the more your variable cost is going to be. That's obvious enough. But here's something else.

## Variable Cost Per Unit Is High When Output Is Low

The *rate* at which the variable cost increases (as your daily rate of output increases) isn't always the same. Suppose you are trying to produce at a very low rate of output—maybe just a few pairs of jeans a day. Then your variable cost per unit of output may be very high. Maybe you must keep a full shift of workers, even if you only produce a few units. So if you produce at a very low rate, your *cost per unit* will be very high. Does that make sense to you? Let me say it another way.

Suppose your plant is designed to operate most efficiently when you are producing ten cartons of hip-hugger jeans per day (two dozen pairs of jeans per carton). So if you operate your plant at less than 10 cartons a day, the operation is not going to be as efficient; if you operate at more than ten cartons a day, the operation is not going to be as efficient. "Efficient," means "low cost per unit." If you produce at less than, or more than, 10 cartons a day, the *cost per unit* is going to be higher.

Now, suppose you aren't producing anything. Then if you decide to start producing at the rate of two cartons per day, your variable cost will go up a lot. The per unit cost of those two cartons a day will be very high. So maybe you decide to operate at the rate of five cartons a day. Your variable cost will go up, sure, but not as fast as your output goes up.

Next, suppose you decide to expand your daily output to ten cartons a day. Your variable cost will get still higher, but it won't increase nearly as much as your output will increase. Your output is expanding faster than your cost is expanding because you are getting closer to your maximum efficiency rate of output. What happens if you decide to expand your output to *more than* ten cartons per day? Suppose you decide to produce eleven, or twelve, or thirteen, or more? Your variable cost will increase much more than your output will increase. Let me say it another way.

What makes up your total cost for any level of daily output you might choose? The "fixed cost," plus the "variable cost." Sure. If you produce nothing, the "fixed cost" is your "total cost." If you decide to produce something, your "total cost" will be the fixed cost *plus* the variable cost. The higher the rate of output, the higher will be the variable cost, so the higher will be the total cost. But the cost *per unit* will be lowest if you operate at your most efficient rate of output.

### Minimum Cost Is Not The Objective

Which rate of output should you choose? Don't worry about that yet; we'll get to it later. But one thing you know already. The output you choose will *not* be the one of *minimum total cost*. The "least total cost" rate of output is where you produce nothing! You pay only your fixed cost. But there wouldn't be much point in running that kind of an operation, would there?

> Exception: If nobody wanted to buy your output, or if the price was so low that you couldn't even make enough to pay for your denim, brass buttons, labor, and other variable factors, then it would be best not to hire any variable factors. Produce nothing. That will minimize your losses. Pay your daily fixed cost as long as you have to, and make arrangements to get out of this business as fast as you can!

## THE TOTAL AND AVERAGE COST CURVES

Now you know a little bit about how your *costs* change as your *rate of output* changes. Wouldn't it be helpful to see how this looks on a graph? We could show the *rate of output* along the horizontal axis and measure the *cost* on the vertical axis. The total cost curve will have a positive (upward) slope—more output, more cost. Right? But the total cost curve won't be a straight line. Why not? Because as the rate of output gets larger, total cost goes up—but it doesn't increase *at a constant rate*. Remember?

What about fixed cost? That will show up on the graph as a straight, horizontal line. Why? Because fixed cost is just as high when you produce nothing as it is when you produce as much as you can! Let's look at all the cost curves.

Here they are, five graphs in a row. Each one shows a different picture illustrating exactly what we have been talking about. Take your time and study and understand all these curves. You should practice drawing them too, just to be sure you *really* understand.

**Fig. 20-1 FIXED COST AND AVERAGE FIXED COST**

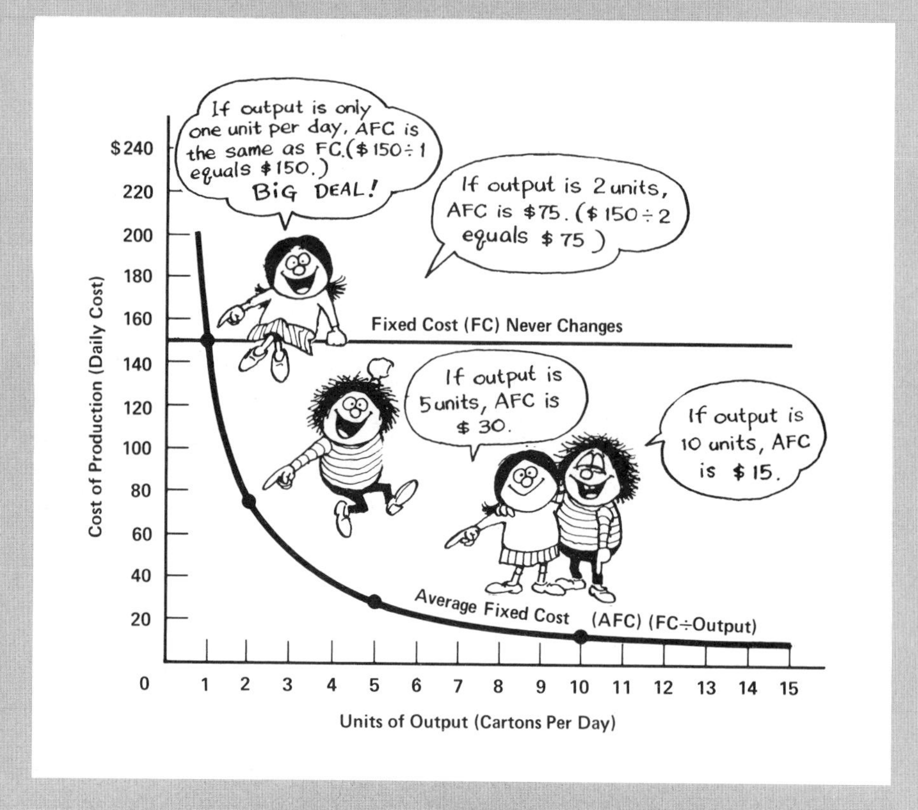


Even If You Don't Produce Anything, You Still Must Pay The Rent, Taxes, And Interest. And The Night Watchman, Of Course!

The fixed cost is always the same, regardless of the rate of output. That's why the total fixed cost curve is horizontal and the average fixed cost curve gets lower and lower.

At low rates of output, the fixed cost is an important part of the cost of each unit. At high rates of output, the fixed cost is only a small part of the cost of each unit. If output is expanded enough, the fixed cost (overhead cost) per unit gets so small that it's hardly worth considering.

A business which has very high fixed costs must produce and sell a large volume to survive.

**Fig. 20-2 VARIABLE COST AND AVERAGE VARIABLE COST**

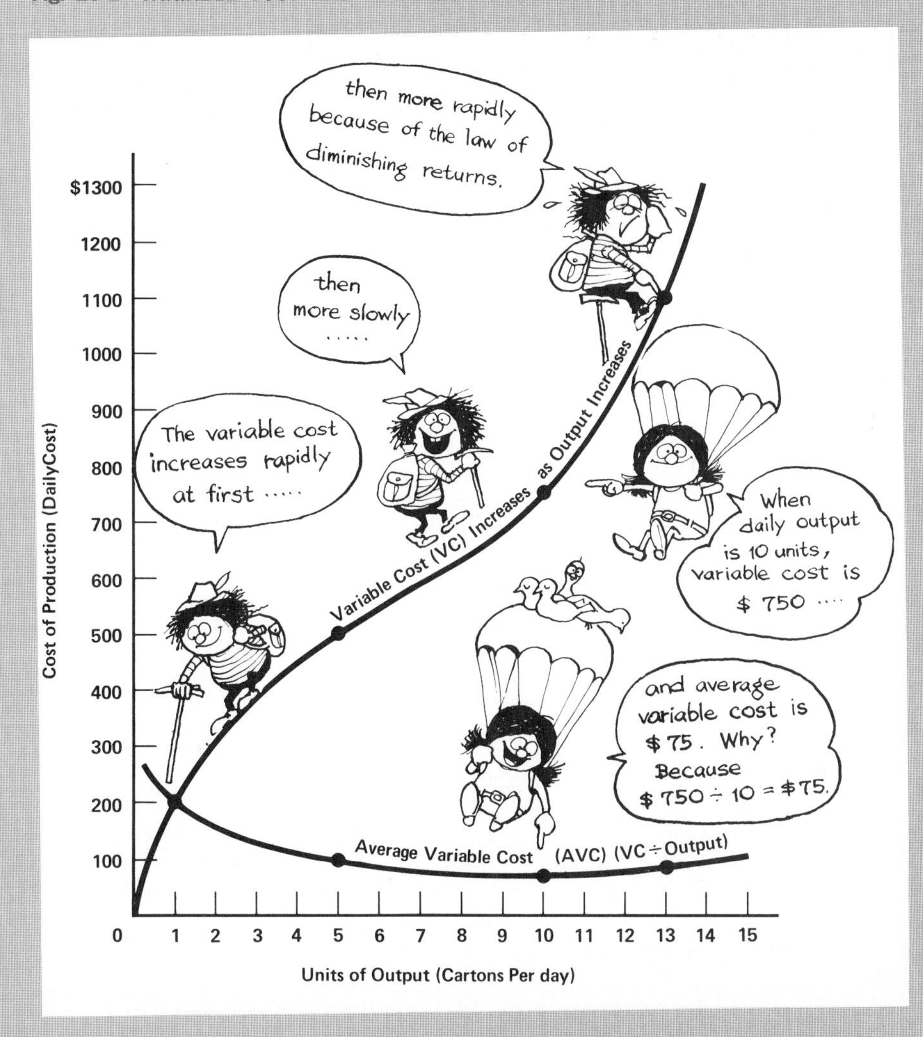


One Thing About Variable Cost Is Sure: The More You Produce, The More You Must Pay.

As the rate of output increases, variable cost increases rapidly at first, then more slowly, then more rapidly again.

(This graph has a smaller vertical scale than the previous figure. We had to squeeze the vertical axis in order to keep this "variable cost curve" from running right off the top of the page.)

As output increases, variable cost increases. The average variable cost goes down for awhile, but after output expands beyond the most efficient range, variable cost increases rapidly and pulls the average up again.

**Fig. 20-3 FIXED COST PLUS VARIABLE COST EQUALS TOTAL COST**

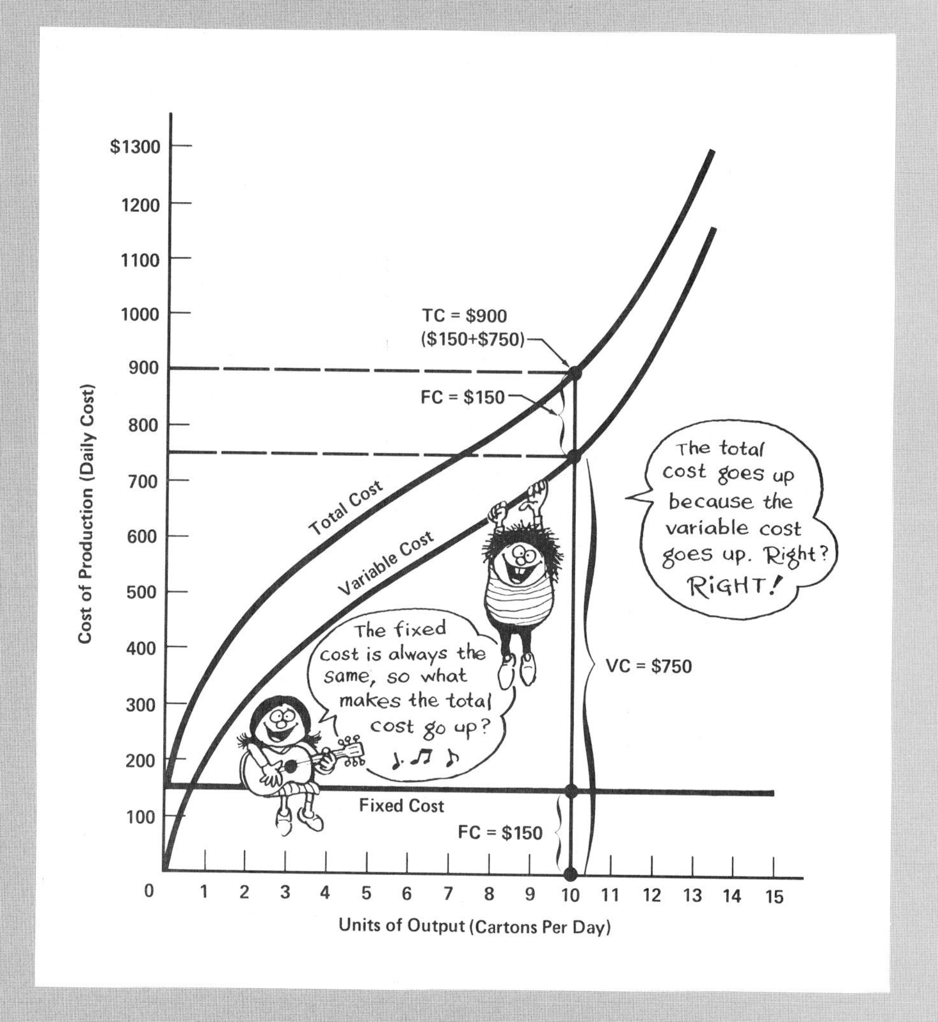


Fixed Cost Is The Cost Of Being In Business. Variable Cost Is The Cost Of Producing Things. Add Them And You Get Total Cost.

When output is zero, total cost equals fixed cost. Then as output expands, total cost is pushed up by the variable cost.

The *vertical distance* (the dollar difference) between the two curves is always $150 (the fixed cost). (Note that the only distances that mean anything on this graph are the *vertical* distances, which show what the costs are at different rates of output.)

When output is ten cartons per day, fixed cost is $150 (as always), and variable cost is $750. So total cost is $900 ($150 plus $750 equals $900).

Fig. 20-4 **TOTAL COST AND AVERAGE TOTAL COST**

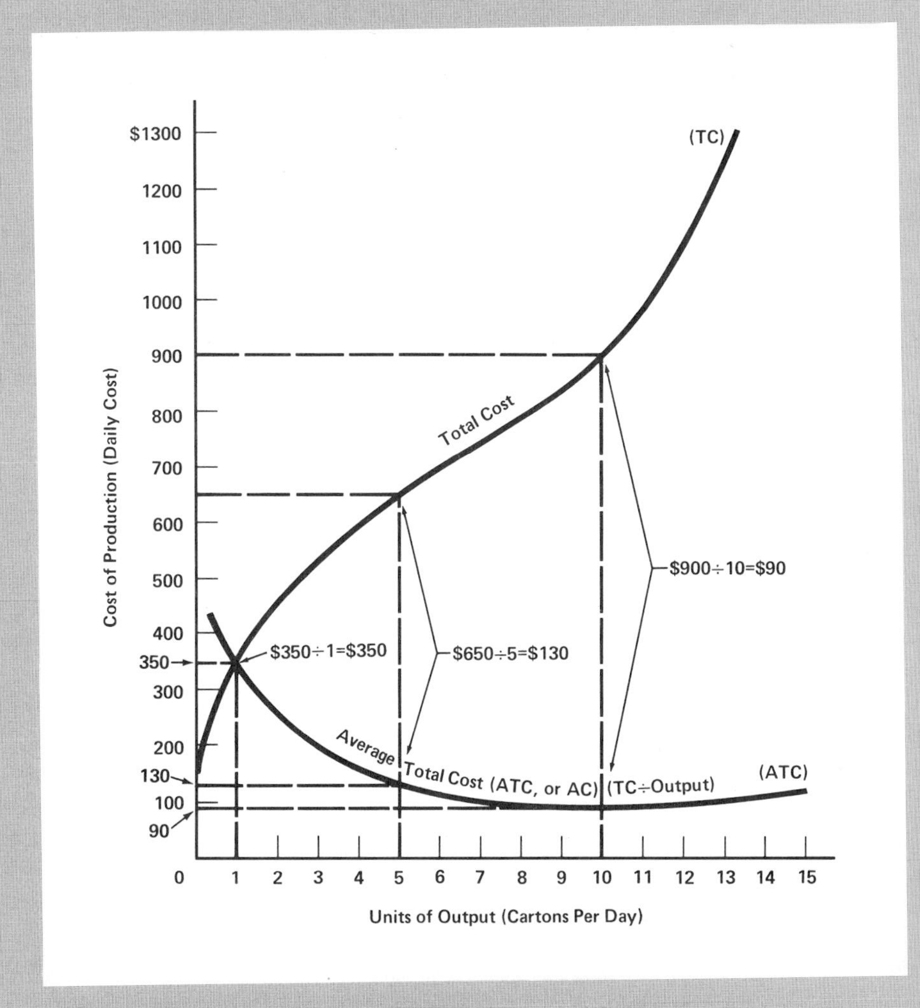


When Output Is Low, Average Total Cost Is High Because 'You Aren't Producing Many Units To "Spread" The High Fixed Cost. If Output Is Too High, Average Cost Is High Because Your Plant Is Running Too Fast.

The total cost curve is shaped exactly like the variable cost curve, except that it begins (when output is zero) at the level of the fixed cost ($150).

You know that as output expands, total cost goes up. But the average cost (the per unit cost) goes down. Average cost keeps going down until your plant reaches its "maximum efficiency" rate of daily output. (10 cartons per day.) Then if output expands beyond that, the average (per unit) cost will go up.

Fig. 20-5 **THE AVERAGE COST CURVES: TOTAL, VARIABLE, AND FIXED**

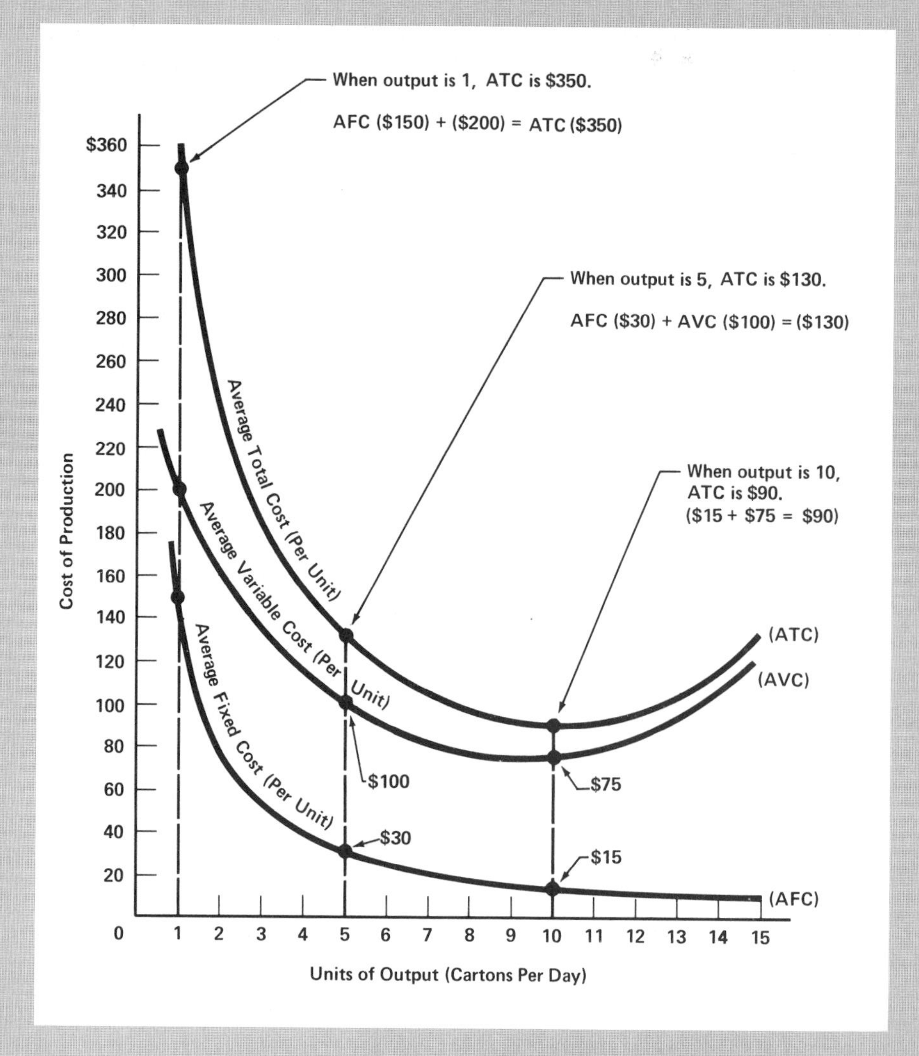


Average Total Cost Is "Total Cost Per Unit." As Output Expands, It Goes Down Because Production Is Efficient. Taken Too Far, Production Gets Inefficient And Cost Goes Up Again.

Average total cost is the average fixed cost plus the average variable cost.

Your per unit cost is lowest ($90 per carton) if you produce ten cartons per day. But you don't know whether or not you will produce at that rate because you don't know what the price is. You'll find out about that, later.

These curves are the same ones you saw on the preceding graphs. They look different here because the vertical axis has been stretched out.

## The Cost Curves Are Hypothetical

Now that you have learned about the different kinds of cost curves, what have you found out? Just this: the curves show what the costs of a business "would be if . . .". If what? If the business was operating at each of the various output rates. But the business is not ever going to operate at all those different daily rates of output. Not ever!

You would never operate your hip-hugger jeans factory at the daily rate of only one unit of output. If you did, your total cost would be $350, and you would have only one unit of output to sell. The only way you could break even would be to sell that unit of output for $350. Think for a minute. Suppose you *really could* sell that one unit for $350 and break even. Wow! Think how much profit you could make by producing and selling a lot more!

## At A Very Low Selling Price You Would Shut Down

No, if the market for your jeans was so bad that all you could sell was one carton per day, then you would just shut down your plant. You would lose the $150 of fixed cost. But the variable cost you would have to pay to produce that first unit is $200. The fixed cost plus the variable cost would be $350. Better for you to lose your fixed cost ($150) than to produce one unit and lose even more! Of course, if someone was willing to pay *more* than $200 for that one unit per day, and if that was all anyone would buy, then you would be better off to go ahead and produce at the rate of one carton per day. But, in the first place, that certainly isn't a likely situation; secondly, even if such an unlikely situation did arise, what you really would do is to produce 10 units a day for a few days and build up some inventory in your warehouse. Then you would shut down and sell one carton per day out of the inventory.

The very low output range of these cost curves is purely hypothetical. It's interesting to know "what would happen to costs if daily output was only one unit." But really, most businesses only operate somewhere fairly close to their "most efficient" operating rate. (Do you see the law of diminishing returns—the law of variable proportions—lurking under the surface here, somewhere?) If the market demand isn't great enough to let you operate your factory near your most efficient daily rate of output, then you will shut down your plant until the market gets better, or you will produce some other product instead of hip-hugger jeans.

What about the very high outputs shown on the graphs? Are the curves only make-believe there, too? Usually, yes. You really can't expand your output very much beyond your most efficient operating range. The curves tell you that. If the demand was very strong and prices very high, you would increase output some. But not very much.

Look back at Figure 20-5 and see how steeply your total cost goes up when output goes beyond eleven units per day. The price would have to be very high to induce you to expand your daily output to as much as twelve or thirteen cartons. Your plant just isn't designed to produce much more than 10 cartons per day.

Why is your "efficient range" of output so restricted? Why can't you just produce 2 or 3 units a day and still make a reasonable profit? And if the demand is great and the price goes up high, why can't you just go ahead and produce 15 or 20 or 30 cartons a day and make lots of profit? You know why not. It's against the law. What law? The law of variable proportions. The law of diminishing returns. Let's talk about that.

## THE LAW OF DIMINISHING RETURNS

The way the variable (and total) costs change at different rates of output illustrates the law of diminishing returns. I'm sure you remember this from Chapter 2. This "law" says that if a factory or a farm (or any kind of productive activity) expands its output more and more without adding any more land or buildings, sooner or later some overcrowding and inefficiencies will occur. The variable inputs won't have enough fixed inputs to work with. Costs will go up.

You can increase your output some by adding more "variable inputs" (say labor) even though the amounts of some other inputs are fixed (say the size of your jeans factory). But there's a limit to how far you can go. Sooner or later the extra output you get when you hire each successive unit of the variable input will get smaller and smaller. The returns (extra product) received from the extra variable input, will diminish.

You can see why the law of diminishing returns is also called the law of variable proportions. It is talking about varying the proportions of the inputs. The reason the returns (extra outputs) diminish is because there is too much of the variable input (labor and materials and all) in proportion to the amount of the fixed input (the factory and machines). Each man has too little of the fixed input (machines) to work with. So the extra output you get when you hire an extra man isn't very big.

### A Flowerpot Wheat Farm?

The law of diminishing returns is obviously true. If it was not, it would be possible for you to grow enough food for the world in your backyard garden (or in a flowerpot!) just by adding enough labor, seed, and fertilizer. Being ridiculous, let's suppose one worker is growing wheat in a flowerpot and another worker decides to spend full time helping him. Chances are that the added output from the added efforts of the second man will be zero. He doesn't have enough land to work with to add any more product at all. The economist would say:

> "The marginal 'physical product' of the second man is zero. Therefore, his marginal 'value product' is also zero. Therefore, the wage he can earn on the 'flowerpot wheat farm' is also zero."

Why is the marginal product (the additional product obtained from the efforts of the second man) zero? Because of laziness? Incompetence? No. It's because of the law of diminishing returns. It's because not enough of the needed inputs are available for him to work with. That's what makes it impossible for him to produce anything. If he could get another flowerpot from someplace, then he could produce just as much as the first worker. But then the "law" of diminishing returns wouldn't be working. Why not? Because nothing is fixed, that's why not.

A silly example, sure. But it illustrates something. It tells us that the more and better land, machines, and other things a worker has to work with, the more productive he can be. Also, the more he produces, the higher his wages can be.

### This Law Explains Why Some Nations Are Poor

Some countries have much good land and resources, and many factories with the best tools and machines. In such countries workers produce a lot, get high wages and have many things to enjoy—cars, furniture, boats, TV sets, refrigerators, good medical care, and all that. In other countries there is not much good land and resources, not many factories with good machines. In those countries, most workers will not produce very much. Their wages will be low. Their living standards will be low. Why is all this true? Mostly because of the law of variable proportions—the law of diminishing returns.

The law applies to all individuals and families, all businesses, and all societies, all nations. It also applies to all productive inputs. Whichever input is used too much (is applied too heavily in proportion to the other inputs) will have a low marginal physical product and a low marginal value product and will receive a low income. The input which is most scarce will be used most sparingly. It will have plenty of other inputs to work with. It will have a high marginal physical product, a high marginal value product, and a high income.

The law of diminishing returns is one of the most important concepts in economics. In our "theory of the firm" it explains the shape of the variable and total cost curves. In the real world, it influences the costs and profits of all businesses. It influences the employment and living conditions of all individuals in all cities, states, and nations. Try to understand it well.

## MARGINAL COST

You know very well the importance of approaching economic choice situations (decision situations) by looking at marginal differences (instead of at the totals). So don't you wonder why nothing has been said in this chapter about marginal cost? How does marginal cost influence your production decisions? How does it fit in the graph, with all those other cost curves? And why have we waited so long to talk about it?

We have waited so long, because I believe the marginal cost curve confuses more students of economics than any other concept, except perhaps the relationships between "savings and investment" and between "interest rates and bonds values"—both of which you have studied and now, I hope, understand. The marginal cost concept is very simple and logical, and the marginal cost curve is just as simple and logical. But the curve is so much easier to understand after you have all the other cost curves clearly in place in your mind. If you don't quite understand all the other cost curves (Figures 20-1, 2, 3, 4, 5), please go back now and study them. If you have command of all those graphs, you will not have any difficulty understanding marginal cost, and the marginal cost curve. But just to be sure, before you look at the marginal cost curve, here's an overview of the marginal cost concept.

### The "Per Unit Rate Of Increase" In Total Cost

"Marginal cost" (as you remember from Chapter 2) means "the addition to the total cost that results from expanding your daily rate of output by one unit." Or you can say it this way: The marginal cost is the "per-unit rate of increase" in the total cost, as daily output expands one unit at a time. As the output rate increases, if the marginal cost is high, total cost rises fast; if the marginal cost is low, total cost rises slowly. Can you figure out how this would look on the total cost graph?

If the *total cost curve* is going up *steeply*, that means that the *marginal cost must be high*. If the total cost curve is going up gently, then the marginal cost is low. What causes the total cost to increase as the rate of output expands? The marginal cost, of course!

If you increase your daily output by one unit, your daily total cost will go up. By how much? By the amount of the marginal cost. Suppose you are producing 10 cartons a day and I tell you that your marginal cost (at that rate of output) is $90. What does that mean? It means that your total cost is $90 higher when you're producing 10 cartons a day than it would be if you were producing only nine cartons a day.

Suppose you want to figure out for yourself what your marginal cost would be if your output was 10 units. Just take the total cost at your 10-unit output, and subtract from it the total cost at your 9-unit output. Then you have it. The marginal cost for any rate of output is found by taking the total cost for that output (output "x") and subtracting the total cost for the one unit smaller rate of output (output "x minus one"). The marginal cost for any rate of output is the *extra cost* required to bring your output *up to that rate — that is, it's the cost of expanding your output from the "one unit smaller" rate.*

### "One More Unit's Worth" Of Variable Factors

What is the marginal cost made up of? It's the cost you have to pay for the extra variable inputs when you expand your output by one unit a day. If you don't have to buy very many variable inputs (that is, if the inputs are highly productive), or if the price of the inputs is low, then the marginal cost will be low. But if you have to buy a lot of inputs (they aren't very productive), or if the price of the inputs is high, then the marginal cost will be high.

Would you expand your output if the marginal cost was high? Or not? That all depends on how much money you are going to get when you sell that "marginal output unit." Right? If you are going to get enough for the "marginal daily output unit" to more than cover the high "marginal daily cost," then you will produce it. You don't mind paying high marginal cost — that is, you don't mind so long as the *marginal revenue* you're going to get is even higher!

### The Marginal Cost Curve

Now you're ready for the marginal cost curve. The first figure (20-6) shows that "marginal cost" is "the rate of increase in the total cost." Then the next figure (20-7) shows how the marginal cost curve is derived from the total cost curve. The marginal cost curve is an interesting, unique, and very important curve. Before you study it, one word of caution.

The marginal cost curve in the graphs that follow is a different kind of curve than anything you've run across yet. You have seen total curves and average curves. But the marginal curve doesn't refer either to *the total* of anything or to *the average* of anything. All it tells you is *how much greater the total cost would be at that daily rate of output, than it would have been at a one-unit smaller daily rate of output.* That's all it tells you. Nothing else.

Be careful. Think carefully about the marginal cost curve. Practice explaining it to yourself. It's really simple, but it's also quite tricky — so watch it. Now you're ready to study Figures 20-6 and 20-7 and learn all about the marginal cost curve. Then Figure 20-8 will show you how the "marginal" curve fits in with the "average" curves.

**Fig. 20-6 TOTAL COST AND MARGINAL COST**

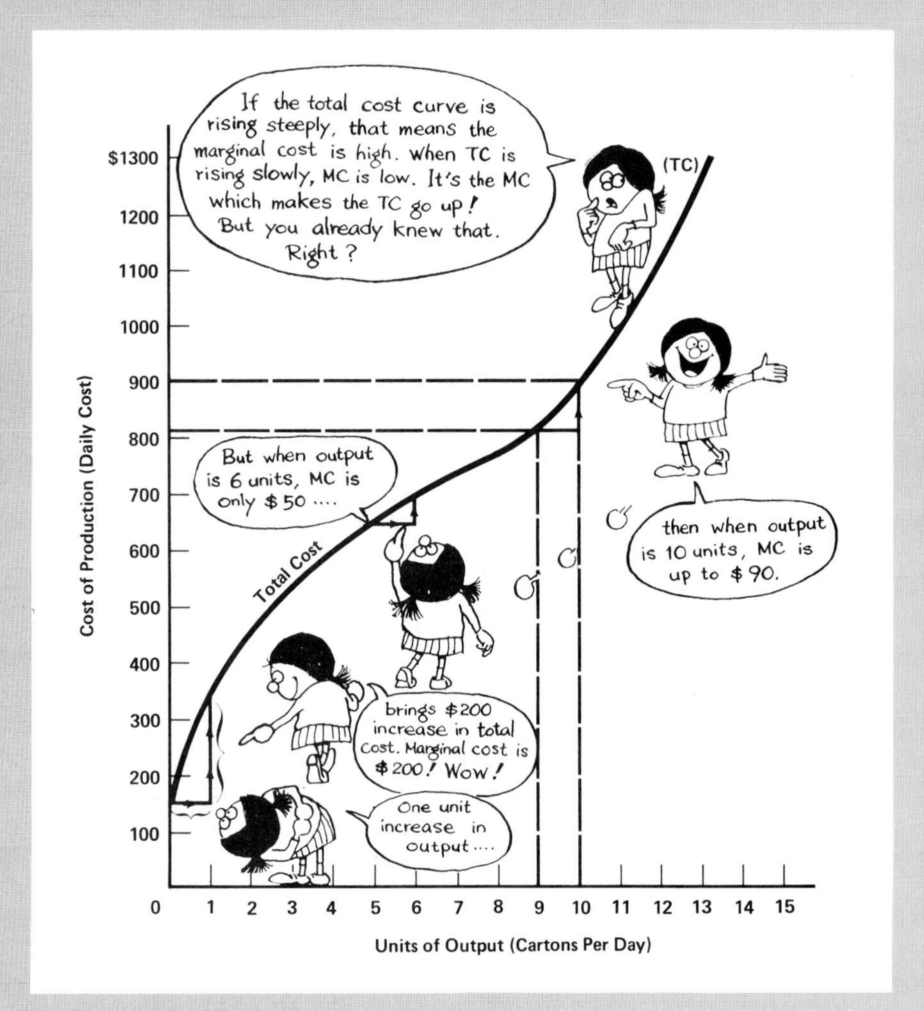


If You Draw A "One-Unit Increase Triangle," The Altitude Of The Triangle (i.e., The Height Of The Vertical Line) Is The Marginal Cost.

The marginal cost is the rate of increase in the total cost.

How much does the total cost go up when the rate of output is increased by one unit? That's the marginal cost!

Fig. 20-7 **DERIVING THE MARGINAL COST CURVE**

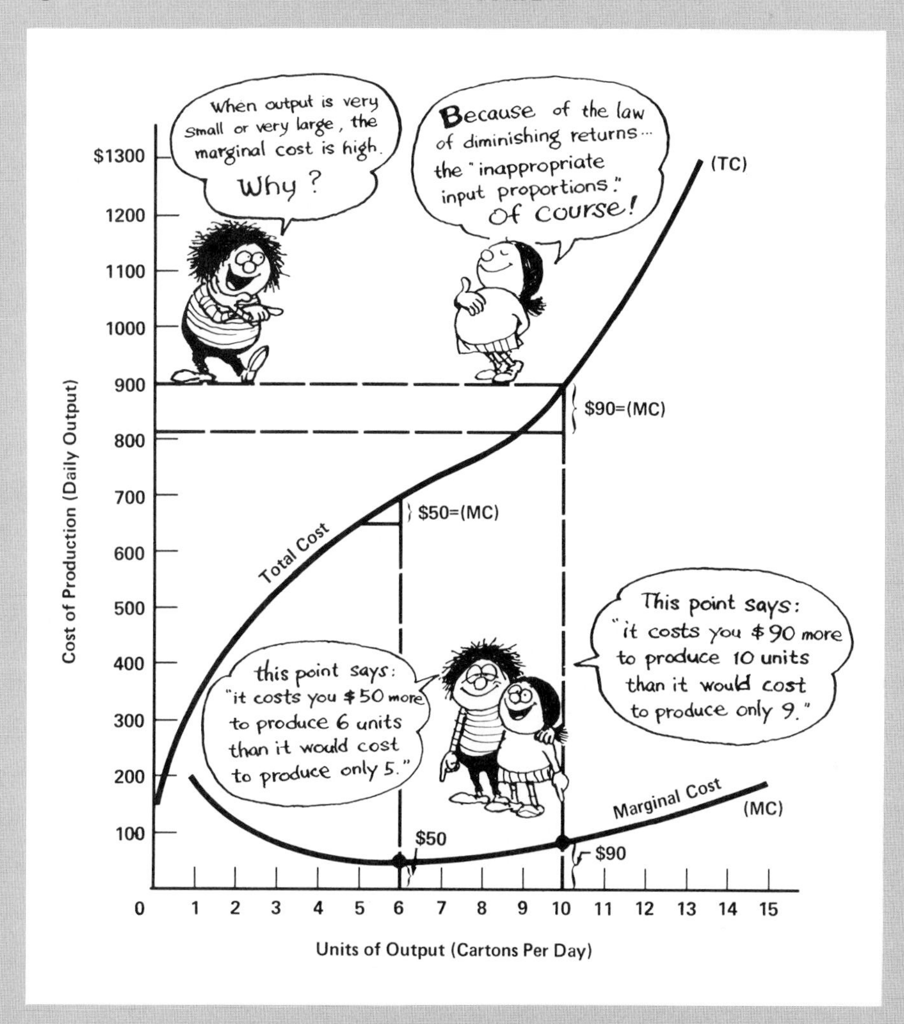


When You Plot The Altitude Of Each Marginal Cost Triangle Along The Horizontal Axis Of The Graph, What You Get Is The Marginal Cost Curve.

For each rate of output there is a "marginal cost." It's the cost of expanding to that output rate, from the "one unit smaller" output rate.

On this graph the marginal cost curve looks very flat, because the vertical scale is squeezed down so that the total cost curve won't run off the top of the page. In the next graph (20-8) you will get a much better picture of the shape of the marginal cost curve, and you will see how it fits in with the average curves.

Fig. 20-8 **AVERAGE AND MARGINAL COST CURVES**

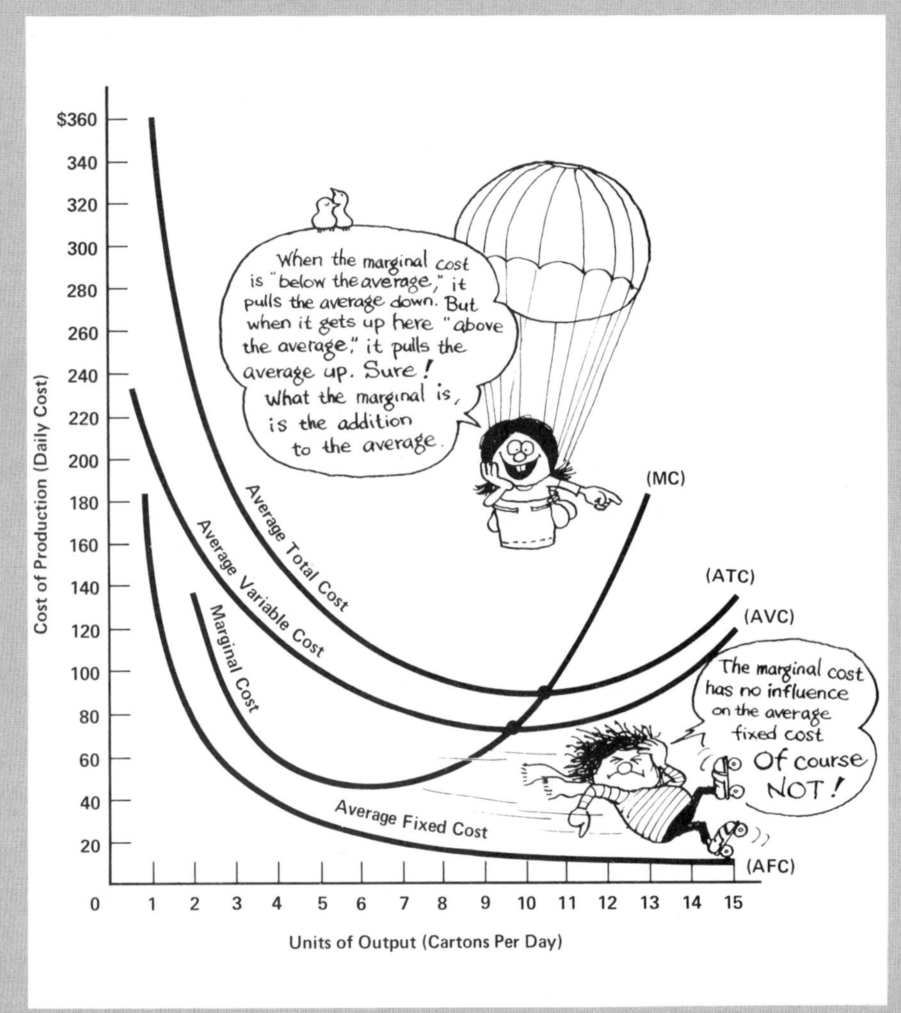


The Marginal Is The Addition To The Average. When It Is Above The Average, The Average Must Go Up. Of Course!

The shape of the marginal cost curve reflects the law of diminishing returns.

The marginal cost curve must go through the average total and the average variable cost curves at the point where the average curves are lowest.

When the rate of output is increased, if marginal cost is below the average, the average must come down. If the marginal cost is above the average, the average must go up.

### Marginal Cost Pulls The Average Cost Up Or Down

The marginal cost curve has a very special relationship to the average variable and average total cost curves. When the marginal cost is below the averages, it is pulling the averages down. When the marginal cost is above the averages, it is pulling the averages up. Why? Because *the marginal is the addition to the average*! If the "addition to the average" is above the average, the average goes up. Here's an example.

Suppose your average in Econ is 80, then you take an additional (marginal) quiz and make 90. The average goes up because the marginal is above the average. But if you make 70 on the marginal quiz what happens to your average? It goes down.

If you expand your daily rate of output by one unit, the cost goes up. How much? That's the marginal cost. If your cost goes up $50, then your marginal cost is $50. If that $50 is higher than the average cost per unit, it's going to cause your average cost to increase. If the $50 is below your average (per unit) cost, then your average (per unit) cost will decrease. High "additional cost per unit" pulls *up* the "average cost per unit." Low "additional cost per unit" pulls *down* the "average cost per unit." Of course!

When you look at the marginal cost at any level of output, it really doesn't tell you very much. It only tells you how much more it costs to produce at that rate of output than it would have cost if you had produced at a rate one unit less. That isn't very much information. It doesn't tell you anything about the overall per-unit cost of your product. It doesn't give you any idea whether you are making a profit or not.

There is much that the marginal cost doesn't tell you. But there is one thing that it does tell you—one very important thing. It tells you how much more it costs to produce at *that* rate of output than it would cost if you were producing at a rate *one unit less*. It's very important to know that. It helps you to decide what rate of output to produce. This will be explained in the next chapter, after you understand *revenues*. But now we need to get this discussion of *cost curves* all tied up.

## ALL THE COST CURVES ARE INTERRELATED

All these cost curves are different ways of looking at the same thing. Suppose you knew what your *total cost* would be, at each rate of output. Could you draw your *total cost* curve? Sure. But what about the other curves? Could you figure out your *total fixed cost*? and your average fixed cost? and your *total variable cost* and *average variable cost*? your *marginal cost*? Sure. No problem. Think about it for a minute and you will see that:

1. Your total cost curve tells you what your fixed cost is. When output is zero, then the total cost *is* the fixed cost.

2. The total cost curve tells you what your variable cost is. If you subtract your fixed cost from your total cost at each output, what you have left is the variable cost.

3. Your total cost tells you the marginal cost, too. For any daily rate of output (output "x"), the marginal cost is the total cost at "output x," minus the total cost at "output x minus one."

It's easy to get the fixed cost, the variable cost and the marginal cost when you know the total cost. But what about the average cost curves? That's just as easy. To convert totals to averages, just divide by the number of units of output. If output is 10 units and total cost is $900, the *average* total cost is $90. If fixed cost is $150, the *average* fixed cost is $15.

Here's another example. Suppose you knew the total fixed cost and the average variable cost for each level of output. Could you figure all the other curves from that? Of course. First get the *total* variable costs by multiplying the average variable cost by the number of units at each output. Then add the total fixed cost to the total variable cost for each level of output. Then you will have the total cost. From the total cost you can get the marginal cost, the average total cost, or anything you want.

Suppose all you had was the average total cost for each level of output. Would that be all you would need to figure out all the other curves? Yes. All you would have to do is multiply the average total cost by the output rate at each daily rate of output, then you would get the total cost curve. Then you could figure out everything else from the total cost curve. (You might have some difficulty figuring out the exact fixed cost, but let's not worry about that.)

Suppose you happen to know what the fixed cost is, and you know what the marginal cost would be for expanding output from zero to one unit, from one unit to two units, from two to three, and so on. Could you then figure the total cost? and the variable cost? and the average costs? Of course! The variable cost, for any output rate, would be the sum of all the marginal costs up to that rate of output. Add fixed cost to that and you have the total cost for that rate of output.

How neatly all these cost curves tie together! Surprising? Not really. They have to tie together. After all, we're looking at the same picture all the time—just in different ways.

### Which Cost Curves Are Most Helpful?

What are all these curves good for? When should you want to look at the total cost curve? And when should you look at the average curves? That depends on what you are trying to find out. The total curves tell you immediately what your total costs would be at each output rate.

If you are interested in finding out the "per unit cost," the average cost curves will show you that, immediately. Also, they will show you immediately what rate of output gives you your lowest per unit cost, how much fixed cost is embodied in each unit at each rate of output, and how much variable cost is embodied in each unit. If these are the kinds of things you want to know, then you should look at the average curves.

Now you know all about your costs, at your Cullowhee jeans plant. So at what daily rate are you going to produce? Will you make a profit, or a loss? and how much? You can't yet say, can you? Of course not. You don't know anything yet about the *market* for your product! You can't answer any of these questions until you know what *price* you're going to get for those cartons of jeans.

### Now You Need To Know The Selling Price

The businessman can't make any decisions on the basis of cost alone. He must also know the selling price. It's pretty obvious that if the price per carton was only a dollar, you wouldn't produce any at all. The best thing you could do would be to shut down your plant and pay the $150 daily fixed costs. As soon as you could you would either get rid of the plant and equipment, or shift into producing something else that pays enough to cover the cost of production.

If the price per carton was $1,000, you would produce all you possibly could. You would strain the production capacity of your little plant until it was highly inefficient. If the price is high enough, you can add to your profit by expanding your output and producing inefficiently! See how important it is to know what the price is, before you start talking about "the most profitable rate of output"?

Before you can go any farther in understanding "the theory of the firm" you're going to have to find out something about *revenues*. But you have already learned enough in this chapter. Take time to go back and review all those cost curves and be sure you understand them well.

The next chapter will begin by explaining your firm's *revenues* at different rates of output. As soon as you understand that, you will be ready to find the most profitable rate of output for your hip-hugger jeans plant, and you will be able to see how much profit or loss you are going to make. After you complete your review of this chapter, and do the exercises, then you'll be ready to go on into the next chapter and find out all about revenues, and about profits and losses.

---

**REVIEW EXERCISES**

**MAJOR CONCEPTS, PRINCIPLES, TERMS (Try to write a paragraph explaining each of these.)**

the law of diminishing returns
the nature and influence of "marginal cost"

## OTHER CONCEPTS AND TERMS (Try to write a sentence or phrase explaining what each one means.)

total cost
fixed cost
"overhead"
variable cost
variable factors
average total cost
average fixed cost
average variable cost

## GRAPHS AND CURVES (Try to draw, label, and explain each.)

Fixed Cost and Average Fixed Cost
Variable Cost and Average Variable Cost
Fixed Cost Plus Variable Cost Equals Total Cost
Total Cost and Average Total Cost
The Average Cost Curves: Total, Variable, and Fixed
Total Cost and Marginal Cost
Deriving the Marginal Cost Curve
Average and Marginal Cost Curves

## QUESTIONS (Try to write the answers, or jot down the highlights.)

1. Draw a total cost curve, and mark off and number both axes. Then derive from the total cost curve (a) the total fixed cost and (b) the total variable cost. Then derive (c) the average total cost, (d) the average variable cost, (e) the average fixed cost, and finally (f) the marginal cost. If you can do all that, you understand the relationships between the curves. If you can't, you don't. (Use graph paper if you have any around. Graph paper makes it easier.)
2. Suppose the wage rate goes up. What does that do to the total cost curve? Or suppose local property taxes are doubled. Or suppose electric power gets more expensive. Or suppose you tear down your old hip-hugger jeans plant and build a new one, twice as big. Can you see how each of these changes would change your total cost picture? Can you draw a new total cost curve to reflect each of these changes? Try.
3. A producer would not produce hip-hugger jeans or any other product at a *very low* rate of output. Explain why.
4. Can you see how important the law of diminishing returns (variable proportions) can be in helping to understand the very low living conditions in the poor, underdeveloped countries, and the very high living conditions in the advanced, highly developed countries? Think about it.

# 21 REVENUES AND COSTS, AND PROFITS AND LOSSES

*If the businessman understands both his costs and his revenues, he can choose his most profitable output.*

Why do businesses produce? To sell things, of course. To bring in money. Hopefully to make profits. When a business sells its daily (or weekly or monthly) output, what it receives in return is its daily (or weekly or monthly) revenue. If the revenue coming in is greater than the cost going out, then the business is making a profit.

You know what determines the cost. It's the fixed cost plus the variable cost. The total cost curve shows how high the cost would be for any rate of output you might choose. You learned all about that in the last chapter.

## HOW REVENUE CHANGES AS OUTPUT CHANGES

What determines the daily (or weekly or monthly) revenue? The number of units the business produces and sells, and the price per unit. Obviously. If the quantity sold increases, or if the price received for each unit increases (or both), the total revenue goes up. If the quantity sold, or the price (per unit) goes down, the revenue goes down.

If you want more revenue, all you have to do is to produce and sell more output. Or sell it at a higher price. But that may not always be easy to do! Also, there's something else to consider: When you produce more, your cost goes up. Remember? We'll talk about that again, later. For now, let's talk about the way the *revenue gets higher as the rate of output and sales increases.*

### The "Constant Price" Case

Suppose you're still operating your Cullowhee, N.C. hip-hugger jeans plant. But you got tired of the grind and the uncertainty of wholesaling your jeans to all those department stores in Asheville and Rock Hill and Texarkana and all over, so you made a special deal to sell all your output to a big retailing company (maybe Sears). They will buy all the output you can produce, and they will pay you a flat rate of $100 a carton. Now you know *exactly* what your total revenue is going to be, for each rate of output and sales.

If you produce one unit a day, your total revenue will be $100 a day. If you produce two units, your total revenue will be $200. If you produce five units, your total revenue will be $500. If you produce 10 units, your total revenue will be $1,000. That's simple enough, right?

### Marginal Revenue Is Always The Same

What about the marginal revenue? How much "extra revenue" do you get when you expand your daily rate of output and sales by one unit? When you expand from zero to one, how much does that add to your total revenue? Or when you expand from one to two, how much does that add? Or from four to five? Or from nine to ten? Or from fourteen to fifteen? The marginal revenue is always the same: $100. If the *price* is always the same, the *marginal revenue* must be always the same.

What about the average revenue per unit (per carton)? You can see right off that it must be $100. If you get $100 for each carton, then the average revenue per carton has to be $100. If you sell your entire output all for the same price per unit, then that price would have to be the average revenue. Of course! So in this case, the marginal revenue and the average revenue are always the same—always $100, at any rate of output you choose.

Can you guess what the *total revenue* and the *marginal revenue* and the *average revenue* are going to look like on a graph? Sure. The total revenue is going to be a straight line with a positive slope. The higher the rate of output and sales, the higher the total revenue. If you want to impress somebody, you can say "the total revenue is a positive and linear function of output and sales."

What about the marginal revenue curve? It's going to be a straight line too, but completely horizontal, at the $100 level. The marginal revenue curve will show that the marginal revenue never gets any higher or lower than $100, no matter what rate of output and sales you choose. And the average revenue curve? It will be the same as the marginal revenue curve, of course! Figure 21-1 shows the picture. Take a few minutes now and be sure you understand that graph.

Fig. 21-1 **TOTAL REVENUE AND MARGINAL REVENUE WHEN PRICE IS CONSTANT**

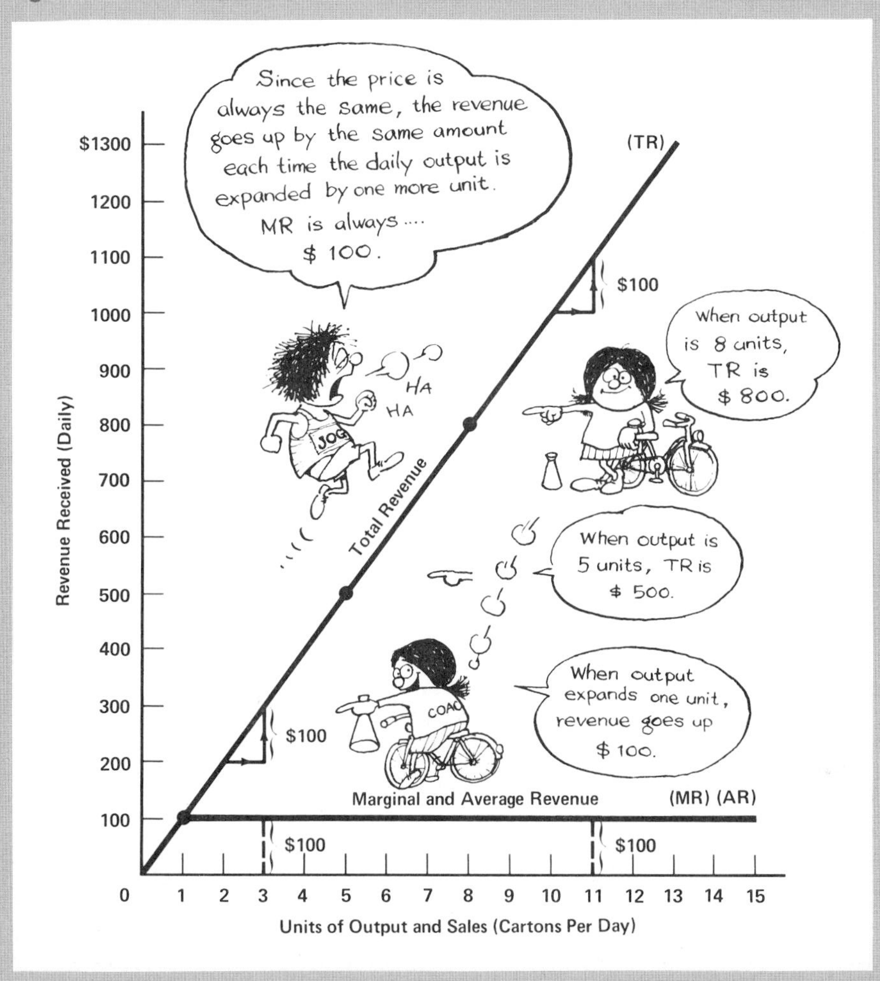


If Price Is Always The Same, The More You Sell The More Revenue You Take In.

If the price stays the same, the total revenue curve will be a straight line. The higher the price the steeper the line.

What makes the total revenue go up as output expands? The marginal revenue from the sale of each extra output unit, of course! So the higher the marginal revenue, the steeper the total revenue curve will be.

If the price suddenly dropped to $50, the marginal revenue curve would shift downward (from $100 to $50) and the upward slope of the total revenue curve would be only half as great. (It would rise by only $50 a unit.)

### The "Decreasing Price" Case

If you're selling all your hip-hugger jeans to Sears for $100 a carton, you don't have to worry about whether or not you're going to be able to sell all of your output. You can sell all you want to, for $100 a carton. That's the easy way. But suppose you thought you could do better by going it alone, and selling your "Cullowhee hip-huggers" to the stores in Asheville and Texarkana and all.

Now things are going to be very different! The lower the price you charge, the more you can sell. If you sell the jeans for $100 a carton, you can sell more than you could sell for a price of $150 a carton. Maybe you could sell a few cartons for as much as $200. (Probably not many.) The lower the price you charge, the more profit the stores can make when they sell your jeans to their customers. So the lower the price, the more they will want to buy. That's what the law of demand says. Right? At lower prices, buyers will buy more. The law of demand works in the wholesale markets, too? Of course!

What a different situation this is! Now you have *two* things to worry about: (1) how many cartons to produce, and (2) how much to charge per carton. If you decide on a high price, then you will be able to sell only a small quantity; if you decide to produce and sell a large quantity, you won't be able to sell them all, unless you lower the price. Can you see how different your "decision situation" is going to be, now?

What does your total revenue curve look like now? If you decide to produce only a few cartons and sell at a high price (maybe $150 a carton) you could do that. Then if you lowered the price and sold more, your total revenue probably would increase. But you can see that the extra revenue you get will be reduced some, because you had to cut the price to get the stores to buy the extra output. *The price cut reduces the extra revenue you get — the marginal revenue.* So the total revenue doesn't go up as much as it would if the price stayed the same. The total revenue curve on the graph is not going to keep on shooting up in a straight line. Not anymore. It's going to go up less and less as you cut the price more and more. Let's take an example.

Suppose that if you set the price at $200 a carton, you could sell only one carton a day. Then if you lowered the price to $190, you could sell two cartons a day. At a price of $180, you could sell three cartons. To make it easy, let's assume that each time you want to increase your daily sales by one more carton, you must cut your price by another $10.

### The Higher The Output The Lower The Marginal Revenue

Now what does your total revenue curve look like? It's very steep at first, when the price is very high. But as the price gets lower, the increase in total revenue gets smaller and smaller. As output expands, the total revenue curve bends and gets more nearly horizontal. What about the marginal revenue? — that is, the extra revenue you get when you increase your daily sales by one unit? That goes down, too. When total revenue isn't going up as fast, that means the marginal revenue isn't as large.

As the price gets lower, the *average revenue* gets lower. Why? Because *the average revenue* (revenue per unit) *is* the price. Always! We are going to look at all these things on a graph in just a minute. But first, take another look at "elasticity of demand."

If you lower the price of your jeans, the stores will buy more. You know what will happen. The law of demand tells you so. But *how much* more? Now, suddenly you find that you are very interested in the elasticity of the demand for your product.

### How Elastic Is The Demand?

How elastic is the wholesale demand for Cullowhee hip-huggers? If the demand for your product happens to be highly *elastic*, then when you lower the price from $200 to $190 a carton, maybe you can sell 3 or 4 cartons a day, instead of only two. Or if the demand is highly *inelastic*, it might be that when you want to sell more than one carton you must lower the price, not from $200 to $190, but maybe from $200 all the way down to $50, to get the stores to buy more than one carton a day.

You remember that when a producer lowers his price to try to sell more, that affects total spending for his product (that is, it affects the revenues he receives) in two ways. The price is lower, so *the lower price* tends to pull his total revenue *down*. But the quantity sold is greater, so *the greater quantity* tends to push his total revenue *up*. These two things are pulling and pushing on total revenue: The lower price pulling it down; the greater quantity pushing it up.

If the demand is *relatively elastic*, the *quantity* changes *more* than the price changes. So with relatively elastic demand, when the price is cut the "upward effect" of the increased quantity brings in enough revenue to *more than offset* the "downward effect" of the lower price. But suppose the demand is *relatively inelastic*. Then the quantity would change *less* than the price changes. The "upward effect" of the increased quantity would not bring in enough revenues to offset the "downward effect" of the lower price. So, with *relatively inelastic* demand, a price cut will pull the revenue down — will actually *reduce* total revenue. (Marginal revenue will be negative!)

See how important the elasticity of demand can be whenever a producer is thinking about lowering his price to try to sell more? Now, study Figure 21-2 and you'll see how different the revenue curves look, now that you must lower your price to get the stores to buy more of your Cullowhee hip-huggers.

**Fig. 21-2 TOTAL REVENUE AND MARGINAL REVENUE WHEN PRICE MUST BE LOWERED TO INDUCE PEOPLE TO BUY MORE**

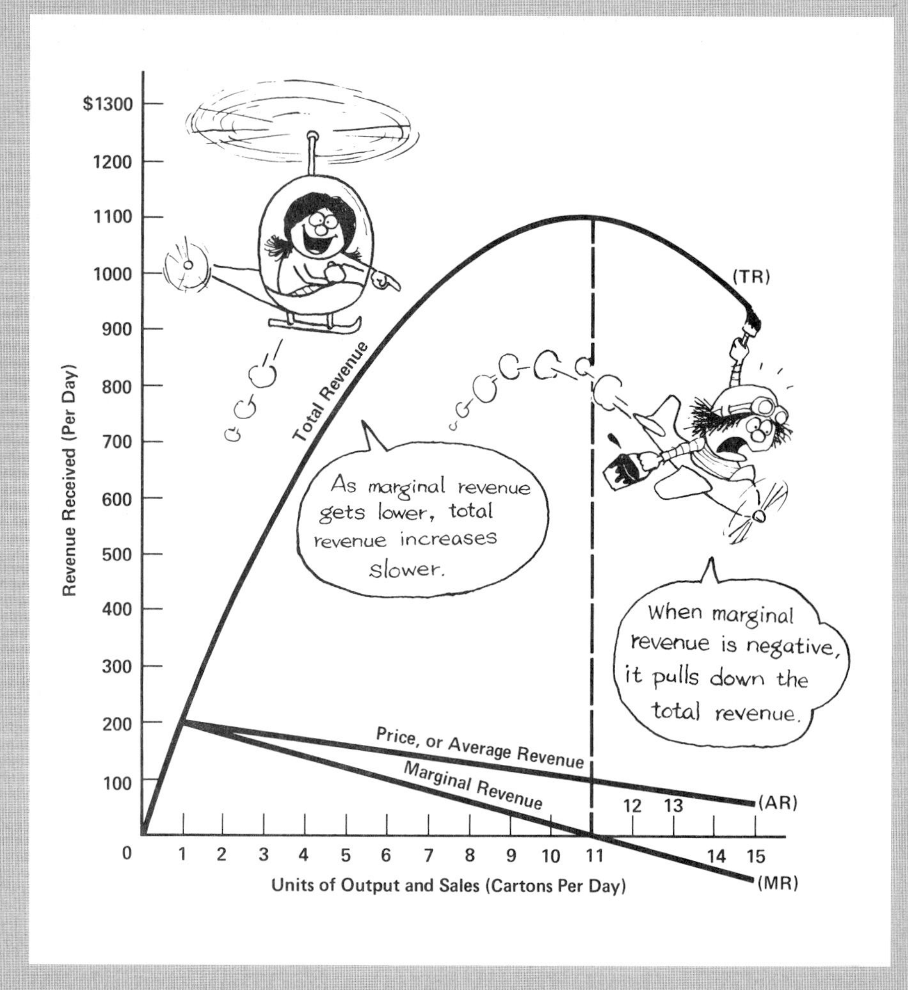


If You Must Lower The Price To Sell More, Selling More May Reduce Your Total Revenue.

As the price is lowered, and more output is sold, total revenue increases less and less. If price-cutting continues, ultimately the price will get so low that total revenue will decrease.

The shapes of these curves depend on the elasticity of demand. When demand is relatively *elastic*, marginal revenue is positive and total revenue is rising. When the elasticity is *unitary*, marginal revenue is zero and total revenue doesn't change. When demand is relatively inelastic, marginal revenue is negative and total revenue is falling.

You can tell from looking at Figure 21-2 that the demand for your hip-hugger jeans is highly elastic when the price is high and output is low. But the demand gets less elastic each time you lower the price and sell more. Between 10 and 11 units of output (prices of $110 and $100) the demand has unitary elasticity. At outputs greater than 11 units (prices less than $100) the demand is relatively inelastic.

### Elasticity Usually Decreases As Sales Increase

A straight-line demand curve (such as the "average revenue" line in Figure 21-2) always has a different elasticity from one point to another. When prices are high and quantities are low, the curve shows relatively elastic demand. When prices are low and the quantities are high, the curve shows relatively inelastic demand. Why is this true?

If you aren't selling much, and you lower the price a little bit, the price cut doesn't reduce your previous revenues much. *You don't have* much previous revenues! So if you aren't selling much, when you cut your price you don't "lose" much, but you might easily *double* your *total sales*! Your increased sales will increase your total revenue. Demand is relatively *elastic*. But what if you're already selling *lots* of units? When you lower the price a little bit, think of all that revenue you're going to lose! Chances are you won't be able to sell enough more to make up for it, so the *price cut* pulls the total revenue *down*. Demand is relatively *inelastic*.

Let's take an example. Suppose a producer is selling ten units a day, and he lowers his price a dollar to try to sell more. The price cut "costs him" $10 of "previous revenues." Maybe he can sell enough more, to more than make up for the ten dollar revenue loss. If so, demand is relatively *elastic*.

Now, suppose the producer is selling a thousand units a day, then lowers the price a dollar. How much "previous revenues" will that "cost" him? A thousand dollars! It's likely that he cannot sell enough more to make up for the thousand dollars of lost revenue. So when the price goes down it pulls the total revenue down. Demand is relatively *inelastic*.

### The "Price Line" (Average Revenue Curve) Is The Demand Curve

You know that the price is always the average revenue. *It has to be.* If you are selling things for $100, then the average revenue per unit is $100. If you are selling things for $20 each, then the average revenue per unit is $20. *Price* and *average revenue* are always the same thing. (Unless you are using price discrimination—that is, charging different prices to different buyers. But let's not worry about that, now.)

Here's another little thing that may need explaining. How is it that the average revenue (price) line is really *the demand curve* for your hip-hugger jeans? The average revenue curve shows you how many cartons the stores are willing to buy from you at the different prices you might charge. If you set the price at $200, then the quantity they will buy is only one carton per day. But if you set the price $100 per carton, they will buy eleven cartons a day. If you set the price at $70, they will buy 14 cartons per day. So sure enough, the average revenue (price) line really is the demand curve for your output. It tells how many they would buy at each price you might set.

What about when you were selling all your output to Sears for $100 a carton? They would buy all you wanted to sell, at that price. Then the *price* and the *average revenue* and the *marginal revenue* were all the same. Did the price (average revenue) line indicate the demand for your output then, too? Yes, it really did. Whenever you can sell as much as you please without lowering the price, that means the demand for your output is "absolutely elastic" or "infinitely elastic." It means that without lowering the price at all, you can sell more and more and more, if you want to. But you won't want to. Why? Because as you produce more, your cost goes up. Let's talk about that, right now.

## THE COST AND REVENUE CURVES

Now you know quite a bit about revenues. And you remember a lot about costs, from the last chapter. So finally, it's time to put the revenue curves and the cost curves together. Then you can figure out what will be the most profitable daily rate of output for your hip-hugger jeans factory. Also, you will be able to see how much profit you're going to make (if any).

Several graphs are coming up now. These graphs will put the revenue curves and the cost curves together. There are no new curves on any of these graphs. You have seen them all before. But please, don't go on to these graphs unless you understand all the cost and revenue curves. If you haven't already learned them, go back and learn them now.

If you understand all the cost curves and all the revenue curves, the graphs coming up will be very simple and very clear. You will be able to see exactly how your costs and revenues change as you adjust your daily output rate. You will be able to see exactly how much daily output you should produce to get maximum profit. And finally, you will see exactly how much profit you will make! All the explanations are there, right on the graphs. Take your time and study each graph as it comes up. I think you'll find it interesting to watch all the pieces finally falling into place.

Fig. 21-3 **TOTAL COST, TOTAL REVENUE, AND TOTAL PROFIT WHEN PRICE IS CONSTANT**

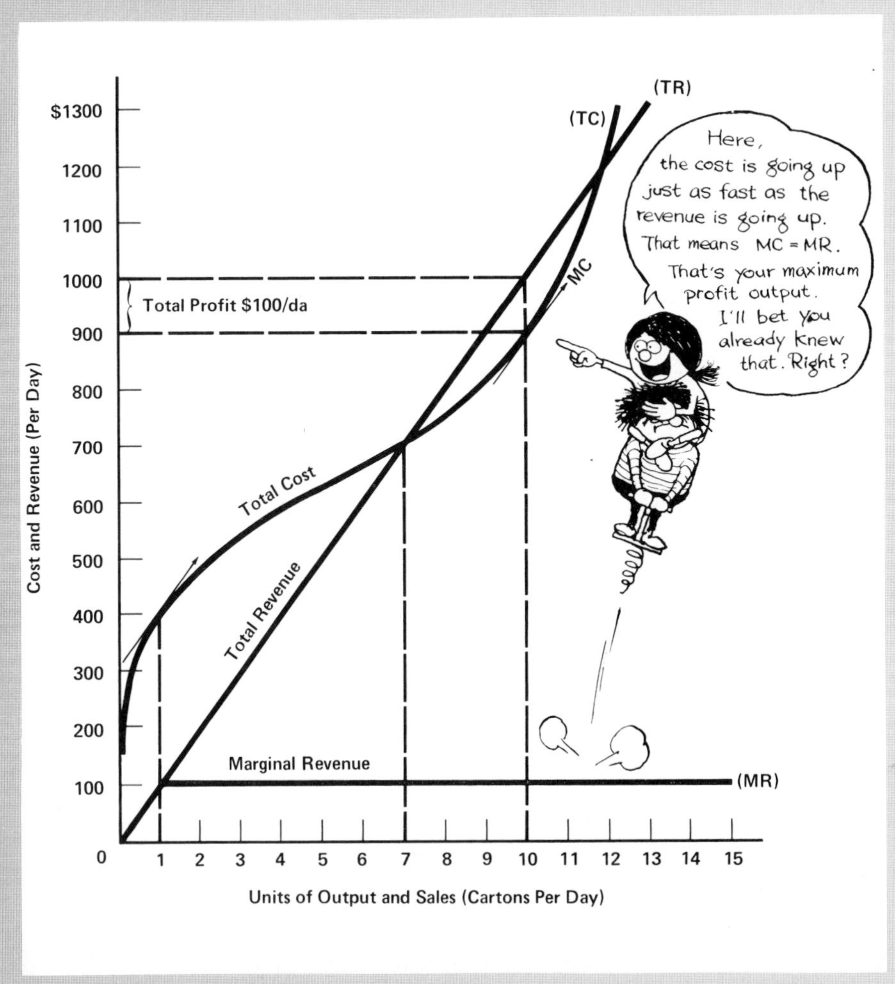


Total Revenue Just Keeps On Going Up, But After Ten Units A Day, Total Cost Goes Up Even Faster!

You can sell all you want to, for a price of $100 a carton.

If you increase your output from zero to one unit, that will add a lot to your cost ($200) but not as much (only $100) to your revenue. But then, as you expand your daily output rate from one unit to two, to three, and on up to ten units, each time you expand your rate of output, your total revenue goes up *more* than your total cost goes up.

Total revenue is rising faster than total cost all the way between the output of one unit and the output of ten units. But beyond ten units, your total cost rises more rapidly than your total revenue. If you produce more than ten units, your profits will shrink.

**Fig. 21-4 AVERAGE COST, MARGINAL COST, AND MARGINAL REVENUE WHEN PRICE IS CONSTANT**

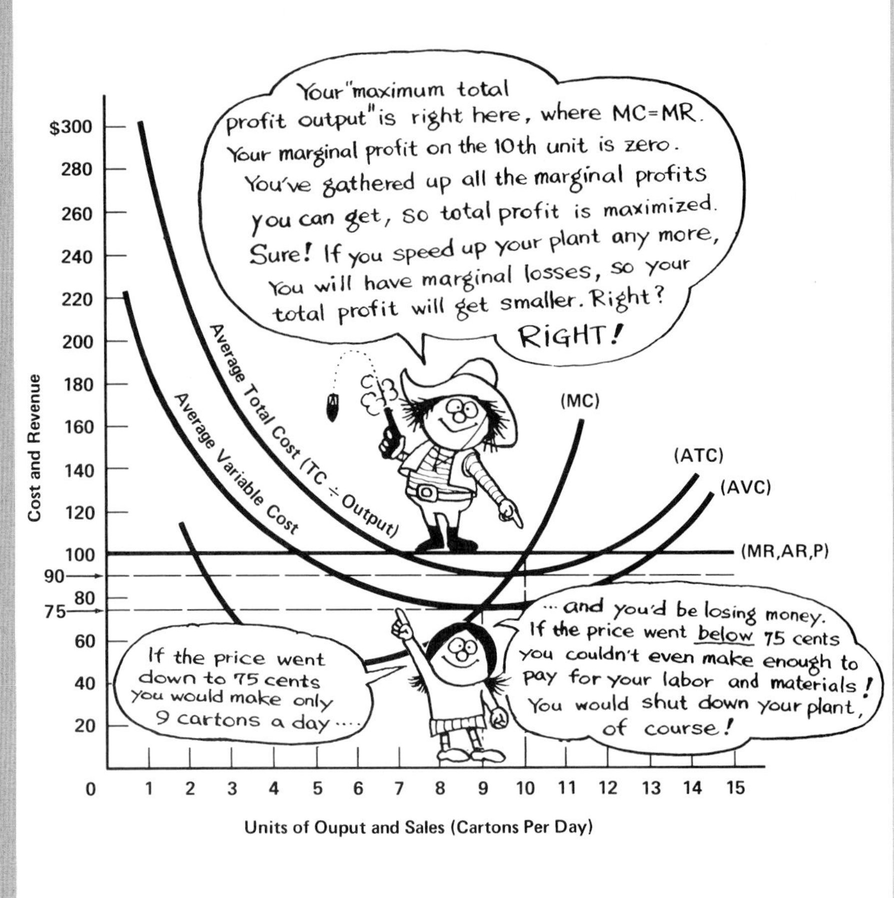


If You Divide Everything On The Total Cost And Revenue Graphs By The Number Of Units Of Output, You Get The Average Cost And Revenue Graphs.

This is just another way to look at the same things shown in the previous graph, except that here everything is on a "per unit" basis. (Note that this vertical scale is much larger.)

At $100 a carton you are making a profit. If the price was $90 a carton, you could only break even. If the price was below $90, you would lose money. At any price above $75, you earn enough to pay for all your variable inputs, plus something left over to help pay for your fixed inputs. At a price less than $15, you couldn't even pay for your variable inputs, so you wouldn't produce anything.

*Note:*
1. Average revenue ($100) times quantity sold (10 units) equals total revenue ($1,000).
2. Average cost ($90) times quantity sold (10 units) equals total cost ($900).
3. Average revenue ($100) minus average cost ($90) equals average profit ($10).
4. Average profit ($10) times quantity sold (10 units) equals total profit ($100).

Fig. 21-5 **TOTAL COST, TOTAL REVENUE, AND TOTAL PROFIT WHEN PRICE MUST BE LOWERED TO INDUCE PEOPLE TO BUY MORE**

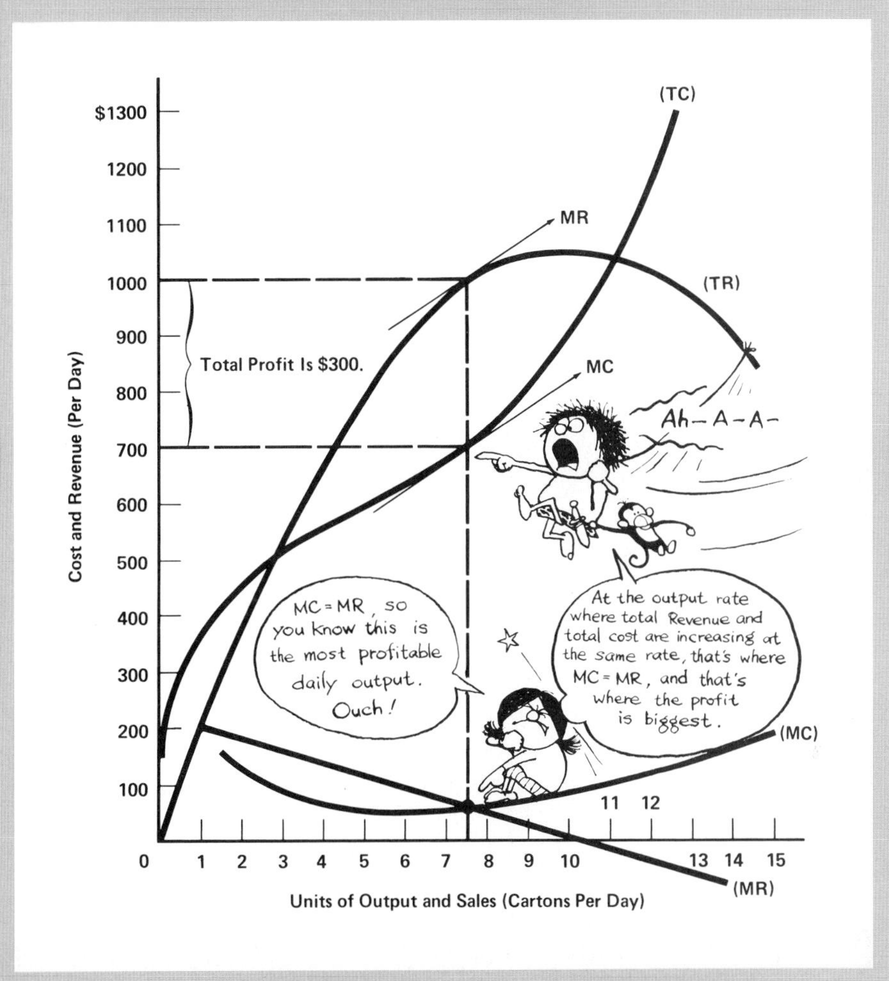


If You Must Cut Your Price To Sell More, You Will Not Produce And Sell As Much. Of Course Not!

To increase your sales by one unit, you must lower your price by $10.

The increased output tries to push your total revenue up, but the decreased price tries to pull it down. When output is small and the price high, the increased quantity more than offsets the reduced price. But when the output is large and the price low, the quantity doesn't increase enough. The price cut pulls the total revenue down.

**Fig. 21-6 AVERAGE COST, AVERAGE REVENUE, AND AVERAGE PROFIT WHEN PRICE MUST BE LOWERED TO INDUCE PEOPLE TO BUY MORE**

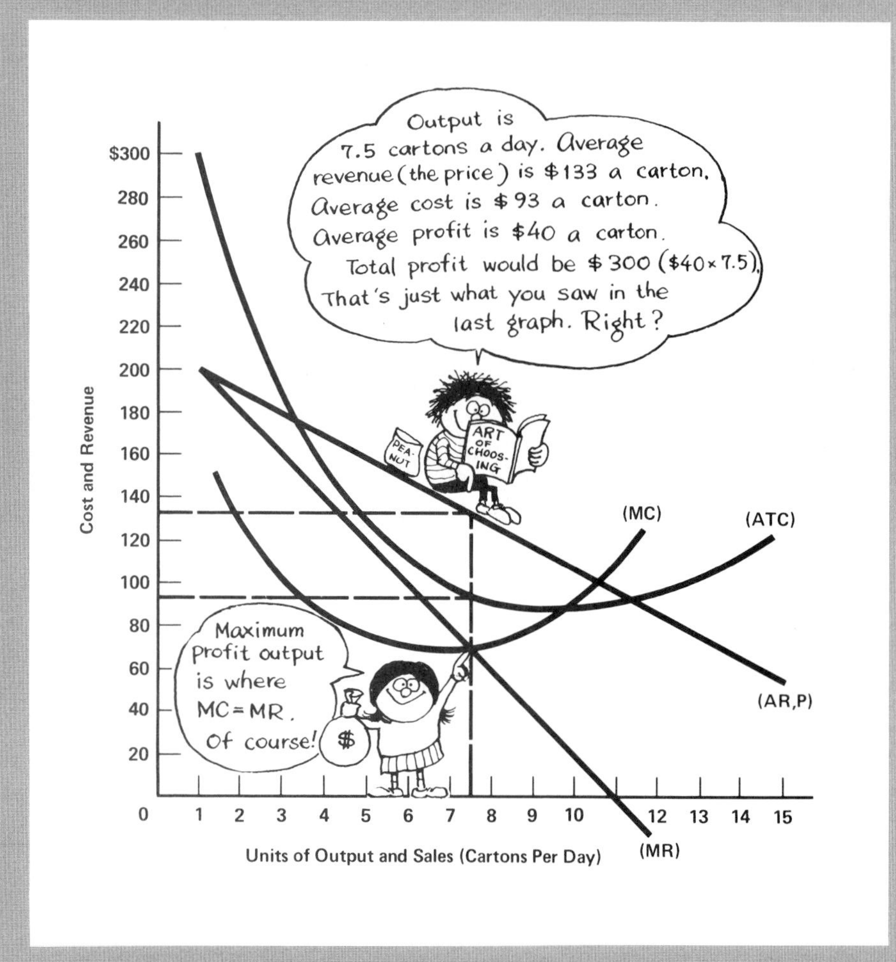


If You Divide The Total Cost And Total Revenue Curves By The Number Of Units Of Output, You Get The Average Cost And Average Revenue Curves.

This graph shows exactly the same picture as the previous graph, except that here, everything is on a per unit basis.

As you lower the price to sell more, the marginal revenue drops rapidly. Soon you reach a point where you lose as much money by lowering the price as you get back when you sell the extra unit of output. That's where marginal revenue is zero and total revenue is maximum. But that isn't the maximum profit output. Your maximum profit is where your marginal cost equals your marginal revenue (7.5 units). If you expand beyond 7.5 units a day you will be adding more to your cost than to your revenue. If you lower the price and sell more, your profits will shrink.

(The AR and MR curves are the same ones you've seen before. Here they look much steeper, because of the larger vertical scale. The average variable and the average fixed cost curves are excluded here to keep from cluttering up the graph.)

Now you know how the costs and the revenues "fit together" and relate to each other. You know how to find your most profitable rate of output. Then, at that output, you can see exactly what your costs and revenues will be, so you can see how much profit you will be making (if any). Yes, now you know quite a lot about "the theory of the firm." But how stable is your "cost and revenue" situation? What might cause it to change?

### How Stable Are Costs And Revenues?

Does a firm's "cost and revenue picture" stay the same? or does it change? You know the answer. It changes. Anything which changes the price of any of your *inputs*, or of your *outputs*, will change the picture. If input factor prices rise, your costs go up and your profits go down. You might even wind up making losses and trying to get out of this business altogether. But if the price you get for your product goes up, that will bring you more profits. Whenever your costs or revenues change, you face a new situation. You have to make some new decisions. These things are changing all the time. Business decision-making is a continuing job.

Now that you know how a firm's cost and revenue picture looks, you probably can see right away that this has quite a lot to do with *supply*. One reason we need to understand the firm's cost and revenue picture is to be able to understand supply. Remember? So now you're ready.

### If Price Is Fixed, "Marginal Cost" Is The Firm's Supply Curve

Suppose you are producing 10 cartons a day of Cullowhee Hip-Huggers and selling them all to Sears for $100 a carton. Then suppose there's a big increase in demand. Maybe Sears offers you $140 a carton and urges you to supply more. Will you increase your daily output, and supply more? Yes, you will!

When your *marginal revenue* goes up to $140, you will keep expanding your output until your *marginal cost* goes up to $140. That way you will maximize your profit. Look back at Figure 21-4 and you will see that you will increase your output to 11 cartons a day. And look how much profit you will be making — more than $50 per unit, or about $600 a day. Wow! With all that profit, this is a pretty good business to be in!

I'll bet you will try to expand your plant as soon as you can, so you can make even more profit. Other firms, producing other textile products, are likely to shift into the production of hip-hugger jeans. Looks like the hip-hugger jeans industry is about to expand! Why? Because people are demanding more of these jeans, and that makes this a very profitable industry.

It's going to take a while for the industry to expand. Meanwhile, in the short run, how much will you supply to the market? You will supply as much as your marginal cost tells you to supply! If the price rises higher, you will produce more. You will expand your output to where the *marginal cost* goes up as high as the new *marginal revenue* (price). If the price goes down, you will produce less. You will reduce your daily output to where your marginal cost equals the new (lower) marginal revenue (price). So, in the short run, see what determines your propensity to sell? Sure. The amount you will produce and offer for sale at each possible price, is determined by your *marginal cost*.

The marginal cost curve of the individual firm tells the producer how much output he should produce and sell per day, for each price that might exist in the market. Of course, if the price ever drops so low that the businessman can't even pay for his variable inputs (labor and materials and all) then he won't produce anything. But for each price higher than the average variable cost, it's the *marginal cost curve* that tells how much the producer will produce and supply to the market. So the marginal cost curve really does show how much the producer will offer for sale at each "possible price" which might exist in the market. His marginal cost curve (at prices above AVC) *really is* his supply curve.

### The Marginal Cost And Marginal Revenue Curve

It might be a good idea to look at this on a graph which shows only the marginal cost curve and the marginal revenue curve. Then you can see this concept, without all the other curves in the graph to confuse you. The next two graphs (Figures 21-7 and 21-8) show you the "marginal cost-marginal revenue" relationships. You will see that the marginal cost really does show how much you will produce and sell—how much you will supply—at various selling prices. So the marginal cost curve really is your supply curve. It shows your "propensity to sell" at the different prices which might be offered for your Cullowhee hip-huggers. Study the graphs carefully, and I believe all this will become perfectly clear to you.

## SHORT-RUN AND LONG-RUN SUPPLY

Once you get the hang of it, it's easy to see that your marginal cost curve (at prices higher than your average variable cost) is the supply curve for your plant. You know that if the selling price goes up you will produce and sell more. If it goes down you will produce and sell less. How much more or how much less? That depends on your marginal cost. Sure. But that isn't quite all there is to it. Here's more.

The marginal cost curve is the supply curve for your *existing* plant. Suppose you build a new plant. What then? That would be a brand new

**Fig. 21-7 MARGINAL COST AND MARGINAL REVENUE, AND MARGINAL PROFIT AND MARGINAL LOSS**

The Producer Who Thinks Marginally Will Keep Adding Output As Long As He Can Add More To His Revenue Than He Does To His Costs.

At a price of $100, you will produce and sell 10 cartons a day. If the price goes up to $140, you will sell 11 cartons a day. If the price drops below $100, you will sell less than 10 cartons. Your MC curve is your supply curve.

This graph doesn't tell if you are making any profit. It only shows you your "maximum profit" (or "minimum loss") rate of output.

The marginal cost and marginal revenue curves only tell you what daily rate of output will be "most profitable" for your plant. That's all. After you find the most profitable rate of output (where MC = MR) then you must look at the other cost curves to see what the cost of your product really is, and how much profit (or loss) you will be making.

Fig. 21-8 **MARGINAL COST AND MARGINAL REVENUE, AVERAGE COST AND AVERAGE PROFIT: THE CASE OF A PRICE INCREASE**

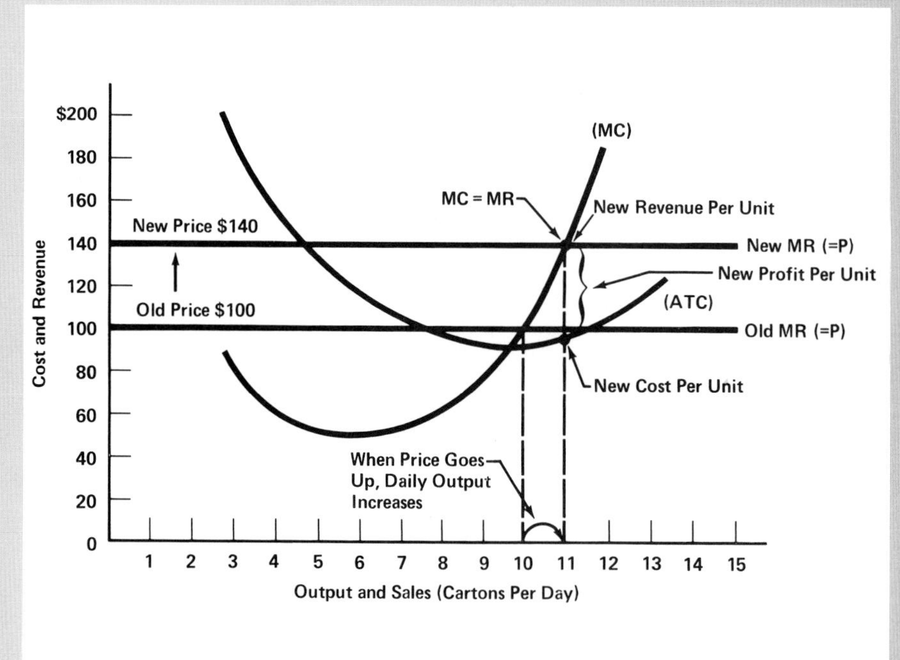


When The Price Goes Up, Everyone Makes Profits And The Industry Expands.

An increase in price brings increased profit. If you know your Average Cost (ATC), you can see how much profit you're making, per unit. So you can figure out your total profit. (Average profit, times quantity sold, equals total profit.)

When the price goes up from $100 to $140 per carton, output expands from 10 units per day to 11 units per day, and average cost per unit goes up from $90 per carton to $95 per carton. Marginal cost increases to $140 where it is equal to the new marginal revenue. Profit increases from $10 per carton to $45 per carton. Wow!

This is going to be a *very rapidly expanding* industry. The high demand for hip-hugger jeans has forced up the price. More competitors will keep entering the industry and pushing more product into the market until the price is back down to the "normal" level.

ball game! Your new plant might be built to operate most efficiently at a daily output of only five cartons. Or it might be made to operate most efficiently at an output of 30 cartons. What a big difference that would make!

The marginal cost curve is only the *short-run* supply curve for your firm. Short-run means that the law of diminishing returns is working. You must operate with a certain plant size and with certain fixed costs, so there are severe limitations on how much you can produce. That's what we mean by "short-run." You're stuck with your existing situation whether you like it or not.

### All Factors Are Variable In The Long-Run

When we talk about the long-run supply for a product, we're thinking about how much would be produced and offered for sale if we would wait long enough for new plants to be built. More businessmen could come into the industry. The firms already in the industry could change the sizes of their plants, or build more plants.

How long is the long run? As long as it takes for the industry to fully respond to a change in the prices of inputs or outputs—that is, to expand or contract in response to a change in demand or cost conditions. The long run is however long it takes to increase or decrease *all* the factors of production—not just the "variable factors." If you wait long enough, there is no fixed cost! There are no "fixed input factors" in the long run.

What would determine whether or not an industry would be expanding "in the long-run"? Suppose someone comes up to your Cullowhee plant and says: "Are you going to stay in this business? Perhaps even expand your plant? Or are you going to try to get out of this business?" Your answer would depend on your present and expected profits. Right? If your business is very profitable, probably you will stay in it, and expand. Chances are that other people are going to start producing hip-hugger jeans, too. In the long-run, the industry will expand. Obviously.

Suppose you're not making any profits at all. Suppose you're losing money. That's different. As soon as you can, you are going to get out of the jeans business. Other producers will too. This is going to be a declining industry. Maybe you will start producing pleated blouses, or sleeping bags, or something else. Or maybe you will get out of the textile industry altogether. Maybe you will sell out and go to the West Coast and set up an almond processing plant in Chico—or an abalone plant in Arcata—or a toy plant in Eugene—or a ski shop in Bellingham.

See the beauty of the long run? You can do anything you want to. *All factors are variable*. But you have to plan it out ahead of time, and work it out as time goes by. Otherwise you just keep sitting in the same old place, making the same mistakes over and over until everything's gone and it's too late to make a new start. That's no way to economize and optimize (but a lot of small businesses seem to do it that way).

The amount of time required for the long run is very different from one industry to another. In some industries, it's only a few days. In other industries it would take years for a major expansion or contraction to occur.

### Supply Is Much More Elastic In The Long-Run

The production-motivating function of price works, *to some extent* in the short run, but it works *a lot more* in the long run. In the short run the production-motivating function of price is limited because the fixed factors can't be increased. Outputs are limited by the law of diminishing returns. In the long run the supply response is much greater—supply is much more "elastic."

In the short run the producers can't produce very much more. Costs go up too fast as output expands. But in the long run, all factors—plant size, machinery and equipment, number of firms in the industry—all are variable. So if the price goes up and you wait long enough, the supply response can be great. For most things, supply is somewhat responsive to price in the short run, but much more responsive in the long run.

### The Very Short Run

Economists sometimes talk about the very short run—a time period so short (maybe one day) that you don't even have time to adjust your daily rate of output. All you can decide is what to do with the products you have already produced. You can only decide whether to hold your inventories, or to sell them. That's all.

In the very short run, the only function "price" can perform is to *ration* the existing goods. There isn't time for the production-motivating function of price to work *at all* in the very short run. The production-motivating function of price works *to some extent* in the short run. But when does it *really* work? In the long run, of course! That's when the production-motivating function really works great!

## THE NORMAL PRICE AND LONG-RUN EQUILIBRIUM

By now you know that if the price of something is above "normal" the industry will be expanding. If the price is below "normal," the industry will be contracting. But what does "normal" mean? How high a price is the normal price?

The "normal price" is that price which is just high enough to bring a "normal profit" to the "average firm" in the industry. A price any higher than "normal" will bring abnormally high profits and will attract new firms into the industry. The industry will expand and the supply flowing into the market will increase. Then the increased supply will create a temporary surplus and push the price down. This process will continue until the price gets down to the "normal" level. Then the industry will stop expanding and the price will stop going down. The industry will be in long-run equilibrium.

A price below "normal" is one which is too low to cover all the costs (including "normal profits") of the "average firm" in the industry. If the price is below "normal," firms will be leaving the industry, seeking a better deal in some other industry. The industry will get smaller. The supply flowing into the market will decrease. This will create a temporary shortage and push the price up. This process will continue until the price gets up to the "normal" level. Then the price and the industry will stabilize. The price in the market will be the "normal price" and the industry will be in "long-run equilibrium."

## Higher Cost Increases The Normal Price

What could happen to cause the "normal price" to change? The only thing that could change the normal price would be a change in the cost of production. Why? Because the cost of production is the only thing that determines the normal price! The normal price covers the average total cost (including a "normal profit"—a profit large enough to keep the producers from leaving the industry). If anything happens to make the "average firm's" costs go up, the normal price goes up. If the cost of production goes down, the normal price goes down. Whenever the "normal price" moves up or down, the industry expands or contracts until the market price is pushed to the new "normal price" level.

Suppose you're going along producing ten cartons of hip-hugger jeans per day. Then for some reason the cost of denim goes down and the cost of labor goes down and maybe some of your other costs go down. Suddenly you're making big profits. You can guess what's going to happen.

Before long the market price is going to go down. Why? Because before long more firms, seeking profits, will come into the hip-hugger jeans industry and increase the supply and push the price down. How far down? Down to where the "excess profits" are gone. Down to the new (lower) "normal price"—the price which reflects the new (lower) cost of production.

Suppose production costs went up? It would work the other way The firms would be losing money, so they would cut back production and start getting out of the industry. Supply would decrease. Prices would be forced up. How far up? Up to the new (higher) "normal price"—the price which reflects the new (higher) cost of production.

### The Buyers Must Pay The Normal Price

In the long run, how much are the people of the society going to have to pay for hip-hugger jeans? The normal price. In the long run the people who buy each product are going to have to pay the cost required to produce that product. If a product requires a lot of the society's most valuable resources to produce it, then a person who wants that product is going to have to pay a high price to get it. He is going to have to produce something very valuable for the society, in exchange, to get the money he needs to buy the high-priced product. See how all the pieces fit together and come out just right, in the pure market model?

In the pure market model, increases in cost always get passed along to the buyers, in the long run. Lower cost always gets passed along to the buyers, too. Each person must pay the true, full cost of whatever he buys. He who "uses up" a lot of society's valuable things, must pay for all those things. In order to get enough income to buy all those expensive things, he must produce a lot of valuable things for the society. See how it all comes out right?

Of course you know that it doesn't work out exactly this way in the real world. Still, even in the real world these market forces are strong. There's a lot going on in the real world that you just can't understand unless you understand these market forces. For example, when someone says: "Let the businesses pay for the cost of cleaning up pollution," what that someone is *really* saying is this: "Let the people who buy the products of those industries pay the cost of cleaning up the pollution." That's just about how it will come out in the long run, anyway, and there's really no way we can help that. (We'll get into the economics of this environmental problem, in Part Eight.)

### The Role of Competition

The "normal price" is also called the normal competitive price. Can you see why? It is competition (new firms coming in when the price is too high) that pushes the price down to "normal." Without competition (new firms moving into the high-profit industries) this "normal price" idea wouldn't work out right. An increase in demand might not generate much new supply. It might only generate a lot of profits for the monopolistic producers!

Of course the price of each product must be high enough to cover the cost of producing it. If the price is too low to cover all the costs, firms will get out of the industry. The supply flowing into the market will decrease, so the price will increase. If the costs were higher than the price and the price *didn't* increase, ultimately all the producers would go out of business and the product would disappear from the market. You can see that competition isn't essential to keep the price up *high* enough— that is, "up" to normal. That would happen anyway. But competition *is* essential to keep the price down *low* enough—that is, "down" to normal.

A producer with *complete monopoly* could keep the price up above "normal," indefinitely. If a producer (monopolist) can keep other firms from entering his industry, he can keep the supply restricted, keep the price up, and keep all the big profits for himself. See how *monopoly* can interfere with the operation of the market process? We'll talk about this issue in detail in Chapter 24.

## DO BUSINESSMEN USE THESE COSTS AND REVENUE CURVES?

In this chapter and the previous one, you have learned a lot of curves which illustrate the costs and revenues of a "typical business firm." Do businessmen actually use, and go by these curves? Generally, no. But they *certainly do* go by the *concepts and principles* which these curves illustrate! *The purpose of the curves is to illustrate the concepts and principles* which guide business decisions.

I doubt if you could go to any businessman in the country and get him to tell you what would be the marginal cost of expanding his daily output by one unit. It certainly isn't likely that he has tried to draw a graph of his marginal cost. But you may be quite sure that any good businessman, when thinking about increasing his rate of production, considers the *extra cost*, and compares it with the *extra revenue* he expects to get.

The economist uses the cost and revenue curves to make *explicit* the ideas and concepts and principles which are "just common sense" to a good businessman. But the curves aren't important. It's the concepts and principles the curves illustrate, that matter. If you are thinking about going into business (or if you're just thinking about "the ordinary business of life"), there are many times when these concepts and principles can help you a lot.

Sometimes businesses actually use the curves. A modification of the economist's total cost and total revenue graph is frequently used. Businessmen call it the "break-even chart." First let's talk about it, then we'll take a look at it.

### The Break-Even Chart

The break-even chart is a simplification of the total cost-total revenue graph. It shows a straight-line total cost curve (starting part way up the vertical axis to reflect the fixed cost) and a straight-line total revenue curve beginning at zero. The total revenue line is going up faster than the total cost line. By increasing its output and sales to where the total revenue is high enough to equal the total cost, the firm "breaks even." At higher outputs than this, the firm makes a profit. At lower outputs it loses money.

Figure 21-9 shows a typical break-even chart. This shows that if the demand for the product is great enough to buy up an output of 500 units per day (or per week, month, or year) the business will break even. If not, they will lose money. If they can sell more than 500 units, they will make a profit. You should take time to study Figure 21-9 now.

Who uses break-even charts? The automobile producers, for one. They use a break-even chart to relate to a "model-year." By the time the company is all set up to produce a new car (say the 1975 Impala) they have already incurred big costs. So they must sell many '75 Impalas before they can break even. But beyond the break-even point, additional sales add to profits.

When General Motors is deciding its next year's model plans, it is very interested in knowing for each of its models: "At our estimated price and cost, how many cars will we have to sell to break even? What are our chances of selling that many?" If the chances don't look very good, then that model will not be produced—unless perhaps the cost can be reduced, or the price can be increased. If the cost was lower or if the price was higher, the firm could break even with a smaller quantity sold.

### It's The Concepts That Count

The graphs economists and businessmen use to illustrate costs and revenues provide a way of seeing more clearly the process by which the business makes its "economic choice" decisions. The cost and revenue curves simply illustrate the "best" choices for the firm which seeks to maximize its profits. The curves show how much the profit-maximizing firm will produce and sell. The curves serve only to illustrate the important concepts.

*The "marginal cost equals marginal revenue" concept is an excellent way of conceptualizing the business decision process.* It is a most valuable framework in which to pose the questions. The businessman who "thinks marginally" as he makes his decisions is likely to be right more often than one who does not, just as the individual who "thinks marginally" in deciding how to use his scarce money, time, and other resources is more likely to achieve his objectives than one who does not. But there's no guarantee. If you guess wrong about what the marginal cost or the marginal revenue will be, you will make the wrong decision, in spite of your good use of a sound concept.

Cost and revenue curves provide excellent tools for conceptualizing the business decision-making processes. But the curves leave out most

Fig. 21-9 **BUSINESSES USE A MODIFIED VERSION OF THE TOTAL COST AND REVENUE GRAPH: THE BREAK-EVEN CHART**

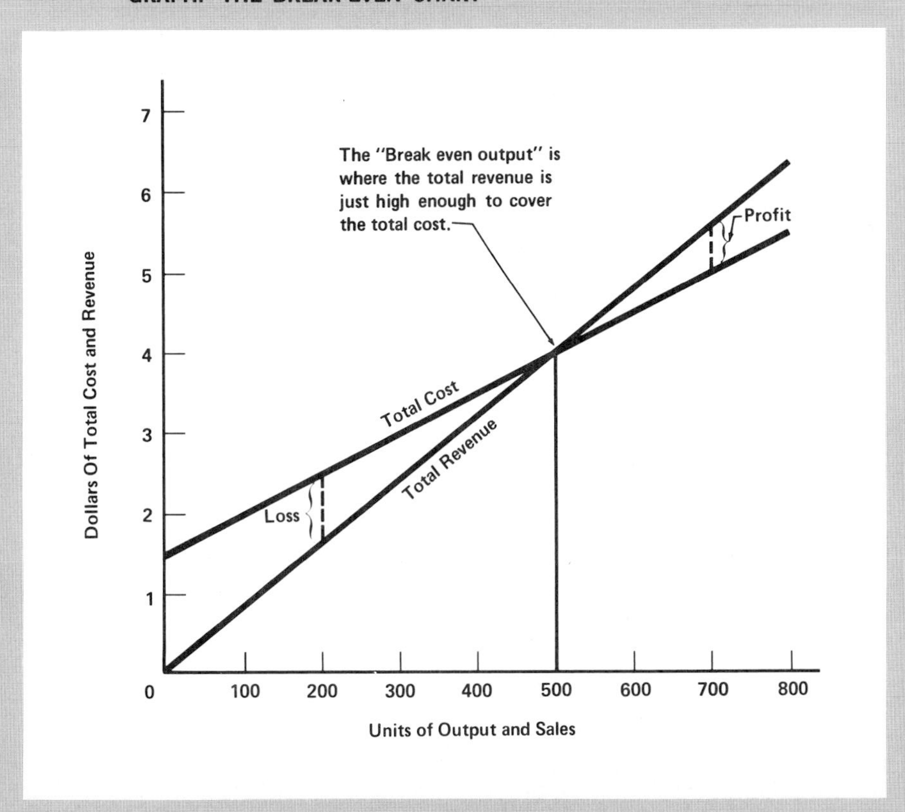


If A Business Is Going To Make A Profit, Its Total Revenue Must Rise Fast Enough To Get Above The Total Cost Before The Business Runs Out Of Customers!

This chart shows that when output is zero, total cost is already high, but revenue is zero. Then the larger the output, the higher the revenue.

If the demand for the product is great enough so that 500 units can be sold, then the business will break even. If more than 500 can be sold they will make a profit.

What would happen if the price was increased? The total revenue line would get steeper and the break-even point would be at a smaller output. A lower price would have the opposite effect. If the cost drops, the break-even point will move to the left, indicating a smaller break-even output. Higher cost would require that more be sold before the business could break even.

of the important practical questions which the businessman must face:

> "How can I reduce my costs?" "What will happen to my revenues if I improve my product?" "What will happen to the cost of my inputs in the future?" "Will the demand for hip-hugger jeans continue to expand?"

These and many other unanswerable questions face the businessman in the real world. The economist's cost and revenue graphs don't help him very much on these important issues. Remember what I said in Chapter 2 when we were talking about "How Businesses Choose"? It takes more than an understanding of economic concepts to be a successful businessman! Still, the economic concepts can help a lot.

## SUMMARY: THE "PRICE-CHASING" PROCESS

Now you understand how the "production motivating function of price" directs the resources in the "model pure market economy." People demand more of something, so there's a temporary shortage. The price goes up. Producers sell more out of inventory and begin hiring more inputs, to produce more output. Each firm expands its output until the marginal cost goes up as high as the new (higher) marginal revenue (price). At the high price, businesses make big profits.

Big profits act as a magnet to pull more resources into the industry. New plants are built. Soon, more supply begins flowing into that market. Little surpluses develop. The price starts moving down. This process continues until the price moves down to the normal level. "Excess profits" are eliminated. The industry is back in long-run equilibrium."

In the pure market system, *each businessman is always chasing the "market" price.* He wants to find a high-priced product he can produce a lot of, at low cost, so he can make big profits. At the same time, each *industry is always chasing the "normal" price.* If the market price is above or below normal, the industry will be expanding or contracting, chasing the normal price, seeking "long-run equilibrium." *The theory of the firm (and much of the entire study of microeconomics, for that matter) is the study of "price chasing"—of producers chasing the market price and of industries chasing the normal price. The "chasers" never quite get there, just as the greyhounds never quite catch the "rabbit"—but they keep on chasing just the same.*

Now you know about supply and demand and costs and revenues, and how businesses behave—so what else is there? What other basic concepts and principles do you need, to complete your understanding of microeconomics? Not very many, really. But there's a major segment we haven't gotten into yet. You don't yet know anything about how *input prices* can change, and force industries to expand or contract. That's the subject of the next chapter.

## REVIEW EXERCISES

### MAJOR CONCEPTS, PRINCIPLES, TERMS (Try to write a paragraph explaining each of these.)

effect of the elasticity of demand
effect of changes in the normal price
effect of time on the functions of price
the "price-chasing" process

### OTHER CONCEPTS AND TERMS (Try to write a sentence or phrase explaining what each one means.)

total revenue
marginal revenue
average revenue
the short run
the long run
long-run supply
normal price
normal profit
long run equilibrium
normal competitive price

### CURVES AND GRAPHS (Try to draw, label, and explain each.)

Total Revenue and Marginal Revenue When Price is Constant
Total Revenue and Marginal Revenue When Price Must be Lowered to Induce People to Buy More
Total Cost, Total Revenue, and Total Profit When Price is Constant
Average Cost, Marginal Cost, and Marginal Revenue When Price is Constant
Total Cost, Total Revenue, and Total Profit When Price Must Be Lowered to Induce People to Buy More
Average Cost, Average Revenue, and Average Profit When Price Must Be Lowered to Induce People to Buy More
Marginal Cost and Marginal Revenue, and Marginal Profit and Marginal Loss
Marginal Cost and Marginal Revenue, Average Cost and Average Profit: The Case of a Price Increase
Businesses Use a Modified Version of the Total Cost and Revenue Graph: The Break-Even Chart

### QUESTIONS (Try to write the answers, or jot down the highlights.)

1. Your marginal cost curve (at prices higher than the average variable cost) is the supply curve for your jeans plant. But the MC curve at prices *lower* than AVC is *not* a part of your plant's "supply curve." Explain why all this is true, and draw a graph to illustrate.

2. Suppose you owned a mineral spring and you were charging people for coming and getting jugs of water. You would like to set the price to make as much profit as you can. Your cost of production is zero, so you want to maximize your total revenue. That will bring you your maximum profit. Where would you set the price? Where the elasticity of the demand is unitary. Right? Sure. Can you explain why? Try.
3. If the "model pure market system" was working perfectly, then it would always be true, of everything, for everybody, that: "You get what you pay for." Also, it would always be true that: "You must pay for what you get." Explain exactly what all this means, and exactly why it must be true.
4. The curves themselves are of no significance. The curves are just a way of explaining and making *explicit* the implicit "common sense" concepts and principles which guide the intelligent decisions of businessmen. Try to state, in your own words, as many as you can of the "common sense" concepts and principles which the cost and revenue curves illustrate. How much do you think the curves help you to see and understand the concepts and principles? Discuss.

---

# 22 HOW AN INDUSTRY RESPONDS TO COST CHANGES

*A change in demand or supply in the input markets affects production costs, so the industry must readjust.*

What happens when the demand increases? It creates a shortage and the price goes up. Then producers produce more. In the long run, the industry expands. You know that so well that it is almost second nature to you now. (I hope.) This chapter talks about it more and goes into it further.

When you expand your output of Cullowhee hip-huggers, or when the egg producers expand their output of eggs, where do you suppose all the extra labor and capital and materials and other inputs come from? They're bought in the "input factor markets," of course! There is a demand for each input factor, just as there is a demand for each output product. In the "model pure market system," when the *demand* for an input factor increases, its *price* increases (just like in the output markets).

In this chapter we're going to talk about what happens when input demand goes up and prices in the input markets rise. Increased factor prices mean increased cost. Increased cost brings a higher "normal price," and losses for the firm in the industry. What happens then? And why? And how? That's what we're going to be talking about.

This chapter goes a little further with, and makes some new applications of concepts you already know about. This chapter will broaden and deepen your understanding of the concepts and principles, but there's nothing very new or different here. You'll be reading more about "consumer sovereignty"—that is, about how, in the pure market system, the consumer exercises ultimate control over the society's economic choices. You'll see how the final output markets and the input factor markets influence each other. You'll gain a much better understanding of the "price mechanism": of how, in the pure market system *the "rationing" and "production motivating" functions of price really do control things.* That's important—really *basic.*

## INPUT FACTOR DEMAND IS DERIVED FROM CONSUMER DEMAND

First, a few more details about how "consumer sovereignty" is supposed to work. Thinking back a few chapters, remember what happened after the announcement that eggs bring "perpetual youth"? The demand for eggs increased, of course. Then there was a shortage of eggs, so the price went up. After that, the factors of production started moving out of other industries and into egg production. *High incomes and profits attracted the factors into the egg industry.* You already know all this. But specifically, step by step, how is it all supposed to happen? Let's look at some examples.

Suppose some of the workers at the Chicago stockyards really would be good workers in the egg business. How do they know that the demand for eggs is high, and that the businesses are making big money producing and selling eggs these days? They are workers. They aren't businessmen and they don't plan to go into business. How do they get the message that they should quit being stockyard workers and go get jobs in the egg industry?

Or suppose you have some land, conveniently located and ideally suited for egg production. How do you know that you should shift the use of your land into the egg business? Or how about the farm equipment companies? They produce and sell all kinds of farm machinery (capital) to be used in all kinds of agricultural enterprises. How do they know that they should produce less of something else so they can produce more (and build up their inventories of) egg-producing equipment?

### Final Demand And Derived Demand

You probably already know enough about how the market operates to figure out exactly how these "factor owners" get the message. What induces them to begin to move their factors into egg production? It all happens through the market—through demand, and price. The businessmen who are in touch with the egg business and who are watching for the opportunity for profits will begin *to demand more egg-producing factors*—more labor, land and capital. This increased demand for the factors of production will have exactly the same effect on the price of the factors as the increase in the demand for eggs had on the price of eggs.

This increased *factor demand* results from the increased *consumer demand*. That is why economists refer to "consumer demand" as final demand and to "factor demand" as derived demand. All consumer demands are "final demands." When people demand things to "consume," this means they want to use the things for their own personal desires, needs, pleasures, or objectives. It is from this *final, consumer demand* that all factor demands are derived. It's easy to see that if there's no *final demand* for something, there will be no "derived demand" for the factors to produce that "undemanded something." The greater the final demand for a product, the greater will be the derived demand for the factors needed to produce that product.

Let's take another example. Suppose I demand a tractor to help me to raise my backyard corn crop. My demand for the tractor results from the fact that there is a demand for corn. If there was no demand for corn, then there would be no demand for tractors to be used in growing corn. There would be no demand for tractor tires to go on the corn-producing tractors. There would be no demand for labor to produce the tires, or for steel to build the tractor, or for iron ore to make the steel, or for steam shovels to scoop up the iron ore.

All demands for all things in the society ultimately result from and reflect the final consumption demand for things. When you think about it, how obvious that is! All kinds of basic and intermediate products are being produced all the time. But if there wasn't any consumer demand for the final products, all this productive activity would stop. Anything which increases the final demand for a product *automatically* increases the demand for the input factors used to produce it.

### Factor Markets Adjust To Changing Demand

In the "model market process," whenever the consumer demand for a product increases, the price goes up. Then the demand for the factors that are needed to make the product will go up. Shortages will show up in these factor markets and will push up the prices of these factors. The higher prices for the factors will pull in the additional factors. That's what brought about the expansion in the egg business!

Increased *consumer demand* always signals the need to expand the industry. Then, it is through the increased *derived demand* for the factors of production that society's wish is carried out. It's "consumer sovereignty" and "the invisible hand" in action!

When the consumer demand for eggs increases and the price of eggs goes up, businessmen will start to pull together the needed factors and go into the egg business. They are attracted by the highly profitable price that eggs are bringing these days. Some profit-seeking businessman will offer a good price to buy or lease your land. He will run an ad in the

*Chicago Tribune* offering good wages for stockyard workers who would like to shift into egg production. He will contact the farm equipment people and place orders for egg-producing machinery and equipment.

As soon as the businessman gets all the input factors he needs, he will start producing eggs, and making profits. Society's wish for more eggs is being fulfilled. See how neatly it all fits together? Consumer sovereignty and the invisible hand. Great!

### Consumer Demand And Derived Demand: A Summary

We can summarize the whole process this way. The increase in the demand for eggs has resulted in an increased price for eggs. The increased price for eggs makes egg production a very profitable business. Because egg production is a very profitable business, businessmen wanting profits try to go into egg production; those already in it try to expand their operations.

The only way businessmen can go into the egg business is to get and bring together the needed factors of production. How do they get the factors they need? They go into the factor markets (the labor market, the land market, and the capital market) and try to buy them. This increases the demand for the egg producing factors. More of these factors move into the egg business to get the higher wages, rents, interest and profits offered there. That's how the egg industry expands, more eggs are produced, and the wishes of the people for more eggs, are fulfilled.

This little summary describes in capsule form a major segment of "the market process in action." It's essentially this: When consumers demand more eggs, businesses seeking more profits demand more egg-producing factors. The higher demand brings higher prices for the egg-producing factors. The higher prices attract more factors into this "shortage" industry and gets them to do what the society wants them to do—produce more eggs.

## THE NORMAL PRICE MAY CHANGE AS THE INDUSTRY EXPANDS

Remember a few chapters ago, when we were talking about the supply and demand for eggs? The "normal price" was 75¢ a dozen. What will happen to the normal price as the egg industry expands? Will the *normal price* stay at 75¢ a dozen? Or increase to 80¢ a dozen? You can see that the normal price might move up. When the price of eggs goes up, the "derived demand" for the egg-producing factors goes up. Then the factor prices go up. So you would expect the cost of producing eggs to go up.

If egg producers must pay more for their labor and land and capital, then it seems likely that their cost of production is going to increase. There is good reason to expect that in any rapidly expanding industry the normal price (the "long-run equilibrium price") of the product will be rising. The higher normal price would simply reflect the increased cost of the factors of production. For any *declining* industry, you might guess that the normal price will be moving down as the prices of the needed labor, land, and capital move down in response to the decreased demand.

## How Big Is The Egg Industry?

Whether or not the normal price actually does rise or fall as the industry expands or contracts depends on many things. No need to get into all the "normal-price influences," but here are two you should know about.

If the egg business is *very large* and the expansion is great, the egg producers may try to hire almost all of the stockyard workers. This would create a serious labor shortage in the Chicago area and the price of labor might go very high. In a case like this the normal price for eggs might rise from 75¢ to 85¢, or maybe to 90¢, or more. But if the demand for labor by the egg producers is only a small part of the total demand for that kind of labor in the Chicago labor market, then the price of labor (the wage rate) may rise only slightly, and only temporarily. (The same idea holds true for land and capital too, of course.)

Another thing which may cause the "normal price" to change as the industry expands or contracts is the change in *productive efficiency* which might occur. As an industry expands there is a good opportunity to bring in the latest technology—the best capital, and the most modern techniques. The modern techniques may be so much more efficient that the cost of producing eggs may go down, even though the price paid for each individual factor may go up.

You can see that it is difficult to be sure what is going to happen to the "normal price" as the industry expands or contracts. In our examples we have been taking the easy way out. We have been assuming that the industry is small (relative to the total demand in the factor markets), and that the normal price doesn't change.

## As One Industry Demands More Inputs, Another Demands Less

As the demand for eggs increases, something else is going on at the same time. The consumers who are eating more eggs are eating *less* of something else. The people who are buying more eggs are buying less cheese or meat or bread or butter or pork sausages or something. The increased demand for eggs brings an increased price for eggs, and increased derived demands for the factors of production needed to produce eggs. But at the same time, the *decreased demand for the other products* is bringing decreased prices and decreased derived demands for the factors needed to produce the other products.

While the egg producers are expanding their operations, demanding more factors of production and paying higher prices for the factors (land, labor and capital), the businessmen in the declining industries are reducing their operations, demanding fewer factors of production, paying lower prices for the factors. They're trying to get out of the "declining demand" industries and into something better as soon as they can.

The whole scene is one of resources moving out of the areas of declining demand and lower prices, into the areas of increasing demand and higher prices. It is the shift in consumer demand (from other things, to eggs) which initiates this process. It is the shift in the demand for the factors of production which carries out, and completes the process. Yes, it's "consumer sovereignty" and "the invisible hand," in action, all right!

You already know that this very neat, very smooth way in which the "model market system" operates is not exactly the way the market process works in the real world. Several of the chapters coming up soon (Chapters 24, 25, 26, and 27) will deal with some of the "real-world imperfections" in the operation of the market process. Those chapters will talk about how well (or how poorly) the "modified market process" really does work—how well it serves the interests of society. But we aren't quite ready to get into that yet. First you need more understanding of how it works in the "model system"—and more about supply.

## The Supply Responds To Cost Changes

Let's go back to the initial egg market, before the increase in demand. The price of eggs was 75¢ a dozen and the equilibrium quantity flowing across the market was 10 carloads per week. Then the demand increased. The price went up to $1.00 and the higher price brought forth more eggs. Remember?

Now let's look at it another way. Suppose the demand for eggs didn't change, but that something happened to *change the supply* instead. What might happen to cause the supply to change? Then what would happen next?

You already know from your hip-hugger jeans business that for the "fixed price" case, your marginal cost tells you the quantity you should produce and sell at each price. Your marginal cost always determines your willingness to produce and sell more. (It determines your "propensity to sell.") So your marginal cost curve (at prices higher than your average variable cost) really is the supply curve for your output of Cullowhee hip-huggers. If the price goes higher, you will expand your output to where the marginal cost is equal to the new (higher) price, and that is how much you will supply to the market. Remember? Sure.

The supply curve in the entire market is nothing more than a composite curve, made up by adding together all the marginal cost curves of all the little firms in the industry. You already knew that. So, implicitly, you already know the two things that would cause a change in the market supply curve: (1) more little firms or less little firms in the industry, or (2) higher or lower marginal costs for each little firm. Let's talk about these two kinds of changes in supply.

### Supply Changes As Firms Enter Or Leave The Industry

If the number of firms in the industry increases, this will increase the supply. It adds more "little marginal cost curves" to the total "composite market supply curve." If more firms are producing and adding their outputs, to the market, then the total amount arriving in the market is greater. Obviously.

You have already seen how the supply increases as more firms enter the industry. Remember how the industry expanded, following the increase in the price of eggs to $1.00 a dozen? As the industry expanded, the supply increased—shifted to the right on the graph—to show that more eggs would be shipped into the market at each price. The extra carloads of eggs came from the new egg producers—the new little firms in the industry.

It works the other way too, of course. If firms leave the industry, the amount coming into the market at each price gets smaller. The supply curve shifts to the left.

### Supply Changes As Factor Costs Change

The other way for the supply to change is for the cost of production to change. Any change in the firm's marginal cost will change the amount the firm will produce and sell, at each price. If the variable input factors get more expensive (or less productive), the marginal cost will increase. As the marginal cost increases, the *supply will decrease*. Less will be produced and offered for sale at each price. But if the input factors get cheaper (or more productive), the marginal cost will go down and the *supply will increase*. More will be offered for sale at each price.

It is obvious that changes in the prices you have to pay for variable factors will change your marginal cost. But when factor prices change, this changes *more* than just the *marginal* cost. It changes the *entire cost and profit picture* for each firm in the industry. The *long-run* repercussions of such changes can be significant!

### How A Cost Increase Affects A Business

Suppose that for some reason there is a big wage increase in the garment industry. Your hip-hugger jeans plant feels the effect right away. The higher wages push your marginal cost way up. This changes the amount you will supply in the short-run, but that isn't all. Now you must pay higher average costs per unit and higher total costs for each level of output. The higher costs wipe out your profits. You are losing money!

What happens to your cost per unit (as shown by your average cost curve)? The per unit cost goes up, so the average cost curve *shifts upward*. The per unit profit you were making (shown by how high your average revenue curve was above your average cost curve) is all gone. Now, even at your "most profitable output," your average revenue curve is *below* your average cost curve.

Your per-unit *cost* is higher than the *price*. You're losing money all right! Perhaps we should look at this on a graph. Figure 22-1 shows the picture and tells the story. You should study it now.

### How To Choose Your "Minimum Loss" Output

Now you know what happens when the cost goes up. Your average cost is higher than the price. You lose money, no matter what output you choose. But the output where your marginal cost equals your marginal revenue is your "minimum loss" output. You produce at that rate, and plan to get out of the hip-hugger jeans business as soon as you can.

Let's suppose that the retailer you sell to, won't pay you any more than $100 a carton, even after the cost increase. Then maybe you will just go out of business. Or maybe you will produce something else, or find another market where you can sell your hip-hugger jeans. Let's talk about how it would work out in the model pure market system.

The higher cost means that all the little firms in the industry are losing money. They will all be trying to find something more rewarding to do so they can get out of this industry. The most flexible firms can get out first. As they leave the industry one by one, this will reduce the supply (shift the market supply curve to the left). The increased cost already reduced the market supply (shifted the curve) once. It caused each "typical, average firm" to cut back from 10 units to 9 units (as shown in Figure 22-1.) Now, as the money-losing firms start leaving the industry, the market supply will be reduced even more.

As more and more firms get out of the industry the market supply is reduced more and more, step by step. Each reduction in supply causes a

Fig. 22-1 **WHEN COST INCREASES, THE COST CURVES SHIFT UPWARD**

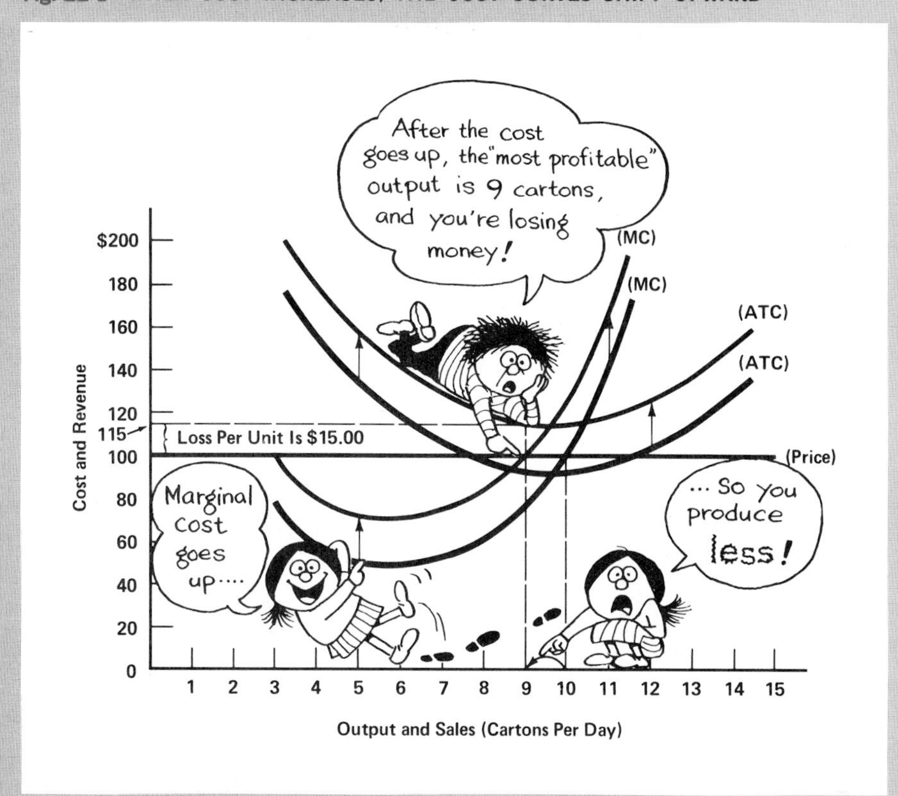


When The Wage Rate Goes Up, Output Is Cut Back And The Firm Loses Money.

When the cost goes up, you cut back your output to where MC=MR, at 9 cartons a day. You are losing $15 a carton, for a total daily loss of $135 (9 units @ $15 equals $135).

If you shut down your plant, you will lose $150 per day (your fixed cost), so it's better for you to produce than to shut down. But you must find some way to reduce your cost or increase your revenue, or else you must get out of this business. Otherwise you will go broke.

little *shortage* in the market and *pushes the price up* a little bit. Each time the price inches up, it cuts down a little bit on the losses of the firms remaining in the industry. You can look back at Figure 22-1 and see what will happen as the price line inches up from $100 to $101, to $102, and on and on.

As long as there are losses in the industry, firms will keep on leaving, seeking "normal profits" (or better) elsewhere. How long will the decrease in the market supply (the leftward shifting of the supply curve) continue? It will continue until the market price moves up enough to cover the increased costs and to once again provide a "normal profit" to the businesses remaining in the industry.

### The Price Goes Up Enough To Cover The Cost

When the market price moves up enough to provide "normal profits," this means that the new normal price has been reached. All the buyers who are still buying the product will be paying the full cost of production, including enough to cover the higher wages, and also including a normal profit for "the average firm" remaining in the industry.

I'll say it again, this way: The wage increase at first *reduced the short run supply*. That caused some shortages, so the price increased some. But the firms in the industry still were losing money, so they were trying to leave. In the long run several firms will leave the industry, reducing the supply enough to push the price up to "normal" again. The firms still remaining in the industry will again be making normal profits.

The "long-run condition" arrives. The industry is stable again. Now this product is more expensive. Society is more carefully conserving its higher-priced, scarcer, more valuable factor of production—labor. What we've just seen is how this adjustment process looks from the point of view of "the typical firm." Now let's take an "egg market" example and watch how this process works itself out in the total market.

## A COST INCREASE AFFECTS THE ENTIRE MARKET

Let's begin by supposing that something happens to push the cost of production up so much that eggs will now have to sell for a dollar a dozen in order for the egg producers to make normal profits—that is, make enough profit to keep them from leaving the egg business. Another way of saying it: Cost of production has increased in the egg business and has forced up the "normal price" from 75¢ to $1 a dozen. A number of things might cause such a cost increase to occur. Let's look at some of the possibilities.

Any one of the factors of production might have become more scarce, or more expensive. A wheat blight or corn rust disease might have made grain very scarce and forced up the price of chicken feed. Something might have happened to cause many of the egg-laying hens to die, thus forcing the price of hens to move up drastically. Transportation costs might have gone up, or electricity, or water, or the interest rates on the money the egg producers have to borrow to finance their businesses—to buy laying hens, egg-packing machines, land and buildings, and all that—might have gone up.

Strong demands for labor in the economy might have pushed up wage rates. Or maybe the poultry workers decided to unionize (create a monopoly for the sale of their labor services) and then strike for higher wages. Or maybe the government passed a minimum wage law, or new health and safety regulations which required new expensive equipment or more skilled personnel or more expensive procedures. Or maybe a new anti-pollution regulation requires more expensive equipment and procedures to eliminate the smell of chickens (or other chicken-related smells). Or perhaps the state and local governments decided to impose higher taxes.

Many things could happen to cause an increase in the cost of producing eggs. But no matter what makes the cost go up, the effect on the supply will be about the same. The supply will be smaller (the supply curve will shift to the left) and shortages will develop in the egg markets. All the egg buyers will start bidding against each other, offering more money to try to get their "normal amounts" of eggs. The price will go up.

At the higher price, the shortage will be eliminated. Once again we will have a *short-run equilibrium* flow of eggs across the market. The quantities shipped into the market by the sellers will be equal to the quantities taken out of the market by the buyers. But the quantity flowing across the market will be smaller than before.

### High Prices Conserve Scarce Products And Factors

Eggs are now more scarce, more valuable in the society. Eggs will be conserved. The factors of production used to produce eggs will also be conserved. If the higher cost results from corn rust destroying much of the corn crop, the higher price of eggs will reduce the amount of corn needed for chicken feed, so that more of the very scarce corn will be available for other purposes. If the higher cost is a result of a higher demand for labor in the economy, the higher price for eggs will conserve the use of labor in egg production and will let more of the labor be used to do some of the other things the society wants done.

The higher price of eggs conserves the (now scarcer) eggs. The higher price also conserves the (now scarcer) factors needed to produce eggs. Figure 22-2 shows how this situation looks in the egg market, and explains what happens. You should study that graph, now.

Fig. 22-2 **INCREASED COST DECREASES THE SUPPLY**

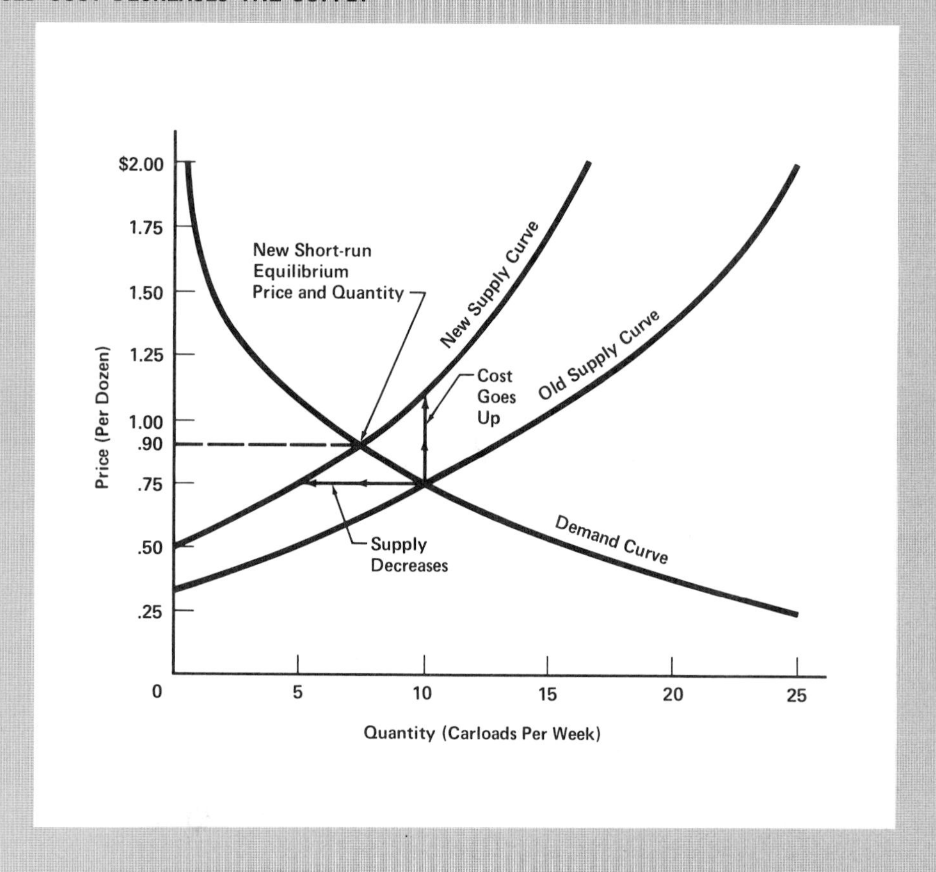


The Supply Curve Shifts To The Left.

At each level of *output* the cost is higher. (The marginal cost curve shifts *upward*.) Therefore, at each *price* the quantity offered for sale is smaller. The *supply curve* represents both these shifts—upward for cost, and leftward for the supply.

The shift of the supply curve to the left shows that after the cost goes up the quantity of eggs offered for sale at a price of 75¢ drops back from 10 carloads to 5 carloads a week. No longer can 75¢ be the equilibrium price.

People are trying to buy 10 carloads of eggs and only 5 are available. There is a shortage. Eggs are more scarce—harder to get. Up goes the price. Hungry buyers keep offering higher prices until the price of eggs moves up to where the quantity offered by the sellers is equal to the quantity demanded by the buyers.

The new price of 90¢ provides a new equilibrium flow of eggs across the market, with a total of 7½ carloads per week being exchanged. But remembering Figure 22-1, you should expect that even at this higher price the egg-producing firms are losing money. In the long run some firms will leave the industry, so the supply will shift farther to the left and the price will rise even more.

## Supply Keeps Changing To Bring A New "Normal Price"

Figure 22-2 shows that the new short-run equilibrium market price is 90¢ a dozen. At this price the producers will ship in more than the 5 carloads a week the 75¢ price would bring. But at 90¢ the producers still supply less than the original 10 carloads they were producing before the cost increase. The figure shows that the new (temporary) equilibrium quantity flowing across the market will be 7½ carloads a week.

The higher price (90¢) partly offsets the higher cost of production, but not completely. The egg producers are still losing money. As soon as they can make the shift, some egg producers will stop producing eggs and start producing turkeys or broilers, or maybe pork or beef, or something else. Or maybe some of them will just go broke, move to the big city and go on welfare. It wouldn't be the first time something like that had happened!

As each producer shifts out of the egg industry, the supply of eggs will decrease a little more and the price will move up a little more. Producers will continue to leave the industry and the supply will continue to get smaller (shift to the left), until the price gets high enough to provide normal profits to all those who are still producing eggs. If we assume that the new "normal price" is $1 a dozen, then we know that the supply will continue to get smaller until the market price moves up to $1 a dozen. When the market price reaches $1 a dozen, the egg producers will find that once again they are making normal profits. There is no longer any reason for them to want to get out of the egg business.

When eggs start to sell for $1 a dozen, the "normal price" (the "long-run equilibrium price," or the "normal competitive price") again exists in the market. The normal price has been brought about by the long-run adjustment (decrease) in supply. The egg market and the egg industry are once again in long-run equilibrium.

Figure 22-3 shows all this on the supply and demand graph. The graph shows how *increased cost* shifts the supply in the short run, then shifts it even more, in the long run. The shifting continues until the market gets back to a new "normal price." Then the egg market and the egg industry are once again in long-run equilibrium. You should study Figure 22-3, now.

## Reduced Cost Brings A Lower Price

You can look at Figure 22-3 and figure out what would have happened if the cost of production had gone down instead of up. Everything would have moved in the opposite way. The supply would have increased and the price would have gone down. But even at the new, lower price this probably would be a very profitable industry. If so, other producers soon

**Fig. 22-3 IN THE LONG-RUN THE SUPPLY DECREASES MORE**

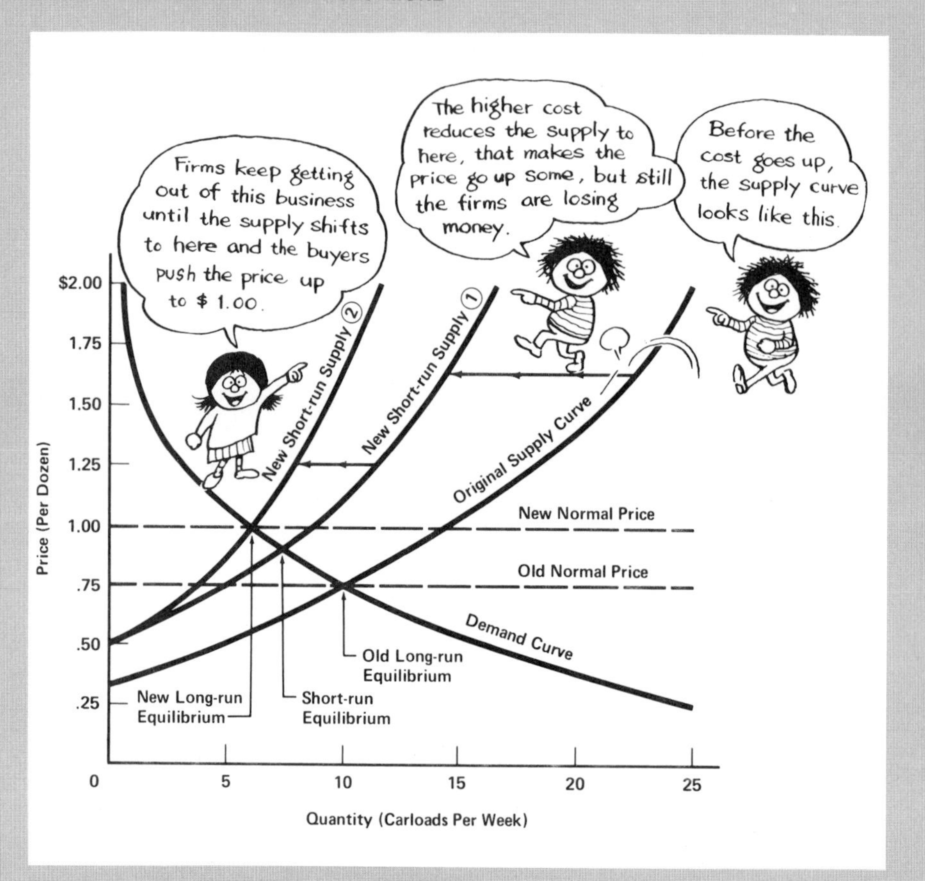


Eventually All The Increase In Cost Is Passed On To The Ones Who Eat The High-Cost Eggs.

The initial cost increase reduces the supply by increasing the marginal cost of the producers in the industry. This decrease in supply creates a shortage. The quantity being shipped into the market drops back from 10 carloads a week to 5 carloads a week. Many buyers can't find any eggs so they begin offering more—bidding against each other—and force the price up.

Initially the price moves up to 90¢. The producers then increase their shipments from 5 to 7½ carloads a week. But at 90¢ a dozen the egg producers are still losing money.

Producers start getting out of the egg business. As the producers leave, the supply decreases and the price inches up. The price finally reaches the new normal price of $1 a dozen. At this price, 6 carloads per week are crossing the market.

Notice that throughout all of this, the demand has not changed. First the people were buying 10, then 7½, and finally 6 carloads per week. They are buying less as a reaction to the higher prices, because of the income effect and the substitution effect. But the demand (the "propensity to buy") has not changed. They buy less, true. But that's a reaction to the market conditions—to the shortages and higher prices. It isn't a change in the "propensity to buy"!

would shift out of the turkey, broiler, pork, beef, grain, and other businesses and start producing eggs. Each time a new egg business added its output to the market, this would increase the supply a little more and inch the price down a little.

How long would new producers continue to come into the egg business? You know the answer: until all the excess profits are eliminated. The supply would continue to increase and the price would continue to inch down until a new "normal competitive price" is reached. The new normal price would reflect the new, lower cost of production in the egg industry.

When the new normal price (perhaps 60¢ a dozen) is reached, the egg market and the egg industry will again be in stable, long-run equilibrium. The price, and the flow of eggs across the market from buyers to sellers, will continue at a stable level until some new change occurs to upset the equilibrium—that is, until something else happens to change either the supply or the demand.

### "Full Cost" Includes "Normal Profit"

The *cost of production* of each product is reflected in the *supply*. Through the supply, the cost is reflected in the *price* of the product. This is the way the market process, in the long run, requires each consumer to pay the full cost (and no more than the full cost) of what he buys. Reduced cost of production increases the supply and lowers the price. The consumer doesn't have to pay as much as before. Increased cost reduces the supply and increases the price. The consumer who still wants to use the higher cost, scarcer, more valuable product, must pay more for the privilege. It is through this process—cost changing the supply, and supply changing the price—that the market requires the people who buy each product to pay the full cost of producing that product, plus a normal profit for the producers.

Whenever the price of a good isn't high enough to provide a "normal profit," then firms will leave the industry and go elsewhere. As the businesses leave, the supply flowing into the market will get smaller, so the price will rise. The price will keep on rising until the businesses stop leaving the industry—that is, until the normal price is reached.

The "normal profit" is really a part of the *cost* of getting the good produced. If "normal profits" aren't being paid, the businessman can't have *his* distributive share. So "normal profit" is the producer's "just reward," in the same way that the wage is the "just reward" for labor, or rent for land. The consumer who gets the goods must pay for the "normal profit," just as he must pay for the "normal wage" embodied in the products. So the "normal profit" really is a part of the cost of producing each good. If the consumers are not willing to pay a price high enough to provide "normal profits" then in the long run the product will disappear from the market.

### Ultimately The Consumer Pays The Full Cost

You understand, now, that in the model market system all the costs (including normal profits) ultimately are paid by the people who buy the products. Even in the real world this result comes very close to being true most of the time. This simple truth needs to be understood by people who suggest that businesses should pay more taxes, or spend more to overcome poverty, or to clean up pollution, or to make automobiles safer, or to give their workers higher wages or better retirement benefits, or whatever. I'm not saying you *shouldn't* argue for all these things. I'm only saying that you should understand the full effects of what you are suggesting.

Ultimately, it is not the owners of the business, but the buyers of the product who must pay all the costs. When the buyer reacts to the higher price by buying less or by ordering from some other (perhaps foreign) producer instead, the business may wind up "out of business." Any cost increase or decrease — for labor, land, or capital, or for taxes, pollution control or anything else — sets into motion an inexorable chain reaction. The effects ripple throughout the economic system, going many places, doing many things. The one thing you can be sure of is that the people who will ultimately pay most of the increased cost, are the people who buy the product. As long as the market system is working, there's just no way around it.

## THE MORE ELASTIC THE DEMAND THE MORE THE SUPPLY MUST RESPOND

You know that when the cost of production changes, the supply changes. When the supply changes, the price changes. The price *must* change. But as the supply changes, *how much* does the price change? A little? Or a lot? If you think about it for a minute you will see that the answer depends on this: How responsive are the buyers, to price changes? Or to say it differently: How elastic is the demand?

If the demand is *highly inelastic*, that means the buyers are *not very responsive* to price changes. They buy almost the same quantity no matter whether the price goes up or down. So a *small change* in the supply would cause a *big change* in the price. If cost increases, and if demand is highly inelastic, a small decrease in supply can push the price right up to the new "normal" level.

What if the demand is *highly elastic?* As the supply decreases the price goes up, the buyers will *buy a lot less*. So a lot of firms will have to leave the industry before the price will move up to the new "normal equilibrium" level. This may not yet be clear, so let's talk about it some more.

Suppose the cost goes up and the supply gets smaller, as you saw in Figures 22-2 and 22-3. How many firms must get out of the industry before the price moves up to the normal level? Will the price go up very quickly to bring a new long-run equilibrium? Or will the price increase come about very slowly, and only after many firms have left the industry? The answers to all these questions depend entirely on this question: "How elastic is the demand?"

## The "Highly Inelastic Demand" Case

If the demand for something is highly inelastic, it means that people really would like to keep on buying about as much of it as they are buying now. If the price goes down they will buy a little more, but not very much more. If the price goes up they will buy less, but not very much less. How far up would the price of salt have to go before you would use less salt on your eggs in the morning? Or how much more salt would you use on your eggs if salt was a penny a pound? Your demand for salt is highly inelastic, wouldn't you say?

When the demand is highly inelastic, and if the cost goes up and decreases the supply, the price will move upward quickly to cover the increased cost. The new normal price will come about very soon. Not many firms will have to leave the industry. Maybe none will.

If salt starts to get more expensive, no matter. You will pay twice as much to get the amount you want, won't you? Let's look at this on a graph. It's so much easier to see that way. Figure 22-4 shows exactly the same picture as Figure 22-2 except that a new, highly inelastic demand curve has been added. With the highly inelastic demand curve you can see how much the price would go up following the initial increase in cost and shift in supply.

Figure 22-4 shows that if the demand is highly inelastic, the producers don't have to be worried about a cost increase. The price will move up to cover the cost increase and the buyers will continue to buy almost as much as before. But what if the cost goes *down*? The supply will increase. What then? Disaster!

When the demand is highly inelastic, a supply increase will create a surplus and the price will move down to try to "clear the market." But as the price moves down, people don't buy much more. The surplus persists. So the price just keeps on falling and falling. How low would the price of salt have to go to induce people to use more salt on their eggs in the morning? Very low, right?

**Fig. 22-4 WHEN DEMAND IS HIGHLY INELASTIC, AN INCREASE IN COST IS PASSED ALONG VERY QUICKLY TO THE BUYERS**

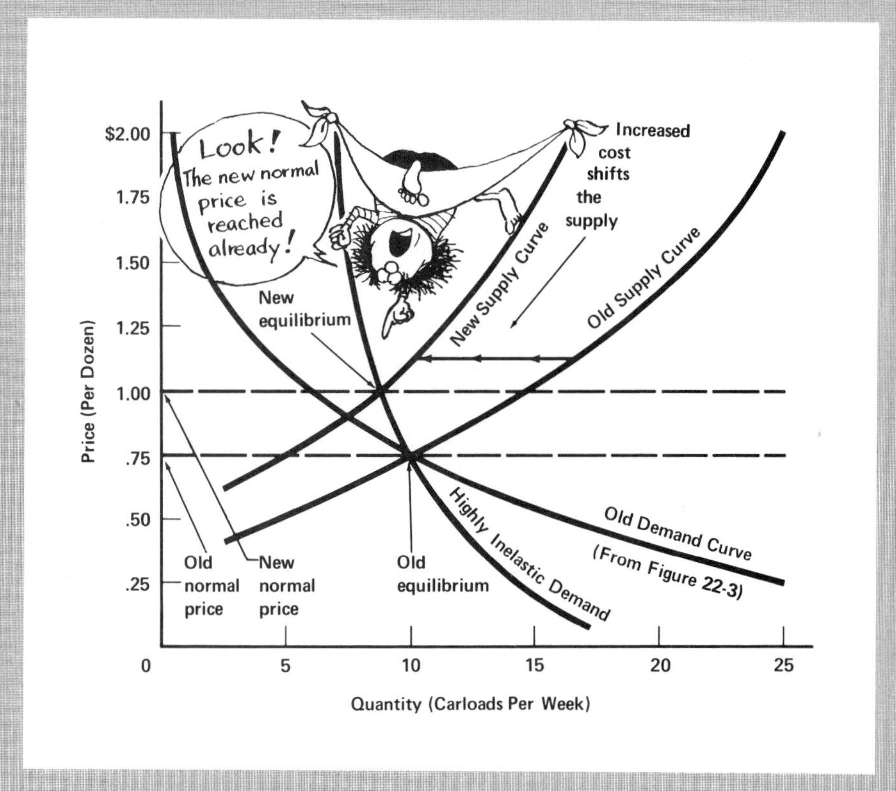


When The Demand Is Highly Inelastic, The Rationing Function Of Price Doesn't Work Very Well. Right?

In this case, the demand is so inelastic that the price immediately jumps up from 75¢ to the new "normal price" of $1 a dozen.

The "inelastic demanders" insist on getting their eggs, even if they have to pay a dollar a dozen! At the initial normal price of 75¢ a dozen, 10 carloads per week were bought. But at a price of $1.00 a dozen, 9 carloads are bought.

The price increased from 75¢ to $1.00, an increase of one-third (33 percent) and resulted in a decrease in quantity bought from 10 carloads to 9 carloads, a decrease of only one-tenth (10 percent). So the elasticity of demand is 10 percent divided by 33 percent (10/33 = .3). Much less elastic than *one*. Highly inelastic demand!

Those who produce and sell a product like table salt, for which the demand is very inelastic, do not need to worry about their costs going up. Increased costs will be easily and quickly passed forward to the consumer. But such producers are really vulnerable to a *decrease* in costs which would increase the supply. The demand for most agricultural products is relatively inelastic. Whenever there is a bumper crop of some farm product, this creates a surplus and causes the price to drop very low. This is a part of the reason why the government has programs to "stabilize" the prices of many farm products. The government sets minimum prices, buys up surpluses, and limits the acreages for certain crops to try to keep prices from falling too low.

It's paradoxical, but the farmers might get more income if half of everybody's crops got eaten up by grasshoppers! Inelastic demand is great for the producer who has enough market power (monopoly power) to be able to keep the supply low and the price high. (Remember the northside chief and his tuba monopoly?) But for the producers in an industry where the supply is likely to increase (or maybe where the supply is likely to fluctuate up and down from year to year), highly inelastic demand can be disastrous. Study Figure 22-4 now.

## The "Highly Elastic Demand" Case

The producer faced with *highly elastic demand* has a problem, too. But his problem is exactly opposite from that of the salt-maker. With highly *elastic* demand, if the supply increases and pushes surpluses into the market, what happens? The price moves down just a little and everybody buys a lot more. Soon the surpluses are all cleared away. So if the cost goes down and the supply increases, the price will move down only very slowly, a little bit at a time.

When the cost of production goes *down*, the firms in the industry make excess profits. The effect is about the same as if the *price* had gone *up*. Each firm produces more in the short run. Then in the long run the excess profits attract more firms into the industry. But if the demand is highly elastic, as the industry expands and the supply increases, the price doesn't seem to want to go down very much. How long will it take for the price to be pushed down to where the firms are making only "normal profits" again? How long will it be before all of the cost decrease is passed along (by lower prices) to the buyers? A long time!

Whenever the demand for a product is highly elastic, if the cost decreases, the producers will enjoy "excess profits" for a long time. A large expansion in the industry will be required to increase the supply enough to push the price down to the new normal level. But what about the opposite case? Suppose the cost had *increased* instead? That's a sad story. Stop now, and study Figure 22-5 and you will really see what a sad story it is!

Fig. 22-5 **WHEN DEMAND IS HIGHLY ELASTIC, AN INCREASE IN COST MAY DESTROY THE INDUSTRY**

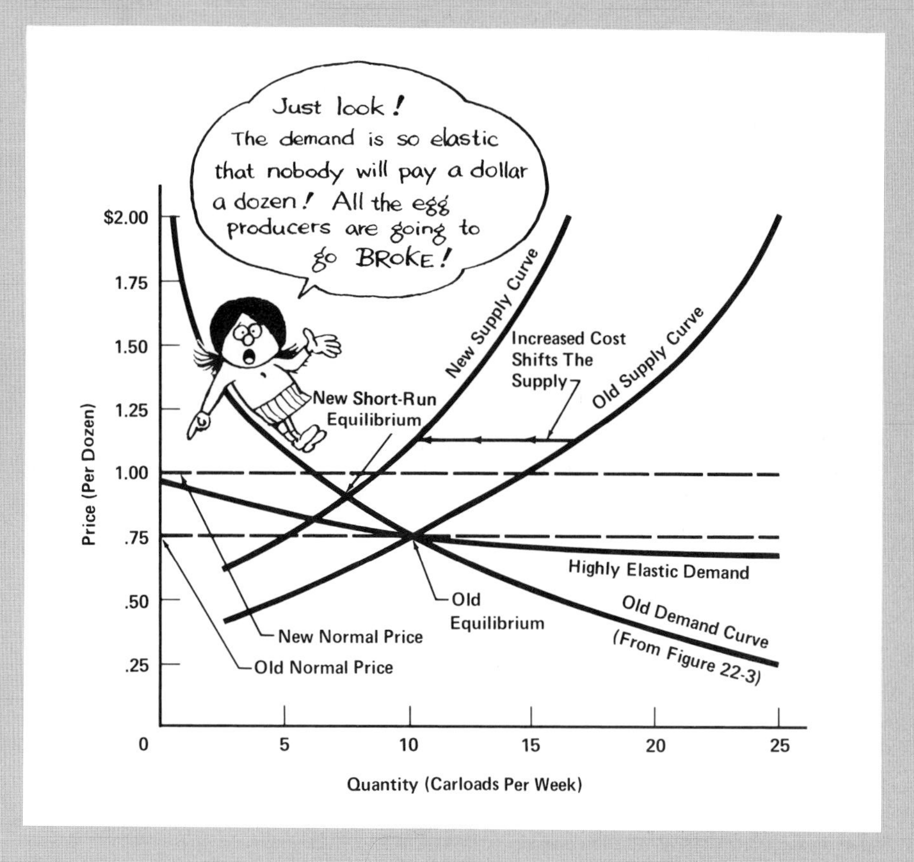


When Demand Is Highly Elastic, The Rationing Function Of Price Works So Well It May Destroy The Industry!

The demand for eggs is so elastic that nobody will pay a dollar a dozen.

At the new normal price of a dollar a dozen, the quantity of eggs demanded would be zero. Unless the demand increases or the cost decreases, all the firms will leave the industry and eggs will disappear from the market.

The society doesn't want eggs badly enough to continue to pay the cost of producing them. Society wants those egg-producing resources shifted into some other industry. The market process will see to it that that's exactly what will happen!

## Substitutes Make Demand More Elastic

What makes the demand for a product elastic? or inelastic? One thing that makes demand elastic is the availability of lots of substitutes. The demand for any brand-name product—like toothpaste or tires or beer or bread or canned peas—usually is highly elastic. Take gasoline, for example.

Suppose the going price of gasoline is about 40¢ a gallon and you're buying 10 gallons of Texaco a week. Will you still buy from your favorite Texaco station if they raise the price to 42¢ a gallon? to 45¢? Or suppose the Texaco station drops the price to 35¢ a gallon. You still might buy only 10 gallons a week, but how many of the regular Gulf and Mobil and Exxon and Arco and Chevron and Amoco buyers will switch to Texaco? Many! The Texaco station will sell a lot more gas. See how highly elastic the demand is for any *one brand* of gasoline?

But suppose *all* the service stations raise their prices or lower their prices by the same amount and at the same time. That's very different! And as you very well know, in the case of brand-name price cutting, everybody plays follow the leader. Why? Because they have no choice.

The demand for any one brand of gasoline is so elastic that if one seller cuts his price very much he will get almost all of the customers. So the other sellers *must* cut their prices or else they will be forced out of business. If the Texaco station cuts its price, you may be sure that the Gulf and Mobil and all the other stations will cut their prices, too. Why? Because the demand for any one brand of gasoline (or any other brand-name product) is likely to be highly elastic.

Businesses improve their products or advertise (or both) to try to create "brand-name loyalty." If you are loyal to Texaco, then you will keep buying it even if some other brands are cheaper. Your demand for Texaco is not so highly elastic. The people at Texaco would like for everybody to be like you!

There are several things which influence elasticity of demand, but by far the most important is the availability of acceptable substitutes. Suppose you're one of those guys who's going to eat two scrambled eggs for breakfast every morning, no matter what. Then, for you, there is no acceptable "breakfast food substitute" for eggs. Your demand for eggs is highly inelastic. But if your sister likes corn flakes and pancakes just as well as eggs, her demand for eggs will be much more elastic. Whenever people consider something to be *essential* to them, that means "no acceptable substitutes are available." The demand for that good will be highly inelastic.

## The "Market Mechanism" Depends On Elasticity

An understanding of the elasticity concept is very important to the businessman. It's just as important to you, if you want to understand how the market system works. Elasticity means responsiveness. We have been talking about the responsiveness of buyers and sellers to price changes. *Price elasticity of demand and supply is what makes the market system work.*

The system flexes in one way or another in response to price changes. The price changes are responses to the changing demands (propensities to buy) of the people, working against the built-in (cost) constraints of supply. As the prices move around, it's the elasticity (of demand and supply) that determines how well the price mechanism will do its job.

The supply curve is a graphic picture of the production motivating function of price. The demand curve is a graphic picture of the rationing function of price. It is the elasticity of supply which determines how well the "production motivating function of price" will work. It is the elasticity of demand which determines how well the "rationing function of price" will work. Surely, "elasticity" is a most vital concept in microeconomics.

## THE FACTOR MARKETS, AND INCOME DISTRIBUTION

Throughout this chapter we have been talking about how factor prices influence production costs, and how cost changes influence business behavior.

It's important to know about the factor markets, because factor prices influence outputs, and product prices. But the factor markets are important for another very significant reason. What reason? Factor prices determine people's incomes. Of course!

If the price of labor (the wage rate) is high, people who sell their labor will receive good incomes. Anything which changes the wage rate (or any other factor price) will change some people's incomes. So to understand income distribution we must understand factor pricing. *Looked at one way, factor prices influence cost, and supply, and price. Looked at the other way, factor prices determine the income shares (the wages, rent, interest, and profits) the individuals in the society will receive.* Be impressed with the importance of understanding how prices are determined in the factor markets!

In this chapter we have been looking at factor prices as *costs* which must be paid out by the business. Now, it's time for us to look at factor prices as *incomes* to be received by the people who are selling the factors. Remember the dollars in the circular flow diagram, flowing (as costs) from the businesses and going (as incomes) to the households? Now we're going to look at it from the "households" side. That's what the next chapter is all about. As soon as you're sure you have this chapter down pat, you'll be ready to go on into the next one and learn all about "how the market process distributes the income."

## REVIEW EXERCISES

### MAJOR CONCEPTS, PRINCIPLES, TERMS (Try to write a paragraph explaining each of these.)

the relationship between cost and supply
the role of "price elasticity" in the market process
how factor markets influence product outputs and prices
how factor markets influence income distribution

### OTHER CONCEPTS AND TERMS (Try to write a sentence or phrase explaining what each one means.)

final demand
derived demand
highly inelastic demand
highly elastic demand
"full cost" of production

### CURVES AND GRAPHS (Try to draw, label, and explain each.)

When Cost Increases, the Curves (ATC and MC) Shift Upward
Increased Cost Decreases the Supply
In the Long-Run the Supply Decreases (Shifts Leftward) More
When Demand is Highly Inelastic, An Increase In Cost is Passed Along Very Quickly to the Buyers
When Demand is Highly Elastic, An Increase in Cost May Destroy the Industry

### QUESTIONS (Try to write the answers, or jot down the highlights.)

1. How does the market process get the factors of production to respond to the society's desires—to move into the industries where demand is increasing and out of the industries where demand is decreasing? Explain the process, step by step.
2. Suppose the consumer demand for eggs really did increase by 30% this year. Would you expect the price of eggs to increase, in the short run? What about in the long run? Would you expect the "normal price" of eggs to go up? Explain.
3. Here's a question you've seen before. Now that you've read more about the concept, see how well you can do on this now:

   If the "model pure market system" was working perfectly, then it would always be true, of everything and for everybody, that: "You get what you pay for." Also, it would always be true that: "You must pay for what you get." Explain exactly what all this means, and exactly why it must be true.

4. If a business is producing a product which has a highly elastic demand, the business is likely to go broke if its costs go up very much. But businesses producing products for which the demand is highly *inelastic*, have more to fear from a cost *decrease* than a cost *increase*. Explain.
5. The "market process" is really nothing more (or less) than the rationing function and the production motivating function of price at work in the output and input markets. How well each "function of price" works depends on the elasticity of supply and demand—that is, the responsiveness to price—in the input and output markets of the economy. Discuss.

# 23 HOW THE MARKET PROCESS DISTRIBUTES THE INCOME

*The price you get for the factors you sell determines your income and your share of the output.*

What determines your income? That depends on what productive inputs you have to sell and how much they sell for. If you own a lot of high-priced inputs, you'll get a big income. If you don't you won't.

**Factor Ownership And Factor Pricing Determine Income Distribution**

In the pure market system, your share of the output would be determined entirely by the value of what you have to sell "in the factor markets"—that is, by what you sell to the businesses that buy in the factor markets. Your share is determined by *how much* you sell, and the *price* you get when you sell it. So the question of income distribution in the pure market system is one of "factor ownership," and "factor pricing."

In the last chapter we talked a lot about the prices of the input factors. The purpose there was to understand how factor prices affect supply and how the output price automatically moves to the "normal" level, where it will exactly cover the full cost of production—no less, and no more. There we were interested in factor prices as *production costs*. In this chapter we are interested in factor prices as *sources of income for the factor owners*. We're talking about the same thing, but now we're looking at it in a different way.

When a business hands out paychecks, that's *cost* to the business and it's *income* to the employees. Obviously. If the wage goes up and the paychecks get larger, that will cause the business to make some changes. That's what you found out about in the last chapter. But the wage increase will also give more income to the employees. That's what this chapter is going to be talking about.

### What Influences Factor Prices?

In the last chapter we were concerned with *what would happen in the industry* if the price of one or more inputs went up. In this chapter we are more concerned with "what might cause" the input prices to be high or low, or to go up or down. If you can find out *why* the labor of a cherry picker is priced lower than the labor of a bulldozer operator, and why the labor of a school teacher is priced lower than the labor of a bricklayer or an accountant or a pediatrician, then you can understand why cherry pickers and school teachers don't get as large a share of the society's output as do bulldozer operators and bricklayers and accountants and pediatricians.

You know that in the real world, many things influence wage rates and other factor prices. The issue is really very complex. So, for awhile, let's just stick with our "very pure model" so we can see what underlying market forces are pushing and pulling on these factor prices. After you understand the market forces, then we can talk about how the real world doesn't come out just exactly that way.

Before, I said that the question of income distribution in the pure market system is a question of "factor ownership" and "factor pricing."

That's the truth. But in micro-theory we don't need to deal much with the question of factor ownership. We can sort of take it as "given." We can assume that everyone owns some "labor" he can sell, and that the hard-working and thrifty people (and maybe some lucky ones) will own some land and capital, too.

With the "factor ownership" question "assumed away," the question of income distribution becomes entirely a question of the *prices* for the labor, land, and capital in the factor markets. That's what we'll be talking about throughout most of this chapter.

## SUPPLY AND DEMAND DETERMINE FACTOR PRICES

What determines the price of what you have to sell in the factor markets? Supply and demand, of course. If you want to understand the pricing of productive factors (or the pricing of anything else) you need to understand what determines the supply and what determines the demand. Essentially, all you need to do to understand the question of income distribution in the pure market system is to understand factor demand and factor supply.

What determines the demand for (the propensity to buy) a productive factor? The value of what it produces, of course! It depends on the *value* the factor *adds* to the output. If the marginal "value product" of the factor is high, the demand for it will be high. The price paid for it will be high. Whoever owns it and sells it will get a high income and will be able to claim a large share of the society's output.

## An Overview Of Factor Pricing

As you know, the price of each factor of production has a special name. But it's only a price, just the same. If labor is being sold, we call the price "the wage rate." For land we call it "rent"; for capital it's "the interest rate." The fourth distributive share—the one the businessman gets to keep after he has paid everyone else their wages, rent, and interest—that's called "profit."

Profit is the only one of the "distributive shares" which is not really a "price." Profit is what's left over after the prices of all the "hired factors" have been paid. Profit is determined by (1) the value of the output and (2) the cost of the inputs (wages, rent, and interest).

## Who Will Be Rich And Who Will Be Poor?

The size of the wages, rent, interest, and profits will determine which people will get large incomes and which ones will get small incomes—which ones will be rich and which ones will be poor. Most people work for wages. If wage rates in a country are low, most of the people in that country will be poor. If wage rates are high, most of the people will enjoy a high standard of living. But what might cause the wage rates in a country to be high? Or low? *That* is the question.

You already know, in general terms, what would make the wage rate high, or low. If there's a high demand and a low supply of labor, then the wage rate will be high. The high demand will keep the price (wage rate) high. Or you could say it another way. You could say that if the *productivity* (marginal value product) of labor is high, the wage rate will be high. Both these statements are true. Either high demand or high productivity will support a high wage rate. So right away you know there must be some kind of close relationship between high *demand* for labor and high *marginal productivity* of labor. Of course there is. Think about it and you will see how obvious it is. Labor is demanded only because it is productive! The more productive it is, the greater the demand for it. Obviously!

The prices of the other factors (land and capital) are determined in the same way as the price of labor. High demand (high productivity) will bring high prices (high rent, or interest). The owner-sellers of the high-priced factors (labor, land, and capital) will receive big incomes and will be able to claim big shares of the economy's output.

You already knew this much about how the distribution question is answered in the pure market system. But that's just the surface—just the table top. Now it's time to lift up the table cloth and take a good look to see what's going on down there, supporting the surface. The "under the table" questions are concerned with finding out *what influences* the demand for (the propensity to buy) and the supply of (the propensity to sell) labor, and land, and capital.

### The Demand And The Supply Of Productive Factors

Why does a business firm demand (want to "hire") *any* productive factor? Because it's productive, that's why! The more productive it is, the more the business will demand it.

The demand for labor is determined in exactly the same way as is the demand for land or capital—by its productivity—its marginal value product (MVP). All we need to do to understand the *demand* for *any* productive factor is to understand what determines it's *productivity.* That's easy enough to see. We will get into that in just a minute. But first, what about supply?

Factor *demand* is determined the same way for all the factors, but not so for factor *supply*. The supply of each factor is unique. Think about it. Do you think the labor supply is determined by the same things that determine the supply of land? or capital? Of course not. For each factor, to what extent will the supply (the propensity to sell) be responsive to the price? That's really the question.

To understand income distribution in the market system, first you need to understand the demand for productive factors. The same concept of demand applies to all the factors. Then, as you look at each factor individually (labor, land, and capital) you need to find out about the unique determinants of the *supply* of each. Since the demand determinants are the same, the difference in analyzing the pricing of productive factors, from one factor to another, must result from supply differences. That's right. But you may be surprised how similar these (obviously unique) *supply* functions can be! We'll get into factor supply soon, but let's discuss factor demand first.

## WHAT DETERMINES FACTOR DEMAND?

This section explains what determines the demand for a factor of production. That's an essential step in understanding income distribution in the market system. But before we go on, one word of warning: *No society distributes income entirely according to the "pure market forces."* Still, it can be very helpful to understand what the pure market forces are, and how they *would work* in a model market system. That's what we're going to be talking about now. Later on we'll get into the question of how all this gets modified in the real world. But for now, just understand the pure theory—the distribution forces at work in the *model pure market system.*

### Factor Demand Reflects Marginal Productivity (MVP)

When you are operating your jeans plant, how many people will you hire—that is, how many "units of labor per day" will you buy? If you think an additional unit will bring you more "value product" than it costs (if it adds more to your revenue than it adds to your costs) you will buy it. So your "propensity to buy" a unit of labor is determined by your estimate of its marginal value product. Here's an example.

Suppose the kind of labor you use in your plant costs $20 a day. You are currently employing 25 people (using 25 "units of labor" per day). Should you increase your daily rate of labor input by one unit? Should you hire 26 people a day? That depends on how much more revenue you expect to get. Right? Suppose with one more worker (without increasing any of your other costs) you could produce enough more to increase your total revenue by $25 a day. Would you hire one more person? At a wage rate of $20 a day? Sure you would!

When you increase your labor input and add more to your revenue than you add to your costs, that increases your profit. That's good business. You will continue to hire more labor and expand your output as long as the marginal value product (MVP) is greater than the wage rate—that is, as long as the marginal "value product" is greater than the marginal "input cost" (MIC). But as you hire more and more labor, the MVP goes down more and more. When the two become equal (when MVP = MIC), that's the "rate of production" where your profit is at a maximum. You won't hire any more.

Suppose you have adjusted your output to the most profitable daily rate. That means you are using the most profitable daily amounts of labor and other inputs. You're making as much profit as you can. At that rate of output marginal cost equals marginal revenue, and for each factor of production, marginal "input cost" is equal to marginal "value product." You are operating at your "maximum profit" rate of output.

Now suppose the market price of hip-hugger jeans goes up from $100 a carton to $150 a carton. What happens to the productivity of the labor you're hiring? and to the productivity of your land? and capital? It goes up! When the price goes up for the product, suddenly the output of each factor becomes more valuable. Suddenly your demand for variable factors goes up! See how an increase in the demand (and price) for your output can bring an increase in your demand for inputs?

Your demand for labor or for any other input factor depends on (1) the amount of "real physical product" added by the marginal factor, and (2) the amount of *money* the extra "real physical product" adds to your revenue. Anything that (1) increases the amount of real product an input factor can produce for you, or (2) increases the price at which the real product is sold, increases the *marginal value* of that factor to you. So it increases your demand for (your "propensity to buy") that factor.

### Factor Demand Is Always Derived Demand

You demand labor and other inputs for your jeans factory because somewhere, consumers are demanding hip-hugger jeans. If consumers were not demanding hip-hugger jeans, you certainly wouldn't be demanding labor (or denim or machines or electricity or anything else) to make hip-hugger jeans. Obviously not. You remember this from the previous chapter. But a brief review won't hurt.

There are only two reasons for demanding something: for *consumption* (to bring "final satisfaction"), or for *production* to be used to produce something else that is wanted). The *production demand* (demand for input factors) always reflects the fact that somewhere there is a *consumption demand*. If there was no *consumption demand* for a product, then you can be sure that there would be no *production demand* for factors to be used to make it.

When we talk about the productivity of an input factor, we are talking about the *value* a unit of that factor adds to the output. *All that value derives from the demand for the final product*. If there's no final product demand, there's no derived demand and there's no value created. Nobody will hire a worker to produce something that no one will buy! Everybody knows that. It's just common sense.

### Marginal Productivity (MVP) Reflects The Law Of Diminishing Returns

What determines how much more real output you will get if you hire one more unit of labor? The most important thing is this: "How much *other factors* is the new worker going to have to work with?" If the worker is going to have lots of efficient machinery and plenty of raw materials to work with, then he will add a lot of extra product. If not, then he won't. He can't.

Suppose all the machines and everything else are already being overused. All there is for the new worker to do is to go around and clean up and straighten up a little bit. Then the extra worker isn't going to add very much product. We're talking about the law of variable proportions—the law of diminishing returns. Right? Sure. The "inexorable law of diminishing returns" has a *very* important influence on income distribution, both in the model market system and in the real world. It really does.

If you expand your output of hip-hugger jeans beyond the maximum efficiency rate of 10 cartons a day, your marginal and average costs are going to rise. Why? Because the extra output you get when you hire an additional worker, is diminishing. The output diminishes because you are trying to produce faster than your plant is designed to produce. The extra workers don't have enough fixed inputs to work with. So they aren't very productive. You probably wouldn't hire extra workers and push your output beyond 10 units, anyway. But then again, you might.

What might cause you to hire an additional worker, even when you know he isn't going to add very much output? Either one (or both) of two things: (1) a very high price (per unit) for the output, or (2) a very low price (per unit) for the input (labor). We could say it this way:

(1) The demand for your product in the market might go up, and push the price up so high that the little bit of extra product added by the extra worker would bring in enough extra revenue to more than cover the extra cost of hiring him; or

(2) The price of labor might go down so low that even though the additional worker doesn't add much output (value product), still he adds enough to more than cover his very low wages.

As the product price goes higher, or as the price of the input factor goes lower, you will hire more of the input factor. How much more? Enough to push down the marginal productivity (marginal value product — MVP) of the factor to where it is no longer profitable to hire any more — that is, to where it costs you as much to hire an extra unit of the factor as the extra output is worth to you. Once you get to that output rate, you don't hire any more.

Can you see that the *law of demand* applies in the factor markets just as it applies in the consumer goods markets? Sure. The lower the price of the factor, the more the businesses will buy. Why does the law of demand hold true in the factor markets? Partly because of the *law of diminishing returns*. As more is hired, the marginal value product goes down; so the price of the factor must go down to get more to be hired!

## Short-Run Factor Demand Reflects Diminishing MVP, And Output Demand

What about the *elasticity* of the demand for labor? or for any other factor of production? That would depend on how fast the marginal value product of the factor diminishes. Get it? If the productivity of the factor *diminishes rapidly* as more is hired, then the demand for that factor (in the short run) will be *highly inelastic*. But if you hire more and more of some factor (say labor) and each additional unit adds almost as much value product as the previous one did, that means that the returns are not diminishing so fast. So the demand for that factor will be more elastic.

You can see that the elasticity of the demand for the final product has a lot to do with all this. Suppose the product demand is *highly elastic*. (Remember Figure 22-5?) If the price of the final product drops a little, people are induced to buy a lot more, so a lot more inputs will be demanded.

It works this way. When input prices go down, cost of production goes down, so supply increases. The increased supply pushes the selling price down and (if the product demand is highly elastic) a lot more output will be bought. So a lot more output will be produced and a lot more inputs will be hired. See how the lower input cost causes a lot more output to be bought? And that causes a lot more inputs to be bought? Sure. That's how the highly elastic *product demand* causes the *input demand* to be highly elastic too. Of course.

## In The Long-Run, Factor Prices Influence Economic Growth And Change

Another thing you can see right off is that industries will respond to factor prices *much more* in the *long run* than in the short run. Suppose the wage rate goes up. You may cut back a little and release two or three workers. But in the short run you can't cut back very much. It would be better to shut down instead. Either way, you're going to lose money. The higher cost of labor pushes you into a loss situation. What are you going to do about it?

All you can do in the short run is try to minimize your losses. But as soon as you can you are going to get out of this business (or else you will install labor-saving machinery and replace a lot of workers). See how the demand for labor is more elastic in the long run? The quantity bought will respond a lot more to the price increase, if you'll wait awhile, until the industry can adjust.

In all the modern nations of the world, wages have been rising for many decades. As the wage rates have gone up, businesses have brought in more capital. With more capital to work with, labor has become more productive. The higher productivity has justified even higher wages. Each *higher wage rate* has made it profitable to bring in *more labor-saving capital;* each increase in the "capital-to-labor ratio" has made *labor more productive* and has supported *even higher wage rates.*

This is an important part of the "economic growth" process which has brought such big increases in the standard of living in all of the advanced economies of "mixed socio-capitalism." Those nations in which this process has been most rapid are the ones in which, now, the people have the most free time and the most things. See how important the *long-run elasticity* of factor demand can be?

## The Factor Demand Curve

Would it be possible to draw a curve to show the demand for a productive factor? No question about it. We could call it a "diminishing returns curve," or a "marginal value product curve," or a "factor demand curve." Either way, it's the same curve. You already understand why. After you finish studying Figure 23-1 you will understand it even better. You should take time to study it now.

## Elastic? Or Inelastic Demand?

The factor demand curve in Figure 23-1 shows an interesting example of unitary elasticity of demand. But it really wouldn't be likely for a firm's

**Fig. 23-1 THE DEMAND FOR A FACTOR OF PRODUCTION**

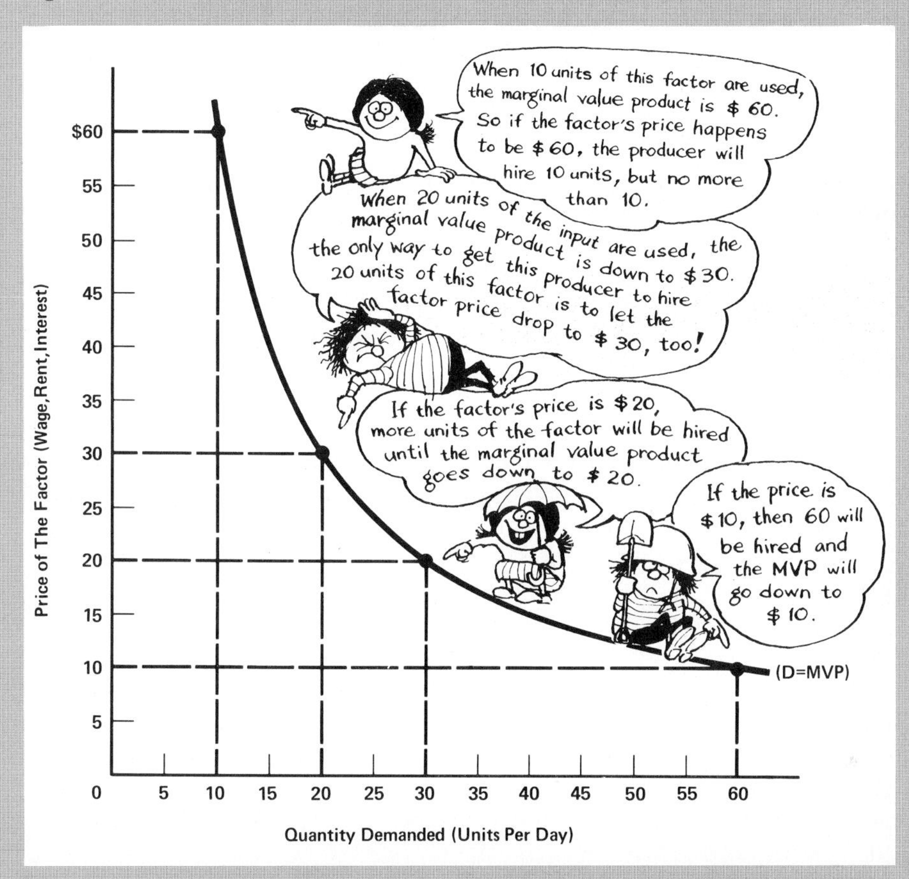


The Law Of Demand Works In The Factor Markets, Too!

As the marginal value product of the factor diminishes, the price must go down to induce the businessman to use more of it. As the price of the factor goes down, the businessman uses more, so its marginal value product diminishes. *How much more* he will use depends on *how fast* the marginal value product diminishes.

If the price is high you will add only a few units of the factor, so the marginal value product of the factor will be high (high enough to be worth the high price). As the price goes lower you will add more units of the factor, so its marginal value product will diminish. The lower the price of the factor the more you will use and the lower its marginal value product will be.

This curve just happens to show a "unitary elasticity" factor demand curve. Any decrease in price would be exactly offset by the increase in quantity bought, so the total amount spent for this factor would always be the same ($600), no matter how high or low the price might be.

This curve will shift only if there is a change in "the marginal value product situation" for this factor. The curve will increase (shift to the right) if (1) the selling price of the output goes up, or if (2) more of the *other* factors become available for this one to work with, thus making this factor more productive.

demand for a variable factor to have "unitary elasticity." It would depend on how critical the factor proportions (the "factor input mix") happen to be for that firm.

Suppose that in order to produce anything at all, each worker's position in the plant *must* be manned. Then the factor proportions of labor and capital are absolutely fixed. The plant's short-run demand for workers will be absolutely inelastic, right up to the "shut-down factor price," (where the factor price gets so high it would be cheaper for the firm to shut down than to operate).

On the other hand, suppose we're talking about a trucking company that can lease more trucks and hire more drivers at a moment's notice. There really isn't much "fixed factor" in this case, so the returns don't diminish much. The trucking company can hire a lot more "variable inputs" with very little "diminishing" of the marginal value product, so the firm's demand for the variable inputs (trucks and drivers) probably would be elastic. The short-run demand for labor and other variable factors would likely be inelastic for most manufacturing plants. Whenever the marginal cost of expanding the output rises rapidly, that means the marginal value product diminishes rapidly. The demand will be highly inelastic.

Most plants are designed to use a certain amount of labor and other variable inputs. If the price of any one of these inputs goes up or down the quantity can be adjusted some — maybe some of one variable input can be substituted for some of another — but it isn't likely that the "input mix" will be changed very much in the short run. But what about in the long run? That's a different matter!

### Long-Run Factor Demand Is More Elastic

In the long run new plants will be built. The new plants will fully reflect the relative prices of all the inputs. When a businessman is designing his new plant he insists on using minimum quantities of the highest-priced factors and much more of the lower-priced factors. This is where the demand for each factor becomes really elastic. So this is where the rationing function of price really works to conserve the scarcest and most valuable of society's inputs.

As time goes on, the higher-priced factors will be conserved more and more. They will be carefully economized — used only for their most highly productive uses, and with lots of other input factors to work with. So what happens? They become even more productive, the demand for them increases (shifts to the right) even more, and their price goes up even more. And the processes continue. The scarcest, most productive, highest priced factors tend to become even more "relatively scarce," more productive, and higher priced as time passes. This is what has happened (and is happening) to labor in all the advanced countries in the world.

In the long run, factor demand is likely to be relatively *elastic*. In the short run it's likely to be relatively *inelastic*. This question of the elasticity of the demand for a productive factor, can be very important. You can see why.

The short-run elasticity of the demand for labor is important in influencing employment in the economy. Suppose there's a lot of unemployed labor. If the short-run demand for labor is highly *elastic*, then a small drop in the wage rate will induce businesses to hire a lot more people. The unemployment will be eliminated. But if the short-run demand for labor is highly *inelastic*, then the wage rate will have to go down quite a lot before it will induce the businesses to hire all of the "surplus labor." See how elasticity (responsiveness of buyers to price changes) can be very important in the factor markets, in the short run as well as in the long run?

The more elastic the demand for the factors of production, the more readily the businesses adjust to changing conditions in the factor markets. Elasticity means responsiveness to price. If the businesses are highly responsive to changes in factor prices then the price mechanism is working just great. But if they aren't very responsive to price changes, then the market process (the price mechanism) isn't working so well.

Suppose nobody responded at all to price changes. Elasticity would be zero. Nothing would shift around in response to price. What about the market process then? It wouldn't work at all. Elasticity is essential if the market process is going to work. But you already knew that. Right?

## SCARCITY AND HIGH PRICES MAKE A FACTOR MORE PRODUCTIVE

The greater the marginal value product of a factor, the greater will be the demand for it. Of course. But did you know that *the high marginal value product of a factor may simply be a reflection of the high price of the factor?* The rationing function of price will cause a high priced factor to be carefully economized—to be used very sparingly. If a factor is used very sparingly, its marginal value product will be high. So a factor's price can be the cause of (not just the *result* of) its high marginal value product. Causality goes both ways! Get it?

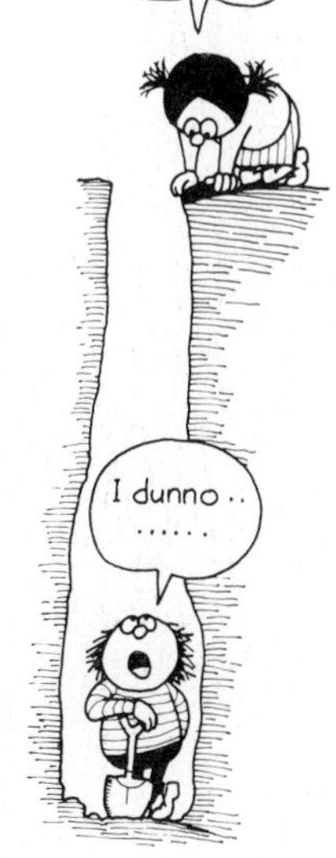


### The Scarcest Factor Has The Highest Marginal Productivity

*The productivity of an additional unit of any factor of production depends on the availability of "other factors."* If a lot of capital and land are available, then an added unit of labor will be highly productive. If a lot of labor and land are available, then an added unit of capital will be highly productive.

As the quantity of any one factor is increased, it becomes increasingly important to get more of the *other* factors to work with it. If somebody is running a big farm and has lots of land and labor but only one tractor, then he will be much more interested in getting another tractor (more capital) than in getting more land or more labor. Why? Because an additional tractor would be much more highly productive—would add much more value to his output than would more land, or labor. Why? Because the tractor is the relatively scarce factor, that's why!

Which factor are you willing to pay the most to get more of? The relatively scarce one, of course. Your demand for the relatively scarce factor is higher because of its higher productivity. An extra unit of the relatively scarce factor would add more to your "value product." That's why you will pay more to hire it, and its owner will receive a high income.

These concepts apply just as much to *specific* factors as they do to "labor, land, and capital *in general*." If the *kind* of capital or land you own is relatively scarce, or if the *kind* of labor you have for sale is relatively scarce, then the marginal value product of your kind of capital or land or labor, will be high. The demand for your inputs will be high and your income will be high. You will get a large share of society's output.

### An Example Of "The Relatively Scarce Factor"

Suppose you live in the "Grape-belt" area of Chautauqua County in western New York State and you're the only person in the county who has a grape-picking machine. You may be sure that when harvest time comes you will have plenty of opportunity to keep your machine running day and night. Your relatively scarce piece of capital will have high productivity. It will be in very high demand. You will be able to make a lot of money with it. One more grape-picking machine would have a high marginal value product in Chautauqua County! Chances are somebody else (maybe everybody else?) will buy one soon. Maybe *very soon*.

Now, suppose every grape farm in Chautauqua County finally has its own grape-picking machine. But suppose you are the only man in the county who knows how to repair one. You may be sure that the marginal productivity of your labor services will be very high. Your efforts will be in great demand. You will make a lot of money.

The marginal value product of an additional repairman in your county would be very high. Chances are that pretty soon someone else will learn to fix those monsters. Or else some skilled mechanic will move in from Canandaigua or Bass Island, or maybe even from St. Joseph or Roseburg or Fresno or Yakima! Why? You know why. Because the "production motivating function" of the high price for his labor will pull him to Chautauqua County, that's why!

What about the marginal productivity of grape-picking machines in Chautauqua County, now that everybody has one? It's much lower. Still, if the cost of buying and owning capital is low enough (as compared with the high price of labor), more machines will be bought. The reason all the farmers in the United States and Canada and in the other advanced nations are using so many of these mechanical harvesters—grape-pickers and corn pickers and tomato-pickers and bean-pickers and cotton-pickers and pickle-pepper-pickers—is because of the relative scarcity (and the high price) of labor. Each time farm workers get higher wages, more "mechanical pickers" take over.

The high-priced factor is conserved. It's used very sparingly, so an extra unit would have high marginal productivity (like skilled grape-picker repairmen in Chautauqua County). The low priced factor is used more freely and abundantly. It is used so freely and abundantly that the marginal productivity of another unit of it would be low (like grape-picking machines in Chautauqua County).

## High Prices Make Factors More Productive

Remember what happened to eggs when the price went up to $2 a dozen? Only the true egg-lovers (or the filthy rich) kept on buying eggs. To them, eggs are really valuable. How valuable? Worth $2 a dozen of course! When eggs are *priced* at $2 a dozen, then you can be sure that eggs are *worth* $2 a dozen to all the people who are still buying them. That's obvious, isn't it? The same principle holds true for the business-man who is buying labor, land, and capital.

In the real world, is it true that you always get what you pay for?" In one sense, not necessarily. But in another sense, yes. Some people spend their money more carefully than others, and they get more "value received." I'm sure you know that. But you also know that whenever a person buys something, he *thinks* what he's getting is going to be worth as much to him as he pays for it. Otherwise he wouldn't buy it. Obviously! Nobody would pay more for something than he thinks will be worth to him!

Let's say it this way: "No one buys anything unless he expects to get at least as much out of it as he pays for it." This says that the "rationing function of price" will restrict the purchase of everything (every resource, every good, every input), so that *nothing will be bought for any purpose (consumption, or production) unless the buyer thinks the use is going to be worth at least as much as the cost.*

In the factor markets, it says this: No businessman will pay a dollar for a unit of input unless he expects the unit of input to add at least a dollar's worth to his output. If a factor is selling for $10, then you can be quite certain that the marginal value product of that factor is at least $10. If not, it wouldn't be hired. The $50 factor has a marginal value product of at least $50. It will be used sparingly enough to *ensure* that its marginal productivity is at least $50.

High-priced factors will be used very sparingly. They will be used only in highly productive uses, and they will be used together with an abundance of other factors. So *a high price for a factor will guarantee high productivity for that factor.* It is true in the model market system and it is true in the real world. (Exception: "Resource administrators" sometimes forget this principle and let a $20,000 a year professor or executive do the work that a secretary or student assistant could do.)

In a nation, a city, or any area where wage rates are very high, labor will be conserved. Lots of labor-saving machinery will be used. The productivity of labor will be high. If the wage was high and the productivity of labor was not high enough to cover it, the labor would not be hired. If it was hired, the businesses would go broke.

**Factor Prices Influence Production Techniques**

The way production is organized, and the production techniques used in a society, always reflect the *relative scarcities* and the *relative prices* of the productive factors. In the United States and other advanced nations we use steam shovels and conveyor belts. In countries where labor is very cheap they use hand shovels and wheelbarrows. If lumber is plentiful and low-priced, while steel and other building materials are very scarce and high-priced, the society will cut down the forests and build things out of wood. But if lumber is very scarce and high-priced and other building materials are plentiful and low-priced, lumber will be used sparingly and other building materials will be used instead.

You can think of all kinds of ways that production would be organized differently if we had different prices on our labor, natural resources, and capital. If you look at a highly developed country and compare it with an underdeveloped country, you will see immediately the many differences in the way production is organized. In each case the production organization reflects the relative scarcities, the relative prices, and therefore, the *relative productivities* of the factors in that economy.

**The LDC's Use More Labor; Labor Productivity Is Low**

If you look at the way production is organized in the less developed countries (LDC's) it's easy to see the difference. It's evident in agriculture, mining, manufacturing, transportation, and even in the way the households operate. The most striking difference is the abundant use of labor and the limited use of capital in the LDC's. If we did things in the advanced nations the way things are done in the LDC's, that would be highly inefficient—very wasteful of our scarce and valuable labor. But if the LDC's organized their productive activities the way we do, that would be highly inefficient for them—very wasteful of their scarce and valuable *capital*.

We're doing it the right way for us; they're doing it the right way for them. But the different "factor scarcities" and the different organization of production has a very big influence on the "average productivity" of labor (that is, on the society's output per man), and on the distribution of income. Where labor is in great supply (relative to the other input factors) wages are low and "the people in general" don't get very much product.

When labor productivity is low, that means there isn't much output per person. So it isn't possible for each person to *have* much output. Obviously! There's just no way! Why is labor productivity so low? Because it isn't combined with highly productive capital and resources. Since labor is so plentiful it isn't carefully conserved. Its marginal value product is low, and the incomes (and output shares) of the workers are low.

(Note: *Highly skilled* labor—maybe tractor mechanics or grape-picker mechanics—may be more scarce, more productive, and higher-priced in the LDC's than in the more advanced nations. The LDC's need to develop capital, but they also need to develop the *kinds* of labor needed to operate and maintain the machines and equipment.)

## Demand For One Factor Reflects The Supply Of Other Factors

By now you are already aware, at least implicitly, of the importance of *factor supply*. We have been talking about factor demand, but *supply* seems to keep creeping into the discussion. The demand for an input is determined by its marginal productivity. But its marginal productivity is determined by how much of it is being used already (relative to the *other* factors it's working with). How much of it is being used already is determined by whether it is a relatively "high-priced" or "low-priced" factor. And its price reflects its *relative scarcity*—that is, its *supply*. See how the supply and the demand are interrelated? all tied together?

It is the productivity which determines the demand and puts the price on the factor, yes. But what is the most important thing that determines the productivity (and the demand) for a factor? The relative scarcity. That is, the *supply*! If one factor is *relatively plentiful*, then it will be used in abundance. Its productivity will be pushed down *low*. Its price will be low. But if a factor is *relatively scarce*, it will be used sparingly, and its productivity and price will be *high*.

Suppose the price of a factor is high. If the supply of the factor is free to expand, then as time passes the supply will increase the price will fall. The marginal productivity of this factor will get lower and lower. That's what has happened in the case of capital in the United States and in the other highly developed nations. Capital has become more plentiful, so labor and natural resources have become more scarce. Labor and resources are now higher priced (and are economized more) than in the past.

You can see that it could be difficult to separate completely the supply from the demand for productive factors. But it is useful (for some purposes, essential) to make the distinction. As we get into this next section, you will understand why.

## THE SUPPLY OF PRODUCTIVE FACTORS

How much of each factor (labor, land, capital) will be available and offered for sale in each market, at each price (wage, rent, interest)? That's the "factor supply" question. The answer is different for one factor or another and it depends on several things. Partly it depends on how we choose to look at the question.

### Factor Supply To The Individual Businesses

Suppose a small business wanted more labor or land or capital. If it offered to pay more, could it get more? Sure. So the supply of any factor to any individual business is likely to be highly elastic. The quantity offered for sale to an individual business (or to any one *industry* for that matter) probably will be quite responsive to a change in the price.

If a firm or an industry is expanding, it must be pulling in more input factors from somewhere. Factors shift from one industry to another. When higher prices are offered in the egg industry, more inputs are supplied. But we can't understand the income distribution question for the society as a whole by looking at the individual firm, or industry. For example, if we want to find out what market forces are setting "wages in general," we can't figure it out by looking at the egg industry. To get at the issue of "factor price determination" for the economy as a whole, we look at the nation's *total supply* and *total demand* for each factor.

A minute ago we were talking about what influences the "total productivity" and "total demand" for a factor. You saw why the "total productivity and demand" for labor in the LDC's is so low. Remember? So now let's talk about what determines the *total supply*. When we finish that, you will be able to begin to see how the income distribution question for the society is worked out.

### Factor Supply For The Entire Nation

For the nation as a whole, the supply of each input factor is *highly inelastic*. At higher prices how much more *total* labor, or *total* natural resources, or *total* capital would become available? Not very much.

The total supply just isn't very responsive to price. In a short period of time the total supply may not be responsive to price at all!

The nation's total labor supply might respond a little bit to increases or decreases in the wage rate, but we aren't sure about that. Higher wages may induce more housewives to join the labor force; on the other hand, higher wages might induce more housewives to quit their jobs and live on their husband's incomes. Higher wages may convince more students to quit college and take jobs; but higher wages may make it possible for students to make it through college by working only part time, or on their parents' high income. Low wages might even force students to quit college and take full-time jobs. (The "production motivating function of price" may work in reverse!)

See how hard it is to say anything definite about how "the quantity of labor offered for sale" responds to *price?* As the wage rate gets higher some people will offer more of their labor for sale, but others will offer less. In the low-wage countries, people work longer and harder than in the high-wage countries. In the United States where wages generally are higher than in any other country in the world, a large number of adults are not even in the labor force (wives, students, retirees, etc.). Those who are in the labor force work fewer "hours per year" than in low-wage countries. So we can't say that our high wages are inducing people to offer more labor for sale. It seems to be working the other way!

### The Backward-Bending Supply Curve For Labor

We really don't know very much about how the "quantity of labor supplied in the nation" would change, if all wage rates went up or down. Some studies suggest that if wages start out very low and then increase a little bit, more labor will be offered. But then if the wage continues to increase, after awhile people start dropping out of the labor force, or demanding a shorter workday or workweek and more time off. This leads to the idea of the "backward-bending supply curve for labor."

Figure 23-2 shows a "backward-bending supply curve." It shows how the wage rate in this "hypothetical nation" would be determined. But don't put too much stock in this backward-bending supply curve. The truth is that we really don't know much about the relationship between "the wage rate in general" and "the quantity of labor that would be supplied." So, remembering this word of warning, it's time now for you to study the graph.

### The Supply Of Land And Natural Resources

It's easy to see that the nation's supply of labor is not going to be very responsive to wage rate changes. It's more or less fixed by the size,

Fig. 23-2 **THE SUPPLY AND DEMAND FOR LABOR**

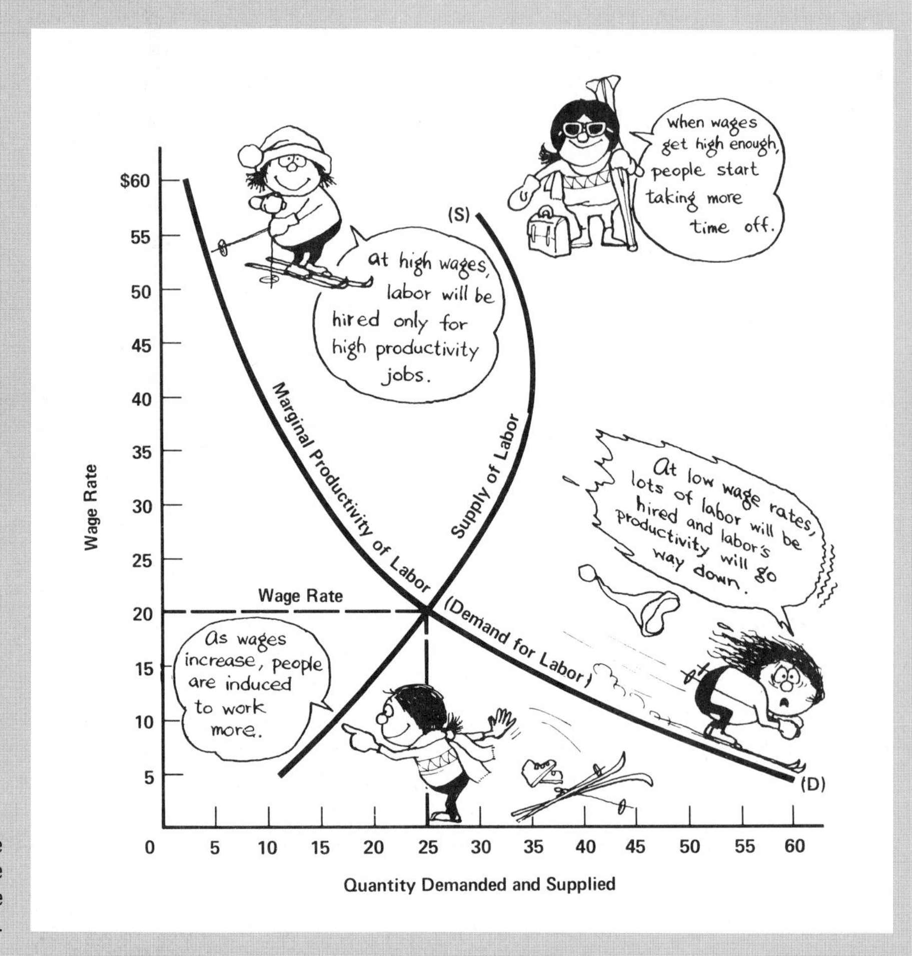


After The Wage Rate Gets Just So High, The Labor Supply Curve May Bend Backward.

If "the wage in general" goes up for "labor in general" this may get more people to take jobs, and workers may be willing to work overtime and to forego vacations. But if the general wage rate in the nation keeps on going up, after awhile some people will start cutting down on their work time so they can take time to enjoy all those things they can buy with their high wages.

age, skills, habits, economic conditions (etc.) of the population. What about the nation's supply of land and natural resources? That's even more fixed and unresponsive than labor. So the production motivating function of price certainly isn't going to have much influence on the *total quantity offered*, either of labor, or of land and natural resources. On labor or land for a *specific purpose*? Yes. But on the *total quantity offered*? No.

What about the nation's supply of capital? Does that respond to price? Capital is made up of *produced* goods. So surely it would seem that a higher price for "capital in general" would soon bring forth more capital. Right? Well, don't be so sure. Better think about that one for a minute, too.

## The Supply Of Capital

How does a nation increase its supply of capital? By saving (not consuming) and investing (building capital). Capital formation requires less consumption. What would make people consume less and businesses invest more? That's the key to capital formation.

If capital is bringing a high return to its owners, that means the interest rate is high. Will that induce more people to "not consume"? To save? Maybe so. Then again, maybe not. A high return to capital, just like a high wage rate, might work either way. Some people might save more (to build more capital to get the high return); others might save less (the capital they already have now brings them enough income, so they're all set).

For capital in any particular industry, or for any one *kind* of capital—grape-pickers or IBM computers or shoe machines or gear grinders—the supply will be very responsive to demand changes (to the production motivating function of price). But for the nation's *total supply* of capital, it isn't at all certain that the price (interest rate) is all that important. Why do people save more (or less) this year, than last? Many reasons. Some of the reasons may be much more important than the "price" they get paid for saving (the interest rate).

## How To Handle The Problem Of Factor Supply

Here we are, faced with a dilemma. If we're going to understand the market forces which influence income distribution, then we must understand factor pricing—that is, factor demand and factor supply. But we don't know much about factor supply. About all we really know is that for the economy as a whole, factor supply is not very responsive to price changes.

Neither the supply of labor, nor land, nor capital responds much to factor price changes. Maybe that's all we need to know! Maybe there's no dilemma after all. Let's see.

Suppose we assume that the elasticity of the nation's supply of each of the "factors in general" is *zero*. That would mean "zero responsiveness of the total quantity to a price change." *Absolutely inelastic demand.* The quantity available would be the same whether the price was high or low. We know this assumption isn't exactly true; still, it isn't so far from the truth, either. If it will help us to understand factor pricing—that is, income distribution in the market system—then it's a good assumption. And it does, so it is. Just watch how neatly all the pieces now fall together.

If the supply of each input factor is absolutely fixed, then what will determine the price of each of the factors? *Demand,* or course! You know that the demand for each factor is determined by its productivity. But the *productivity* of each factor (as you so well know) is determined by the availability of *other* factors.

The productivity of *labor* is determined by the amounts and kinds of land and capital available. The productivity of *land* is determined by the amounts and kinds of labor and capital available. And the productivity of *capital* is determined by the amounts and kinds of labor and land available. See where all this is leading us? It's taking us directly to the answer to the distribution question!

The productivity of (and therefore the demand for) each factor, is determined by the supply of the other factors. So if one factor is relatively scarce, that means that the other factors are relatively plentiful. The productivity of the relatively scarce factor (and the demand for it) will be high. And the price? High, of course! The strong demand and the limited supply will keep the price high. Whoever owns and "sells" some of this relatively scarce factor will receive a large income and will get to have a large share of the output of the society.

### Demand And Supply Curves For Factors Of Production

Figure 23-3 shows a graph of the factor demand curve and the "absolutely inelastic" supply curve. This graph could apply to any one of the factors. You should study it now.

From studying Figure 23-3 you can see why many people in the less-developed countries are so interested in building capital and developing their natural resources. If the people of those countries (or of any country) are going to receive high wages, then there must be an abundance of good land (resources) and/or capital for the people to work with. Economic development, which the less-developed countries are so relentlessly striving for, is the process of building capital and developing and making use of the nation's resources. With more capital and resources to work with, labor will be more productive. Output per person will be higher. The people will be able to have more things.

Fig. 23-3 **THE SUPPLY AND DEMAND FOR A FACTOR OF PRODUCTION**

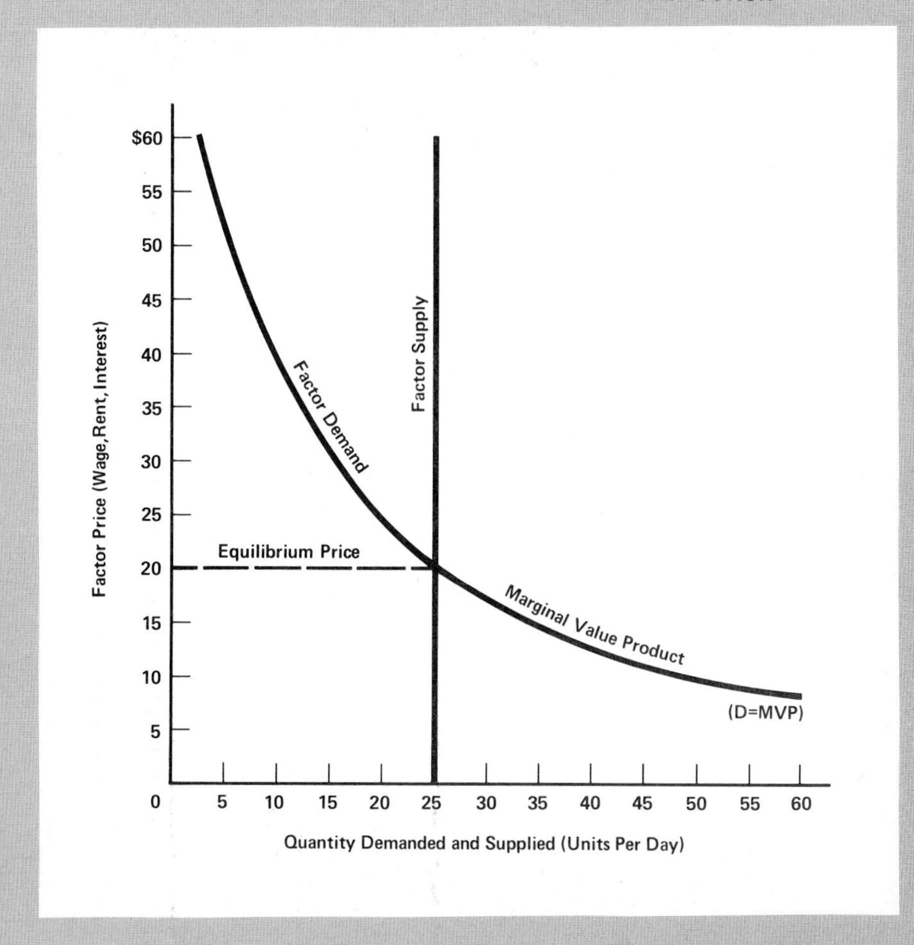


The Total Supply Of A Factor Of Production Usually Isn't Very Responsive To Price. That Is, It's Highly Inelastic.

The factor supply wouldn't really be a straight vertical line, but this is a useful approximation, at least for prices not too far from "the equilibrium price." (At very low prices the factor owners might not offer any of their factor for sale!)

Any increase in the amounts of other factors available would cause this factor's demand curve (MVP) to shift to the right. The price of the factor would increase.

Anything that would cause the supply of this factor to increase (shift to the right) would lower the price of this factor, and would automatically increase the productivity of the other factors.

Anything which would cause the supply of this factor to decrease (shift to the left) would raise the price of this factor, and would automatically decrease the productivity of the other factors. For the economy as a whole, the productivity of each factor depends on *the supply of the other two factors.*

## THE MARKET MODEL AND THE REAL WORLD

Figure 23-3 can help you to understand why the general level of wages, rent, or interest in one economy or another, would be high or low. But you know that this doesn't answer all of the "real world" questions, not by a long shot. It just gives us a place to start.

### There Are Many Kinds Of Labor, Land, And Capital

Each particular kind of labor, land, or capital has its own relative scarcity, its own marginal value product, and its own "supply and demand situation." One kind of labor, or land, or capital may be highly productive, while another may be of no value at all in the production process. While the total supply of labor or land or capital may not respond at all to a change in demand and price, *each kind* of labor, land, or capital may be quite responsive to price changes. If the price of one kind of labor is unusually high (like grape-picker mechanics) this will "motivate the production" of more of this kind of labor. But in the meanwhile, the owner of the high-priced kind of labor will enjoy an especially high income.

It is in this "constant shifting about" within the various factor categories that the production motivating function of price really does its work. It entices the factors to do the kinds of things most valued by the society. It stimulates the development of new and better labor skills, and new and better capital. It even stimulates the discovery of more of the most productive kinds of natural resources. The person who is most successful in developing the kind of labor which is highly valued in the market (or the person who gains ownership of the kinds of land or capital which are highly valued) will receive a high distributive share of the society's output.

### The Normative Question Of Income Distribution

In this chapter, most of the time we have been talking about income distribution in the pure market system. The theory really does relate to what goes on in the real world but it doesn't answer all the questions. It doesn't get into the issues of extreme affluence or poverty. It doesn't talk at all about how the society's income and output *should* be distributed.

Most people, in every society, seem to agree that the distribution of income in their society is something less than perfect. How bad is it? What should be done about it? Why isn't a better distribution arrangement being worked out? Most societies are working on this problem, constantly. Each price change or wage change makes a difference. Each tax or social security or unemployment compensation or "medical program" change is a change in the distribution system. Each manpower training program, minimum-wage law, and anti-poverty program exerts some influence. Yes, people are always trying to work through the political process to change the distribution system. And sure enough, it's being changed all the time.

The issues we're talking about now could take us far beyond the pure theory of income distribution. This book will not be able to go very deeply into all of them. But in the final Part (Part Eight) there's a chapter that deals with real-world income distribution problems. For now, be sure you understand what the "distributive forces" are and how they work in the "pure market system."

## MICROECONOMICS, PURE COMPETITION, AND THE REAL WORLD

For the past five chapters you have been studying and learning about "the pure theory of microeconomics." You have found out how demand and supply work to set prices, and how prices work to ration things and to motivate the production of things. You know how the *costs* of a business firm change as the *output* of the firm changes. You know quite a lot about "the theory of the firm." You also know a lot about how the forces of the market process influence income distribution in the society. In fact, you know quite a lot of microeconomics!

Almost all of the principles explained in these five "micro-theory" chapters rest on the assumption of pure competition. That means there must be many buyers and sellers, all responding to price, freely entering and leaving all markets—all the product markets and all the factor markets. The prices respond to demand changes and cost changes. Then the buyers and sellers respond to the price changes. Everything works out just right. That's the way it works in the model. The real world is different.

The model system, made up of markets with "pure competition," is different in many respects from the real world. The four chapters of Part Seven will explain several of the differences. First, be sure you understand the pure theory. Then go on into Part Seven and find out about the things that can go wrong when we try to take our "theoretical results" into the real world.

---

**REVIEW EXERCISES**

**MAJOR CONCEPTS, PRINCIPLES, TERMS (Try to write a paragraph explaining each of these.)**

how MVP influences factor demand
how the law of diminishing returns influences factor demand
how time influences factor demand

how "output demand elasticity" influences factor demand
how factor prices influence economic growth and change
how relative scarcity influences factor prices
how factor prices influence factor productivity
how factor demand is determined by other factor supplies

## OTHER CONCEPTS AND TERMS (Try to write a sentence or phrase explaining what each one means.)

marginal value product (MVP)
marginal input cost (MIC)
the normative question of income distribution
pure competition

## CURVES AND GRAPHS (Try to draw, label, and explain each.)

The Demand for a Factor of Production
The Supply and Demand for Labor
The Supply and Demand for a Factor of Production

## QUESTIONS (Try to write the answers, or jot down the highlights.)

1. In the real world, each "factor of production" (labor, land, capital) is really made up of a great number of different factors of production. Each *kind* of labor is really a different "factor of production." How different is your "labor" today than it was five years ago? How about five years from now? What are you doing these days to try to develop a more productive, more highly valued kind of labor, so the "price" you can get for your labor will be higher in the future? Discuss.
2. Explain as much as you can about what influences (1) the demand for a factor of production, and (2) the supply of a factor of production.
3. Suppose we're talking about a completely planned economy—the market process has no influence whatsoever. In planning the production activities for each industry, should the planners try to use each kind of labor, land, and capital in the same proportions as the market process would direct? or not? Also, do you think the planners should try to distribute the income among the members of the society in the same way the market process would direct? Explain.
4. Now that you know something about how the "prices" of labor and of capital are influenced by relative scarcity, think back over the past 100 years or so in the history and growth of the United States. What do you think was the effect of the "open immigration" policy on the wage rates? and on the returns to the owners of capital? and on the rate of economic growth? And today, what do you think are the effects of the *more restrictive* immigration policies? Discuss.

# PART 7

IF IT'S SUPPLY AND DEMAND THAT SETS PRICES,
HOW COME SO MANY PRICES TURN OUT TO BE $4.98?

In the real world the market process never works exactly the way it does in the model, but you may be sure that in a modified way it does work, and that the price mechanism really is a powerful force for economic motivation and control.

# 24 SOME INHERENT FLAWS IN THE MARKET SYSTEM

*In the real world the market process is hampered by instability, inequality, and sometimes the failure of market values to reflect true social values.*

Do you know of anything *perfect* in the real world? I don't think so. There is no such thing as a perfect car, house, job, parent, son, daughter, wife, husband, sweetheart, ball-point pen, or economics professor. If you look at something carefully enough and long enough you will find imperfections. Whenever you think you have discovered something "perfect," either stay away from it so you will never see the flaws, or else look out for a let-down! That's the way life is. The "beautiful market process" is no exception.

This chapter tells you about several of the flaws in the market process. I don't think the let-down will be too big a shock for you. Even with the imperfections the market system has some good things going for it. In this chapter we're going to be talking about the problems of instability, of inequality, and about some special situations in which the market can't accurately reflect the society's wishes. Then the following chapters will get into monopoly and price fixing and some other things. But first let's talk about instability.

## THE LONG-RUN EQUILIBRIUM IS NEVER ACHIEVED

Did you ever go down on the beach and watch the waves come rolling in to the shore? The booming surf keeps frothing and pounding, the water surges far up on the beach, then rushes back. You say to yourself, "What a restless thing this ocean is!" As you stand there watching the ocean frothing and foaming, the waves peaking, curling, breaking, you might say to yourself, "If that ocean was in equilibrium it would be completely flat and smooth. I could build sand castles here at the edge of the water and they would last all day. As it is, my sand castles only last until the next big wave comes up." Can you see where this story is leading?

Equilibrium is the condition we would have if everything "came to rest." There are no disturbing influences. Have you ever seen the ocean in "long-run equilibrium"? I don't think so. The ocean always "tends toward" equilibrium, but never gets there. The same is true of markets. In real-world markets, *long-run disequilibrium is really the "normal" thing*!

### Demand Is Constantly Changing

When you go into the supermarket to buy some paper plates for your picnic, what do you bring with you when you come out? Oh, a jar of pickles, a new paring knife you thought you might need, a bag of charcoal, some Polish sausages and horseradish to go with the lunch you packed, and a few other things. (Of course, you forgot the paper plates.) Sort of unstable demand, wouldn't you say? Suppose that when you get back to your car in the parking lot you find that your left rear tire is flat. Your demand for tire service increases immediately. And so it goes with all of us.

People's demands for things are changing constantly. Almost every moment your desire for one thing is increasing while your desire for something else is going down. With this happening all the time to all the people, how could we expect the demand for anything to be stable and constant long enough for a "long-run normal competitive equilibrium price and quantity" to ever really come about? Actually, we don't. And it doesn't.

### Equilibrium Is A Concept, Not A Condition

The "normal equilibrium price" is like the normal equilibrium of the ocean. It's important to know which way things are tending. You know that a high wave is going to go down; a deep trough in the ocean is going to get filled up. It's important to know that. But don't get confused about the equilibrium concept and think of it as a description of a real world *condition*. Certainly it isn't. It couldn't be. Demands for all things are changing constantly, costs are changing constantly, and prices and quantities are moving in response to these demand and cost changes. The movement of demand, supply, and price goes on forever.

The concept of equilibrium shows which direction things will be changing, and it may set some limits within which fluctuations will occur. Because you know about the equilibrium ocean level you know how high to build the fishing pier to be sure no wave will ever break over it; you know how deep to dig the channel crossing the bar to be sure that the trough between two waves will never be so low that boats will hit the bottom. But you don't ever expect to see a perfectly placid ocean. This is the way it is in the real world with the free operation of competitive markets.

With the ocean it is the wind and currents, temperature differences, and the earth's rotation which prevent "normal equilibrium" from being achieved and maintained. In markets, it is the constant change in demands and supplies of outputs and inputs which keep equilibrium conditions "chased after but never caught." The concept of "normal equilibrium" sometimes confuses students of economics. Please, don't think of this "normal equilibrium" concept as something that is supposed to describe the real world. That isn't the idea at all.

Suppose you built yourself a fragile little canoe to paddle in the ocean along the edge of the beach. Maybe you're planning to look down through the clear water and spear some fish. You know that water seeks its own level and that in its "normal equilibrium state," the ocean must be flat and smooth. So you build your canoe. You take it to the beach and wait for equilibrium so you can launch it. While waiting, you grow old and die. Well, that's about how long you would have to wait for the long-run equilibrium to arrive in the competitive markets of the economy!

Now that you know that the market process (like the ocean) never really reaches a smooth, stable equilibrium, perhaps you are ready for an even more serious problem. Market fluctuations can be much more violent and unpredictable than ocean waves!

## SUPPLY ADJUSTMENTS MAY HIT THE MARKET LIKE A TIDAL WAVE

As soon as the peak of a wave rises above the "normal equilibrium level" of the ocean, the force of gravity starts trying to pull it back down. This "pulling back toward equilibrium" begins immediately. The force of gravity goes to work and pulls the wave back down. But when the market process is working, the "corrective action" may be slow in coming.

When the demand or the supply shifts and forces the price to a level above or below the "normal equilibrium" level, it may be a long time before any visible "corrective response" will start to move the price back to the normal level. The markets for agricultural products provide the best real-world examples of how slow and uneven the supply response can be. Suppose you are producing corn in your backyard plot. Then suppose, just about harvest time, there is an article in Consumer Reports, reprinted in *Reader's Digest*, saying that fresh corn-on-the-cob steamed in the husk is not only delicious, but also has great and heretofore unrecognized nutritional value!

### The Demand For Corn Increases

What happens to the demand for corn? That's obvious. But what happens to the supply of corn? Nothing. Nothing *can* happen to the supply of corn on such short notice! It was planted months ago. It has been growing all this time. Now it will be harvested, and it will be sold at whatever prices the market will offer.

As you sit looking out your back door watching your corn tassels wave in the breeze and looking at the article in *Consumer Reports* and the condensed version in *Reader's Digest*, you feel real good. You remember when, back in February when the snow was on the ground, you flipped a coin to decide whether to grow all corn or all tomatoes. It came up "tails"—all corn. You knew it had to be all corn or all tomatoes because you have an in-sagging, convex production possibility curve. You get the greatest total output by specializing in one product or the other.

But why are you so happy as you look at the articles in the magazines and watch your corn tassels waving in the breeze? Because you understand enough about the basic concepts of economics to know that there's going to be an increased demand for corn; that the increased demand is going to support a higher market price for corn, and that *lucky you* are going to be one of the beneficiaries of this fortuitous turn of events!

Yes, lucky you. The demand is going to increase for sure. The price is going to go up from 75¢ a dozen for those fresh, succulent ears of corn, to $1.50. Then to $2, or $2.50, or maybe even $3. You are *really lucky*! But what about all those corn-loving people who live in the cities and dream all winter long of the pleasure of having fresh corn-on-the-cob in the summertime? They aren't so lucky. "But," you say, "this high price will bring forth more corn. Then the price will go down to 'normal' and everyone will be happy again." Will this happen? Maybe. But not this summer.

### The Supply Response Is Slow In Coming

How high must the price go, to bring forth more corn this summer? The answer to that is obvious. We aren't going to get any more fresh corn this summer than was planted in the spring. We have already planted and grown all the corn we are going to plant and grow this season.

We can't import the corn from other countries because those magazine articles have already been translated into seventeen different languages, and the increased demand for fresh corn is world-wide. In fact, the price of corn in some foreign countries is increasing so much that it looks like some of our corn is likely to be exported to the countries where the prices are higher. You yourself are considering making an air-freight shipment to Stockholm!

### The Rationing Function Works Just Fine

The market process is supposed to work by bringing forth more output in response to an increase in price. But what is happening in the corn market and in the corn industry? As the price goes up, this rations the available supply of corn. Those who are willing to pay a high price are the only ones who will get to eat any of it. A person who doesn't like corn very much will stop eating corn and leave it for somebody else.

The high price guarantees that people who are having corn-on-the-cob will not cook a dozen ears when all they really need is seven. At the higher price they probably will buy only four ears, and cut each one in half. You can be sure that some of the chickens and cows that usually eat corn are going to have a quick change in diet! The rationing function of price is working fine. But the price keeps going up and no new supply comes forth.

You are a lucky corn producer. Everything is going great for you. Every night you sit out with your shotgun, guarding your corn patch. Several times each day someone comes by and asks if he can buy your entire corn crop. (Everybody is trying to make a buck!) One fellow has a big truck, and wants to harvest your crop and take it to New York City where the price is really going wild. You're tempted to sell, but you decide to sit tight, to harvest each ear at just the right time and take your chances at your little roadside stand.

As it turns out, it seems that you can do no wrong. Everything you do seems to make more money for you. The longer you wait, the higher the price goes. What a lucrative summer! You can hardly believe it. You finally sell your last dozen ears for $3.75. That's *five times* as much as the "normal price" you expected to get when you were planting the corn, last spring. What a summer! What a profitable business corn farming can be!

### Getting Ready For The Supply Response

After your corn is all sold you feel very wealthy. You sit around thinking for a while, then you call the owner of the seven acres of land which adjoins your backyard plot. You ask him if he would like to sell. After some haggling, you make a deal. Then you start contacting agricultural equipment dealers to see if you can get a tractor, a cultivator, a planter, a harvester, a sprayer and all the other things you need. It's obvious to you that if you are going to grow seven more acres of corn in your spare time, you need to use a lot of capital to increase your productivity.

You contact various fertilizer dealers to find out the most productive kinds of fertilizers for growing corn. You call the county agricultural agent to ask about how to increase the output per acre. You are really planning to cash in on the gold mine you discovered this summer! But wait just a minute. What do you suppose the wheat farmers out in Kansas are doing? And the egg producers in New Jersey? And the cattlemen in Texas? And the potato growers in Idaho? And all the others who have land which could be used to grow corn? They are doing the same thing you're doing.

### All The Farmers Respond At Once And The Price Collapses

Everyone is planning to cash in on the gold mine—the high price and high profits from producing corn. You know what's going to happen. This market for corn is not going to move itself slowly, step by step, back to equilibrium. There will be no "step by step increases" of the supply and "inching downward" of the price. Next summer when the corn begins to be harvested, there is going to be so much fresh corn—the supply is going to be *so much greater* than it was this year—that the market is going to be glutted. There will be thousands of carloads of surplus corn. The surpluses are going to force the high price to collapse. It's going to happen all of a sudden.

The demand for corn is not very elastic. As the price plummets downward people still aren't going to eat but just so much corn. Even at a price of 10¢ a dozen not many people are going to eat five ears of corn a day! The quantity people eat does increase as the price goes down, yes. But it doesn't increase very much. Before the summer is over, lots of hogs and cows and chickens are going to get to enjoy some high-quality corn!

Well, after it was all over, you lost money on your seven acres and all that equipment and fertilizer, didn't you? But you are no fool. You won't get caught making the same mistake again! So you curse the day you planted all that corn, and you resolve that from now on you will grow tomatoes instead. So next summer, guess what? Fickle fate has done it again! Corn is very scarce. The price is very high.

This is no way for an economic system to serve the society! This thing is out of balance! When does all this "disequilibrium" work itself out? When does all the "disequilibrium" in the ocean work itself out? You already know the answer. (Sorry about that).

This example is a little over-dramatized. But it isn't *completely* unrealistic. Recurring fluctuations in prices and quantities do happen, especially in farm products. The problem can be very serious if the weather happens to be especially good, or bad, to accentuate the fluctuations.

The important point for you to recognize is that in competitive real-world markets, the supply may not adjust smoothly to a change in demand. The price may go very high or very low following a change in demand. Then it may be some time before supply responds. If the market for something keeps fluctuating violently from year to year, or from season to season, or just from time to time, this is not a healthy condition. An economic system of violently unstable markets would be intolerable.

### The Government Stabilizes Prices Of Farm Products

Do you begin to understand that there may be some logic behind government stabilization of some agricultural outputs and prices? Not that the government always does it right—far from it! But there are times when anything the government might do would be preferable to the violent fluctuations which free markets might generate.

The markets for agricultural products are the markets in which the "classical picture" of "pure competition" comes closest to reality. These are the markets in which the market process is supposed to work best. What a shame that these are the very markets in which we can't let the process work! Often the supply response is not smooth and gentle, but violent and fluctuating. Such fluctuations hurt both the consumers and the producers. So the political process (government) asserts its ultimate power to influence the economic choices. The government sets minimum prices, establishes acreage restrictions, and buys up surpluses and pays subsidies to farmers. Why? Because the people of the society are not satisfied with the way the market process works things out. The "automatic adjustment process" described in the pure market model, doesn't work out right in the real world.

## How Elastic Is The Supply?

Not just agricultural markets, but *all* markets are constantly in the process of adjustment. Demands change. Costs change. In markets controlled by competition, each change in demand pushes prices upward or downward; each change in cost influences supply and also pushes prices upward or downward. The responsiveness (elasticity) of supply, both immediately and in the long-run, will determine how smoothly and easily this adjustment process—this movement toward equilibrium—will come about.

If the short-run supply is elastic, then the quantity offered for sale will respond readily to a change in demand and price. The price will not rise very high. Adjustments toward equilibrium may be easy and smooth. But if the short-run supply is highly inelastic (as in the case of corn), an immediate response to the price change is impossible. The price may go up very high before any new output comes to market. Then when the response comes, it may flood the market.

In the case of corn (or for farm products in general) the problem is simply this: The "very short-run" supply situation lasts for such a long time. How much output is available for sale from one harvest time to the next? Only the amount which has already been produced. That's all there is! The total supply cannot respond to an increase in price, so if the demand goes up, up, up, the price just goes up, up, up, too.

By the time the "very short run" (with no output variability) is over, suddenly the long run (with *complete* output variability) is here. See the problem? Each farmer must make his long-run production plans on the basis of very short run (often very temporary and largely irrelevant) market supply, demand, and price conditions. Whenever this kind of "delayed supply response" situation exists—in farm production or in anything else—recurrent fluctuations of price and output are inevitable. Why?

Because the market process can't work things out smoothly, the way the "model" says it's supposed to.

When it takes a long time for the supply to respond, then the government is likely to overrule the market process and do something to stabilize prices and outputs. Why? Because recurring violent fluctuations of prices and outputs are socially intolerable. It really can be a big problem. But it isn't the *only* problem with the market process as it works in the real world. Let's talk about another problem.

## THE PROBLEM OF INEQUALITY OF WEALTH AND INCOME

Over the past 200 years or so, the market process has been playing a very important role in the world. It has been the major force influencing and directing the economies of most of the modern nations. But all this time, some people have been criticizing and attacking it. Why? On what grounds? Usually because of the way *the distribution question* is answered.

The most serious failure of the market process to satisfy the wishes of most people of the society, has been in the way the *distribution of income* works out. No society has ever permitted the market process to have *complete* control over the distribution question. Of course not! It could never work. All the little children would starve to death! Some of society's output must be distributed to the unproductive children. It must be done by some "process" other than the market process—either the political process or the social process must do it.

### The Family Is A "Social Process-Income Distribution Unit"

Throughout history *the family unit* has played an essential role in answering the distribution question. In modern times, even in the advanced nations the family continues to play an important role in modifying the harshness of the pure market process. The family takes care of the very young, the very old, and the other unproductive members of the family. In this way, many people who produce nothing still get to have a "distributive share" of the output. In the capitalistic countries this "family distribution function" has played an important role in making the harsh market system tolerable. Most of the unproductive people have been cared for by their relatives—their family.

### Governments Redistribute Wealth And Income

In recent years, as many families have sort of drifted apart, governments have taken over more and more of the "income redistribution" function. For many centuries, most of a person's "economic security" has come from his family unit. But now, in "modern society," traditions are breaking down.

The family unit is no longer so strong. Many individuals feel that they belong—not to a family, with responsibilities to the family—but to themselves. No longer can the family perform the "income redistribution," "economic security" function for everyone. The babies (and maybe wives) still get fed, but a lot of kinfolk get left out. Private and religious charities help some, but they have never been big enough to do more than supplement the family's redistributive role. So they can't take over the job.

What's the answer? There's only one possible answer. The political process must get involved. If we're dissatisfied with the way the market process distributes the output and income, then we must look to the government to do something about it.

The only way the unacceptable inequality—the poverty, malnutrition, starvation which the market process would leave in its wake—can be overcome, is by some kind of government action. The unproductive members of the society—young, old, infirm, others—must not be left to starve, just because there is no family to take care of them! So to meet the problem, governments undertake "socialist reforms" such as the U.S. Social Security program.

### The Rich Make The Choices And The Poor Go Hungry

Just as the pure market system is intolerable because some people don't get enough, it is also intolerable because some people get too much. Suppose you inherited a million dollars. How much influence could you have on your society's choices? on the production and distribution questions? Quite a lot! But suppose there's a fellow down the block who didn't inherit any money at all. He's hungry and broke. How much influence do you suppose he's going to have on whether or not you will plant corn or tomatoes? You know the answer. In the market system, people with no money don't influence *any* choices.

You might say: "If he's broke, then he *should* be broke. He isn't producing anything for the rest of us, so there's no reason why we should be producing anything for him. If he will do something productive he will have some income. Then he can go out and buy what he wants and influence some of society's choices, just like the rest of us can." Of course you can say that. That's the basic justification for the inequality which (at least to some extent) is an essential condition in the market system.

No economic system has ever operated on the principle of "equality." It isn't likely that one ever will. So that isn't the real issue, anyway. The real issue is the question of *how much* inequality there will be in the society. How many will be how poor? What opportunities will they have to improve their circumstances? Will people have something approaching *equal opportunities* to be productive? to earn good incomes? Or not?

In the real world, equal opportunities don't exist, either. The problem of "unequal opportunity from birth" appears to be a basic fact of life. One person is born to an uneducated, poverty-stricken, unwed mother. Another is born into a family of millionaire economic and political leaders. Equal chance? Of course not. But what can the market process do about it? Nothing. That's the problem! Let's talk about it.

## People Don't Have Equal Opportunities

A person who has wealth can have the income he receives from that wealth, and he can also have the income he receives from the sale of his own efforts—his "labor." Furthermore, one who has wealth to begin with, can get a high-priced education and increase the value of his "labor" greatly—like from janitor to corporate lawyer! If your family is wealthy enough to let you finish high school without having to stop and go to work, you are lucky. Not all people are that lucky. In almost all countries such extravagance is out of the question.

Any person who has enough wealth behind him to let him go to college without having to work to support his aging grandmother and his two high-school age sisters, really is lucky! Suppose your family is even wealthy enough to lend you the money you need to buy some land and capital so you can go into business. Wouldn't you say you had more than an "equal chance"? You have an excellent chance for a good income. In a wealthy nation, many people have such good breaks. But not *all* people have this kind of head start toward economic success in life.

If you start out with such a big head start, and if you work only reasonably hard and keep your capital and land producing to help you, you may be a very wealthy person when you die. Then your children may have so much wealth that they will not need to sell their labor at all. They just need to keep their wealth (land and capital) invested and working for them and they will get richer and richer. If your father leaves you a million dollars inheritance, and if that million dollars is invested in productive capital earning a return of only five percent, then the income on your inheritance will be $50,000 a year—about $1,000 a week. Unless you work at it, you will not be able to fritter away all that money every week. A thousand dollars a week? That's a lot of fritterin'! So what happens? You'll just automatically get richer and richer.

## Wealth Generates Income And More Wealth

Suppose your father leaves you a *five million dollar* investment, paying a five percent return. That will bring you an income of $250,000 ( a quarter of a million dollars!) every year. That's about $5,000 a week. Almost a thousand dollars a day! You'll have to spend it all up (for nothing of value) in order to keep from getting even richer!

If you spend it for capital or any other "valuable property," you will just keep piling up more wealth. To keep from getting richer, you must spend it *just for fun*! I don't think you could do it. So unless you just give the money away, you can't help getting richer and richer.

### The Process of "Snowballing Enrichment"

Each time you get richer, your income goes up more. So you get even richer. And richer, and richer, and richer! Through no fault of your own, you are going to gain more and more control over society's choices. You can't help it. Then your children (or other heirs) are going to be very wealthy and they will not be able to keep from getting more and more and more wealthy. That's the way the market process makes it turn out.

Your heirs really will be in charge of a lot of society's economic choices! Should they be! Is this just? If so, on what grounds? If not, should the political process intervene? Maybe the government should take away some of your big income. ($5,000 a week? WOW!) Then, when you die, maybe the government should take away a lot of your wealth. People disagree about whether it should, or shouldn't. But you may be quite sure of one thing: It does—on both counts.

The process of "snowballing enrichment," where the rich keep getting richer and richer, is something like the process Karl Marx was talking about when he said that capitalism had within itself "the seeds of its own destruction." He said that the richest people (the bourgeoisie) would get constantly richer and more powerful, and that more and more people would be pushed into the ranks of the poverty-stricken. Marx predicted that economic (and political) power would become concentrated in the hands of a very few people and that someday the poverty-stricken masses (the proletariat) would rise up and take over.

### A "New Deal"?

In the century which has passed since Marx was observing and writing, the political process has proved its ability to soften the harshness of the market process. The "total concentration" of wealth and power in the hands of a few people, has not occurred. Most people are enjoying *more* wealth, not less. Also, many of the wealthiest people decided to give away much of their great fortunes. For example, look at the Ford, Rockefeller, Carnegie, and many other foundations, and at Duke, Vanderbilt, Carnegie Tech, and many other universities which have been financed by gifts of private wealth.

The government has become increasingly demanding of the wealthy, and increasingly helpful to the poor. The society, through its democratic processes, has demanded these changes in the government's attitudes toward the rich and toward the poor. Governments have been increasing taxes on incomes, sales, property, inheritance, and other things; while increasing the payments for welfare, social security programs, medical care, job training, and other aids to the poor. These reforms have worked together to redistribute some of the wealth and income among the people of the society. This "income redistribution" has gone somewhat farther in most other economies of "mixed capitalism" than it has in the United States. But there's been quite a lot of redistribution in the United States, too. Some people say there's been much too much. Others say there's been much too little.

When too few people are holding all the "high cards," maybe it's time to call in the cards and have a "new deal." This is sort of what President Franklin D. Roosevelt was talking about during the depression of the 1930's, when he announced his "New Deal" programs. Of course we didn't really "pull in all the cards and start over." But several "income redistribution programs" were started at that time, and we have continued to move farther in that direction ever since. Each time the government takes some of my "property" (income, or wealth) and uses it to feed a fatherless child, or to subsidize college education for a worthy student, or to put picnic tables in a national park, isn't that sort of a "new deal"? It really is.

### What Happens When Property Rights Get "Bunched Up"?

Society's resources and other valuable things are controlled by the people who hold the *property rights* to those resources and things. If the property rights get concentrated in the hands of a few people, those people are the ones who can make most of the choices for the society. If this happens, we can no longer think about the market process as offering a sort of "economic democracy" where each person has an *equal opportunity* to enjoy society's bounty and to influence society's choices. The few who are lucky enough to be wealthy, with high incomes and government-protected rights of private property, will enjoy lavish shares of the society's bounty and will make most of the choices about what will be done with the society's resources.

In any market-directed society, if the land and capital (the wealth) get concentrated in the hands of a few, it becomes *essential* that there be some kind of a "new deal." Sometimes the political processes of the society are sufficiently flexible and responsive to the needs of the society so that the "new deal" can be accomplished through the democratic process. But if not, then you may be sure that (as Marx predicted) the ultimate result would be revolution.

What about the problem of poverty and inequality of wealth and income in the United States, today? Specifically what is being done about it? And what are the ultimate answers? These are important questions—unsolved problems, really. We'll talk about them in the final part of this book (Part Eight) which deals with current problems.

For now, just be sure you understand that the market process could never be left in *complete charge* of the distribution choices for any society. The results would be intolerable. In all market-oriented societies, both the social process and the political process are called on to modify the unacceptable results which the market process, acting alone, would bring. In the past few decades, in all the economies of "mixed socio-capitalism," the political process has become increasingly involved. This trend seems likely to continue.

Now we need to take up a new kind of problem. This new problem is one you may have been vaguely aware of for many years. What problem? The fact that sometimes the choices made "through the market" do not really reflect the best interests of the society—even when the market seems to be working perfectly in every respect. How can that be?

## SOME SOCIAL COSTS AND BENEFITS ARE NOT REFLECTED IN THE MARKET

Sometimes important *social values* are not reflected in the market. If the *true value of the social benefit* of something is not reflected in the market demand, or if the *true value of the social cost* of something is not reflected in the *market supply*, then the market price will not reflect society's true values. So the "production motivation" and the "rationing" which such a price accomplishes, will be misdirected! Production will be motivated too much (or too little); goods will be conserved too stringently (or not stringently enough).

If the price is "too high," uses will be more severely limited and production more strongly motivated than would be required for the "optimum social good." If the price is "too low," resources will not be conserved enough, and new production will not be motivated enough to achieve the "optimum social good."

Whenever the price of something is "too high" or "too low" for the optimum social good, then the results will not reflect the society's best interests. The resources which move in response to a "socially erroneous" price will not move to where society, all things considered, would most *want* them to go. Why not? Because some of the true social benefits or true social costs of the product *are not being reflected in the market price* of the product. These extra social benefits or extra social costs are outside of (are external to) the market process. These "unreflected benefits" or "unreflected costs" are called externalities.

### Cost Externalities And Benefit Externalities

Many people are talking about "externalities" these days. Industrial pollution is a cost externality—a social cost which does not show up in the market price of the product. But when you plant trees on your land, there are benefit externalities. The whole neighborhood gains some benefits from your "reforestation project," but those external (neighborhood) benefits are not reflected in the demand for your land or in the price you will get if you sell it. There are many kinds of "cost" and "benefit" externalities. Let's look at a make-believe example.

### An Example Of Cost Externalities: Cutting Timber

Suppose you lived in a village in a beautiful green valley. The hills surrounding the valley are covered by a virgin forest of fir, spruce, and hemlock. These forest lands around the village are owned by the heirs of old Sam Johnson, the man who founded this village, many years ago. Everyone here kindly remembers his name. There's a small monument to his memory in the village square. But all his heirs have long since moved to Seattle.

One day the logging crews arrive with their heavy equipment and begin clear-cutting the forest on the hillsides around the village. Everyone in the village is terribly upset. It is only the presence of those timberlands which make this valley a pleasant place to live. But more important, those timberlands prevent the heavy spring rains and melting snows from turning the surrounding hills into a mud-slide that would bury the village.

To the village, those stands of timber may mean life or death. But to the present owners — the heirs of Sam Johnson — the timber is simply an economic resource, ready to be used in response to market demand and price. The timber owners mean no harm to the village. They are just responding to the market demand for timber in the way every good businessman should.

From the social point of view, this stand of timber is a very valuable asset. At stake is the property (and maybe the lives!) of the people of an entire village. How, then, can the market process, which is supposed to restrict the use of things which are very scarce and valuable — how can it permit the clear-cutting of the timber on these hills? The price of the lumber isn't nearly high enough to cover the great social cost of cutting the timber and destroying the village! How can the market permit such a thing to happen? I think you already know the answer.

### External Social Values Do Not Influence The Price

The *values* we are talking about are "external" to the market. The "external social cost" of cutting the timber is not reflected in the price of the lumber. The "external social cost" doesn't affect the supply of the lumber; therefore it doesn't push up the price and *conserve* the lumber the way it should.

If the owners of the timberlands valued their timber at its *true social value*, then they would not cut it unless the market price of lumber went very, very high. But since the high "external value" of the timber is not reflected in the "economic value" of the timber, the present owners see no reason why they should not cut it and sell it. The timber owners have the "property rights" to do whatever they please with their trees. They have chosen to do what the market has *directed* them to do—cut the timber and sell it. What a classic case of externalities this is!

What can the villagers do? What would you do? The villagers meet in the town hall and decide to talk to the foreman of the lumbering crews. Then, if that does no good, they will try to sabotage the equipment. At the same time, one member of the village will be rushing to the county seat to try to get a court order prohibiting the cutting of the timber. Perhaps the property rights of the owners to cut the timber can be taken away "by due process of law."

The lumbering foreman has a job to do and he doesn't want to get fired. He appears sympathetic to the pleas of the villagers. He says: "I would like to help you. But you know I can't. My hands are tied." Then the people of the village decide to take the law into their own hands. They pour sugar in the fuel tanks of the bulldozers, shoot holes in the tires of the logging trucks and chop up the ignition wires on the draglines. These (usually) law-abiding citizens decide to take the law into their own hands to protect their village. They are forcibly denying the "private property rights" of the timber-land owners. They are breaking the law.

If the system and the law are right, then what the villagers are doing is very wrong. If what they are doing is right, then the system and the law in this case are very wrong. Luckily, we live in an open, flexible society in which, if the system is wrong, it can be changed. That's why they sent a man to the county seat. But change takes time. If the timber is cut, no amount of "system changing"—of restricting the owners' property rights—will bring the timber back! And it may be that the law will not agree to protect the timber anyway. The law doesn't always do what every group of dissatisified people wants done! What a mess. How could such a thing happen in an economic system directed by the market process?

### Externalities Require Government Intervention

This is just one of those cases in which the *true values* of the society are not reflected in the market. Most of the social cost is external to the market. The villagers bear the cost. The timber market (made up of the sellers and buyers of timber) is not influenced by this "villagers-cost." In a society in which the market process plays a major role, these "other values," these "external social costs" must be protected either by the government or by social traditions and customs. Otherwise wasteful, "non-optimal" choices will be made through the market process.

Whenever there are large "externalities"—either external social *costs* or external social *benefits*—the society usually demands that the government modify the results which the market process would bring. Everyone knows that it isn't going to do your friend any good or society any good if your friend gets hooked on heroin. But if your friend offers to pay *enough*, someone will supply the heroin to him and make the profit. The people who produce and sell illegal drugs are responding to the market process. The fact that the *social cost* doesn't show up in the market is not a matter of their concern. So what must happen? The political process—the government—must come in and *overthrow the market* and change the result to a more socially desirable one.

In the case of Sam Johnson's valley where the wooded slopes are in danger, legal action probably will be taken to prevent the owners from cutting the timber. Then the owners will be given some court-determined "just compensation" to make up for the loss of their "property rights" to cut and sell the timber. In the case of illegal drugs, the government tries to *abolish* the market—to make it impossible for the drugs to be bought and sold. In both of these cases the society is using its political process to overthrow the market. Why? Because in these instances the market does not reflect the true values of the society. The most important considerations (social values) lie outside (are external to) the market.

### An Example Of Benefit Externalities: Urban Housing

Externalities (social benefits and social costs which are not reflected in the market) show up in many places in every market-oriented society. Let's take another example. In housing, such as apartments in the big cities, there can be externalities. It is sometimes said that people who live in decent houses in decent neighborhoods, are more likely to be stable, law abiding citizens. They will be more serious about educating their children, better contributors to the society, and less likely to be on welfare. It is said that slum housing contributes to crime and violence and social disruption. If this is true, then there are externalities—external social benefits—in urban housing.

### Investors Respond To (Internal) Market Conditions

Consider the real-estate investor who is thinking about building a new real estate development to serve poor people in the urban area. He cannot afford to make the investment unless he expects to get his money back, plus a reasonable interest and profit for himself. The return he can get in this investment must look at least as good as what he could expect to get in any other investment opportunity he might be looking at elsewhere in the economic system. Obviously! It doesn't matter that the new real estate investment would result in reduced crime, better education for young people, growth of new businesses and employment opportunities, better social harmony for the people in the area, and all that. All these benefits are *externalities*. These are *benefits which the investment will bring to the society, but for which the investor will not get paid.*

A prudent businessman (unless he has some money he wants to give away) cannot consider external social benefits as he is deciding where to invest. So the chances are the urban housing development will never happen. From the point of view of the *society* the benefits would justify the costs. But from the point of view of the market—that is, the return the investor can expect—the investment is not justified. It will not pay itself off and provide the investor a "normal profit," so he will invest in something else—something which will.

What should the society do about urban housing? Should we call on the government to modify the market process? Should we get the government to pay subsidies or give special tax breaks or otherwise support the construction of housing to serve the urban poor? Perhaps so. Under the circumstances, there is no way that the market, operating by itself, can produce as much housing as the "social benefit" calls for. Why not? Because much of the social benefit is external to (is not reflected in) the "housing investment market."

If you look around you, you will see many examples of "externalities." Suppose the local coal mine spreads dust in the air and over the countryside for an area of twenty miles. If this effect does not show up as a *cost* to the coal producers, then this is an "externality." It's a social cost, but not a cost that shows up in the cost (and price) of coal.

If I decide to plant Christmas trees on the barren hillside so that I can make profits from selling the trees, my tree farm provides externalities. The trees stabilize the soil so the spring rains will not wash topsoil into the stream, muddy the water, and kill all the fish. But do I get paid for this external benefit? No. The only way the *market* can take these externalities into account (that is, the only way I can get paid) is for some way to be devised to "internalize" the externalities. That means, to get market *costs* and *prices* to reflect the *full social cost* and *full social benefits*.

### How To "Internalize" The Externalities

The obvious way to "internalize" a cost externality (say, pollution) would be for the government to require that the business pay an additional cost as a penalty for polluting—or else have the business pay the full cost of eliminating the pollution. If the business must pay for the *full social cost*, then the "cost externality" becomes internalized. The cost will be reflected in the market supply, so the price the consumer will have to pay will go up. Only those consumers who are willing to pay the higher price, covering the *full social cost* (including the pollution cost) will get the product!

In the case of the timberlands surrounding the village in the valley, after internalizing the externalities, the total cost of the timber would reflect the full social cost of wiping out the village. If all that cost is going to have to be paid for by the people who buy the lumber, then nobody is going to buy the lumber! If nobody is going to buy it, then nobody is going to cut it and sell it. Obviously! The village will be saved.

In general, how can the government "internalize" externalities? External social *costs* can be internalized by levying *taxes* or by imposing other government charges to *push up the cost* (and the price). This will discourage the production and limit the use of the product. External social *benefits* can be internalized by paying government *subsidies* which will *push down the cost* (and the price). This will encourage the production and use of the product.

The "village destruction tax" the government levied on the cutting of Sam Johnson's valley timber is very high! The loggers all go home. Can't you just hear the village mayor's speech when he says: "The timberlands have been saved by the timely internalization of the externalities!"

Let's say it this way. Whenever the "cost to society" is greater than the cost the *producer* has to pay, then there is some "external social cost." Whenever external social cost exists, the market process cannot lead to the best choices for society, because the supply (which is supposed to reflect the full social cost) does not reflect these "external" costs. So the price of the product doesn't reflect the full social costs. The cost paid by the producer is too low and the price paid by the consumer is too low. The cost (and the price) must be pushed up to "internalize the externalities." For benefit externalities, the cost (and the price) must be pulled down.

### There Are Many Kinds Of Externalities

The term "externalities" has broad meaning in economics. It is used to refer to any kind of "outside the market" influence. If a new typewriter repair shop opens up in my town, this may let my local business get its typewriters repaired faster and cheaper. This is a beneficial externality for my business. Ultimately my customers will benefit. Of if a new processing plant opens up in my area and causes more overcrowding of the already overcrowded highways, railroad connections, and telephone services, this will increase my costs. I would see this as an undesirable externality. Ultimately my customers will have to pay more. In both these examples the ultimate result is for the society to receive extra benefits or to pay or suffer extra costs which do not show up in the "profit and loss statements" of the new typewriter shop or the new processing plant.

Whenever you are producing something which provides benefits to society greater than the value of the money you receive, then there is an external social benefit. You are not getting fully paid for the good things you are doing for the society. Whenever this situation exists, the market will not direct you to produce as much as the society really would like for you to produce. The price doesn't reflect the *true value of your output* to society. The social benefits are not reflected in (do not increase) the market demand. Therefore the demand and the price are not high enough to pull forth as much supply as the society would like. In the case of the Christmas trees, if the local villagers would give me a subsidy of 50¢ a tree I would plant more trees. I would reclaim and stabilize more of the barren hillside and provide more benefits for the society.

### The Political Process Must Influence The Choices

When externalities exist, frequently the political process comes in to subsidize, or provide special tax breaks, or otherwise aid those who are producing things which have external social benefits, like Christmas trees, or urban housing. The government also comes in and penalizes or regulates or sets up special taxes or other extra charges for producers whose processes or products generate external social costs. Throw-away beer bottles may be the next to go!

Whenever externalities exist, the entrance of the government into the market process is perfectly understandable. Sometimes it is absolutely essential. Whenever there are external costs or external benefits, the only way society can arrive at the best economic choices is to modify the results which the market would bring. One approach is for the government to subsidize the ones providing external benefits, and place extra costs on the ones generating external social costs. Another is to absolutely prohibit the production or use of some things.

In markets where large externalities exist, the market process cannot protect and serve the best interests of the society. As population expands and more people get crowded together, each action by anyone has more effect on all the others. Externalities multiply. It seems that the problem of externalities is likely to get worse before it gets better.

## THE MARKET ALONE CANNOT ACCOMPLISH LONG RANGE GOALS

There's another kind of "social consideration" external to the market. It's this: How does the market handle the issue of long-range objectives and goals for the society?" The answer: It doesn't. It can't.

If the society is going to establish long-range objectives and work out long-range plans for achieving the objectives, it must do these things through the political process. It cannot do these things through the market. Markets operate essentially in a short-run framework. When we are trying to decide what kind of world we would like to have to live in by the year 2000, we are not dealing with the kinds of decisions which the market process can take care of. We know that during this next quarter-century, the market process, with some modifications, can generally provide for the efficient use of our resources. In that way it can help us to reach our objectives. But the objectives and goals and the general plan for their achievement must be worked out through the political process. The market can't do that for us.

Is it possible that we might visualize a better kind of world for the year 2000? Might it be possible to work out some plans and achieve something better than would happen if we just "let nature take its course?" Should the society, through its political process, decide on some objectives and develop some programs to move us toward those objectives?

During the early 1950's, some people in the Bureau of Public Roads and some members of Congress and some others worked up plans and legislation for the construction of a 41,000 mile "interstate highway network." The planned network was to link all the major cities of the United States and to be completed over a 20-year period. Soon the plans were worked out and the engineering and construction of the interstate highway system was begun. Today most of the system is complete. Why? How did it happen? Because the objectives and the plans were made by the government, some twenty years ago.

Another example is the U.S. space program. The decision was made to put a man on the moon in the decade of the 1960's. This was possible only because the objective was established and the plans were made by the government. This kind of long-range planning and programming is not a function which the market process can be expected to perform.

## Market Decisions Reflect Expected Demands And Costs

Most business and consumer decisions are made on the basis of future expectations, based on existing (or recent) demand, supply, and price conditions. Businesses make long-range plans based on projected costs and demands for things. All businesses react to changing demand and cost conditions as they see them now, and as they expect them to unfold.

The problem of setting long-range objectives and goals for the society is really a special case of externalities. The market can only reflect what the businesses guess the people are going to want. The businesses which make the best guesses will succeed and make the most profits. You already know that. But how can this process, efficient though it is (and it really is)—how can it get at "social goals?" It can't.

How could the market process get at such things as eliminating poverty? Or developing the Tennessee Valley? or developing (while protecting the natural resources of) Appalachia? or the Upper Midwest? or the Ozarks? or the other chronically depressed areas of the nation? Or how could it solve the vital problems of the nation's inner cities? or work out integrated national systems of transportation and communications? or develop and maintain clean lakes and rivers? and clear air? and a junk-free landscape? Or how can it produce a cure for cancer? Or open new economic opportunities to underprivileged people, while providing them "a living wage?" Or support nationwide systems of public education, health, welfare, housing? It can't.

### The Market Process Can Implement The Plans

The market process just isn't designed to choose for the society *which* of these "long-range social objectives" we will seek, or how much "opportunity cost" we will accept as we seek the chosen objectives. But once the goals are decided and the plans made, most of the day-to-day implementation of the plans can be carried out very efficiently by the market process—by letting prices pull the resources into the "chosen" places to do the "chosen" things. But the objectives and the plans—to reach the moon, to revitalize the cities, to build a new and exciting educational system, to eliminate poverty, to protect the environment, to save Lake Erie, to find a cure for cancer, or whatever—must be the brainchildren of the political and social processes. The market process can't make these choices for us.

The objectives and programs needed to meet the long-range needs of our society (and our world) will have to be approached, not through the market, but through the political process.

Where does this leave us? We've spent several chapters talking about the market process and how it works. Do we scrap it, now, and take off with the political process to try to solve some of the big problems of our nation and our world? When we come out of the "model market system of pure competition" and move into the real world, what do we have left? Anything of value? Yes. Nothing pure or perfect, but certainly valuable.

Chapters 26 and 27 look the real world squarely in the eye to try to see what's going on out there. But before that, there's Chapter 25. More disillusionment, I'm afraid. It talks about monopoly power and price-fixing; about how market prices in the real world don't behave the way they do in the model. As soon as you're sure you understand the problems of instability and inequality and externalities and long-range goals, go on into Chapter 25 and add the problem of *monopoly* to the list.

---

## REVIEW EXERCISES

**MAJOR CONCEPTS, PRINCIPLES, TERMS (Try to write a paragraph explaining each of these.)**

disequilibrium is the "normal" thing
need for farm price stabilization
the role of supply elasticity
the need for some "income redistribution"
the essential role of inequality

the process of "snowballing enrichment"
the need for some "new deal" changes
the basic problem of externalities
the problem of long range goals

## OTHER CONCEPTS AND TERMS (Try to write a sentence or phrase explaining what each one means.)

the delayed "supply response"
income redistribution
social cost
social benefit
externalities
cost externalities
benefit externalities
internalizing the externalities

## QUESTIONS (Try to write the answers, or jot down the highlights.)

1. If "long-run equilibrium" doesn't ever exist in the real world, then exactly what good, if any, does it do to learn how it looks in the model and to describe how the "model market system" would move into "long-run equilibrium" following a change of demand, or cost? Do you think your understanding of the "long-run equilibrium" concept helps you to understand anything about what goes on in the real world? Discuss.
2. The markets for agricultural products are "triply treacherous," since:
   a. the very-short-run lasts for many months, then suddenly the market jumps to the long-run;
   b. the weather may smile on everyone and bring a bumper crop, or it may do everyone in and create shortages; and
   c. the demand for most agricultural products is inelastic so the price must move drastically for people to buy a lot more (or less) of the product.

   Explain what all this means, then talk about what you think ought to be done about "the farm problem."
3. I don't suppose anybody likes inequality "for its own sake," but some inequality seems to be essential to permit "rewards" to be given to those who produce the things wanted by the society. No economic system has ever been able to function satisfactorily without rewards, and the resulting inequality. So *inequality* is not the question. The questions are: "How much inequality do we want to permit?" and "In the real world of economic uncertainty and practical politics, how do we achieve the 'desired amount' of inequality—no less and no more?" Discuss some of the difficulties you see in trying to get this complex socio-economic problem worked out.
4. In our complex, densely populated modern societies, almost all activities have some "spillover effects" which have either favorable or unfavorable influences on other people. If the "spillover effects" occur in some kind of "market-directed" activity, then that's what we mean by "externalities." List some "social benefit externalities" and some "social cost externalities" that you are personally familiar with, and explain how each one might be "internalized."

# 25 HOW MONOPOLY POWER CHANGES THE MARKET PROCESS

*The various kinds of competition and market structures have a lot to do with the way the market process really works.*

The neatness and beauty of the pure market process is a delight to behold. We are all free to do as we please, yet just the right amount of everything gets produced. Each of us gets his share of the output as determined by his productivity. It's so great it's just unreal. And that's true. It really is unreal. The last chapter told you some of the reasons it's unreal. This chapter and the following ones will explain more about it.

## "PURE COMPETITION" AND "MONOPOLY POWER"

You know a lot about how the model market system works. You know how changes in the demand and the supply change the market price. You know how prices move freely, responding to shifts of demand or supply. But did you know that all those things won't happen that way unless there is pure competition in all the markets? We haven't yet said much about "pure competition." But now it's time to recognize that unless the market has something like "pure competition" the adjustment process can't work out the way the pure market model says it will.

### Competition Protects The Buyer

Back in Part One of the book, in our "island economy," there was a big demand for tuba (coconut-palm booze). Remember? The northside chief was the only seller of tuba. In response to the great demand for tuba, he might have simply raised the price (instead of increasing the output). But suppose he had lots of competitors, all trying to sell tuba. Then he couldn't just raise the price and make lots of profits. Right? Of course not.

Competition is the thing which prevents businessmen and factory owners from limiting their output and raising the price sky high. With competition, if one seller limits his output and holds up the price, what happens? Nobody will buy from him! Everybody buys from the lower-priced sellers. If there are *many* buyers and sellers of a product, we say the "market structure" is one of pure competition.

### What Is Pure Competition?

"Pure competition" means that there are enough sellers and enough buyers in the market so that no one of them thinks he is big enough to create any surplus or shortage. The idea is that each producer knows he is so small that even if he stopped producing entirely, this would have no noticeable effect on the market price. *No producer thinks of cutting back his output, to hold up the price.*

Would you cut back your "backyard plot corn production" to keep the price of corn from falling? Obviously not. Then you are producing and selling under conditions of pure competition. Would any one of your roadside stand customers stop buying corn to force the price to go down? No. So the buyers are buying under conditions of pure competition. Both buyers and sellers simply "react" to the market price. They don't try to change the price. But *as they all react, "the market" changes the price.*

When the conditions of pure competition are met, the price moves upward only if many buyers are trying to buy more, or if many sellers are producing less. With pure competition existing in the market, the price is a true reflection of society's *demand*, and of society's *cost* of production. But if any buyer or seller gets so big that he can create a shortage or surplus in the market, then pure competition no longer exists.

### A Producer With Monopoly Power Can Set His Own Price

Without pure competition, no longer can we say that the price is a true reflection of society's demand for and cost of the product. If one buyer or seller gets large enough to influence the total demand or total supply for a product, he can then influence the *price* of the product. He can restrict his output and keep the price up high and make big profits. This power to restrict output and hold up the price (as you probably already know) is called market power, or monopoly power.

Is pure competition the usual thing? You know it isn't. In all the modern economies in which "free markets" play an important role—the United States, Canada, western Europe, Japan, and most of the other nations of the modern world—"pure competition" usually exists only in the markets for farm and fishery products. Most manufactured products are made by a few large producers. Sometimes there is strong competition among these big manufacturers, but the competition is very different, and brings different results than "pure competition" would bring. The outputs still adjust to changes in demand, but the *adjustment process* is different. Also, the equilibrium price and the quantity flowing across the market may be quite different from that described by the "pure competitive market model."

### Producers Can Respond Directly To Demand Changes

Manufacturing firms usually set the selling prices of their products. They try to set and hold the prices at levels they believe will cover their costs and bring in some profits. Once the price is set it will not change in response to every little change in demand, or cost. The price stays constant. This does not mean that the buyers have no influence over the amounts which will be produced. If the buyers demand a large amount of something, a large amount of it will be produced and sold; if they demand a smaller amount, production will be cut back and a smaller amount will be produced and sold. But the adjustment doesn't come about through the "price mechanism," as described in the "pure competition" model.

In the markets for almost everything you buy, outputs respond *directly* to demand. Prices stay about the same. Instead of production adjusting itself in response to a price change (as was the case in our egg market example) the production of manufactured products (toothpaste, TV sets, automobiles, ball-point pens, economics books, and almost anything else you can think of) responds directly to the strength of the demand. The manufacturer sets the price, and he doesn't change it very often. Of course, if the demand for some product turns out to be especially strong the producer probably will find some excuse to move the price up to increase his profits.

### There Are Subtle Ways To "Raise The Price"

No producer likes to be thought of as a selfish and ruthless profiteer. But if the demand for some product is really strong he probably will find a way to raise the price. There are all sorts of ways to do this. Perhaps he will stop producing the cheaper line and only produce the more expensive line. Then soon, the expensive line may begin to look more and more like the old cheap line. Finally he will announce the introduction of a new, even *more* expensive line. If a textbook really catches on, soon you will see a "revised edition"—at a higher price!

There are all sorts of ways of "raising the price" without coming right out and announcing that you are raising the price. But suppose the demand falls? If it turns out that much of what has been produced cannot be sold at the set price, what does the manufacturer do? He will cut back production, but he will also find some special way to reduce the price to get rid of the surplus. He might use a "special sale," or larger trade-in allowances, or perhaps he will remove the brand labels and sell the entire surplus to a discount outlet. There are many ways to lower the price without announcing that you are lowering the price!

Figure 25-1 illustrates the situation in which the individual producer is producing a brand-name product—perhaps Ford Pintos or Bobbie Brooks knitwear, or whatever. He fixes his price and then adjusts his output in response to the quantity people are buying. You should study Figure 25-1, now.

### Competition Always Limits Monopoly Power

What keeps the businessman from setting his "usual price" very high, so that he can make large profits? The strength of the demand is an important factor. Competition, although not "pure," still cannot be ignored. There is *always* someone out there, looking for a chance to take away your customers. Or the businessman might fear charges of "monopoly," and governmental intervention. Or perhaps he fears bad publicity from consumer groups. He doesn't want a bad public image! Or maybe he is considering the fact that high profits will have to be openly reported and that the labor unions will probably strike for higher wages if they see such high profits.

There are many forces in the real world which work against individual firms setting exorbitantly high prices or making exorbitantly high profits. Probably the most important force is the role of actual and potential competition—that is, the "threat" of acceptable substitutes from other producers. If any one steel company, automobile manufacturer, television producer, or other manufacturer sets prices which are very much above the others in the industry, his sales and profits are likely to drop. Customers will stop buying from him and buy other (substitute) brands instead. So why don't the producers get together and decide to all raise their prices at the same time? This could be a *very* profitable move. But as you know, there's a law against it. Several laws, really. The "antitrust laws" try to protect the consumer by prohibiting that sort of thing.

### Administered Prices? Or "Competitive" Prices?

How different the markets are in the real world, from the market model of pure competition! In markets of pure competition, the price moves freely in response to changes in demand and supply. It is the

Fig. 25-1 **THE "ADMINISTERED PRICE" (HORIZONTAL) SUPPLY CURVE**

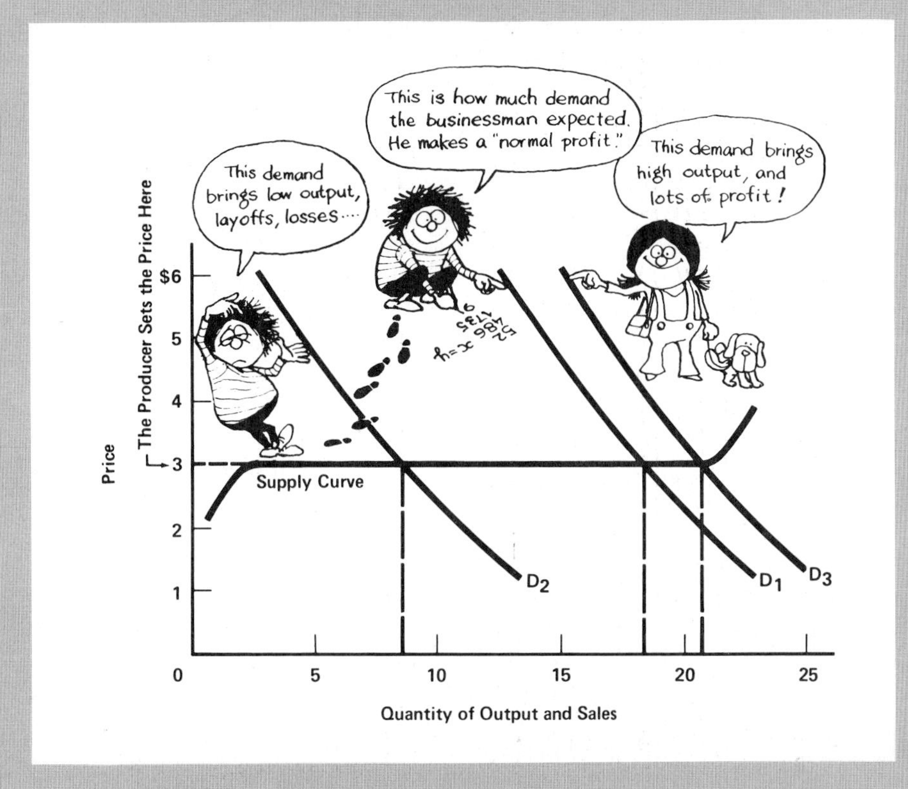


The Price Is Set By The Producer And The Amount Produced And Offered For Sale Responds Directly To Changes In Demand.

This illustrates the supply of a manufactured product marketed under normal circumstances in an economy like the United States. For a broad range of output, the *price* will stay the same. If the quantity people are buying is large, production will be stepped up; if the quantity people are buying is small, production will be cut back. The producer would go on a four-day work week, or perhaps close down for one week each month rather than lower the price.

If the demand is low, the output is low; if the demand is high, the output is high. The price stays the same. But notice that if the demand goes very high (shifts far to the right) the price *will* go up. The firm will find some way to move the price up to take advantage of this golden opportunity for high profits. Or if the demand goes very low (shifts far to the left) the price will go down. Sales and discounts and other gimmicks will be used to get rid of the surpluses. If the demand does not soon pick up, the businessman will stop producing this unprofitable product.

*price,* as it moves up or down, which carries the message from the consumers to the producers, and from the producers to the owners of the factors of production. It is the *price increase* which pulls the factors into the industries which society wants expanded, and it is the *price decrease* which moves the factors out of those industries which society wants contracted. In the real world, very few prices are free to move like that.

*The prices for almost everything you can think of are set by someone.* The economist calls "set" prices, administered prices. Manufacturers, wholesalers, retailers, barber shops, doctors, lawyers, taxicab companies and airlines all sell their products and services at "administered prices." The wages received by most people are "administered prices." Interest rates are "administered prices." Even many of the farm products are now sold at prices which are "administered," either by the government or by farmers' organizations.

All administered prices reflect demand and cost conditions, at least to some extent. The administered prices for manufactured products generally are set on the basis of cost and demand estimates. Then the prices are held firm and the output is expanded or contracted, depending on how much is being sold. Yes, administered prices *are* influenced by demand and cost conditions. But the extent of the influence and the results obtained are not the same—and in some cases are greatly different—from the "pure competitive market model."

### The Effects Of Administered Prices

Economists have spent a lot of energy and effort and have written many volumes on the issues of "administered prices" and "market (monopoly) power." Yet we still don't know for sure exactly how to assess the effects of administered prices in the real world. We know that administered prices reduce the immediate responsiveness of the market price to a shift of demand. But that isn't always bad. Violent fluctuations resulting from sharp changes in demand don't do anybody any good. Administered prices do introduce a degree of stability, which is desirable. But *administered prices permit the seller to bend the economic choices of the society toward his own interests.* This would not be possible in the pure competitive model.

Another problem is that *administered prices tend to inch upward,* all across the board, year after year. Workers want bigger paychecks. Businesses want to show increasing profits. Everyone who is selling anything would like to get more and more money, year after year. When prices are administered, the seller has the opportunity to inch up his price—so he does. This results in a tendency for *a continual upward movement of prices.*

Now you have an overview of the issues of pure competition, market power, and administered pricing. But there's a lot more to it than that.

Not that it's difficult, or complicated, or hard to understand. It isn't. In fact it's all very "real-worldly." When we dig into it you will recognize many familiar things—advertising, for example. So let's go back and start at the beginning and talk about "market structure."

## "COMPETITION" IS DETERMINED BY MARKET STRUCTURE

The market is not a place or a thing, but a concept. You remember that from Part one of the book. Right? I'm sure you remember what "market structure" is, too.

The "structure" of something is "how it's put together." The *structure* of a house is *how it is built*. When you study anatomy you learn about the structure of the human body—how it is put together—how many bones there are, and how many muscles, and where they are and how big they are and all that. You learn what all the parts are and how they fit together. When we are looking at the *structure* of something we are not concerned with how it works—only with how it is built, how it is put together, what the pieces are, and how they all fit together.

You already know that each market is "made up of"—that is, built of, or structured out of—buyers on one side and sellers on the other. All the buyers and all the sellers are not standing there glaring at each other, of course! But they are there, just the same. No more than two people—one buyer and one seller—may meet face to face at any one time. But if each buyer and each seller knows that there are a thousand other buyers and a thousand other sellers just waiting to appear, then we have a "market structure" of pure competition.

### Pure Competition Is A Form Of Market Structure

Whenever there are many small buyers on one side of the market and many small sellers on the other, the market structure is "pure competition." In a situation like that no *one buyer* and no *one seller* can go into the market and throw his weight around. With a thousand buyers and a thousand sellers, if one seller gets mad and goes home, there will be no noticeable effect on the market—on the price, or on the quantity flowing across the market.

Suppose one buyer decides to buy twice as many eggs this week. That won't have any effect on the price or quantity either. With pure competition, the "natural market process" determines the price and the price regulates the quantity flowing across the market from the sellers to the buyers.

When we talk about "pure competition" we are talking about one form of "market structure"—a structure of *many buyers and sellers—so many that no one of them, acting alone, could have any noticeable effect on the market—either on the price, or on the quantity flow.*

### Bilateral Monopoly Means One Seller And One Buyer

What kind of market structure would be the complete opposite of pure competition? Complete monopoly, of course! It would be a market where there is only one buyer and one seller. This market structure is sometimes called bilateral monopoly. Other times it is called monopsony-monopoly (one buyer-one seller). When the steel workers' union bargains with the steel industry we have a labor market of "bilateral monopoly," or "monopsony-monopoly." The union has complete control of the supply. It's the only seller. The steel industry has complete control of the demand. It's the only buyer. You can see how ridiculous it would be to assume that the wage rate for steelworkers will be set by "the natural and impersonal forces" of the pure competitive market!

Whenever a market structure of bilateral monopoly exists, neither the buyer nor the seller can ignore his market power. Anything he does will change the condition in the market. He must take this important fact into consideration. So the monopolist's decision-making process is more complicated than that of the "pure competitor." The monopolist must consider costs and revenues, just as a "pure competitor" would. But *the monopolist must also figure out what effect everything he does will have on the market price*!

Now you have seen the two most extreme forms of market structure. All other market structures are variations of these extremes—such as one seller and many buyers (monopoly), one buyer and many sellers (monopsony), or a few on each side of the market, and so on. All kinds of variations in the sizes and numbers of buyers and sellers are possible. Let's talk about some of the variations.

### Most Product Markets Have Few Sellers And Many Buyers

Many (most) consumer product markets in the real world consist of a few sellers and many buyers. Each *buyer* knows that no matter how much he buys, it won't force the price to go up. But each *seller* knows that if he increases his output very much he will have to reduce the price to sell all of his output. He can set his prices, and as long as his prices are not higher than the prices of the few other producers (his competitors) he can continue to hold a share of the market.

How many products can you think of which are produced and sold by "a few sellers"? How about metals—steel, aluminum, copper, etc.? Or coal, oil, and natural gas? Railroads, trucking companies, bus companies, airlines? Producers of automobiles, washing machines, refrigerators, TV sets, tires, paints, chemicals? Local building contractors, shipbuilding firms, computer companies? Toothpaste, deodorants, hair spray, after-shave lotion, headache pills? About everything you can think of, right?

### Each Producer Must Keep His Price "In Line"

Most of the products you buy are produced and sold by a few large sellers. Each company determines the price it will charge for its product. But each company knows that the price it charges must not be very far out of line with the prices other companies are charging for similar products. If one company's price is so high that it's out of line, no one will buy that company's product.

If one producer is selling only half as much as he would like to, one way he could sell more would be to lower his price. But he knows that if he lowers his price he will take customers away from his competitors. And he knows that his competitors would be very unhappy about that. So what would they do? They would lower their prices too, just like in a gasoline "price war." In the end, they will all wind up in about the same place—except that everybody will be receiving lower prices and probably losing money. So what good would it do to lower your price? No good. You would wind up hurting your competitors and yourself, too. Every producer already knows that. So usually, nobody lowers his price.

### Each Producer Limits His Output

In the kind of market structure we are talking about (where there are a few sellers and many buyers) does each firm produce as much as it can? Not on your life! The rate at which each plant operates is determined by the rate at which people are buying the output. Each businessman keeps his production regulated so that no unwanted surpluses will be produced, and no price cutting will occur.

This is very different from what would happen in a "pure competition" market. Here, each firm is responsive to demand, but their responsiveness does not result from *price movements* in the market. Each firm responds *directly*. If their customers are buying more, they produce more. Isn't that the way you would operate if you were selling Cullowhee hip-huggers to the department stores in Asheville and Knoxville and Nashville and all?

In the egg market, the only way the producers knew that they should start producing more eggs is because the price of eggs went up, offering excess profits. But the way that General Motors knows to cut back on the production of the Olds 88 and increase the production of the Vega station wagon is by seeing how many of each the people are buying. The response of production to demand is *direct,* rather than through the movement of price. If Bethlehem Steel finds that people aren't buying as many hot rolled sheets as before, Bethlehem doesn't wait around and let a glut on the market push the price down. Production is cut back immediately.

## OLIGOPOLY IS A MARKET WITH FEW SELLERS

The kind of market structure we have been talking about is one in which there are only a few sellers. Each seller *knows* that he will push the price down if he is not careful. So each one is very careful. Economists call this kind of market structure oligopoly. "Oligopoly" is an important kind of market structure in the world today. We need to talk about it for a little while so you can get a "comfortable feel" for what it is and how it works. So let's resurrect our island example and talk some more about "palm-tree booze." What do we call it? Tuba! Of course.

### The Island Example, Again

Remember in the beginning when the northside chief had the monopoly on tuba? That was a market structure of complete monopoly on one side and several buyers on the other. The northside chief (monopolist) had the power to restrict the output, raise the price and make profits. Any profit-seeking businessman who finds himself in a situation like that will take advantage of it. Why should he produce a lot of tuba and force the price way down? His maximum profit will come if he limits output and sells his tuba at a high price.

The thing about a market structure of monopoly is that, once he has it, a good businessman *cannot ignore the fact* that he has it. He cannot ignore the effect of his actions on the market—on the price and quantity flow. If he does ignore his monopoly position, he loses the opportunity for more profits. What kind of businessman would he be if he didn't take advantage of his profit opportunities? That's the problem of monopoly. Once it exists, the seller (if he is a good businessman) *must* restrict his output to keep from pushing his price down.

In order to introduce *oligopoly* to our island, we really need more people. So let's assume there are several more islands just like our island, all scattered within easy trading distance. Soon trade develops among all these islands. Also, let's suppose that the only island on which the people know how to make tuba is *our* little island. And suppose that as the years have passed, all of the families on our island have learned how to make tuba. Now there are four major tuba-making operations on the island. All the operations are approximately the same size, and all four families are selling tuba to the people on all the other nearby islands.

### Oligopoly Discourages Price Competition

The northside chief always sells his tuba at the "going price." He is wise enough to know that if he tries to charge more than that, no one will buy from him. They will buy from the eastside, westside, and southside islanders, but not from him. He realizes also that if he tries to sell more by cutting his price, his "oligopolistic competitors" will lower their prices too, to keep from losing their customers. In the end, all the producers would be worse off. So he doesn't cut his price, either.

Is there "competition" in the tuba market? Yes. Oligopolistic competition. But notice how different this "oligopolistic competition" is from pure competition! Remember, in the case of the egg market, one egg producer doesn't care if his neighbor goes into the egg business or not. He knows that one more or one less producer isn't going to have any effect on the market. But with oligopolistic competition, each seller is acutely aware that everything his competitors do, will affect him; everything he does will affect his competitors. Each seller is very careful to restrict his output to that amount which he can sell at the going price. You can understand why.

### What Happens If Someone Cuts The Price?

Suppose the westside islanders are not careful enough about limiting their production of tuba. One day the westside chief realizes that they have produced so much tuba that the only way they can sell it all is to cut the price. So he decides to pull a quick price cut and sell off his surpluses before the other tuba sellers know what's up. He decides to cut the price from eight fishsticks (FS 8.00) to six fishsticks (FS 6.00) per jugful. He sends out a boatload of tuba. As the boat approaches a neighboring island the crew unfurls a huge banner saying: "SPECIAL PRICE! TUBA FS 6.00 A JUGFUL!" The minute the boat gets to the beach, the people begin to clamor for the tuba. In less than an hour all the tuba is sold. The boat heads back home for a new supply.

It just so happens that about two hours after the westside boat leaves, an eastside boat arrives and tries to sell tuba for FS 8.00. No one will buy. They explain that they can now get tuba at FS 6.00 from the westside islanders and that they are not going to buy any more tuba from anyone else unless they can get it for FS 6.00 a jugful. The eastside islanders realize immediately that they can't sell any tuba on that island unless they lower their price to six fishsticks. Those stupid westside islanders! Trying to get away with a trick like that! They should know that *with a market structure of oligopoly, price competition only lowers the profits for everybody*!

### Everyone Must Cut The Price

The eastside islanders reluctantly agree to sell their boatload of tuba for FS 6.00 a jug. They would rather do this than to take back a boatload of tuba or to try to make the long run to the next island before nightfall. By the time they arrive home the eastsiders are in the mood to convince the westsiders of the error of their ways—by fair means or foul! But by the time they get there, the westside chief has already been thinking about this thing. He begins to realize that it would be more profitable for him (and for everybody) if he would just dump his surplus in the lagoon and keep the price at FS 8.00. So before any conferences are even called, he sends around a messenger saying that he made a mistake, and announcing that from now on, all westside tuba will be sold for no less than FS 8.00 a jugful.

Since the demand for tuba is not very elastic, the total amount people would buy at a price of FS 6.00 probably isn't very much greater than the total amount they would buy at a price of FS 8.00. As the price moved down to six fishsticks, the producers would be able to sell a little more tuba, but not much more. They wouldn't sell enough to make up for the loss (FS 2.00 a jug) caused by lowering the price. Everybody would make less profits, so no one is going to cut the price of tuba again.

### Oligopoly Generates Price Leadership

Will anyone ever *raise* the price of tuba? Maybe. If one raises his price, all the others must also raise their prices or else the initial price increase will be rescinded—that is, it will be withdrawn in a hurry! Suppose the northside chief raises the price to nine fishsticks a jugful. Unless everyone else immediately does the same, the northside chief will not be able to sell any tuba.

Here's a new idea. Perhaps the northside chief can become recognized as the "price leader" in the tuba market. If so, he can raise the price to FS 9.00 and be confident that all the other tuba makers will immediately follow. Then all will get more profits at the higher price. Can you see why price leadership usually develops in oligopolistic industries? Sort of obvious, isn't it?

In the United States, if Bethlehem Steel announces a ten percent price increase for hot rolled plate and cold rolled strip, unless the other steel companies soon announce similar price increases, Bethlehem will probably announce that it isn't going to raise the prices after all. Each price will go back to where it was before. Bethlehem doesn't want to lose all of its customers! They know that no one is going to buy steel from Bethlehem, when they could get the same product from USS, Republic, Inland, J & L, or any other steel company, for ten percent less!

### Characteristics Of Oligopoly Markets

All oligopoly markets have three special characteristics:

1. It is essential that prices be identical (or nearly identical) in oligopoly markets.

2. No producer can profit by using price competition—by cutting his price to less than the "going price."
3. The only way any producer can raise his price and then stick to the higher price is for all the other producers to follow along and move their prices up, too.

If you understand why these three simple things are true, then you have a good basic understanding of oligopoly.

A market of "pure oligopoly" consists of a few big sellers of an identical product. Each seller is big enough so that any move he makes will affect everybody. He cannot make any move without considering what effect it will have on the market. And he must always be mindful of what those other sellers are doing, and of how they are responding to each of his moves. It's a careful, watchful, nervous kind of a market situation!

### Oligopolists Would Like To Plan And Work Together

Producers don't like the limitations and restrictions imposed on them by this "pure kind" of oligopoly. No producer can do anything with his price or quantity without considering what the other producers are going to do. Producers like to escape from such restricting influences. One way to escape is to have a little conference with all the other oligopolists and agree on what prices to charge. You might even agree which market territories each one will sell in. A "cartel" agreement like this surely would relieve some of the pressure!

An agreement among the oligopoly sellers would let all of them know just how everything is going to go. Everybody can have high prices and carefully restrict outputs and make high profits. How? By eliminating the nervous, watchful kind of "oligopolistic competition" and replacing it with an agreement which, in effect, creates a kind of "pure monopoly." When all the big sellers get together and form a cartel and plan market strategy, they become, in effect, "one seller."

An "oligopolistic conspiracy" is great for the big producers, but expensive for all the rest of us. This is the kind of agreement "in restraint of trade" which the U.S. antitrust laws are designed to prohibit. You already knew this was an "illegal conspiracy"—one of the things prohibited by the Sherman Antitrust Act of 1890, didn't you?

### Product Differentiation Makes Each Product "Special"

There is another way—a legal way—which producers may be able to escape from the nervous, watchful conditions of "oligopolistic competition." One day the northside chief decides to order from Japan some beautifully designed, multicolored tuba bottles with interesting figurines on the bottle caps. Then when the new bottles arrive he fills them with tuba, and labels them with the slogan THE ONLY MONEY-VAULT-AGED AND DECORATIVELY-BOTTLED TUBA IN THE WORLD. Then he makes a ceremonial trip out to the other islands and donates one bottle to the chief of each village.

On each island the northside chief makes a speech. He says: "Of all the very, very finest things in the entire world, by far the most outstanding and regal is this. This is the only Money-Vault-Aged and Decoratively-Bottled Tuba in the world. This tuba should be used only by those who have that highly developed sense of taste which will enable them to appreciate the very finest. You should never give this tuba to anyone unless you care enough to give the very best!" The northside chief then announces that the price of this special money-vault-aged tuba will be 22 fishsticks a jugful and that he can supply only 12 jugfuls per month.

You wouldn't believe the reaction! Everyone wants some vault-aged tuba. The high price appears to be no deterrent at all. There is a long waiting list of people who want some of this special tuba. In actual fact, the tuba is no different than the other tuba. But in *economics*, it is different. If the people think something is different, it will carry a higher price. So *from the point of view of economics, it is different*!

The northside chief is making beautiful profits. How did he swing it? He escaped from the "oligopolistic competition" by product differentiation—by making his product different from those of his oligopolistic competitors. Now he has escaped from that nervous, cautious oligopolistic competition.

## PRODUCT DIFFERENTIATION GENERATES MONOPOLISTIC COMPETITION

The northside chief is still selling ordinary tuba in the "oligopoly" market. But in his special "vault-aged tuba," he has himself a little monopoly, all his own. He has hit upon a truly profitable scheme—a scheme recognized by businessmen all over the world—"differentiate your product, and advertise!" But how long do you suppose he is going to be able to enjoy this great special advantage? Can he keep this "separate little monopoly" all his own? Forever?

The other tuba producers are not stupid. They, too, are going to think up some special brand names, get some special bottles, differentiate their products, advertise, and create some little monopolies of their own and start cutting in on the northside chief's little monopoly. Just watch.

Soon the westside chief gets a bright idea and orders some pink food coloring from Japan, mixes it with some of his tuba, bottles the tuba in smaller containers of clear glass so the pink can be clearly seen, puts on a label showing a strong man with big muscles, and the words: WEST-SIDE TUBA FOR HEALTH, STRENGTH, VITALITY. He runs his own advertising campaign, complete with two giant-type strong men borrowed from a traveling carnival. He prices his bottles at 27 fishsticks each and announces that because of the great difficulty of producing this special tuba he can only supply 14 bottles a month. Soon he too has a long waiting list of customers for his special "healthful-type" tuba.

You can guess what happens from now on. Soon the eastsiders develop their own brand; then the southsiders develop their own brand. As all this is going on, the northside chief asks himself, "Why have only one brand? Why not one brand to sell for FS 100.00, in a gold-plated jug, and only sell two each month? Maybe have another brand to sell for FS 15.00? And several others?"

### Soon There Are Many Brands Of Tuba

Before long each family has five or six or eight or ten different brands of tuba, all sort of competing with each other, but no two brands exactly alike in the minds of the buyers. The northside chief still has his monopoly in "money-vault aged" tuba and the westside chief still has his monopoly in "health, strength, and vitality" tuba. Each producer has, in effect, a "little monopoly" in each of his own brand name products. What the producers are doing now is *competing by advertising*.

Each producer is trying to convince the people that his kinds of tuba are the best kinds. But by now the island people are beginning to get a little immune to all the advertising and "medicine shows." Several people are beginning to suspect that the different kinds of tuba really aren't all that different after all. Some strong-willed and pioneering people are beginning to buy the brands that are priced lowest, and without even being apologetic about it!

### Price-Consciousness Must Be Carefully Undermined!

The producers do all kinds of things to try to stop this intelligent "price consciousness" on the part of their customers. They pay people to put on fancy clothes and go around drinking expensive tuba. They spread the word that only "low-class" people drink low-priced tuba. They put on "educational programs"—movies and guest speakers and free wall posters and all that—for the school children to convince them early in life, about such "truths" as:

> "You get what you pay for"—"expensive things are good things and show good breeding and good taste"—"low-priced things are shoddy and only the very poor or the 'country clods' would buy such things"—"a truly refined person never asks about the price"—"anyone who is really intelligent will be glad if the price is high because he will know he is getting *good* goods."

So goes the propaganda campaign. You would be absolutely astonished how many otherwise intelligent people fall for it! The sales of the higher priced brands of tuba increase quite a bit, and then stabilize.

Each family is now competing with all the other families in selling tuba at each "price level." Each family continues to have its own little monopoly in each of its brands, but each brand is in close competition with other brands. This kind of market structure should not seem strange to you. You've been living with it all your life! You have been bombarded with it by radio and TV ever since you were old enough to hear and see!

The kind of market structure we are talking about now, is called monopolistic competition. "Monopolistic competition" is simply competition among "little monopolies." Each "little monopoly" is a "brand name," or "differentiated product" monopoly, just as in our island example. With monopolistic competition, most of the "competing" is done through advertising and trying to improve (or differentiate) the product.

### Monopolistic Competition Is Everywhere

In the United States and throughout the world, monopolistic competition exists almost everywhere you look. We find it in toothpaste and cigarettes and soaps and detergents and candy bars and beer and cola drinks and gasoline and swimming pools and TV dinners and almost everything else. It exists with big, nationally known products, and also on the main street of every town and village. Each store is trying to advertise a little better, trying to compete for customers, trying to display its products better, to run attractive sales and to do all kinds of things to get more customers to come in and buy. Each of these businesses, just like each of the brand name products, is really a "little monopoly." The amount of "monopoly power" or "market power" each one holds is really very small. There are so many "little monopolies," all with similar products and services, always trying to take customers away from each other.

Is "monopolistic competition" a kind of market structure? Sure. But it isn't quite as neat and easy to describe as the market structures of pure competition or pure monopoly or pure oligopoly. Why? Because with monopolistic competition, instead of having one product we have many *slightly different* products. The competing products are very much alike. Still, the products aren't *exactly* alike.

### With Monopolistic Competition, Each Producer Has Some Freedom

A "monopolistic competitor" has a little bit of freedom to move his price around without losing all his customers. Once you differentiate your product, you will have customers who will stay with your product even though you raise the price some. All producers would prefer to have "brand name loyalty" among their customers, because this frees them from competition. It lets them raise prices without losing much business.

Just think how profitable it can be to advertise effectively. How long will people continue to pay as much as 30¢ or more for a dozen aspirin, when they could buy a different brand and get a *hundred* aspirin for 19¢? For a long time, I suppose. The difference which advertising can create in the minds of your customers is as real to them as if the difference *really was* real! So it pays to advertise. That's the way it is, with monopolistic competition.

## HOW BAD IS MONOPOLY POWER IN THE REAL WORLD?

How does the market process really work, when we get away from pure competition? Does the market still get the choices made *in response to the wishes of society*? Or does the market process *serve the producers, at the expense of the rest of the society*? This is a difficult question. Many aspects of the "real world" answer aren't even known—not by anyone.

No producer, not even one with *pure monopoly*, can completely escape the influence of consumer demand. At prices which are too high, the monopolist will lose customers. Also, at high prices, if the monopolist is making high profits, the magnetism of these profits to potential competitors is very strong. So we might say that no producer, no seller ever has complete, *absolute* "monopoly power." Some kind of actual or potential competition—either in the same product or in some subsitute product—seems to always be lurking there, somewhere.

### "Bigness" Is Not All "Badness"

Another fact to consider is that "bigness" is not all "badness." Specialization is limited by the size of the market. You can't specialize in something unless the market is large enough to use up all of your specialized output. If the market will only buy half an acre of corn from you, you certainly can't specialize and produce seven acres of corn. It takes large markets to support large firms. Large firms can introduce new technology and can spend money for research and development, and for new, more efficient machines and techniques and processes.

A big firm can look more to the long run than to the short run. It can smooth out some of the ups and downs in its production operations by stockpiling things during times of slack demand and then selling out of inventory when demand picks up. There are several very real economic advantages of bigness, both in production and in marketing. But the problem of "market power," or "monopoly power" always exists when a few big firms make up the seller's side of the market.

The modern societies of "mixed capitalism" are faced with this dilemma:

> Each society wants all the advantages which bigness can bring: good management, effective planning, production efficiency, stability, sound financing, industrial growth and all that. Yet each society wants each business to have many competitors, so that no one will have much monopoly power.

See the dilemma? What should the public policy be? How far should the government go in trying to pass and enforce laws and regulations to prevent businesses from getting big, from joining together, from doing things which the businessmen themselves claim are necessary for increased efficiency? Where and how should the lines be drawn? This is the dilemma.

### We Want To Have Our Cake And Eat It Too

The "market model" tells us that the best interests of the society will be served by markets in which there are many competitors. With pure competition, each producer will be forced to respond to market prices which reflect (a) the wishes of the buyers in the market, and (b) the true cost conditions. Any producer who doesn't respond to market prices soon goes out of business.

If we absolutely prevent bigness, we will force inefficiency on our economic system. The standards of living of all the people will suffer. There isn't much logic in that! On the other hand, we can hardly afford to let the markets become monopolized. The market model tells us that sellers with a lot of monopoly power can (and will) guarantee themselves large profits at the expense of society. So what do we do?

### The "Monopoly Power" Question Is Still Unresolved

Are we certain that the market model really is close enough to the real world to be trusted as our guide on this vital matter? The honest answer is that we don't know. A good understanding of basic economic concepts can be of great help in approaching the problem. But the truth of the matter is that we don't yet have a body of "real-world theory" which is capable of taking us all the way to clear and indisputable answers on these vital public policy issues.

This is but one example of the many kinds of problems facing modern society which are simply unresolved at our present state of knowledge. If you want to do something important in the world while you are making your journey through life, and if this kind of thing appeals to you, perhaps you will help us to work out answers to some of these questions. The farther you go in economics (with your mind turned on), the more you will realize how much help we could use!

## GOVERNMENT POLICIES TOWARD BIG BUSINESS AND MONOPOLY POWER

We have gone into quite a bit of detail on this problem of monopoly power, and on two real world variations: oligopoly, and monopolistic competition. If you have a general understanding of how the sellers behave in these two kinds of market structures, and if you can tie all this in with the idea of administered pricing, then you understand the concepts. Also, you understand some of the public policy issues and problems. But before we leave this interesting (and sometimes confusing and controversial) subject, let's take a brief look at some governmental attitudes toward monopoly power.

### All Nations Of "Mixed Socio-Capitalism" Limit Monopoly Power

What has been the public policy toward monopoly power, in the nations of mixed socio-capitalism? No nation permits the unlimited creation and exercise of monopoly power. Of course not! You can understand why.

Even before the first anti-monopoly ("antitrust") laws, if sellers got together and agreed to limit their output and keep their prices up, this was held to be illegal by the courts. It was considered to be a "conspiracy" against the public—a *conspiracy to restrain trade*.

### U.S. Antitrust Policy

In the latter 1800's in the United States, there was rising sentiment against the great economic power of big business. In 1887, the *Interstate Commerce Act* was passed to limit the monopoly powers of the railroads. Then three years later (1890) the *Sherman Antitrust Act* was passed to prohibit all businesses from trying to monopolize any market, either by joining together in some sort of "trust" arrangement, or by merger, or by any other means.

The Sherman Act was not very strictly enforced. In fact, it was hardly enforced at all. Only a few of the most obvious and flagrant violations were prosecuted. By the early 1900's, Standard Oil had managed to get control of most of the oil companies in the country, and American Tobacco had managed to monopolize the tobacco industry. Then, in

1911, the U.S. Supreme Court forced both these companies to split up—but not until after the Rockefeller family (Standard Oil) and the Duke family (American Tobacco) had made many millions of dollars in profits!

In 1914, two new laws, the Clayton Antitrust Act and the Federal Trade Commission Act, were passed to strengthen the government's position against monopoly. After that, enforcement was somewhat better. Since 1914 there have been several additional laws and a great many court cases. Some have strengthened the government's antimonopoly position. Some have weakened it.

A part of the inconsistency in U.S. antitrust policy results from political pressures which are so strong that they are certain to have some influence. But perhaps the major cause of the inconsistency is the uncomfortable fact that we really don't know what we want. We want all the advantages of bigness and all the advantages of smallness. Which to give up? and which to have? That's the question. Sound familiar? The economic problem—the problem of choosing. Right?

Antitrust enforcement became much more effective during the 1950's and 60's. New laws were passed, and the courts prohibited several mergers between competing businesses—mergers about which, in earlier decades, the courts would have said: "Not too bad. Let's let them go ahead."

**What About "Conglomerate" Corporations?**

During the 1960's, a new kind of "big business" problem arose (or exploded!) on the scene. What problem? The problem of the rapid growth of giant conglomerate corporations—corporations producing everything from oil and chemicals and building materials and aircraft engines and agricultural machinery, to soft drinks, record albums, cake mixes and shampoo! And each "conglomerate corporation" was always looking for more and more, new and different partners to merge with.

What to do about the problem of "conglomerate mergers"? Is it a problem? Each merger is between *non-competing* firms (aircraft engines don't compete with record albums!) so why worry? But what about all that financial power and economic control, falling into the hands of one corporation? Controlled by one board of directors? Isn't that bad? And if so, what should be done?

The courts soon took the position that conglomerate corporations can't be permitted to keep on merging with and absorbing more and more other corporations. The courts began prohibiting some of the attempted mergers, and some of the conglomerates were required to split off some of their merged units. So in the early 1970's the "urge to merge" has been much more restrained. Still, a lot of mergers continue to occur.

## Big Business Is A Fact Of Life

Today multi-billion dollar corporations (conglomerates, and others) make up a major segment of the U.S. economy and of all the other economic systems of "mixed socio-capitalism." It's very likely that that's the way it's going to be from now on.

That's enough on the "monopoly power" issue, for now. It's time to move on into direct confrontation with the question: "How does the market process work in the real world, anyway? After all is said and done, do we wind up where the 'model' says we will? or not?" That's what you'll be thinking about as you're reading the next two chapters—how all this really does work, in the real world.

---

## REVIEW EXERCISES

**MAJOR CONCEPTS, PRINCIPLES, TERMS (Try to write a paragraph explaining each of these.)**

pure competition
pure monopoly
oligopolistic competition
monopolistic competition

**OTHER CONCEPTS AND TERMS (Try to write a sentence or phrase explaining what each one means.)**

market power
monopoly power
anti-trust laws
administered prices
bilateral monopoly
monopsony-monopoly
oligopoly
cartel
price leadership
product differentiation
conglomerate corporation
Sherman Antitrust Act (1890)
Clayton Act (1914)
FTC Act (1914)

**GRAPHS AND CURVES (Try to draw, label, and explain.)**

The "Administered Price" (Horizontal) Supply Curve

**QUESTIONS (Try to write the answers, or jot down the highlights.)**

1. Explain why pure competition is essential to the functioning of the "model market system."

2. Discuss some of the advantages and disadvantages of administered prices.
3. Can you think of any specific ways you have seen businesses "raise the price, without *announcing* that they were raising the price?" Discuss.
4. Explain why oligopoly tends to eliminate price competition and to generate price leadership.
5. Explain why, when oligopoly exists, there is a strong incentive for product differentiation to arise and for the market structure to gravitate toward monopolistic competition.
6. Discuss some of the advantages and disadvantages of business "bigness."
7. Discuss the highlights of "the evolution of public policy toward big business" in the United States.

# 26 HOW "REAL-WORLD" MARKETS REALLY OPERATE

*How well the market process works in the real world depends on how the prices are set and how the buyers and sellers respond.*

What do you know about *real-world* markets? You know how the *model* works. Sure. People demand more, or less. That pushes the price up, or down. Then producers and factor owners respond to the price. As they respond, they are "following the wishes of the people"—and at the same time, they are pushing the price "back to normal." It's a neat system! But what about in the real world? It doesn't work quite that way.

You know that in the real world, such things as monopoly power and externalities and instability and such, keep interfering. Real-world economic systems just don't work the way the "pure model market system" works. But then comes the *tough* question. How do real world economic systems *really* work? How do markets *really* function in the economic systems of mixed socio-capitalism? That's what this chapter and the next one are all about.

The "model market process" lets you see *the forces at work* in all markets. Once you understand the market forces, you can see how such things as monopoly power and instability and inequality and externalities would interfere—would result in something less than the best (the "socially optimal") results. There's no mystery about that. But that's not enough. It helps, sure, but it really doesn't answer the question.

What's the answer? After all is said and done, how do the results really come out in the real-world markets? No professional double-talk, no issue-dodging, no beating around the bush. What's the answer? That's really a tough one. Each product, each market, each purchase, each sale is a "special case." Each market transaction is unique—not identical to any other. It's really difficult to generalize.

Whatever you say about *real work markets* can't be very precise. If so, it would be *wrong* more often than right! Whenever you generalize about the real world, it's impossible to know how often your generalizations are going to hold true. Every rule is bound to have many exceptions. See the problem?

Generalizing about the thousands (*millions,* really) of unique, "special cases" which make up the real world, is treacherous business to be sure! That's why most economics books don't try to generalize about the real world. But this book is going to try. In this chapter and the next, I'll give you as much insight as I can into how real-world markets operate. Finally, at the end of the next chapter, we'll get back to the model and ask the question: *"How well* does the market model *help to explain* the workings of real-world markets?" But now: On to the real world!

### Now The Initial Focus Is On The Real World

How do real-world markets operate? Essentially, that breaks down into two questions:

1. How are prices *really set,* in real-world markets?

   Who really sets the prices? and how? and with how much freedom? and how much restriction? from competition? from government? or from where? How does it *really* work?

2. How do the buyers and sellers *really respond* to the prices that are set?

   Does the rationing function of price really work, to discourage the use of (to conserve) the most scarce, most valuable things (resources—outputs, inputs, and all)? And does the production motivating function of price really induce more resources, more factors, to flow into the production of the higher-priced things? Is the real world price *response* like the price response in the model?

If you knew the answers to both of these questions—how prices really are set, and how buyers and sellers really do respond—and if you knew the answer for *all* of the output markets and *all* of the input markets in the economy, then you would know exactly how real-world markets work. No doubt about it! And you would know for sure how the "real-world economy of mixed socio-capitalism" was working. But can I tell you all that? About the thousands of transactions going on in all of the thousands of markets in the world? Of course not.

The price-setting and price-responding processes in each market, at each moment, are in some ways unique. How impossible it would be for anyone to see all the little details! Still, there are some "general conditions" at work in most real-world markets. Those "general conditions" are what we'll be trying to get at in this chapter.

You have been seeing and living in "real-world markets" all your life. You will recognize the characteristics of the real-world markets the moment I begin to flash them at you. How different they look, from the pure market model! And yet, looked at more closely, how surprisingly similar! So, now, first a quick review of the key role of the price mechanism, then we'll get right into "real-world markets."

### The Price Mechanism Is The Key

You know that the "model market process" works its magic through the price mechanism. The demand of the people, acting against the restraint of supply, sets the price of each good at the level which reflects society's wishes. Then both the producer-sellers and the buyer-consumers respond to the price. What induces them to respond, to follow society's wishes? The two "functions of price"—the *rationing function* and the *production motivating function*, of course!

Suppose we want to look at real-world markets and see how they're working. What do we look at? At the price mechanism? Sure. Are demand and supply setting prices? If not, what (or who) is? And how?

Once the prices are set, what then? Do the prices in real-world markets induce the proper responses? That is, do the rationing function and the production motivating function of price really work? How well?

Suppose "real-world prices" are set the way "model prices" are set. And suppose "real-world responses to price" are the same as "model responses to price." What then? Would "real-world markets" work out exactly the same way the "pure market model" works out? That's right. But everybody knows that in the real world, both the *setting of* prices and the *responses to* prices are different than in the pure model. So the "real-world" results are different than the "model" results (as you know). First, let's talk about how product prices are set in the real world.

## HOW OUTPUT (PRODUCT) PRICES ARE SET IN THE REAL WORLD

You've bought lots of things in your life. How do you suppose the prices were determined, on all those things you've bought? Do you suppose somebody *set* most of the prices? Probably so. You've seen enough "output pricing in the real world" to know a good bit about how it actually happens. You know that businesses set their prices, and advertise, and sometimes run sales, and all that.

### Most Output Prices Are "Administered Prices"

Almost anywhere you look, almost any price you see, has been "set" by somebody. That is, almost all prices are "administered prices." A few output prices may be set by the free, unmodified influences of the "pure market forces." But not many. Sometimes agricultural prices may be set by pure market forces. When prices for agricultural products are above the "government-supported minimum prices," then these prices reflect demand and supply. But there's usually some kind of "price administering" going on. There are many kinds of "producer cooperatives" and other associations for marketing farm products. Sometimes they try to restrict supplies to keep market prices up.

"Producer co-ops" are government-approved monopolistic arrangements for stabilizing (keeping up) the prices of farm and fishery products. Some farm product prices are set (fixed) by government. I expect that in your state, milk is marketed by dairymen's co-ops and that the prices are set by the "State Milk Commission" (or whatever the "milk price control board" is called in your state). So even in the "pure competition-type" farm product markets, prices are pushed around (administered) by monopoly influences, and by the government.

When you get right down to it, almost every price you can find is either set or manipulated or influenced by somebody or something *other than* "the pure market forces of supply and demand." Does this mean that supply and demand are irrelevant in the setting of real-world prices? Of course not. That's really the issue we ought to be thinking about. So let's do.

### The "Administered Price" Must Reflect Reality

How far are the "set" or "administered" prices, from the prices which "the pure market forces of supply and demand" would set? That's the important question. What's the answer? It's different from one industry to another, from one product to another, from one market to another, and from one time to another. Generally speaking, in the "market oriented" economic systems, for most things and at most times, most prices probably aren't *very* far from what the model would call "the normal price." That is, in general, most real-world prices of produced goods are pretty close to "the cost of production, plus a normal profit for the producer."

### Most Prices Really Do Reflect Production Costs

In the real world you'll seldom see anything like pure competition in most markets. Still, there's usually *enough* competition to keep most prices pretty much in line with the cost of production. What competition? Competition from substitute goods, from both *domestic* and *foreign* producers. It's obvious that if any producer's price gets very far above the cost of production, somebody will try to come into the market and start underselling him. This threat of actual and potential competition really does influence the pricing decisions of most sellers. It convinces them not to set their prices "too high"—or not "*very much* too high," anyway.

It is likely that in most real world industries, prices ride somewhat higher, and production is somewhat less efficient than the "long-run-equilibrium-pure-competitive-model situation would bring. But is that relevant? "Long-run equilibrium" is a very fictional thing—like a waveless and ripple-free ocean. The idea of "long-run equilibrium" was never meant to describe "an existing condition," anyway. It's just "a direction toward which things will tend to move."

Real-world markets, with administered prices and scheduled production plans and all that, are much more controlled, and therefore much more stabilized than markets of "pure competition" could be. The increased stability of prices, outputs, and employment certainly is of some social value. Maybe the social benefits of increased stability are greater than the social costs of the higher prices and the lower efficiency which real-world markets (may) bring. Nobody knows, really.

### Some Prices May Be Exorbitantly High

What about exorbitantly high prices? Surely some producers sometimes manage to "corner the market and soak the public." Maybe a company has a patent on something, or maybe a secret process, so nobody else can produce it. Then, if there's no good substitute the producer can really set the price high, far above the cost of production. That way he can make a lot of profit for a long time! Yes, that's true. There are always some sellers "making a killing." That's one of those "exceptions to the rule." Some of those are going on all the time, of course.

What about some of the "high prestige" stores, and the "brand-name" products which sometimes have prices many times higher than the cost of production? With heavy advertising and promotion expenses burning up the society's scarce resources and sometimes fooling the people into making unwise choices? Is this a good way for an economic system to run? Of course not. But it's one of the ways real-world markets *will* run and *do* run until the government comes in and forces the producers to tell the truth about their products.

### How Advertising May Distort The Market

Here's an example. There is no question that many millions of dollars each year are "conned" from the people by deceptive advertising. Recently the Federal Trade Commission (FTC) ordered several of the companies to stop their deceptive advertising. The FTC is even trying to get some of the companies to make public statements *retracting* some of the deceptive and misleading statements they have been making in the past.

But the companies aren't going to give up this profitable activity without a fight—and they have a lot more to fight with (and for) than the FTC!

A problem of deceptive advertising really does exist in real-world markets. Deceptive advertising—with exorbitantly high prices for the deceptively advertised products, and with most of the buyer's dollar going to pay the advertising cost, and with the continuing flow of society's valuable resources into this socially harmful use—this is one of the most obvious and most often criticized failings of "the market system" as it works in the real world. Yet, it's hard to get at. It's hard to separate good advertising from bad advertising. Informational advertising is very helpful—essential, even—in the functioning of real-world markets. Deceptive advertising is the scourge of real-world markets. Let's all hope it won't be long before we learn to handle this problem. The system will work better when we do.

### "Public Utility" Services And Prices Are Regulated

In some markets the "tendency toward monopoly" is so strong that it can't be resisted; competition just can't work to ensure the consumer a quality product at a reasonable price. When that happens, there's no way the market process can work. The government must come in and perform the "consumer protection function." That's what happens in the "public utility" industries.

When you buy electricity or gas, or telephone service or an airline ticket or bus or subway tokens, the price you pay is regulated by the government. Why does the government set the prices for these "public utility type" products and services? Because it's impossible for competition to work to protect the consumer in these industries. The public utilities are just naturally "monopoly-type" industries. That's why the public utilities are called natural monopolies.

Free, open competition among the public utilities would be highly inefficient. Can't you just see several electric companies setting up poles and stringing wires, or several gas companies digging trenches and burying pipes, or several bus companies running city buses, down your street? No way! So the government (sometimes the city, sometimes the state and sometimes the federal government) has regulatory commissions to decide which company will provide what kinds of services, where, and how much they will be permitted to charge.

What do the regulatory commissions do? They establish "required standards of services," then they set the prices for these services. How do they set the prices? Just exactly the way you would guess. First, they try to figure out what the average cost is "per unit of service" (including a "normal profit" for the firm, of course). Then they try to set the price just high enough to cover the "per-unit cost, plus normal profit." The regulatory commission tries to do what competition would do, that's all.

## Regulatory Commissions Try To Do Competition's Job

Whenever a natural monopoly exists, it has to be regulated. Otherwise the buying public would have no protection from low quality and high prices. In competitive markets, suppose a producer doesn't produce a good product and sell it for a reasonable price. What happens? He loses his customers to his competitors! Of course. But in "natural monopoly" markets, there aren't any competitors! That's why the government must come in and regulate.

Do the regulatory commissions do a good job of figuring out "average costs" and "normal profits"? Then do they really set the prices and standards of services where they "should be"? Sometimes they do and sometimes they don't. It usually isn't as easy to make it work as you might think. The history of public utility regulation in the United States provides plenty of proof that government price-setting doesn't always work out so well in the real world.

Consider this: A small increase in the average electric bill could add thousands of dollars of profits to the electric company. The electric company can afford to put a lot of effort into convincing the regulatory commissioners that the price needs to go up a little. What a great opportunity for bribes and graft! There have been several scandals about crooked deals by regulatory commissioners. Probably some crooked deals have been pulled and never discovered. But in view of the great opportunity to make millions and not get caught, I suppose it's surprising that the regulatory commissions usually have been so honest!

How close do public utility prices come to where they "ought to be"? They're probably not too far out of line, in most cases. But it's a constant struggle between the government regulators and the utility companies. The firms are busy trying to convince the commissions to let them reduce their services (to cut their costs) and to push up their rates. The government regulators are busy trying to ensure "adequate services" and to hold the rates down. It would be so much simpler if we could just let competition take care of these things. But with the natural monopolies we can't do that. So the regulatory commissions keep on struggling with the problem.

## Real-World Price Setting Is A Constant Struggle

So how are output prices set in the real world? Mostly, they're administered—set by somebody. And mostly, they're pretty close to "the cost of production plus normal profit." What about the *cost of production*? Costs may be higher in the real world than in the model. True. "Inefficient plant size" may be responsible—or there may be other reasons. Still, it seems to be *generally* true that real-world markets usually do generate a high degree of efficiency.

Real-world markets are scenes of constant struggle—opposing and balancing forces, offsetting each other. Businesses trying to push down costs, to expand their markets, to raise their prices—but being hemmed in, by various kinds of competition, by government restrictions, by labor and consumer organizations—all pushing, pulling, squirming—each trying to get things to be a little better from *his own* point of view.

Somehow, real-world markets seem to work out pretty well—not at the "optimal condition" described by "long-run competitive equilibrium." Of course not. But for the most part, probably not *too far* from it. There are lots of improvements that could be made. But even now, price-setting in the output markets usually doesn't work out badly, all things considered. But what about the "price responses" in the output markets? That's what we need to talk about now.

## HOW PEOPLE RESPOND TO OUTPUT PRICES IN THE REAL WORLD

In real world markets, do prices really ration (cause people to conserve, and limit the use of) things? Sure. That's obvious. What about the production motivating function? Do prices really motivate people to produce? Of course. What else could induce people to work so hard, and to produce all the many things you see around you everywhere? The production motivating function of price does it.

### The Rationing Function Of Output Prices

The rationing function of prices really does work in the real world. The "rationing response" isn't as complete as in the model, but usually it's pretty close. Sometimes businesses use misleading advertising to try to overcome the rationing function of price—to get people to buy higher-priced things than they really want to.

Generally speaking, the rationing function of price really does work. As prices go up, people really do shift to other products. And you know what people do when things "go on sale"! Right? Higher prices squeeze out some buyers and lower prices get people to buy more. Yes, in almost all of the markets in the real world, the rationing function of price works all right. Not perfectly, but adequately.

### The Production Motivating Function Of Output Prices

The production motivating function of price really does work in the real world, too. But it doesn't work quite the way it's explained in the model. To be sure, if one product is selling at a very high price (much above it's average cost), then pretty soon some competitors will come in and start selling acceptable substitutes. This is the production motivating function of price at work, all right.

Usually, though, in the real world the "motivating process" works in a different way than in the model. The market gets into balance and everything comes out looking a lot like the condition described in the model, but real-world markets get into balance sort of "backwards" from the way the model explains it. Here's an example.

Suppose Polaroid has developed a new kind of camera. The price is set at $19.95, and Polaroid is going to produce and sell as many as people will buy. They think they will be able to sell enough to go beyond the "break-even point," and make profits. They put on a nationwide advertising campaign and introduce the new camera. So what happens?

The new camera sells just great! The price stays at $19.95, and Polaroid keeps producing and selling more and more and more. Is Polaroid responding to the production motivating function of price? Sort of. But really, they're responding *directly* to the demand. They're responding to the "production motivating function of *sales*." As more cameras are sold, more are produced. The price doesn't change. You saw a graph illustrating this "administered pricing" concept, back in the chapter on "monopoly power." Remember?

Polaroid sells more and more cameras and makes high profits. What's going to happen next? You could guess. Kodak wants in. Leica and Nikon want in, too. These other companies come out with competing models as soon as they can. Are Kodak and Leica and Nikon entering this market in response to the "production motivating function of price"? Sort of. What they're after is some of the high profits, that's for sure.

But it wasn't an upward movement in *price* that attracted Kodak and Leica and Nikon into this market. It was the high volume of *sales*. It's the high volume of sales that makes this little camera so profitable. *In the real world, the "production motivating function of price" usually works out to be the "production motivating function of sales."*

## The Buyer Gets A "Multiple Choice"

In the "model," when the demand increases the price increases, so the producers make excess profits. The profits attract more firms into the industry, so the supply increases. That pushes the price back down to "normal." But in the real world it usually works differently. In the real world the producer chooses a product he thinks will sell, sets a price he believes will be profitable, then produces the product and pushes it out to let it "stand the test of the market place."

Of course the producer doesn't introduce his product and leave it out there lying all alone in the market, ignored! He advertises and does everything he can to convince the buyers to buy it. So really, the buyers are offered a "multiple choice." All the competing products are out there in the market, vying for the buyer's favor. The buyers will choose some and ignore others. The chosen ones will continue to be produced. They will be imitated by other producers. The ignored ones will disappear from the market.

Those products which nobody thinks about producing, or which nobody thinks would be profitable, will not be offered. A product can't be "chosen" if it isn't offered. The consumer usually doesn't "generate" new products. Businesses do that. Each business is always trying to develop profitable products. Some producers guess right. They succeed, make profits, and get more chances to develop new products and to offer more "multiple choices" to the buyers. Some producers guess wrong. They don't make it.

### The "Chosen Ones" Survive And Prosper

This process goes on all the time. You see it working in all kinds of products and services. The well-managed businesses, with good products and services and good customer appeal, will survive and grow. The ones that can't stand "the test of the market place" will fail and disappear.

This process works with all products, and with all services. It works with drug stores, department stores, clothing stores, variety stores, candy stores, grocery stores, appliance stores and record stores; with beauty parlors, barbershops, gas stations, launderies and cleaners, taverns and bars, restaurants, hotels and motels, dance halls, movies, campgrounds, recreation areas; with auto repair services, building contractors, plumbing shops, hamburger drive-ins, and just about everything else you can think of.

Consumers keep making their "multiple choices." The "unchosen ones" fail the test of the market place. They just fade away. The "chosen ones" survive and prosper. They expand their outputs. Soon other producers imitate them. That's how "the production motivating function of price" works in the real world. It's quite different from the model, but most of the time it seems to come out pretty close to the same place.

Most of the time, in most real-world markets, the things people are demanding are being produced in about the right quantities. Most prices are pretty close to the cost of production. Anyone who is ready to pay the price can get the product he wants. There are always some businesses making good profits while others are going broke, and there's always a movement of input resources toward the profits and away from the losses. That's the way it works in the output markets. What about the input markets? How do the input markets work in the real world?

Are the real-world input markets like the real-world output markets? —similar, yet different from the model? You could guess. Sure they are. So we need to get into these questions: "How are input prices really set?" and then, "How do the 'rationing' and 'production motivating' functions of price really work in the input markets?" That discussion is coming up right now.

## HOW INPUT PRICES ARE SET IN THE REAL WORLD

Are real-world wages, rents, and interest rates set by pure market forces? You know the answer. *Nothing in the real world is determined completely and purely by the "pure market forces" described in the model.* Surely everyone knows that by now!

### There Are Many, Many Input Factors

In the model we talk about three factors of production. We talk about "labor, land, and capital" as though all "labor" was identical, and all "land" and all "capital" were identical. But really there are all *kinds* of labor, land, and capital. One kind of labor may be very scarce, very productive, and very highly paid. Another kind may be very plentiful, unproductive, low-paid, and maybe unemployed.

So in the real world, instead of having only three factors of production, we really have maybe thirty, or three hundred (depending on what kind of business you're in and how you choose to look at it). Each businessman will decide about each *kind* of capital, each *kind* of land, and each *kind* of labor, in exactly the same way he would decide *between* labor, land, and capital.

Each decision to hire or not to hire, which to hire and how much to hire, reflects the principles illustrated in the model. The productive kinds of factors will be hired; the more productive a factor is, the more the business will be willing to pay for it. It really does work that way in the real world—except that there's a lot of guessing going on—and there are lots of pressures influencing the decisions.

### Most Factor Prices Are "Administered Prices"

You probably won't be too surprised to find out that most *input* prices (like most *output* prices) are administered. Can you think of any real world wage rates that are actually set by "pure market forces"? Probably not. In a minute you'll find out that you can say the same thing about interest rates.

In the case of land and natural resources, the market forces of supply and demand really do sometimes set some prices. But almost always there's some kind of monopoly power or government regulation or something, and the prices wind up "administered" after all. Let's talk about each of these factor prices, in turn—about how wages, interest, and rents are set in the real world.

### How Wage Rates Are Determined

I suppose everyone knows that wage rates aren't set by the "free market forces of supply and demand." Everybody knows about "minimum wage" laws. The "minimum wage" started out several years ago at less than $1.00 an hour and has been increased several times. Now it's up around $2.00, and Congress is always talking about increasing it even more. You can be pretty sure that it will keep going up. Woe be unto the congressman who votes against an increase in the minimum wage! The minimum wage surely isn't set by the pure market forces of supply and demand. What about other wages?

Most big companies bargain with labor unions to set wages. The sellers of "professional labor" (doctors and lawyers and public accountants and such) usually set their charges (wages) by following the pattern recommended by their "professional association." Government employees usually have some sort of salary scale, approved by the legislature or the civil service commission or somebody. It's hard to find any "unadministered wages."

Are the administered wages somewhere close to where the "free market forces of supply and demand" would set them? Maybe yes, maybe no. But if the "administered wage" is pushed up very much higher than the "free market wage" would be, there's likely to be some unemployment. Why? Because businesses will not hire people who can't produce enough to pay their way. That's almost as true in the real world as it is in the model.

Suppose the "administered" wage rate is set very high. What will the producer do? What *must* he do? Either he must respond (a) by hiring less labor, or (b) by bringing in more or better capital to make labor more productive, or (c) by raising his price enough to cover the higher wage. Prices will be increased and/or more capital will be added and/or employment will be decreased to where those workers who are still employed will have a high enough "value product" to be "worth" the high wage.

See what happens? *The value product adjusts itself to the wage rate.* How about that! When the wage goes up, those businesses which *don't* do something to increase labor's "value product," soon will be out of business. We'll talk more about all this in a minute, but first, look how interest rates work—sort of in the same way.

### How Interest Rates Are Determined

Everybody knows a lot about wages, but most people don't know much about how interest rates work. Really, there's nothing mysterious about it. Most people can't understand interest rates because they don't understand about money. But you can, because you *do*!

Banks lend money to consumers and businessmen who want to borrow. If the banks have lots of excess reserves and want to make more loans, interest rates go down. It really does work that way. If the banks are all loaned up and consumers and businessmen are still clamoring for more loans, interest rates really do go up. So is the market model working perfectly? with supply and demand setting interest rates in the real world? No, not exactly. Why not? Because in the real world the market is "rigged."

### The Fed "Rigs" The Money Market

What determines if the banks are all "loaned up," or if they have lots of excess reserves? Many things. But the very biggest of these "many things" is what the Federal Reserve decides to do. The government is always trying to follow a monetary policy "easy" enough to let the economy be prosperous and growing, but not "easy" enough to let the economy slip into rapid inflation. Remember? Sure!

The Fed is always manipulating bank reserves by using open market operations and adjusting its discount rates. That means the Fed is "rigging" the supply side of the money market. That's how it makes interest rates high, or low. Also, its open market operations influence interest rates directly. How? Through the effect on bond prices. When the Fed sells bonds, that pulls money out of the banks' reserves and it also pushes down the prices of bonds and that automatically pushes up interest rates. Remember?

Interest rates really are determined by supply and demand, more or less, but not "free" supply and demand! The Fed always has its hand on the "supply control knob" and can make the money market as "tight" or "easy" as it wants to. Then the big banks set their interest rates to reflect the "tightness" of the money markets, and all the little banks follow along.

If lots of businesses suddenly decided to borrow lots of money and do a lot of expanding, would this push interest rates up? Sure it would. How far up it would push interest rates would depend on how much the Fed would permit the money supply to expand—that is, how much new bank reserves the Fed would be willing to generate. If the Fed would permit the money supply to expand a lot, interest rates would rise only a little.

Suppose the Fed didn't want the money supply to expand very much. Then bank reserves wouldn't be allowed to expand very much. The money market would get very tight and interest rates would go very high. The high interest rates would discourage businesses from investing. You found out about that from your uncle's Chincoteague oyster-breading plant, and from your roommate's father's feasibility study of the Isphem-ing pelletizing plant. Remember?

It wouldn't be wrong to say that in general, interest rates are set by "supply and demand." But since the Fed controls the supply, it can and does control the general level of interest rates in the country all the time.

### The States Have "Usury Laws" Setting Maximum Interest Rates

In addition to the controls by the Fed, the states have various kinds of "usury laws" limiting the maximum interest rates banks and other lenders can charge for various kinds of loans. For example, many states limit interest charges on "revolving charge accounts" and "overdue credit card charges" to 1½% per month. That works out to a maximum rate of 18% per year (12 months, times 1½% per month = 18% per year). The State of New York and some others limit the interest rate on home mortgages and other loans. The highest rates permitted by state laws usually are for pawn shops, which, for very small loans and for very short periods of time, usually charge interest rates which amount to several hundred percent (sometimes more than 1,000%) per year.

### What About The Productivity Of Capital?

Doesn't the demand for (the productivity of) capital have anything to do with the setting of interest rates? The productivity (marginal value product) of capital makes up the *demand side* of the "money market" (where interest rates are determined), doesn't it? Sure. If businesses expect capital to be productive enough to be profitable, they will be trying to borrow more money and buy more capital. The borrowing will push upward on the interest rate. But will the interest rate actually go up? That depends on what the Fed decides to do.

Suppose the Fed "holds the line" on money expansion. Interest rates will go up more and more. Does that mean the businesses think the productivity of capital (the marginal value product of the new machines and things) will be high? Of course. Only when businesses expect a high productivity of capital will they borrow and buy the capital and pay the high interest. Obviously!

If the Fed is trying to curb inflation (using a "tight money" policy) interest rates will be high. Only those businesses which have *highly productive* uses for capital will borrow and invest. Only that capital which is expected to be productive enough to pay the high interest cost will be bought. See how a *high interest rate can force an increase in the productivity of new capital investments*? Just as a *high wage rate* can force an increase in *the productivity of employed labor*. Labor that isn't productive enough to be worth the high wage, won't be employed. Capital that doesn't look productive enough to be worth the high interest, won't be invested in. See what's happening? In the model, factor productivity determines factor prices. But here in the real world it looks like *factor prices are determining factor productivity*! More on this interesting issue in a few minutes, but first, what about *rents* in the real world?

### How Rents Are Determined

Suppose you own a piece of land, or some natural resource — maybe a puddle of oil about half a mile down in the ground. How much money can you get from your land, or from your oil? That depends on how high the demand happens to be. You can set the land rent or the oil price as high as you want to, but if the price is too high, no one will buy. There are many "substitute pieces of land," and "substitute puddles of oil" which people can buy instead of yours. If for some reason people want very much to have *your* land, or *your* oil, then they will pay an especially high price to get it. But if they don't, they won't.

Perhaps the prices of land and natural resources ("rents") are somewhat more responsive to "pure market forces" than wages or interest rates. But not necessarily. Some monopoly power usually exists on both sides of most real-world land and natural resource markets. Still, it's fairly safe to say that the price which is paid for each bit of land or natural resource usually reflects the strength of the *demand* for it.

If you happen to own forty acres in central Florida near Disney World, you're going to make a lot more money on your land than if you owned forty acres someplace in the Everglades. Why? Because there's more *demand* for Disney-World land! If you own a pool of oil down in your land and I own a pool of salty water down in mine, you will get rich and I will stay poor. Why? Because there's more demand for oil than for salty water.

There are many reasons why the price in the real world doesn't work out exactly as it does in the model. You may sell your oil too cheap, to the first fellow who comes along. Or you may sell your forty acres near Disney World for half a million dollars when someone else would have been willing to pay you five million. But if the demand for some kind of land goes up, pretty soon the word gets around and the price responds.

Most land and natural resource prices are "administered" to some extent, either by the buyers or the sellers. Governments sometimes regulate these prices, too. But in general, prices of land and natural resources really do come fairly close to reflecting the strength of the demand. And the demand for land really does generally reflect its marginal productivity (its marginal "value product").

### A Factor's Productivity Adjusts Itself To The Price

In real-world markets input prices (like output prices) are usually "administered prices." Somebody "sets the price" almost all the time. But, as you might guess, the price of each input factor really is pretty close to its "productivity." Why? Because the business can't afford to pay a factor much *more* than its "value product." If it did, the revenues wouldn't be high enough to cover the costs, and the business would go broke!

If the factor's price is much *less* than its "value product," then as time goes by, more of the factor usually will be hired. As more of the factor is hired, its "value product" really will go down. So in the real world the *ultimate results* look very much like what you would expect from looking at the model. But just as in the case of the output markets, the input markets seem to get there via a different route—sort of "backwards" from the way the model works.

In the real world, the usual thing is for the price to be set first—that is, "administered" by the seller or the buyer or the government or somebody. Then *the productivity of the factor adjusts itself to the price* (instead of the *price* adjusting itself to reflect the *productivity*, as in the model). After you read the next section, about how the real-world input markets respond to prices, you will understand this a lot better.

## HOW INPUT MARKETS RESPOND TO FACTOR PRICES

Do input markets respond to price changes? Of course. The rationing function of price *conserves* the higher-priced factors; the production motivating function of price *pulls in more* of the higher-priced factors.

### The Rationing Function Of Factor Prices

What happens when the wage for some kind of labor goes up? Everyone using that kind of labor tries to use less of it. When less of the higher-priced labor is used, its productivity goes up. (But unemployment is likely to go up, too.) See how a high *price* of a factor can force the productivity of the factor to move up, in line with the price? Producers really do buy less of the high-priced factor, so that causes its productivity to increase—that is, *the units which are still employed* are more productive. The rationing function of price causes it to work out that way.

The rationing function of price works for all factors of production—for every kind of labor, capital, land, and natural resource. The rationing response is much smaller in the short run than in the long run, of course. It takes awhile to automate the plant to cut down on the high-priced labor. The longer you wait after the price goes up, the greater the response you're likely to see.

### "Featherbedding" Practices Impede The Rationing Function

There are many real-world restrictions which keep this rationing function from working out the way the model says it will. Restrictions can be severe in the labor market. Suppose you were working in a plant and your union had just gotten a big wage increase. Then suddenly you found out that you were going to be "automated out of a job." You wouldn't like that much. Right? Nobody else does, either. So labor unions fight hard to keep high-paid workers from being replaced by machines.

Ever hear of "featherbedding"? Labor unions sometimes get employers to agree not to lay off people, even though their jobs no longer need to be performed. The classic example is the fireman (the man who is supposed to stoke the steam boilers) who always rides along on the diesel locomotive (which has no steam boilers, of course). The idea of "featherbedding" is that the worker might as well take along a featherbed and sleep, since there's no work for him to do.

There's no way of knowing just how much the efficiency of the economy is reduced by "featherbedding" practices. Certainly it's significant. It's more serious in the old, established industries (railroads, steel, coal, etc.) than in the new and growing industries (petrochemicals, computers, etc.). Also it's more serious in the oldest industrialized centers than in the newer industrializing areas. (In the United States, for example, it's more serious in New England, the Northeast and Midwest than in the South, Southwest, and far West.) The newer industries and the newer areas can develop without the restrictions of obsolete techniques and practices to impede them.

### Waiting For The Long Run

The owners of the other inputs (land and capital) are less likely to be able to interfere with the rationing function of price. If the cost of investing in capital goes up, some businesses will invest less. That's true in the real world, just as in the model. The only serious restriction in the response is that of "waiting for the long run."

After the investment is made—the note is signed at the bank, the money is spent, the concrete is poured, the machines are all in place and all that—it's going to be awhile before all this process can be "undone." So that's sort of a natural limitation on the responsiveness to a factor-price change. You learned about "waiting for the long run" from the model. Remember?

What about featherbedding and other such restrictive arrangements? Will the long-run work these things out, too? Sure. Either one way or another: either *employment* will be adjusted (reduced) so labor's productivity can increase, or *competitors* from other parts of the country or from other parts of the world will come in and take over the markets and force the "old-line producers" out of business.

Many kinds of things can be done to slow or stop the adjustment process (for example "protective tariffs" or other import restrictions can be imposed to prevent foreign competition), but the "pressure to adjust" will be there, constantly "pushing against the dike" until something gives and lets the adjustment occur. Usually, when all the opportunity costs are considered, the *harm* done by trying to prevent the adjustment is greater than any *good* effects that might result.

Even with all the restrictions and limitations, the rationing function seems to work fairly well in most real-world input markets. As factor prices for different kinds of labor or land or capital go up, these higher priced factors really are more carefully conserved. There's some effect in the short run, then the response gets much greater as time goes by. In the model, the demand for a factor of production is *much more elastic* in the long run. Remember? That's how it is in the real world, too.

What about the production motivating function? Does a high price for a certain kind of labor bring in more of that kind of labor? What about land? or capital? Sure it does. Here's an example.

### The Production Motivating Function Of Factor Prices

Suppose you heard that people who can read English are being recruited and paid $250 a day to go overseas to read to people in some underdeveloped countries. Do you think anybody will sign up? For $250 a day? Count me in! How about you? Sure. See how labor really does respond to a high price?

What about land? Suppose a wealthy "big shot" just moved into your neighborhood. He doesn't want to park his car on the street overnight, but he has no driveway. He's looking for some little place—some land—where he can park his car. He's willing to pay $20 a night. Do you think somebody will supply him some "land"? I might. How about you? Yes, land also responds to high prices.

What about the interest rate? Suppose you have several hundred dollars in savings, drawing 5% at the local savings and loan. Then you hear that one of the local banks is paying 7% interest on savings accounts. Do you suppose anybody will shift his savings to the high interest bank? Will you? Me too! Does money flow to where the interest rate (price) is highest? You bet!

Now you know the answer to the general question: "Does the production motivating function of price work in the factor markets?" Of course it does. Whenever the price offered for a factor is *high enough* you may be sure that factors will shift from other uses, into the high priced use. It really does work that way in the real world—sometimes. But that isn't *usually* the way it happens. Just as in the case of the output markets, the production motivating function of price seems to work itself out "backwards" in the input markets, too.

### Inputs, Like Outputs, Respond Directly To Demand

Suppose a businessman wants to hire more workers. Does he offer a high wage rate (price) to attract the people he wants? Not usually. He just advertises to let people know he has jobs available at the "going wage." Then do workers who are already getting the "going wage," quit their jobs and come to work for him? No.

It works like this: In a "dynamic real-world economic system" some people are always changing jobs. Some businesses are going broke; others are expanding. Jobs are constantly being eliminated; new jobs are being created. The "structure" (make-up) of the economy is always changing. New technology eliminates old jobs and creates new ones. These structural changes in the economy are constantly displacing workers, who (if they are young and flexible enough) can shift into the new kinds of jobs that are opening up.

### There Is Constant Turnover In The Factor Markets

There is a constant flow of people becoming available and looking for jobs. Some are people coming out of the high schools and training schools and colleges and entering the labor force for the first time. Others are people who have lost their jobs for one reason or another. All these people are watching the ads and looking for jobs. At the same time, new and expanding businesses are hiring people. This constant turnover is always going on, not just in labor markets but in all factor markets.

Factor prices always are being set by the various "administering forces" (within some reasonable distance of the price which *would have been set* by "supply and demand"). Then, at that "set price," labor and other inputs flow across the markets from sellers to buyers. If the set price of a factor is "too high" then businesses will start using less of that factor and more of the other (less expensive) factors. That's how it works its way out. At a higher price, less of a factor will be used, so its *marginal productivity* will be higher (high enough to cover the high price). It really does work that way in real-world markets.

Anyone who wants to hire people at the "going wage" usually can find workers. He doesn't have to offer to pay a higher wage. The same is true in the land and capital markets. There is a constant flow of all kinds of inputs, moving across the real-world factor markets at the "going prices." Usually anyone who wants more of an input can get it at the going price.

So does the production motivating function of price actually work in the input markets? Yes. But just as in the output markets, *the inputs usually respond directly to demand.* If people are demanding more Polaroid cameras, more of them will be produced. The price doesn't have to move. The same is true in the input markets.

If a producer is demanding more of a certain kind of labor, or land for a certain purpose, or a special kind of capital, usually the response will be direct and almost automatic. No change in price is required. That's the way it really does work in the real world.

## WHAT ABOUT FACTOR PRICES AND INCOME DISTRIBUTION?

How is the income distribution question really answered in the real world markets of "modified capitalism"? What really determines who gets to have how much of the output? Is each person's share determined by the value of what he, his land, and his capital produce? Do owners of the scarcest, most productive factors really get the highest incomes?

### Income Really Does Go To The Productive Ones

You may be surprised how closely income distribution in the real world really does reflect what the pure model predicts. Think about it. People who own and sell the high-priced factors really do receive the high incomes. People who "sell" highly skilled or professional or managerial labor receive high incomes. People who own a lot of highly productive capital or land or natural resources receive high incomes. People who work overtime, or those who go out of their way to seek high paid jobs, receive more income than those who don't.

If the demand for your kind of labor is going up, then your income will be rising. You will be able to get more of the society's output. If you own a very scarce and productive kind of land or capital, you will receive a high income. It really does work that way.

You know (in the model) what creates a high demand and a high price for an input. High productivity? Of course! And you know what makes a factor's productivity high, too: lots of other good factors to work with. Is that the way it is in the real world? If a society has an abundance of good capital and land and natural resources and a relatively small population and labor force, will the wage rate be high? Will "the working man" enjoy a good income and a high standard of living? Right! It really does work out that way.

## The Scarcest Factor Gets The Highest Return

When labor is the relatively scarce factor, then labor is highly productive. Workers receive high wages, and most people enjoy high standards of living. But what if there's a great abundance of labor? Suppose there are great hordes of people, but very little capital or land and very few natural resources. What then? The productivity of labor will be very low. Wages will be very low. Most people will be poor. That's the way it works in the model and that's the way it works in the real world, too.

Compare different countries in the world. Or look at different areas within any country. What do you see? In places where there is an abundance of highly developed land and natural resources and highly productive capital (so that each worker has a lot to work with) wage rates are high. Most of the people enjoy a good living. In places where there is lots of labor and not much capital or land or resources, wages are low. Most of the people are poor.

In the areas where labor is more scarce, labor is more carefully conserved. Each worker has more of the other factors to work with, so each is more productive and receives a higher wage; each receives a larger income and a larger share of the output. It really does work out that way. Here's an example, from the history of the United States.

## An Example From History

In the latter 1800's and early 1900's, millions of people moved to the United States. Why? Jobs were available. Lots of land, natural resources and capital were available, so labor was more productive. As more land and resources were opened up and as more capital was built, more labor was needed. More and more labor came in to meet the demand. All the labor coming in, kept wages from rising much, kept the productivity of capital high, kept the profits of the capitalists high, and kept the economy growing by leaps and bounds!

Suppose the massive flow of immigrants had been stopped. What then? Labor would have been more scarce. Wages would have gone up. Living standards would have gone up, and profits would have gone down. There would have been less savings and investment. More consumer goods and fewer capital goods would have been produced. The growth of the economy would have slowed down.

Today, the rapid inflow of labor has been stopped. Capital is now more plentiful and labor is relatively scarce. Wages are high; consequently, profits are low. Most of the people are enjoying high incomes. Why? Because of the high labor productivity which results from the great amount of highly efficient capital.

The owners of land and natural resources have been receiving higher and higher incomes. Why? Because with more people and more capital, land has become relatively more scarce. The more labor and capital there is, the more productive and higher priced the land becomes. Of course! So the landowners get a bigger "distributive share." That's the way it works in the model and that's generally the way it works out in the real world.

### Some Kinds Of Labor May Be Very Scarce While Others Are More Plentiful

In general, incomes in real world markets seem to be fairly close to the results described in the model. But for any *particular* worker, or piece of land, or machine, the "in general" condition may not be very relevant. In any economic system, or in any area, *some kinds* of labor or land or capital are relatively scarce and some kinds are not. (If you're not careful, generalizations can be dangerous.)

If "labor on the average" is scarce, that isn't going to help a person very much if he is uneducated, unskilled, inexperienced, unproductive, or unwilling. If "capital in general" is relatively plentiful in a country, that doesn't mean that some new kind of highly efficient machine isn't going to bring to its owner a very high return. Be careful of the generalizations! You can always "beat the average" if you're smart enough, or lucky enough—or both.

### A Normative Question: Do Real-World Markets Bring "Distributive Justice"?

Do real-world markets bring a "socially just" distribution of the society's income and output? Since it appears that income distribution in real-world markets really does reflect factor productivity, does that mean the distribution is equitable? Does each person actually get the amount of income he really *should* get? the amount he really *deserves*?

This "normative" question of the "social justice" of income distribution can't be answered by the concepts (the "laws") of positive economics. Each person has a right to his own philosophy on issues such as this. Each person can have his own opinion about what is and what is not "socially just."

As each person decides what he thinks, he can try to influence real-world conditions to change in the ways he considers "desirable." If he follows the laws of positive economics (like: "you can't have your cake and eat it too") he may succeed. Each person is free to work through the market process, the social process, and the political process to try to cause the society to move toward the conditions he considers to be "more just," or "more desirable."

### Everyone Is Trying To Redistribute Some Of The Income

Some people acquire great fortunes and then give them away. Why? Some people dedicate their lives to trying to improve the conditions of the poor. Why? Some people try to get the government to collect more taxes from the income-earners and to give more of this money, and more and better services and things, to the poor. Why? Almost everybody seems to be doing something to try to redistribute some of the income—from contributions to the Red Cross and to the church, to trying to get the government to increase payments for unemployment and social security. Why? *To try to change the society's distribution of output and income*, that's why.

Most people seem to agree that the income distribution which results from "the ownership and sale of productive factors" is not socially acceptable—not entirely. It seems to be generally accepted that some redistribution is essential. One problem results from the fact that the ownership of wealth (capital and land and natural resources) is so unevenly distributed among the people. A few people are very wealthy so they receive very high incomes from their property (land and capital). Another problem is that some people and some families are incapable (either temporarily or permanently) of producing enough to live on. Yet society wants them to have a share of the income and output.

The government does many things to redistribute income from the people whose "factors" produce it to those who need it. Taxes are collected from those with more wealth and income, and some of the money is paid to the poor. The government offers various kinds of public facilities and services—parks and playgrounds, schools and hospitals, and many other things. All these "income redistributions" change the results generated by the real world market forces. But there's another approach that can be used. The government can "rig the markets" so that larger incomes will go to "the disadvantaged people"—to those the society feels *should* receive more output. How does the government rig markets? By doing things to push prices up or down—by "input price-rigging and price-fixing."

### Minimum Wage Laws And Other "Input Price-Fixing"

Governments sometimes set input prices to try to improve the conditions of low-income people. Minimum wage laws try to guarantee that everybody who works will earn enough for a "decent living." Maximum interest laws try to prevent lenders from "gouging" the poor. Rent ceilings are sometimes used to try to prevent landlords from overcharging. What are the effects of all this "input-price-fixing"? You already know enough about "the functions of price" to be able to guess. Let's look at some examples.

Suppose the legal limit is 7½ percent on home mortgages, and the market rate on government bonds goes up to 8 percent. Who will lend you money at 7½% so you can buy a house? Nobody. So what does the law accomplish? It keeps the lender from "gouging you," all right! It also keeps you from buying the house you want to buy. Too many people are trying to borrow the low cost loans; too few banks are willing to lend. *Holding prices down always creates shortages.*

Maximum rents work the same way. Suppose the rent ceiling on a certain type of apartment is $80 a month. Then suppose maintenance and repair costs and interest and taxes and other costs go up so that landlords can't break even. What happens? Landlords start to cut corners and let the apartments run down. More people want to rent the low-cost apartments, but fewer and fewer are available. Ultimately these apartments will vanish from the market. So what does the rent ceiling do? It creates a shortage of appartments. Holding a price *down always* creates a shortage.

Minimum wage laws create problems, too. Suppose the minimum wage goes up to $2.50 or maybe $3.00 an hour. Will businesses use less labor and more capital? Of course. Which workers will be hired? The highly skilled, most productive ones. Who will hire the thousands of inexperienced young people who are constantly joining the labor force? At $3.00 an hour? Nobody. So how do all these "surplus workers" get jobs and get a start? They can't. The price of their unskilled, inexperienced labor is too high. *Holding a price up always creates a surplus.*

### Input Price-Fixing Always Generates "Reflex Effects"

The indirect effects of "factor price-setting" by the government can create serious problems. Does this mean that the government shouldn't set minimum wages? or rent ceilings? or maximum interest rates? Maybe so, maybe not.

What it does mean exactly, is this: Whenever the government decides to regulate any input price, some undesired, indirect effects are going to be generated. Maybe the price-fixing will do more good than harm; maybe it will do more harm than good. But you can be quite sure that unless preventive action is taken, the price-fixing will do *some* harm. Why so? And what preventive action might be taken?

Price manipulation *always* produces some indirect effects—some "reflex actions" in the market. Why? Because both the buyers and the sellers respond to the "manipulated" price. How *great* the response will be depends on the *elasticity* of the supply, and of the demand. In the short run, the elasticity of both the demand and the supply is likely to be low, so the response is likely to be small. But as time passes, the response will become greater and greater.

If rent ceilings are imposed this month, there isn't much the landlord can do now but go along. But in the long run he can let his apartments run down. Then, ultimately, he can just abandon them. In some big cities, many apartment buildings actually have been abandoned because they wouldn't "pay their way."

What can be done to prevent the undesired effects of "input price-fixing"? It's very simple: If the government is going to try to *hold a price down*, then it must do something to *prevent shortages*—that is, it must *guarantee an adequate supply*. If rent ceilings are imposed, the government must ensure a supply of rental units. It can build them itself, or subsidize private landlords, or it can take other approaches. If the government is going to hold the interest rate down to 7½ percent for home mortgage loans, some arrangement must be made to guarantee the supply of "loanable funds" to those who want to borrow to buy homes.

If the government is going to try to *hold a price up*, then it must do something to *prevent surpluses*—that is, *to guarantee an adequate demand*. If a minimum wage law is passed, the government might subsidize businesses to induce them to hire inexperienced workers. Or it might establish job training and placement programs to increase the workers' productivity, and help them to find jobs. Or it might take some other approach. It's much easier to pass minimum wage laws and rent ceilings and maximum interest rates than it is to design effective programs to take care of the total problem!

Real-world markets for both outputs and inputs *can be* (and all the time *are being*) manipulated. Sure. Prices are "administered" or "set" or "fixed" by somebody just about everywhere you look. But then what? The "reflex effects" seem to take over—to find some *indirect* way to bring things into balance again. Eventually we wind up with conditions that look a lot like the results you would expect to see in the pure market model.

In the next chapter (27) you'll get a better look at how these "reflex effects" work in real-world markets. Then there's an overview evaluation of how the market process functions in the real world, and then the question: "How useful is our theoretical microeconomic model—of supply and demand and the price mechanism—in helping us to understand what happens in the real world?" But please: don't go on to the next chapter until you're sure you understand this one. The next chapter picks up just where this one leaves off. Then as soon as you're ready . . .

---

**REVIEW EXERCISES**

**MAJOR CONCEPTS, PRINCIPLES, TERMS (Try to write a paragraph explaining each of these.)**

how output prices are really set
how buyers and sellers respond to output prices
how inputs prices are really set

how factor buyers and sellers really respond
how the "Fed" rigs the money market
the "reflex effects" of input price-fixing

**OTHER CONCEPTS AND TERMS (Try to write a sentence or phrase explaining what each one means.)**

public utilities
natural monopoly
minimum wage laws
usury laws
"featherbedding" practices
structural changes
distributive justice
input price-fixing
administered prices
market structure
freedom of entry
elasticity of demand
Federal Trade Commission (FTC)
regulatory commissions

**QUESTIONS (Try to write out answers, or jot down the highlights.)**

1. To understand how real-world markets operate, it's necessary to find out (a) how real-world *prices* are set, and then (b) how the functions of price work their way out in the real world. Can you explain why this is true? Try.
2. Mention and explain several "especially important" conditions which would hold the "administered price-setter" under control—that is, the conditions which would keep him from setting his price "much too high"—much higher than the cost of production, so he could make an exorbitant profit.
3. Price-setting in the real world is almost always a constant struggle among offsetting and conflicting forces. True. Try to give some examples of this, from your own personal experience. In each case, explain how it all worked out.
4. When you're thinking about the "reflex effects" of input price-fixing, the real danger usually doesn't come from what might happen immediately, but from what is likely to happen in the long run. First explain, then use some examples to illustrate.

# 27 AN EVALUATION OF "THE MARKET," IN FACT AND IN THEORY

*An assessment of the market process as it works in the real world, and of the usefulness of microeconomic theory in understanding it.*

What do you think of the market process now? How beautiful it was when we were looking at how everything worked out so neatly and surely, in the model! But in the real world, nothing seems to work just right. Everywhere you look something seems to be going wrong.

### The "Model Pure Market System" Is Make-Believe

The model of the "pure market system" is really a beautiful thing. But much of its beauty results from its many make-believe conditions: freely-moving prices and freely-moving factors of production, plenty of competition among great numbers of small buyers and sellers in all the markets, everybody getting a fair and equal chance to produce and to consume, everybody free to follow his own interests, demand directing the resources of the society to do and make the right amounts of all the chosen things, and all the adjustments coming about smoothly and easily. With all the "assumptions in place," it's really neat!

With all the make-believe conditions, the "model pure market system" looks like a most attractive, most appealing kind of economic system. The demand reflects the wishes of the people. The supply responds to the demand. Any person can choose to have or to eat any of the things produced, just as long as he will pay the honest cost of what it takes to make it, including a fair profit for the producer. Society's most scarce, most valuable things are most carefully conserved, automatically. The most wanted things are the things that are produced. All the "economic choice" questions are answered, and with freedom and justice for all. Beautiful!

## In the Real World, The Market Process Faces Many Obstacles

Unfortunately much of the beauty of the market model fades away when we see it trying to work in the real world. Nothing seems to work right anymore. For the last three chapters you've been reading about the many real-world conditions which keep the market process from working out the way the model says it will. Here's a review of the major problems.

**Monopolistic Interference.** We talk about competition, but there always seems to be some kind of monopolistic interference. Prices are supposed to move around in response to supply and demand conditions, but everywhere you look prices are being administered by somebody. Most of the time these prices probably aren't "too far out of line," but sometimes they are. People with enough economic and/or political power sometimes can keep their prices far above their costs. That way they can enrich themselves at the expense of other people of the society.

**Unequal Opportunities.** Almost everybody is born with a "more than" or "less than" equal share of the breaks. Some have "got it made" right from the day they are born. Others are born "behind the eight ball" with a long, hard struggle ahead of them and very little chance to get their fair share of what society has to offer.

**Instability, Unemployment, Inflation.** Sometimes things are unstable. In individual markets, supply doesn't always adjust smoothly to changes in demand. Prices can go much too high or too low, causing the industry to over-respond. In the economy as a whole, there seems to be a cyclical tendency, for recurrent prosperity and recession, boom and bust. These ups and downs of the economy bring the great social costs of unemployment, and inflation.

**Externalities, And Social Goals.** Sometimes markets don't recognize the true social value of things. The market process places no restriction on the wasteful use (and pollution) of our "free goods," like air and water and virgin forests and roaming buffalo. The market pulls a lot more of society's resources into fancy restaurants and night clubs and whiskey and wine and beer and cigarettes and other "worthless things," than it does into education and public health and criminal rehabilitation and other "good things." It doesn't necessarily keep all the choices aimed toward the best long-run objectives for the society.

**Selfishness Is Approved Of, And Encouraged.** The market process is built on the idea of selfishness—the idea that each person is *most* interested in looking out for himself (and maybe his family and friends), and that he will work a lot harder and take better care of things if he's going to be rewarded for it. Everyone knows that selfishness isn't the most virtuous of human motives. Yet it is this "individual self-interest" which provides the incentives which make the market process work. The market gives material rewards for individual selfishness.

There's a lot to criticize about the market process. No doubt about that. So why do we put up with it, anyway? The market? Who needs it? You'd better believe *we* do. We all do. Even with all the flaws, all the imperfections, all the many vulnerabilities to valid criticism, it's still true that the market process has been the world's greatest single tool, or institution for improving the economic condition of mankind.

## The Market Process Generated The "Modern World"

Without the tremendous driving force of the market process, it is not likely that the "modern world" could ever have emerged. If the market process had not "taken over" in Europe and America and elsewhere, what would life be like for most of us? That's an irrelevant question. Most of us would have died as infants—or, more likely, our fathers and mothers would have—so we just wouldn't be here at all! Look at the poverty and disease and high mortality rates in most of the "tradition-bound" societies and you will get some idea of what the market process has done for the rest of us.

## Market Forces Are Powerful In the Real World

Most of this chapter talks about the power of the market forces as they work in the real world—about how, even in the real-world markets of monopoly power and administered prices and market-rigging and all sorts of "un-model-like" conditions, the market forces still exercise great power in directing the economic activities and choices of the society.

Prices and wages can be "administered," and markets can be "tampered with," or "rigged" in various other ways. But the "tampering," or "rigging" will set off a chain reaction which will immediately begin to try to bring everything back into balance. This indirect, rebalancing effect is what I was referring to in the last chapter as the "reflex effect" of real-world markets.

This "automatic leveling force" really does exist. It really is there, at work, in all real-world markets. *Always.* This "automatic leveling force," or "reflex effect" is very sure to happen. It's a most significant real-world phenomenon. So real and sure to happen and so significant, in fact, that it probably should be called an "economic law."

In the model, the "direct effect" is in control, and brings things into balance. In the real world, more often than not, it's the "reflex effect" that brings things into balance.

To an economist, this "reflex effect" is so obvious that no one ever bothered to *identify* it and call it an "economic law." Yet, this "reflex effect"—this indirect, "automatic rebalancing action"—really explains how the market forces work in the real-world markets. You cannot understand real-world markets without understanding this "automatic releveling force." To emphasize its importance, I believe it should have a special name, and be called an "economic law." So, for the want of a better name, I'm going to call it: *"Bowden's Law" of the Reflex Principle of Real-World Markets.*

If I call it an "economic law," you might remember it better. Maybe it will help you avoid the trap that people fall into when they forget about it! Of course I didn't discover the idea. All economists and lots of other people already understand and respect the "reflex principle," the "automatic releveling force" of real-world markets. I just thought if I called it "Bowden's law," that would emphasize the importance of the concept. Call it anything you want to, but be sure you understand it. It's very important. The next three major sections explain it in detail and give many real-world examples so you can see how it works.

## "BOWDEN'S LAW" OF THE REFLEX PRINCIPLE OF REAL-WORLD MARKETS

What, exactly, is this "automatic releveling force" that's so important? Exactly what is "Bowden's Law of the Reflex Principle of Real-World Markets"? It goes like this:

1. Almost all real-world markets are "controlled" or "influenced" or "rigged" to some extent, by the buyers or the sellers or the government, or maybe by all three.
2. Whenever these "market-riggers" force *any change* in a market (in supply, demand, or price), that puts the market in an unbalanced condition. Automatically and immediately the "natural releveling forces" of the market will go to work to try to put the market back into balance.
3. The leveling forces will exert their influence either (a) to overthrow the "rigged" condition, or (b) to bring all the other conditions into line with the "rigged" condition, or maybe (c) both.
4. Unless the "reflex effect" is foreseen and planned for, the "market riggers" probably won't achieve their objectives. They might even wind up with *worse* conditions than existed before.

That's "Bowden's Law" of the "Reflex Principle of Real-World Markets." You can be certain that it really does work in the real world.

Can you see how this "law" might be very important? Nobody likes to start out to do something to help himself (like rig a market) and then wind up worse off than he was before! Of course not. But people who don't understand this "reflex principle," do things like that. After you've read the next several pages of explanations and examples of how the "reflex principle" works, you'll be able to see it work, almost everywhere you look!

### The Reflex Principle Brings "Rigged" Markets Into Balance

It's the reflex principle which causes real-world markets to look so much like the markets described in the pure market model. In the real world the cause and effect roles often get reversed—what the model describes as the "cause," in the real world becomes the "effect." In the model, supply and demand set the prices. In the real world, businesses or labor unions or producer co-ops or professional associations or governments or other organizations usually set the prices; the reflex principle then brings the market into balance. The "automatic leveling forces" work in various indirect ways to push everything into balance.

### Administered Prices Cannot Ignore Market Conditions

In real-world markets, the "administered price" can't be set too far out of line with basic supply and demand conditions. The price can be set "a little too high," or "a little too low." But the price can't be set "much too high," or "much too low."

Suppose the price is set "much too high," or "much too low." What happens? The automatic releveling forces of the market (that is, the "reflex principle") either will force the price to change, or will destroy the market. If the price stays "much too high" or "much too low" for very long, the market will cease to exist. At a price "much too high," nobody will buy. A market can't exist without buyers! At a price "much too low," nobody will sell. A market can't exist without sellers! Of course not.

If the price is set "a little too high," the market will adjust to the high price; if it's set "a little too low," the market will adjust in the opposite direction. But, for any price to be "administered"—that is, "set" by business or controlled by government—certain conditions must be met:

1. Setting the price up "too high" requires the power to restrict supply (and perhaps also some power to stimulate the demand). Otherwise, what will happen? The high price will pull forth a large supply, but buyers won't buy very much. The market will be flooded! The price will break loose and go down.
2. Setting the price down "too low" requires the power to restrict demand (and perhaps also some power to stimulate the supply). Otherwise, what will happen? At a low price, buyers will be trying to buy a lot, but the sellers won't supply very much. There will be severe shortages! The price will break loose and go up.

### How The Reflex Principle Works

Suppose the administered price is set just "a little" too high or too low. You might say the price is set "within the permissible range," as determined by basic supply and demand conditions in the market. Then if the price-setters have enough power (monopoly, monopsony, or governmental) to control the demand or supply (or both) they can make the price stay where they set it. What happens then? The "automatic reflex forces" will go to work and start to bring everything into balance again.

As time passes, both supply and demand will become more and more elastic—that is, buyers and sellers will respond more and more to the "administered price." The releveling forces will keep pushing to try to rebalance the market. *The reflex forces will become stronger and stronger and push more and more, either (a) until all the other conditions in the market are brought into line with the administered price (so that the administered price becomes the "right price") or (b) until the set price is overthrown.* That's *what* happens. Now here are some examples to show *how* it happens.

## The "Monopolistic Competition" Product Markets

In the output markets, setting the price "too high" on your product will attract more competitors. Everyone will be trying to cut in on your profits! Do the competitors flood the market and force the price to go down? Not usually. It works the other way. As more competitors enter the market, prices don't come down. Costs go up instead. Up how high? Up to where the "administered price" becomes the "right price" and the market is in balance again.

Remember what happened when Polaroid introduced a very successful new camera for $19.95? Many other producers entered the market. Soon there were so many different models competing against each other that no one was making very much profit anymore. But the price didn't go down; the costs went up.

When everybody is trying to sell $19.95 cameras, nobody will be able to sell very many, so no producer can produce very many. That means each producer's average (per unit) cost will be higher. The "reflex principle" eliminates the excess profits in the markets of "monopolistic competition," not by pushing the price down, but *by pushing the cost up.* It pushes up per-unit costs and pushes down the profits. When the excess profits are all gone, the market will be in balance. The industry will stop expanding.

**The Agricultural Product Markets.** In the agricultural product markets (for corn or wheat or cotton or tobacco or whatever) a "too high" price pulls in more supply than the buyers want to buy. As time goes on, more and more people will stop growing other things and start growing the high-priced product. If the cotton price, for example, is held up high enough, long enough (and unless the government makes it illegal to produce so much cotton) one day the pile of surplus bales of cotton will reach to the moon! But before that happens, the "reflex principle" will either force the price to break, or force the government to change its policy.

**The Public Utilities.** Suppose the regulatory commissions set public utility rates "too low." People will use lots of electricity and all, but the low-paid companies won't invest much in new equipment. The quality of the service will decline. If the regulatory commissions set the rates "too high," the utility companies are likely to expand their services too much — to make "wasteful" expenditures. Their costs will rise. There will be excess capacity because of too much investment, and because the people will be careful to conserve (not use so much of) the high-priced services.

See how the costs tend to go up or down, to make the "set" price become the "right" price? And see what happens if the society tries to get its utilities without paying the full cost of production? The society winds up with inadequate utility services. The reflex principle strikes again!

## The Reflex Principle Depends A Lot On The Elasticity Of Supply And Demand

Whenever supply is *highly elastic*, that means sellers are *highly responsive* to price changes. So if supply is highly elastic it's difficult (maybe impossible) for the "price-setters" to set the price either "too high" or "too low" and make it stick. The suppliers respond too much!

Suppose the price is set "too high." Great surpluses will be generated and the high price will be forced to collapse. If the price is set "too low," very little will be produced. There will be *severe shortages*. The price will break loose and go up. It's very difficult to administer the price if the supply is *highly elastic*! If the demand *also* is highly elastic, that would make it *twice as difficult* to administer the price.

Suppose the supply is highly inelastic. Then the price can be set "too high," but the sellers won't offer much more for sale. If the demand is inelastic too, then there won't be much of a surplus. Or the price can be set "too low" and the sellers will still offer almost as much for sale. If the demand is inelastic too, there won't be much of a shortage. What then? Can the "price-setters" set the price "too high" and make it stick? Yes, they can.

So what about the reflex principle, when in the real-world markets, supply and demand are highly inelastic? Do the automatic releveling forces still work? You bet they do. It may take longer, and sometimes the reflex principle has to work itself out a different way. If so, then that's just what it does. (You don't think I'd call it "Bowden's Law" if it wasn't going to work!)

### The Reflex Principle In The (Inelastic Supply) Input Markets

The input markets give an excellent example of how the reflex principle works when supply is not very elastic. You remember something about that from the last chapter. Here's more.

Almost all input prices are administered. Yet, most input prices in real-world markets really do turn out to be fairly close to the "productivity" (the value product) of the input. How can that be?

If the factor's price is set (fixed), then it can't move into line with the factor's productivity. So how is it that the factor's price winds up approximately equal to its productivity? The *productivity* adjusts to the *price*! Of course! If the factor's price is high, less is used. When less of the factor is used, its productivity goes up. That's exactly what happens. That's how the reflex principle works its way out in real-world input markets.

### A Labor Market Example: Productivity Adjusts To Wages

Suppose the union bargains with the company and gets a big wage increase. Does this mean that labor productivity will go up? Maybe not right on that day, but pretty soon, yes. When the wage goes up, a labor-saving machine suddenly becomes more valuable to the firm. It will save the firm more money than before.

When the wage goes up, businessmen will buy more machines. There will be less labor used, per unit of output. Some workers may be fired. So what happens to the productivity of labor? It goes up. Each worker has more capital to work with. How long does this process (this "reflex effect") continue? Until the productivity of labor is high enough to cover the higher wage rate, of course.

Here's another way the high wage may push up the productivity (value product) of labor. After the wage increase, the business may *raise prices*. When the price of the product goes up, suddenly the "value product" of each worker goes up. The reflex effect did it again! But this "price increase" approach can cause problems. Watch.

### The Reflex Effect Of Inflation

Suppose all wages go up, and all businesses raise their prices. That way the "value product" of each worker goes up automatically. But is anybody any better off? Everybody gets more money when he sells his labor, but everybody has to pay more money when he buys the things he wants. We're right where we started! Right back to "square one," as the mathematicians like to say.

Is this another example of the reflex effect? Sure. Frustrating, isn't it? Whenever anybody does something to try to rig the market, it seems that the automatic releveling forces of the market, one way or another, go to work to push things back into balance again. When it's all over, things may not be very different from the way they were in the beginning. Or things may be *very different*, but perhaps in ways the "market-riggers" never expected.

## SOME REAL EXAMPLES OF THE REFLEX PRINCIPLE

Examples of the reflex principle are everywhere. In the United States and other modern nations, wage rates have gone higher and higher. As this has happened, productivity has increased more and more. But what has happened in the industries where labor-saving capital has not been developed? What has happened in the U. S. watchmaking industry, for example? That industry no longer exists!

### High Wages Destroy Low Productivity Industries

Whenever an industry's wage rate gets higher than the value product of labor, then that industry can't continue to exist. Watchmakers wanted wages as high as the next guy. They insisted on it. But then the total value of the product wasn't big enough to cover the wage bill, plus all the other costs (including a normal profit), so the U. S. watch producers went out of business. So where do U. S. watch companies get the "movements" to go inside their watches? From lower-wage foreign countries. High wages destroy low-productivity industries and occupations. The reflex principle? Of course.

Suppose some special arrangement could have been made to keep wages low, for U. S. watchmakers. Would the watchmaking industry still be thriving in the United States? I'm afraid not. Why not? Because of the reflex principle! What would you do if you were a low-paid watchmaker? Get out of watchmaking and go into something else, of course.

No economy can go for very long with great wage differences between people of similar skills. Workers would move out of watchmaking and go into something else. New workers would not enter the watch industry. The industry would die out as more and more of its skilled workers moved into the production of computers and precision instruments and other things.

### Migrant Farm Workers Are Being Replaced By Machines

Here's another, more immediate example. Right now there's a strong push to unionize and increase the wages of farm labor, especially the migrant workers who move from place to place, harvesting crops. So what else is happening in the agricultural inputs markets? More and more capital being demanded? and less and less labor? Of course.

There's a rapid increase in the development and use of agricultural harvesting equipment. The reflex principle again? Of course. The higher wages stimulate more investment in labor-saving capital. More capital is used, and that brings the market back into balance again. But there's likely to be a lot of unemployed (surplus) farm workers, until they can develop new skills and move into other occupations.

What do you think will happen in U.S. farm labor markets? Will farm workers get higher and higher wages? Sure. Then will more and more farm workers be replaced by machines? Of course. If the migrant worker works for a low wage, a farmer can afford to have his crop picked by hand. But if the migrant worker insists on a "living wage" (comparable to what a more highly-productive, semi-skilled worker would receive in a factory) then the farmer cannot afford to hire him.

### As Wages Rise, Mechanize!

So what will happen as the farm worker's wage goes up? Mechanical equipment must be developed to take over the harvesting task. If not, the farmer will have to go into something else (maybe poultry or dairying, or growing wheat or corn) where mechanical equipment *can be* used. Is this another example of how the reflex principle works itself out? Sure. As wages rise, mechanize! (If you can't mechanize, then either prices must rise to cover the higher wage, or you must get out of that industry. It's one or the other. You have no choice.)

During the next few years, hand labor may be virtually eliminated from crop-harvesting in the United States. The crops which can't be harvested mechanically, simply won't be produced in this country. Perhaps they will be produced in Mexico or Central or South America (or somewhere else, where wages are not so high), then shipped to U. S. markets. Either that, or the prices of the hand-picked crops will have to go up very high.

Whenever *wage rates* go up very much, then either (1) the *output per man* must go up, or (2) the *price of the product* must go up. If neither of these can happen quickly enough, then (3) the firms in that industry must go out of business, or shift into the production of something else. You can be absolutely sure that one of these three things will happen. The reflex principle will guarantee it.

As wages rise, if capital cannot be introduced to reduce the use (and increase the productivity) of labor, then output prices must go up. If the market won't stand the higher prices, or if *foreign producers* ship their products in and undersell the domestic producers, then the domestic industry will decline. Unless the trend is reversed, ultimately that industry will die out completely. Under no circumstances can workers receive more in wages than the value of what they produce. There's just no way.

### The Special Problem Of The Service Industries

In an economy which has a rising wage rate, industries which can't mechanize and automate have a special problem. People who buy the products of the nonmechanized industries have a special problem, too. They pay high prices! Look at the repair and service industries, for example.

**The Private Services.** As wages rise, some of the service industries can't mechanize very much. The productivity of the workers can't be increased by adding labor-saving capital. But workers (and maybe the government) will insist that their wages go up. Then how can the "value product" of the worker go up enough to justify the higher wage? There's only one way. The price of the service must go up. That's exactly what has been going on in the U. S. economy.

Can you begin to understand why it's so expensive to get your car worked on, or get your shoes shined, or your plumbing repaired, or a broken windowpane replaced, or anything else done that requires hand labor? We can afford to buy new cars, but we can't afford to have very much work done on our old ones! It really pays, these days, to "do-it-yourself." If you can fix your car, or your roof, or your TV, or your plumbing, you can save yourself some money. It's going to get even more that way in the future. Can you see why?

As the wages of the service workers have gone up, the prices of the services have gone up. That has helped to push up the cost of living for everybody. As the cost of living has increased, everybody has pushed for more wage increases. Is this an example of the reflex effect of *inflation* working to wipe out the "artificial productivity increases" (the wage increases) in the service industries? Right! When "prices in general" go up as much as the service workers' wages went up, where are they? All level again? Yes. Right back to "square one."

How much of the present inflation problem in the United States results from this kind of see-saw action? from each group trying to get ahead, while the reflex effect of inflation constantly erodes their gains? We don't know. But a lot. Maybe most of it. But even if we knew, that wouldn't help much. This kind of problem is not easy to handle!

**The Public Services.** The public services (financed by government) face the same problem as the private services. Ask any local or state government official what has been happening to the cost of providing public services and he may scream and go into convulsions. The salaries of school teachers, policemen, firemen, administrators, sanitation workers, social workers and everybody else have gone up, up, up. But their productivity? It hasn't changed much.

The public service employees feel underpaid, yet the taxpayers who support them, feel pressed to the breaking point. In several states the financial problem has reached crisis proportions. Why? Partly because wages have been going up faster than productivity, and partly because of the rapid increase in the demand for more and better public services. This financial problem is not going to go away until better ways are found to increase the efficiency (productivity) of labor in the public services. When we're talking about teachers and social workers and such, that isn't going to be easy to do!

### Wages Rise But Productivity Doesn't: Inflation

How does the reflex principle work in the service industries? Primarily by inflation. The service workers get their wage increases, but prices go up, so they don't get much of a "real increase" after all. Unless "the people in general" really do produce *more output*, they can't have *more income*. We can't give ourselves more than we're producing! Unless we're producing more, we can't have more. There's just no way. Whenever we try to pay ourselves more than the value of what we're producing, the *reflex effect of inflation* levels things up for us.

Manufacturing workers (and others with increasing productivity, and perhaps with bargaining power) lead the way in gaining wage increases. Then everyone else demands similar wage increases. Ultimately they get them, so prices must rise. Then just about everybody winds up no better off than they would have been if the whole thing hadn't started.

Those whose wages run ahead of the price increases get some increase in their standards of living. For the others, the wage increase is only an illusion. The reflex effect (in this case, inflation) reduces people's real incomes enough to wipe out the effects of inflationary wage increases— increases which go beyond the real increases in output. Inflation is sort of the "last in line" of the automatic leveling forces in real-world markets. If something else doesn't get the markets pushed back into balance, *inflation* probably will.

In real-world markets, almost all prices are set by somebody. Once each price is set, the market forces go to work to bring things into balance. What things? costs, productivity, employment, outputs, consumption patterns, prices—everything.

If the administered price is "just right" (not too high and not too low) then there will be no reflex effect at all. But if the price is either "too high" or "too low" the automatic leveling forces will go to work immediately, either (1) to *break the price* and force the price-setters to set a "more nearly normal" price, or (2) to *adjust other market conditions* (costs, productivity, inputs, outputs, etc.) *to bring them into line with the administered price.*

The adjustment to a "too high" or "too low" price is likely to result in *perpetual surplus or shortage*, unless supply or demand can be controlled enough to prevent it. It isn't impossible to hold the price and still maintain a sort of "artificial balance" between quantities offered and quantities bought. But it usually isn't easy. It's especially difficult if the set price is *much* "too high," or if either the supply or the demand is very elastic.

## Everything In The Market Influences Everything Else

Now I think you understand "Bowden's Law of the Reflex Principle of Real-World Markets." It's almost impossible to rig a market without being hit by some "reflex effects." Why? One simple reason: in any market, everything is "tied in" with everything else. You can't play "pick-up-sticks" in a market, and expect to win. In a market, the pieces don't just lie there, loose. Everything is tied to everything else. Grab any one thing and move it and soon you'll find that you're moving everything else. Here's more about an example you're already familiar with: labor productivity and wage rates.

Suppose the wage rate is high. *Because* the wage is high, labor will be carefully conserved; it will be used only in highly productive uses, and with lots of good capital and natural resources to work with. *Because* it is carefully conserved and used only with lots of good capital and resources, its productivity will be high. *Because* its productivity is high, the business can afford to pay the high wage rate. But what about the productivity of land and capital? It will be low because of *the relative scarcity of the high-priced labor*! See how it's all tied together?

It isn't likely, but suppose something would happen to cause the wage rate to go down. What then? The lower wage rate would work its way out, all the way around the circle. More labor would be hired. The productivity of labor would go down. The productivity of land and capital would go up. It all works out! Any change, anywhere in the circle, will work its way out all the way around the circle. The "reflex principle" is always there in real-world markets, lying under the surface ready to go to work.

Whenever some "external power" comes into the market and pushes the price up too high or down too low, the reflex principle comes into play. The "leveling forces" go to work, usually in some indirect way, in some unexpected direction, to push things back into balance again.

We pass minimum wage laws. Then we see high unemployment of young workers, and of minorities, and in the ghettos, and in depressed rural areas. At the same time, we see inflation. Perhaps we begin to wonder if we are seeing reflex effects of the minimum wage laws. Is it possible for minimum wage laws to do more harm than good? If the reflex effects are serious enough, yes.

We hold down the prices of public utility services to be sure that the companies don't make too much profit. Then we see the dwindling standards of service and the lack of sufficient growth to satisfy the needs of the economy. Reflex effects? Sure. Could it be that our policy of setting a "too low" administered price, is a little bit shortsighted?

I'm sure that when Aesop told his fable about the man who killed the goose that laid the golden egg, he must have had in mind the idea of the reflex principle of real-world markets! If you kill the goose to get all the golden eggs today, the eggs stop coming. That's a pretty powerful "reflex effect" all right!

To rig a market without considering the reflex principle is as ridiculous as to kill a golden-egg-laying goose. It would be a good idea to keep that in mind, as "appealing shortcuts" seem to appear.

Usually, in reality, there's no "easy way." The opportunity costs, the trade-offs, are always lurking there beneath the surface, ready to zap you! That's the "reflex principle." Look out for it.

The indirect effect of "market rigging" is always there. Sometimes we make believe it isn't, but it's there just the same. Each time we manipulate the market, there are *indirect* consequences—reflex effects. It's always a good idea to try to figure out what the reflex effects will be, and to make plans for handling them. Otherwise we may wind up killing the goose that lays the golden egg—or maybe jumping out of the frying pan into the fire.

## PUBLIC POLICIES CANNOT IGNORE THE REFLEX PRINCIPLE

"Bowden's Law of the Reflex Principle of Real-World Markets" doesn't mean that we mustn't rig markets. Of course not. You know already that society's ideas of "social justice" demand that we rig some markets. What the law *does mean* is this: Whenever anybody sets out to rig a market, all the "releveling forces" should be considered. Otherwise the planned objectives are not likely to be achieved.

The "market riggers" must try to foresee the direct and indirect effects of each action they're planning to take. They need to know about all the automatic releveling forces which will come into play following any price-fixing or supply-rigging or demand-rigging action they may take.

Suppose the market-riggers succeed in foreseeing all the reflex effects of their actions. That's only half the battle! Foreseeing is one thing; forestalling is something else.

### The Reflex Principle Is Very Powerful

It is usually difficult, sometimes impossible, to prevent the reflex actions from occurring. Suppose the legislature wants to help the poor cotton farmers, so it sets a high price for cotton. What happens next? The textile producers start using more nylon and other synthetic fibers, and they start buying more Egyptian and Mexican and Brazilian and other foreign cotton. At the same time, the corn and wheat and tobacco and chicken and beef and lettuce producers decide to produce cotton. Soon there are big surpluses of cotton. Finally, the high price collapses.

To be sure, a system of controls could be built up to make the high cotton prices stick. Cotton production could be carefully limited, and government "policing agents" could go around checking to be sure that nobody was growing any more cotton than he was supposed to. Imports of foreign cotton could be severely restricted or prohibited, and the government could police all the borders to keep people from smuggling cotton into the country. Special taxes could be levied on nylon and other synthetic fibers to make them more expensive so the textile producers would not substitute synthetic fibers for cotton. The government could buy up all the surplus cotton and burn it, to keep the price from collapsing. Or the government might build a lot of warehouses and store up the cotton, just in case we should ever want to stack up cotton bales all the way to the moon!

Can the reflex principle be prevented from working? Sure, if we're willing to exert enough effort and energy and resources and money. But it isn't easy. The longer we "hold the line," the more difficult it gets to keep on holding the line. As time passes, some enterprising citizen always seems to figure out some way to get around all the controls and restrictions, and to profit from the artificially high (or low) price. As people do this more and more, the artificial price breaks loose. The reflex action of the market overthrows the controls and brings things back into balance.

### The Reflex Principle May Not Be Irresistible

The more the "rigged condition" departs from the "natural condition" in the market, the more difficult it will be to block the reflex effects. But many techniques can be used to prevent the reflex principle from working. Attempting to keep it from working is like building a dike to keep water from reaching its own level.

There is no question that dikes can be built and maintained indefinitely. The Dutch have shown us that! The real world shows many examples to prove that markets can be rigged successfully, too. But the reflex forces which occur in real-world markets are not quite as easy to predict and control as are the forces of the waters of the North Sea! Containing or blocking the reflex effects can present quite a problem.

### Rigging A Market Requires "Building Some Dams"

If those who have the power to rig markets really are good "theoretical and real-world" economists, they may succeed in rigging the markets and producing the results they seek. They must know the limits of what they can do (as imposed by the natural market forces). Also, they need to be able to foresee reasonably well the total pattern of reflex effects, and to figure out some way to contain, block, or get around each undesired effect.

Even though water seeks its own level, everywhere you go you see artificial lakes. They range in size from little farm ponds, to lakes many miles long, holding billions of gallons of impounded water. How is that possible? Dams, of course.

Who makes the dams? The engineers. The ones who understand the most about the "automatic leveling forces" of water. But sometimes even the engineers figure wrong. Sometimes the automatic leveling force of water is too great. The dam breaks. Disaster! If a dam is going to break, we'd be better off not to build it in the first place. The same is true of market rigging.

"Market rigging" works a lot like "water-level rigging." Only, as you know, we know a lot more about how to control water than we know about how to control markets. That's why "rigged price levels" are much more likely to break loose than are "rigged water levels."

Who rigs the market? Everyone who has (or who can get, and exercise) some monopoly power. Many buyers and sellers have "farm-pond size" market-rigging power. Sometimes big businesses and big labor exercise much more market-rigging power than that. But the really big "economic dam-building market-rigger" is the government.

### The Reflex Effects Are Not Easy To Foresee, Or Forestall

Many of the government's economic policies are specifically designed to rig markets. But market-rigging is much less scientific than dam-building. Even the best economists know less about the releveling forces of markets than a mediocre engineer knows about the releveling forces of water. And worse than that: Everyone thinks he is a good economist, so he argues for whatever he thinks looks good—progressive income taxes, tariffs on foreign cars, more aid for dependent children, minimum wage laws, tuition-free universities, tax incentives to stimulate investment, higher unemployment benefits, etc., etc.—without any regard at all for the reflex effects.

So what kinds of economic policies does the nation wind up with? Not perfect. Not by a long shot. But it's surprising how good the economic policies work out to be, much of the time. Why? Because outstanding economists who really understand the reflex effects are in charge of things? No. Economists don't know all that much, anyway. Then how does it happen to work out reasonably well, much of the time? Through trial-and-error? pragmatism? Right!

### The Reflex Effects Usually Are Handled After They Appear

First, people decide that every worker should earn a "living wage," so the government passes a minimum wage law. Then pretty soon we see high unemployment of young, and unskilled workers. So then the government sets up job training and economic development programs. One thing leads to another. Each time some "reflex effect" shows its ugly head, the government takes a swat at it. Not quite as scientific as dam-building, right? Still, we will keep doing it this way, more or less, for many years to come. Why? Because it isn't easy to find any other workable approach.

We might hope that all of our policymakers will be learning more about the reflex principle of real-world markets. That might improve the government's economic policies. More of the indirect effects might be foreseen and contained or circumvented or offset, right from the beginning. That would be good — but don't count on it.

An idea such as "Bowden's Law of the Reflex Principle of Real-World Markets" is sort of theoretical. Politics and economic policy have a way of being coldly practical. What's the legislator-politician's attitude? "When people begin to scream about the problem, I'll try to do something about it. Until then, don't bother me with 'Bowden's Law.' Bowden is probably some kind of a nut, anyway."

## Policymakers Need To Understand The Reflex Principle

No one will ever be able to foresee all the reflex effects of any government economic policy or program. Still, the better the policymakers understand the reflex principle, the more realistic their programs are likely to be. The less likely the policymakers will be to fall into the trap of offering quick and simple answers to chronic and complex problems.

The market process in the real world is headstrong and powerful. It tries to insist on working things out its own way. Government policymakers need to be very careful about how they choose to manipulate markets. Otherwise, you know what can happen. Out of the frying pan, into the fire!

In view of all the things we have been talking about, perhaps we should junk the market process. Then what would we do? Governmental planning and resource administration would be the only answer, of course. All the market-oriented economies of mixed capitalism are using a good bit of governmental planning and resource administration, already. Why not just get rid of the market, and use governmental planning and resource administration, all the way? No, we'd better not try to do that.

If you think the market creates problems, you're right. But compared with the task of trying to administer and direct all the resources and production activities and everything in the economic system? Forget it.

By now you are well aware that the people in the world simply aren't going to be able to make it without the market process working for them. Manipulated? Yes. Controlled? Sure. But somehow we must use it. The inefficiencies of trying to get along without it would be intolerable.

### The Market Process Is Mankind's Most Valuable Tool

Try very hard—now, and always—to stay aware of the many real-world flaws and imperfections in the workings of the market process. But please, don't throw out the baby with the bath water! The market process—intelligently and realistically modified and, when social objectives and goals require, manipulated—has been, and is, the world's single most valuable tool for serving the material needs of mankind. The personal feelings of freedom which it permits, the careful conservation of most of society's resources which it accomplishes, the stimulation of production which it generates—all these are going to be essential in building the kind of world the people of tomorrow will want and demand.

The needs of the world simply cannot be served through the inefficient process of "resource administration." The idea of the control of each input and output being handled by many "experts" who spend all their time working on such things? Ridiculous. It's just too inefficient, too wasteful, as compared with the neat way the market process could handle it. The needs of the people of the world will never be met if we waste our resources and talents trying to "administer" everything.

The market process doesn't always work right. It presents many problems and has many flaws. But the people of the world would always be poor, except for the market. Even the communist countries are beginning to find that out. Maybe we should talk about some of the good things about the market process.

## SOME ADVANTAGES OF THE MARKET PROCESS

The "market model" certainly isn't a true picture of how things happen in the real world. But the model can help us to understand some of the things that go on in the real world. Why? Because most of the forces which operate in the model, really do operate in the real world, that's why!

People work hard to get more money and things—more private property. They work, study and struggle to get more income. Usually the ones who work at it *really do* get more income. And most people do work at it!

By now, you know many good things about the market process. You know that it is very efficient in stimulating production. For most things, most of the time, it really does a great job of conserving and optimizing the society's things, too. We need to talk about these "real-world functions of the market process" a little bit.

### The Market Process Works Automatically

The efficiency of the market process as a "resource-directing mechanism" is truly impressive. Somehow the market finds the needed factors and moves them to the right places and gets the right things produced. How could you get all the factors to do the right things, without prices, wages, and profits, as incentives? Sure, resources can be "directed" by government administrators. But this usually is an inefficient (and often distasteful) way to do it.

It's sort of miraculous how prices can move factors from one place to another and get them to do one thing or another. The price mechanism is far more efficient in directing resources than any other method yet discovered or devised by man. It is because of this great efficiency that the communist governments are finding it sometimes to their advantage to use the (manipulated) market process (the price mechanism) to direct their resources.

### "Private Property" Gets Us To Conserve Society's Things

The market system appeals to the natural self-interest of each person. By a sort of "neat trick" it gets all of us to become full-time custodians of society's resources. A big part of this "trick" depends on the concept of private property. Look how neatly it works: A patch of forest can be "mine," a house can be "mine," a truck can be "mine," and I will work like a slave (harder, really!) to take care of these things. I will put forth great effort to care for these pieces of the society's land and capital. Which pieces? The pieces which happen to be "mine."

Why will I fight the forest fires and stay awake all night to keep the wolves from the cattle? Fear of punishment? No. Offer of high wages? No. Then why? Because these things are *mine*. I will fight and struggle and exert every effort I can to protect those things which are *mine*! No one needs to order me to care for these resources. No one needs to watch me to see that I do it. I will do it for myself because these things are *mine*. Do you see the great custodial, economizing role of private property? How efficiently it is performed!

The truck will be lubricated when it needs to be, because it's *mine* and I want to take care of it. It will be used to haul the bricks to build the new factory because that's where it's demanded. It's my truck, sure. But it's society's truck too. I take good care of it and I use it to serve the wishes of the society. See how neatly the market gets things taken care of, and used for the desired purposes?

Suppose the truck was owned by the government and I was just the driver. Would I take such good care of it? Maybe so, but probably not. Would you be lubricating, fixing, washing, or otherwise taking care of a government truck during your time off on Saturday afternoon? Not likely.

My forest will be protected from insects, from fire, and from disease so the timber can grow. My forest is one of the natural resources of the nation. Really, it's *society's* forest. But I'm taking care of it because it is also "mine." No "government supervisor" must waste his energy watching over me to get me to take care of the forest. I do it automatically because it's mine. The "private property" thing tricks me into taking care of it. If I'm lucky, someday all my efforts will be rewarded by profits; then I will have the opportunity to become society's custodian for *even more* of its property!

### Efficiency And Freedom Are The Big Advantages

The great efficiency—this custodial role of all the people, and the automatic responsiveness to the price mechanism—is one of the two greatest advantages of a market-directed economic system. The other great advantage is that each individual really does have a high degree of freedom to choose his own way of life.

In the real world the harshness of the pure market forces can be modified to give each person a reasonable chance. Then each is free to decide where he wants to live, what kind of work he wants to do, and how hard he is willing to work. Each person knows that he can work hard, produce more, and get more things; or he can work less, own fewer things, and have more free time. It's his choice.

The relatively high degree of freedom of each individual to choose his own work pattern and life style is often considered to be the greatest advantage of a "free enterprise" or "market-oriented" economic system. The same feelings which support democracy as a form of government also support "free enterprise" as a form of economic system. Both strongly emphasize the importance of the individual and his "right" to choose for himself. Of course you recognize that neither political democracy nor economic "free enterprise" operate quite as neatly in the real world as they do in theory. But, praised be, both do work!

### A Summary Overview Of The Market Process In The Real World

You know that there are several ways in which the model market process doesn't fit the real world. Administered pricing and monopoly power result in major changes. There are many socially desirable conditions which the market cannot bring about. The market permits over-use of some natural resources and pollution of the environment, while letting desirable projects (reforestation, urban housing, etc.) go wanting.

The wide disparity in the distribution of wealth and income lets a few people have a lot of control over society's resource choices, while many people have almost no influence at all. The market process does not respond to the needs of the poor. The market works on a day-to-day basis, responding to each new set of conditions, but it has no way of working out long-range objectives and goals and then making resource choices aimed toward those objectives and goals.

No society would permit the market process to have full control over its resource-use choices. Every society departs from the market process and, through the political process, forces the people to turn some of their private property into public programs: police and fire protection, highways and streets, hospitals, urban renewal, education, manpower training, nuclear submarines, trips to the moon.

All societies practice some forms of income redistribution. When a person's welfare level drops below that level which the society's conscience will permit, the society tries to do something about it—either through the social process, or through the political process. Somehow, arrangements are made so that more of the society's outputs are transferred to the poor.

It isn't likely that anyone would ever suggest that the market process be given *complete charge* over all of the society's economic choices. On the other hand, history has shown the tremendous power and efficiency of the market process. As we continue to modify it, perhaps we should take care not to destroy it completely—at least not until we are sure we have found something at least as good to take its place.

## HOW RELEVANT OR HELPFUL IS THE "PURE MARKET MODEL"?

The model of the "pure market system" certainly doesn't describe the real world. That's for sure! Nothing in the real world works exactly like the model. So why have you had to go to all the effort to understand the model?

Why don't we forget about the concepts of pure competition and easily moving prices, and about supply and demand determining prices, and about "productivity theories" of wages and rents and interests? Why don't we just study the things we can see going on in the real world? What good is the model, anyway?

### The Model Is Not Supposed To Describe The Real World

Everybody knows that the model of the pure market system really doesn't describe the real world. *It's not supposed to do that*! What it's supposed to do is to illustrate some of the forces at work in the real world.

The law of gravity can explain how things would drop in a vacuum, but that sure doesn't describe what's going to happen if you drop a fluffy little feather out of that little window at the top of the Washington Monument! Get the idea? It's a good idea to know about the force of gravity, but that doesn't describe what will happen when you drop the feather. Not by a long shot! That's sort of the way it is with the "pure market" forces.

The "pure market model" can't describe the real world. Of course not. But can't you look at the world, now, and see things you couldn't see before? Could you understand "Bowden's Law of the Reflex Principle of Real-World Markets," if you didn't know anything about the workings of supply and demand, or about the rationing and production motivating functions of price, or about the effects of surpluses and shortages, and all that? Could you? Probably not, right?

### The Model Shows The Market Forces "In A Vacuum"

In the real world, the results of the pure market forces are always modified by special circumstances. Still, it's very helpful to know what these forces are, and how they work. The model gives us a "solid rock foundation" to start from, as we begin to try to understand real-world markets. It lets us see the pure market forces at work. And it gives us a way to "test out," the various market forces, and to "pre-test" various proposed policies to see how each might work out in a model setting— "in a vacuum," so to speak.

An overemphasis on the model sometimes hides the real world from people. Try to be careful about that. The neatness and beauty of the model is so great that it's easy to get carried away. Try not to. But on the other hand, do try to be always aware that in real world markets, the supply and demand forces, and the two functions of price *really are* there, working all the time.

Sometimes the market forces are in *direct* control; more often, prices are "administered" and the "reflex principle" keeps things moving into balance. But either way, market forces are always influencing things.

You can't learn the economics of the real world just from studying the model of the market process. You can't learn about the political or social processes in the real world just by studying social and political models and theories, either. A model is a little story we tell. It isn't true, but it illustrates some of the tendencies—some of the forces that are at work.

### The Model Is True, Just As Aesop's Fables Are True

The economist's model of pure competition is sort of like Aesop's fables. Surely Aesop didn't want us to believe that there was a fox looking up a tree and saying that the grapes must be sour because he couldn't get to them! But Aesop did manage to put his finger on a very human tendency, right? If all you knew about humans and their tendencies was what you could learn from reading Aesop's fables, you would have a very distorted view of the world and its people!

If all you knew about real-world economics was what you learned from the economist's pure market model, you would have a very distorted view of real-world markets. But everybody has some awareness of what's going on in the real world. By studying the economist's model, you can learn to see and understand it so much better. (But don't forget what you *already* knew. That's an essential input, too!)

Is the model necessary? Helpful? You decide, for you. For me, I have already decided. I need it. Without it, I really wouldn't be able to understand how a modern economy of mixed socio-capitalism "gets it all together." Could you? And I probably couldn't do nearly as well at distinguishing between "sound economic policy" and "shortsighted political demagoguery." Could you?

I hope you've done a good job of learning about how the market forces work in the model—about the functions of supply and demand and price and all that. And I hope you've really learned about the real-world limitations and modifications, especially about administered prices and the reflex principle. If so, I believe you will always have a better understanding of what's going on in the world around you. And if so, then it's been worth the effort. Right?

In the four final chapters of this book (Part Eight) you will be reading about current real-world problems. With your new understanding of how the market forces work, you will be able to look at each of these problems with greater insight than you ever could before. But first, before you go on, try to spend some time reviewing and re-thinking this chapter, and all of Part Seven. How deeply you can see into the issues in Part Eight will depend a lot on how well you understand Parts Six and Seven—especially Part Seven.

---

## REVIEW EXERCISES

### MAJOR CONCEPTS, PRINCIPLES, TERMS (Try to write a paragraph explaining each of these.)

real-world obstacles to the market process
the reflex principle and how it works
how labor productivity adjusts to wage rates
the reflex effect of inflation
how high wages destroy low productivity industries
how private property "tricks people"
the efficiency of the market process
the purpose and helpfulness of the "pure market model"

### QUESTIONS (Try to write out answers, or jot down the highlights.)

1. Can you explain in detail what is meant by "Bowden's Law of the Reflex Principle of Real-World Markets"? Try.

2. What would happen if the government came in and set the price of *anything* (gasoline, or shoes, or houses, or economics books, or you name it) *much too high*? Or what if it set the price *much too low*? Explain exactly what would happen, and why, in each case.
3. Why do the reflex effects get stronger and stronger as more and more time goes by? (What happens to the elasticity of supply and demand as times goes by, and does that matter?) Explain.
4. The "reflex principle" helps to explain why producers who sell in markets of monopolistic competition often have high average cost of production, and a lot of excess capacity. Can you explain why?
5. If you were a farm worker, knowing what you know about the reflex principle and how it works, would you be pushing hard to get your wages up higher? or not? Try to look at this from all sides and explain the various possible points of view.
6. Do you really believe that the market process is mankind's most valuable tool? Explain, in as much detail as you can.

# PART 8

THE WAY THINGS ARE GOING ON EARTH,
MAYBE WE SHOULD TRY VENUS, OR MARS!

Today the people of the world are very concerned about domestic and international micro- and macro-problems: pollution, population, poverty, unemployment, inflation, trade balances, underdevelopment—and about how the world's economic systems are changing as we try to solve all these problems.

# 28 POLLUTION, POPULATION, URBANIZATION, POVERTY: MICROECONOMIC PROBLEMS OF MODERN SOCIETY

*These urgent problems have been ignored, but now they must be solved and all of us must help to pay the costs.*

Finally, here it is. This is the last part of the book—the part that talks about some real-world problems—problems you have already heard a lot about—problems you will keep on hearing about (perhaps sometimes worrying about) all your life. These four chapters in Part Eight will be talking about microeconomic problems: pollution, the population explosion, poverty; about macroeconomic problems: unemployment, inflation, wage-price stabilization; about the "international balance of payments" problem; about some special problems of the underdeveloped countries; and about the "challenge to capitalism"—the increasing role of government in the "free economies" of "mixed socio-capitalism."

### The Focus Is On Economic Factors And Opportunity Cost

Many books have been written by economists and others describing the seriousness of the problems you'll be reading about here. It's worthwhile to describe these problems and to present statistics showing how serious they are, but that isn't the purpose here. My purpose is to help you get a good basic understanding of the *nature* of each problem and of the *economic influences* at work.

I want you to gain a realistic awareness of the choices available to the society as we tackle these problems. I want you to be able to see the difficulties involved, and the opportunity cost which each choice is likely to require. If you can come out with that, you'll be far ahead of most people. But even more. You'll be much more aware of the basic economic concepts which are always there, always working in the real world.

For now, don't worry about the detailed statistics. If you ever need detailed statistics on pollution or population or poverty or unemployment or inflation or the balance of payments deficit or such things, the statistics (those which exist) are easy to find. You might start with the annual *Statistical Abstract of the United States*. (I'm sure your friendly librarian would be glad to help.)

### All Of Us Are To Blame For The Major Problems Of Our Society

When something is going wrong we usually try to find someone to "blame it on." It's frustrating to discover there's *nobody* to blame it on! But in this case you're going to find out that's the way it is.

If the problem is poverty or population or pollution, or unemployment or inflation, or balance of payments deficit, or whatever, you're going to find that you can't "blame it on" the rich people, or the labor unions, or the landowners, or the government, or big business, or the foreigners, or the southerners or the northerners, or the younger generation, or "the system," or any other easily available scapegoat. Why not? Because mostly, for all of these problems, *all of us are to blame*, that's why.

What I'm trying to say is this: There aren't any easy answers. The answers are *tough* answers. The opportunity costs are high. You can't have your cake and eat it too, remember? Even if we all decided that we were willing to pay the opportunity costs—to make the sacrifices needed to overcome all of these problems—most of the problems still couldn't be solved very fast. The problems are *too big*. Quick solutions are *physically impossible*! (As this chapter unfolds you will see just how true this is.)

## WHICH OBJECTIVES SHOULD THE SOCIETY PURSUE?

Mankind has always faced problems. Now that the age-old problems of hunger and disease have been pushed back, new problems are coming to the surface. So now, society's objectives—the directions of progress—must change.

### Progress Uncovers Or Creates New Problems

As mankind solves one set of problems, new problems are uncovered or created. Take the population explosion, and the environmental crisis. We solve the "hunger" and "disease" problems. So what happens? People stay alive longer, live better, consume more. Population explosion?

Environmental problem? Yes! Who's to blame? High technology and high productivity have done it. So let's get rid of high technology and high productivity? We'd better not! High technology and high productivity offer our only hope for working out solutions to these problems!

Without our high technology and high productivity, we wouldn't have a chance of handling the environmental effects of the world's exploding population! Not nearly enough progress has yet been made in bringing the population explosion under control, or in dealing with the environmental crisis. Everybody knows we've barely scratched the surface. But we've made more progress than we ever could have made without our high technology and high productivity to help us.

Sometimes people suggest that because of the environmental problem, we should try to stop economic progress — have "zero economic growth." But surely zero growth isn't the answer. Even if it was a good idea, it wouldn't work. People always try to change the things they don't like — that is, to improve things. That's what progress is all about — making things better — moving toward our desired objectives.

## Progress Means Moving Toward Your Objectives

What "progress" means to each person depends on what that person wants most. Picture an unemployed worker whose family eats nothing but dandelion greens and corn meal mush. He is a lot more interested in seeing a new factory open up where he can get a job than he is in protecting the clean air in the valley or the pure water in the stream. To him, a new factory would bring progress. But what about the retired stockbroker who lives in the valley, or the local school teacher, or the wealthy family that owns most of the land and raises horses? A new factory that might mess up the water and air? That's progress? Forget it!

Suppose there's a public hearing on the issue of re-zoning some land to permit the new factory to be built in the area. Will all the local people agree? Of course not! Do you see what a tough situation this creates? Each of us will work for the things that are most important to us. We will try to change those conditions which we see as problems. But is a problem to *me* a problem to *you*? That all depends on *your own* point of view!

Many choices have to be made by "the society as a whole" — by the government. Whenever this happens, disagreements are inevitable. In every society, no matter what the economic or political system, people disagree about issues such as these. Only when a problem becomes very obvious and very serious will everyone agree that something must be done. Even then they still will disagree about exactly *what* should be done, and *how much*, and *how*.

There are many kinds of economic issues and "public policy" problems which fit into this discussion. This chapter talks about three: pollution, population, and poverty. I suppose everyone agrees that something ought to be done about each of these, but there sure are wide disagreements about *what* and *how much* and *how*! Now, a quick look at each of these problems—these serious microeconomic problems of modern society. First, the pollution problem.

## THE ENVIRONMENTAL CRISIS: THE ECONOMICS OF THE POLLUTION PROBLEM

I suppose everybody knows that the pollution problem is bad and it's getting worse. All of us are dumping cans and bottles and garbage and trash and junk cars and all kinds of junk ("solid wastes") into the environment. And that's only the beginning. We're dumping liquid wastes (effluents) from all the industrial plants and processing plants and from all the sewage systems of all the cities all over the country, and all over the world. All these effluents wind up in our streams and lakes and rivers and oceans and everywhere. We're spreading all kinds of chemicals all over the landscape to control insects and weeds and all. We're puffing out smoke and all kinds of "gaseous emissions" into the air from the smokestacks of all the factories and from the exhaust pipes of all the automobiles and trucks, from city incinerators, from the furnaces of all the houses, and from burning steak fat from the backyard barbecue grills.

We're pulling fresh water out of the rivers and streams, to drink, to bathe in, and to use for industrial purposes. So now, less fresh water is flowing into the salt water estuaries—Chesapeake Bay and Galveston Bay and Puget Sound and Pamlico Sound and all the others. So the water in the estuaries is getting more salty. Maybe it's getting too salty to continue to serve as a breeding ground for many kinds of marine life. That's bad. At the same time, the power plants and other plants are dumping warm water into the streams and increasing the temperatures. That may be upsetting the natural balance of life there.

It would take a lot of polluting to pollute all the land and water and air of the earth, and kill off all the wild life and marine life—and human life. But some scientists say we are moving in that direction at a rapid clip! Everyone knows that the trend can't be allowed to continue. Something must be done. But what? And how? These are tough questions.

### We All Contribute To The Pollution Problem

Who pollutes? Everybody. Every living person, every organization, every business, every school, every factory, every store, everybody. We are all responsible for this pollution problem.

All pollution results from either final consumption activities, or from production. If it results from final consumption activities, the final consumers are the ones who are *directly* responsible. If it results from productive activities, then the final consumers are the ones who are *indirectly* responsible. You might say that all pollution is either *final pollution* or *derived pollution.*

"Ask not for whom the industrial plant pollutes. It pollutes for thee!" And who is going to make the sacrifices required to "pay the cost" of overcoming the problem? Make no mistake about it. "Thou art."

All industrial activity and all of the normal processes of life require some "environmental destruction" and some "waste disposal." Until recently, everyone has been thinking of the "environmental destruction effects" and the "waste disposal task" as sort of trivial aggravations—to be ignored—to be dealt with only when absolutely necessary.

Throughout history the task of environmental protection and waste disposal has been treated as "an afterthought in the total economic process of life." But from now on that isn't going to be good enough. Not anymore. With so many people on the earth, consuming so many things, environmental protection has become critical. It's becoming more critical every day.

### We Are "Living High" By Not Fully Paying Our Way

In the past, most of the environmental costs—the costs of waste disposal and resource destruction and all that—have been external to the market. In adding up their costs, businesses haven't had to pay much attention to the resource destruction and waste disposal effects of their activities. Consumers have been able to buy all kinds of products without having to pay for the "full social cost" of producing them. Also, people have been getting by without having to pay the full cost of disposing of their trash and garbage and sewage. All of us have been enjoying the temporary pleasure of "living high" at the expense of our environment. But it won't be long before we aren't going to be able to do that any more.

What does all this mean? It means that some things which have always been thought of as "cheap" are suddenly going to become more expensive as the cost of environmental protection begins to be included in the prices of things. It's already beginning, but it still has a long way to go.

In recent years, paper manufacturers have been putting many millions of dollars into environmental protection equipment and processes. You helped to pay for some of that when you paid the price of this book! Get the point? Already we're beginning to feel the pinch. But it's only just begun. We will all grumble, but we will pay. Really, we don't have any choice!

### Cleaning Up Is Part Of The Job

The environmental protection task must become an important part of every productive activity — really, *of every aspect of life.* A larger and larger percentage of the national output will be directed toward environmental protection. A lot more labor and equipment and power and other resources will have to be re-directed toward this task. Maybe, before it's all over with, we will have to learn to get along with less—fewer cars, less electricity, less gasoline, less plush college buildings, less of a lot of things. Maybe. We can't have our cake and eat it too. Remember?

How much more of the GNP will have to go for environmental protection? How much are we going to have to transform "consumer goods" into "environmental protection goods"? Nobody knows. But we do know that the cost will be high, and that all of us are going to have to give up some things (things we otherwise could have had) to help to pay for it. There's just no doubt about it.

### How Much Pollution Should Be Allowed?

How much "gaseous emissions" or "liquid effluents" or "solid wastes" should be permitted? And by whom? And under what circumstances? Absolutely none? Zero? That's really going to be tough! Be careful. Don't exhale and pollute the atmosphere with your used breath! Don't drive your car because if you do you'll pollute the air with exhaust fumes and you'll support the "derived pollution" of all those gasoline and tire companies! Don't eat or drink anything out of cans or bottles or paper bags. And please! No trips to the bathroom!

Zero pollution? No. That can't be the answer. So how much, then? That's a tough question. Nobody really knows the answers. The biologists and chemists and other scientists are working on it.

Luckily, the environment has much ability to rejuvenate itself. It can give up a lot of resources and absorb a lot of wastes before the effect becomes cumulative. But everybody knows that in the big cities and in the industrialized areas we have gone too far. We've got to cut back. But to zero? No. That would be an impossible, unnecessary, ridiculous objective. But we must cut way back from where we are now. Some scientists are saying that if we do the very best our technology will permit, that will be barely good enough.

### The Political Process Must Solve This Problem

The environmental protection problem is "external" to the market. You already know that's why the problem must be solved through the political process. There's just no other way it could be done. But that doesn't mean that the market mechanism can't be used to help. Of course it can, and it will.

Essentially, there are two ways the government can approach this problem: 1) by a system of direct regulations, with constant policing to see that everyone follows the law, or 2) by charging "prices" for the use of the society's environment, and adjusting the prices up high enough to keep the environment adequately protected—that is, by letting *the rationing function of price* take care of the problem. Either of these approaches will internalize the externalities, and will pass along the "environmental protection costs" to all of us who buy the products.

As you drive along the highway you see signs saying: DO NOT LITTER. $50 FINE. That's the first approach. It's against the law to throw beer cans out of your car window. Another approach might be for the government to add a nickel "deposit" to the price of each can of beer. Then when you turn in the used can for proper disposal or recycling, you get your nickel back. If you throw the beer can beside the highway or in the lake, it costs you a nickel. If people still throw away too many beer cans, raise the price to a dime, or a quarter, or to a dollar, or two, or five. Can you see how effective this approach might be?

## Internalizing The Environmental Externalities

Which approach should the government use? regulations and policing? or the price mechanism? Probably both. Whenever the price mechanism can be used, it's likely to cut down on the "policing cost." But no matter which approach is used, the results are going to be far from perfect. There will be all kinds of special interest groups pushing for more protection or less protection, for tighter or easier regulations, for stricter or more lenient enforcement, for higher or lower "environmental protection prices." That's the way the political process always operates.

Remember the unemployed father of five who feeds his family on dandelion greens and corn meal mush? The one who wants a new local factory so he can get a job? He detests the environmental protection program and writes nasty letters to the mayor and to the governor and to his congressman. The local gas station operator and the local real estate salesman do the same. They want to see business pick up. What about the businessman who wants to build the factory? He writes letters, too. They all want to kill (or soften) the environmental protection program.

What about the retired stockbroker and the civic-minded school teacher and the wealthy landowner who raises horses? They get together and visit city hall and the state capital and Washington. They're urging that all environmental protection programs be tightened!

Who will win out? Who knows. If we're lucky, the final decision will be made with an eye to *what ought to be done*. But the final decision is almost certain to depend partly on who is a member of which political party, who contributed how much to the last political campaign and such things as that. That's too bad, I suppose. But that's the way it is. Such is the nature of the political process.

### How Fast Should The Government Move On Environmental Protection?

What ought to be done? And how quickly? Nobody knows, really. The "doomsday prophets" say it's already too late. We've blown it. Human life on earth is on the way out. Others say that we must move as fast as we can because there's no time to lose. At the other end of the range are those who say that this whole problem has been blown completely out of proportion — that we are working on it too fast already. They say we're distorting the progress of our economy by turning too much effort toward environmental protection.

Many people seem to agree that we should work on the environmental protection issue as fast as we reasonably can. But how fast is that? We could impose enough controls to absolutely "shut down" the economic system! We might say: "Okay. Let's do it." But after we've been going without food or water for a few days we might all decide to change our minds. This is a tough question, with opportunity costs staring us in the face every step of the way. As long as we recognize that, and go after the problem realistically, I think we'll find the ways to handle it. We'd better all hope so!

### The Problem Provides The Means For Its Own Solution

One thing is sure. The great human population is going to have to become compatible with its limited environment — with its city, its nation, its "spaceship earth." Solutions will not be easy to find. The cost is going to be higher than most people would like to admit. But *the cost must be paid*. Soon, everyone will realize that we don't have any choice. Can we do it? Sure. The high technology and economic growth which have let the problem arise, will now provide us with the means to solve it.

So is the solution to the environmental problem at hand? No. Far from it. We have the means to develop the solutions, though. Now it's time for the economic system, through the political process to redirect much more energy and effort toward environmental objectives. "Progress" in the decades ahead is going to have to be redefined — aimed much more toward achieving *environmental objectives* and less toward *other objectives*.

It's good that we have the means. Let's hope we now will be wise enough to pay the opportunity costs and take care of the problem. Try not to grumble too much as more and more of the costs come to rest on you. Sorry, but there's just no other way. Let's all pay the cost and be glad we got around to it in time! (I hope.) But unless we can solve the population problem, the environmental problem will *never* be solved. So what about the population issue? It's time to talk about that now.

## THE POPULATION CRUSH: THE ECONOMIC PROBLEM OF WALL-TO-WALL PEOPLE

Man has been on earth for thousands of centuries. But it was "only yesterday" that suddenly there were *lots of people* on the earth. At the time of the first Crusade (about 1100 A.D.) the population of the world only amounted to about 300 million people. Then in the 4 centuries between the first Crusade and the time Columbus discovered America, the world's population expanded by about 50 percent—to about 450 million. Then in the next 3 centuries (by the time the United States was created as a nation) the world's population *doubled*—to about 900 million. Then in the next *one* century (the 1800's) it almost doubled again! Then, in the next sixty years (1900–1960) it doubled again! Now it appears that the number of people in the world may double again (may increase by *three billion*) in only 40 years (1960–2000). Wow!

### The World's Population Is Exploding

Just think. In 1900, when the world entered this century, there were about 1.5 billion people on earth. By 1960 there were about 3 billion. By the end of this century (by the year 2000) the number could reach 6 billion. Three fourths of the total (4.5 billion) is being added just during this century. Explosion? You bet. Is there any wonder we have a pollution problem? And lots of other serious problems?

It has been said that "of all the people who have ever been alive on earth since the beginning of time, half of them are alive today." That's a sobering thought. Here's another. The number of children who will be born during *your* lifetime will be greater than the total number of children who have been born before you, since the beginning of time.

It's just hard to conceive of the dimensions and seriousness of the population problem. Unless mankind can work out the means for drastically curtailing the number of births, then nature will take care of the problem for us—probably by arranging for most of us to die off!

### People Are Pouring Into The Cities

Where are all these people? these three billion, going on six billion, most of whom are in their teens and younger? They're all over the world.

They're in the high-income advanced countries, they're in the low-income advancing countries, and they're in the poverty-stricken underdeveloped countries. They're everywhere. They're in India and China, in Mexico and Brazil, in Japan, western Europe, the United States and Canada, in Russia, Australia, Africa, South America, Indonesia—everywhere.

Wherever they are, masses of them are moving to the cities—thousands every day—a constant, steady, increasing stream. Why? Why do they pour into the cities? Because there's nowhere else for them to go. The greatest migration the world has ever known is the one that's going on right now—the millions of people, pouring into the cities. It's happening everywhere the world over. Every day it gets larger.

Suppose a peasant farmer has enough land to get by on. Then he raises six or eight or a dozen children. Maybe all of them can eke out a semi-starvation existence from the land. But then by the time the children are in their mid-teens, the girls start having babies, raising families of their own, perhaps each of them having six or eight or a dozen children. Where can all these people go? What can they do? They go to the cities to try to find something to do—some means of existence. But wherever they go, whatever they do, they keep having children. Babies, babies, babies, everywhere babies.

Yes, it's true that more than half of the people alive on the earth right now are in their teens or younger. In Communist China, the number of children under ten is greater than the total population of the United States! In most countries most of those in their teens are already having more babies. And the flood of surplus population keeps pouring faster and faster into the already overburdened cities. How long can it continue? Not very long, now. Not very long at all.

### Population Growth Is Very Difficult To Control

The flood of babies pouring into the world is going to have to be stopped. But how? Every individual has his own feelings about this. Every society has its taboos, its restrictions, its religious beliefs about the "right" or the "obligation" to bear children. Getting around all of that is going to be a tough task.

Here's another tough one. Some people, some organizations, some groups, some nations see *strength in numbers*. Having lots of children can provide a kind of "social security"—the children can care for their aging parents. Or if a "good Democrat" or a "good Episcopalian" raises a dozen children, chances are most of them will turn out to be good Democrats or good Episcopalians. If the French-Canadians or the Irish Catholics or the Southern Blacks want to gain political control of some area, if they just keep having enough babies, sooner or later they "tip the balance"—especially if all the other people are all trying to have small families and slow down the population boom. Even the leaders of some nations still see "strength in numbers" and hesitate to curb the baby boom.

"Everybody just like me, who agrees with me, who wants what I want, should have lots of babies. Everyone else should stop!" See the problem?

### Cooperative Action Is Essential

Several of the nations of the world are aware of what must be done. Many of them are trying. The technology of birth control is just now getting good enough so that a solution is becoming *technically* possible. But technology is not enough. The socio-cultural problems are going to be really tough to overcome. Multi-billion dollar programs—a massive thrust of energy and effort to bring down the birth rate—are going to be required. Nobody knows where these billions of dollars' worth of energy and effort are going to come from.

What's the ultimate answer? First a lot of people are going to have to become aware of the imperative need for population control. The international community, all nations, must work together to convince the people of what must be done. Then the governments must provide the assistance and the incentives required to do the job. The *political process* must direct resources toward this objective.

The tidal wave of new babies will be brought under control, because it must. The only questions are: how soon? and in what ways? In the meantime, what happens to the ones who already have been born? and to the ones who will be born? What about today's little babies, young children, teenagers? They will continue to grow up, and many of them will pour into the already overcrowded cities. The population crush in the metropolitan areas—just from those who are *already* born—will continue to get worse. It's in the cities where the full force of the population explosion wreaks its havoc.

## FOCUS OF THE POPULATION CRUSH: THE URBAN AREAS

We don't have to wait for a serious problem to appear in the urban areas. That problem is already here. Acute. Demanding attention *now*. As the *chronic* problem of the population explosion continues to infect more and more areas and further threaten the future of man on earth, the acute problem of the urban areas will continue to become more acute, more urgent, more undeniable.

In the urban areas, the population crush and all of the other problems of modern society come into focus. We need to spend a while talking about these urban places—these places where the *acute* symptoms of the *chronic* "population disease" weigh so heavily on so many people.

## The Explosive Growth Of Urban Areas

About three fourths of the people in the United States are living in urban areas now. More are pouring in all the time. Some forecasters estimate that in the United States, by the end of the present century another 100 million people will be living in urban areas.

That's enough people to make up ten metropolitan areas as large as New York City! Can you picture that? Or it would make up fifteen metropolitan areas the size of Chicago, or Los Angeles-Long Beach. It would make twenty more Philadelphias or Detroits; thirty more Bostons or San Franciscos or Washingtons or Pittsburghs or St. Louises; fifty more Atlantas or Baltimores or Clevelands or Houstons or Minneapolis-St. Pauls, or Newarks. It would make up between fifty and a hundred Buffalos or Cincinnatis or Dallases or Denvers or Indianapolises or Kansas Citys or Miamis or Milwaukees or New Orleanses or San Diegos or Seattles. If you live in a city of only 100,000 population, it would take a *thousand more* cities like yours just to accommodate all these people! Get the picture? All this is likely to be happening just during the next four decades of your lifetime!

All over the world the same thing is going to be happening. In many countries it will happen much faster and be much more severe than in the United States. The United States and other advanced countries have the technology, skills, education, governmental stability, everything needed to somehow handle the problem. But many nations don't have what it's going to take. Much human suffering is likely to result, and there doesn't seem to be much that anybody can do about it.

## The Problems Of The Urban Areas

The problems of the urban areas are the problems of modern society. Not that all the people with problems are located in the urban areas. Of course not. Many people live in rural areas and some of them face serious problems, to be sure. In the United States there are several government programs aimed toward overcoming the poverty of people who live in Appalachia, the Ozarks, and in other rural areas. The rural area problems can be very stubborn and difficult to solve. But the really tough challenge to the wisdom, the ingenuity, and the technology of modern man is the challenge of the exploding urban areas.

It is in the urban areas where the serious problems of modern man are concentrated and magnified—expanded into crisis proportions. It is in the urban areas that the environmental crisis is most acute. It's there that the problems of unemployment, inadequate education, inadequate medical and health facilities, inadequate housing, high crime rates—all the things which are heralded as the "major problems of our society"—come into focus. And it is in the urban areas that the problems are so frustratingly resistant—so difficult to overcome.

One problem of the urban area is that it is made up of such different "parts." Urban areas contain healthy places and sick places. There are the wealthy suburbs and there are the poverty-ridden ghettos. It is in the ghettos—these "isolated cities within the cities" where all the problems come most sharply into focus. It is in the ghettos where the solutions to "the urban problem" will be most difficult, and it's in the ghettos where much of the effort is going to have to be applied.

### The Problem Of "Population Spillover"

Yesterday, there was a city. It had its boundary and its people lived within the boundary. Then suddenly, almost overnight the population of the city exploded—spread itself out all over the countryside. But the boundaries of the city didn't move out fast enough to keep up with the people. The "city" spread out beyond its own boundaries. When that happened, the city lost control of itself. It lost the means for governing itself, for working out its problems.

All over the country and all over the world we see metropolitan areas sliced up into many local government jurisdictions—people living in the county, working in the "central city," visiting their friends in another "little city" nearby—when really it's all one metropolitan area. Usually there's not much coordination among all these local government units. Each "local government unit" has its own leadership, its own tax system, its own local programs, its own little problems to take care of. But the metropolitan area as a whole really doesn't have a "government." Nobody is in charge of the whole thing!

Most of the problems of a metro-area can't be solved in little pieces. The problems are area-wide. It's obvious that the transportation problem, or the problems of education, law enforcement, fire protection, welfare, housing, air and water pollution, employment and such things just can't be solved piecemeal. So what can be done? In the United States, the Federal government tries to force some coordination and area-wide program planning in the metropolitan areas.

The Federal government now requires that "councils of government" (COG's) or "regional planning councils" be set up to coordinate the programs and plans for the entire metropolitan area—either that, or else the Federal government will cut off the grants for the sewer systems and the urban transit systems and such things. The "COG movement" has helped quite a bit—more in some places than in others—but the problem still is far from being solved.

## The "Metro-Government" Approach

The city of Toronto, Canada, and a few other cities in the world have established "super-governments" that encompass the entire metro-area. The results appear to be very good—but apparently not good enough to convince the "small town" residents of any U.S. metro-area to give up their own independence and toss in with a "metro-area government."

While the urban areas have been growing and their problems multiplying, a financial crisis has been emerging. The "financial base" of the city has been eroding from two sides: (1) the more wealthy taxpayers have moved outside the city where they no longer pay city taxes, while (2) the costs of providing city services have been going higher and higher. So now the cities can't make it on their own. They just can't! It's as simple as that.

## The Federal Government Tries To Help

In the 1950's, as things got worse the Federal Government began to get involved. The Department of Housing and Urban Development (HUD) was set up, and various programs for "urban renewal" were undertaken. Ghetto areas were demolished and low cost housing and urban thruways were built. Then as time passed, the programs were expanded and new approaches were tried.

Still, most of what has been done so far has been more "panic relief" than carefully planned, long-range problem-solving. Some progress has been made, but we haven't yet aimed enough resources, enough effort at this "urban problem" to get very many things really "solved." In the future, more permanent solutions are going to be required. Many people are worrying about it and working on it. That's a good sign, I suppose.

The U. S. Advisory Committee on Intergovernmental Relations has recommended (1) that some kind of effective coordination among the local governmental units—between the cities and villages and counties and townships and all the others which make up each "metro-area"—be worked out; (2) that the cost of public education be taken over entirely by the state government; (3) that the cost of the welfare program be taken over entirely by the Federal government; and (4) that the Federal government give "no strings attached" grants to help out the local governments in the urban areas.

## Developing "New Towns"

Another recommendation of the Advisory Committee (and of others) is that we plan for and develop many more "new towns" to take some of the population pressures off the existing cities. This means that we go out into the rural areas, find suitable undeveloped sites, and develop new cities. In recent years several "new towns" have been developed in the United States and in other countries. It's likely that many more "new towns" will be developed, and that these new towns will absorb some of the (perhaps 100 million) people who will pour into U. S. urban areas (and some of the thousands of millions who will pour into urban areas throughout the world) during the next few decades.

New towns will help. But they won't solve the problem. For one thing, "new towns" won't necessarily be "better towns." But more serious than that, it's the already overburdened *existing* cities which are going to feel most of the pressure. They are going to have to be greatly strengthened, both politically and economically, if they are going to be able to handle the task without worsening "the mess" which we have already allowed most of them to get into.

### There Are Many Non-Economic Impediments

The problems of the urban areas will not be solved by economics alone. Some of the major impediments are political, social, traditional, psychological, cultural. The key individuals in each little borough or suburb or county government don't want to give up their positions of status or power. Racial, nationality, religious, and cultural groups within the metropolitan area don't want to give up whatever "local government control" they may have and toss in with some distant, impersonal "metro-government." Many people fear change. The people who feel at home in the ghetto may be very wary about those "social planners" who are out to tear down their neighborhoods and disrupt their familiar pattern of life—the only pattern of life they have ever known.

To all these non-economic impediments, now add the economic ones: the many billions of dollars worth of resources it's going to take just to get the "physical plant" in shape: all the tearing down and rebuilding; the development of transit systems; greatly improved waste disposal and anti-pollution services; better housing, schools, hospitals—and what about the "people-oriented" programs? Better education, training, manpower development, health, recreation, job development and all that? And with the constant, relentless flood of new people pouring into the area, while the costs keep going higher and higher? This is going to be a tough one, all right!

The opportunity costs of solving the problems of the urban areas, like the opportunity costs of solving the pollution problem, are going to be higher than most people have been willing to admit, even to themselves. But just as in the case of environmental protection, the problems of the urban areas are going to have to be solved.

Ultimately, it's the *population problem* that's going to have to be solved. That's a major cause of the urban problem and the pollution problem, and of most of the other problems of our society. It would be difficult to find a problem anywhere in modern society to which the population explosion is not an important contributing factor. When the flood of births is brought into check, so many of our current problems will suddenly become much more manageable.

While we are waiting to get the flow of births slowed down, we must keep working on the "brushfires"—the continually erupting, acute symptoms of the chronic population problem. That's what we're seeing in the urban areas and in the environmental crisis. Another place it shows up is in the *poverty problem*. We haven't talked about poverty yet. Now it's time to do that.

## THE POVERTY PROBLEM: HOW DO WE CHANNEL ENOUGH OUTPUT TO THE POOR?

What is poverty? Poverty is not being able to have what most people have—not being able to live like most people live. It's like hearing the jingle of the ice cream truck and peeking out from behind the curtain, watching the other kids buying ice cream and knowing that you can't have any. Not today. Not ever. It's like hearing your mother in the bedroom crying because she has no money to give you for lunch. It's like having to be always pleading with people not to disconnect the electricity or the telephone or to repossess the car or the TV set—like being forced to move from one shabby apartment to another because you can't pay the rent—like trying to hide from your "well-off" friends so they won't see how poor you are and feel sorry for you.

The kind of poverty I'm talking about is the poverty of an affluent society, like the United States. In less developed countries, poverty means something quite different. It may mean having no home, sleeping in the streets, getting wet when it rains—going for days with no food—having no access to medical care—perhaps dying of sickness and malnutrition.

### Poverty Is Always Relative

Poverty is always "relative to the people around you." It means "being deprived of the things the people around you have." It may be defined as: "not having enough to meet basic needs." It's easy for everyone to agree on that. But what do we mean by "basic needs"? And how much money does that take? That's where people disagree. In the United States in the early 1970's the government defined the "basic needs" income as being about $2000 for a person living alone, or about $4,000 for a family of four, with about $1,000 more for each additional person over four in the family.

You can see that these minimum basic income figures couldn't be very exact. Picture a family of six living on a farm, growing much of their own food and living in their own house and receiving a cash income of a little less than $6000 a year. The government's figures might put them below the minimum basic income level. By the government's definition, they might be living in poverty. But they might think of themselves as quite well-to-do. Probably they're living as well as (or better than) most of their neighbors and friends!

Now put that same family in a tenement house in the heart of a big city where they must pay for everything, with no relatives or friends to share vegetables and hand-me-downs. Suddenly, they're poor. Where you live and how you live and who your neighbors are can make all the difference! Also it makes a lot of difference whether the family includes young children, or teenagers, or grandparents, and whether or not any of them have chronic illnesses.

Any income definition of "the poverty line" is bound to be too high for some families and too low for others. Still, if we're going to set up any practical programs to cure poverty, we must have some way of deciding who "qualifies" and who doesn't! We must have some measurable definitions. Since there is no "right" definition, let's hope our programs can be flexible enough to take care of the injustices which will result from the definitions we choose.

### How Many Americans Are Living In Poverty?

Using these (admittedly inexact) "minimum income" definitions of poverty, how many people in the United Stares are "living in poverty"? More than one family in ten. More than five million families. A lot of people, to be sure. If you look just at the black and other minority families, you see about one family in four with incomes below the "poverty line." Also, if you look at families headed by very young men or very old men, or by women, you will see a high incidence of poverty.

About a third of all the Americans living in poverty are living in fatherless households. Most of them, of course, are children. There are about ten million children and teenagers under 18 living in households with incomes below the government's "poverty line."

None of these measures or indicators of poverty is very precise or accurate. Still, there's enough evidence to be convincing. Yes, we can be quite sure that a very real poverty problem does exist in our "affluent American society."

## Profiles In Poverty

It's difficult for most people to visualize a poverty situation. If you're from a middle or upper income family, how can you picture a family in poverty? If you begin by picturing a family like *your* family, you'll never make it. A family like your family wouldn't be in poverty.

Try to picture what your family would have been like if your father had died before you were old enough to remember. Or suppose he had contracted some kind of disabling disease that required a lot of medical expense. And suppose there were no well-to-do relatives to help out. And suppose there were already three young children ahead of you, and another due in a few months. What would life have been like for you, then? Very different. Right?

The most frequent, most serious profile of a poverty family is the profile of a fatherless household where there are small children. Other "profiles in poverty" would include families with disabled or otherwise unproductive fathers, and families of high school dropouts who get married in their teens, start producing children and try to make it on their own.

For another profile, you might picture a household headed by a middle aged or older man, perhaps not too bright, not too flexible, whose skills have been made obsolete by changing times. If the man happens to be a member of a minority race or has a poor employment record or a prison record or some other "undesirable characteristics" (or even if he doesn't) he may not be able to get another job. After months or maybe years of looking, he may just give up. Many do.

Other profiles in poverty would show you families living on seasonal employment, working at part-time jobs, temporary jobs, low-paying jobs. Often the poverty is worsened by a large number of people depending on one low-wage earner. If the family includes aging relatives, plus several children, the paycheck just won't meet the needs.

## Poverty Is Self-Perpetuating

An important *cause* of poverty is the *existence* of poverty. Poverty begets poverty. Many people are born and raised in poverty, in neighborhoods where poverty is a way of life. All the influences point toward the poverty pattern of life rather than toward ways to get out of poverty. Avenues of escape are difficult to find. Many people feel beaten before they start — so why try?

If a person is bright, strong, healthy, capable, he will work himself out of poverty. But only a certain percentage of the people born in any generation are going to be bright, strong, healthy and capable. The others, if they had been lucky enough to be born into affluent, middle class families, would have been able to do okay. They would have been educated and trained and "conditioned to succeed." They would have succeeded. But when born in the ghetto and conditioned to the poverty way of life, their chances for success are very slim. They will just stay in the ghetto and be the "perpetuating group" left behind when the strongest and most capable ones break out and leave. The ones left behind are trapped by, and they are the ones who perpetuate the "vicious circle of inherited poverty."

### What Can Be Done About Poverty?

When you look at the poverty profiles it's easy to see that simple solutions are not going to solve the problem. No longer is it possible to quip: "If people are poor, let them go to work. If they are unskilled, offer them training and then let them go to work. If they won't take the training and won't go to work, then let them go hungry."

How simple this solution seems to be! But how do you apply it to the widowed or unwed or deserted pregnant mother of four small children? How do you apply it to the aged couple, over 65? Or to the man who is so old no company wants to hire him? Or to the men and women who are physically or mentally incapable?

Much, probably *most* of the poverty in the United States can't be "cured" by getting people to go to work. In so many poverty families there just *isn't anybody* to go to work! Often, even if there is someone who possibly *could* work, there's no *feasible* way to train them and fit them into an available job.

We need to emphasize manpower training and job development. Obviously. We need such programs as worker education and skill development and on-the-job training, supplemented by job placement services, and subsidies to stimulate new employment opportunities. Of course! Many such programs are already operating, and progress is being made in overcoming some of the "curable" kinds of poverty. But no matter how far we go in job development, this approach cannot *completely eliminate* the poverty problem. Obviously not. So what can be done?

What's the most serious "profile in poverty" we see in this country? The fatherless home with small children. Do you suppose the population problem is lurking under the surface here, somewhere? You know it is. If the number of children born into these homes could be reduced, that would help a lot. Improved birth control technology and liberalized abortion laws may lessen the seriousness of this "profile in poverty." But still, the problem isn't going to be completely solved that way.

The only way we can really hope to handle the incurable kinds of poverty is to have some programs which will provide money to the families who have no way to generate their own incomes. There's just no other way. What kinds of programs? And how much money? To whom? Under what circumstances? Those are the tough questions. The choices usually reflect more political and social considerations than economic ones. Let's talk about some of the programs—the ones we have now, and the ones we might have.

## PROGRAMS FOR OVERCOMING POVERTY

Essentially, there are two ways to approach the poverty problem. One way is to try to "cure the disease"—that is, to help the poor people become more productive so they can lift themselves out of poverty and be self-sufficient. The other is to redistribute some of the income from those who earned it to those who need it. Government agencies, private charities and various religious and other organizations are working on this problem, both ways. I'm sure you know that. In this section we'll be talking about government programs, but don't forget that the efforts of the private charities and religious organizations also are important.

### Curing "The Disease Of Poverty": Manpower Development

The United States has a long tradition of manpower training and development through public education, vocational education, extension services and many other programs. Since the "New Deal" days of the 1930's, the U. S. government has become increasingly involved in these programs. During World War II the government undertook a massive manpower training effort, not to cure poverty, but to increase the output of the economy. But the wartime manpower program cured a lot of poverty. You may be quite sure of that!

In the latter 1940's and throughout the 1950's, "traditional" kinds of manpower training were continued. Then in the early 1960's the Manpower Development and Training Act increased the size and scope of this effort. But it wasn't until the mid-1960's that the *Economic Opportunity Act* was passed, setting up the "anti-poverty program"—that is, the "war on poverty."

**The "Economic Opportunity" Program—The "War On Poverty."** The thrust of this new program was to get local people, including poor people, involved in working out innovative ways to overcome poverty. Many things were tried—neighborhood information and help centers, legal aid for the poor, "black capitalism" programs to help small businesses, job development activities, volunteer counseling and help for the poor, and of course, education and training and manpower development.

Several of the efforts supported by OEO (the U. S. Office of Economic Opportunity) failed miserably. Others, including some of the legal aid and counseling and education programs, were generally thought to be very successful. The OEO-sponsored "headstart" program — pre-schooling to help "culturally deprived" children to get ready for school — was a most popular and apparently very successful program. The OEO also sponsored educational programs — Sesame Street, the Electric Company, and several programs offering special help (in primary and secondary schools and in college) for students from poor and minority families.

There's no question that governments have been making increased efforts to "cure the disease of poverty." Progress has been made, sure. But much more could be done. It seems likely that more *will* be done, and that more success will be achieved. But one thing is certain. Complete success in "curing the disease of poverty" will never be achieved. It just isn't possible.

**Some Poverty Is Not "Curable."** There will always be some people who can't be productive enough to stand on their own. Some will be "down" only temporarily. They will need only temporary help. Others will be permanently incapable of self-support. They will need continuing help for as long as they live. For the people who can't stand alone, their only chance for a share of the society's output is for the society to donate something to them — a gift from the productive people to the unproductive people of the society. No matter how effective the "poverty curing" programs may become, there always will be a need for some income redistribution to take care of those who are either temporarily or permanently incapable of self-support.

### Lessening The Discomfort Of Poverty: Income Redistribution

In the United States we have three big "income redistribution" programs: (1) the "social security" program, (2) the "unemployment compensation" program, and (3) the "welfare" program.

**Social Security, And Unemployment Compensation.** The Social Security program is the largest income redistribution program in the country. It was set up during the depression of the 1930's and has been growing ever since. This is the "Old Age, Survivors, Disability and Health Insurance" (OASDHI) program.

Under the Social Security program, taxes are automatically deducted from people's paychecks and payments are automatically made to people who qualify — by retirement, disability, or poor health. If the wage earner dies, payments automatically go to his dependents. You can see that this program would help a lot to soothe the hurt of lost income. This is a Federal government program and the taxes and benefits are the same, nationwide.

The "unemployment compensation" program gives automatic payments (for a limited number of weeks) to anyone who loses his job. This is a Federal-sponsored, state administered program. The amount of money you get per week and the number of weeks you can get it, depends on rules adopted by your state. People "sign up" for their unemployment payments at the state employment office in their area. Usually the only requirement is that they be ready to take any appropriate job that might become available. If you're interested in how the program is working in your state, go down to the local employment office and ask. I'm sure someone there would be glad to explain it to you.

There are many things to complain about, about the social security and unemployment compensation programs. The payments aren't high enough to provide for "basic needs," and what does a person do after his unemployment payments run out? There are other questions, but generally these two programs seem to be functioning reasonably well. It's in the welfare program where the serious problems arise.

**Welfare, Or Public Assistance.** The welfare program includes several kinds of aid. "Aid to dependent children" (to families without a father's support) takes most of the "welfare" money, but some is also paid to others who can prove that they need it and can't get it from anywhere else. This "welfare program" is supported with Federal and State money and some local money. Within Federal guidelines, each state sets up the rules for its own program. The procedures and the benefits are a lot different from one state to another. (I suppose everyone has heard stories about poor families moving from one state to another so they can get more welfare benefits.)

The costs of the "welfare," or "public assistance" program have been growing very rapidly in recent years. In the early 1970's more than seven million children (about eight percent of all the children in the United States) were being supported under this program. In New York and California this program supports about *twelve percent* of the children in the state. In other states, the percentages are smaller.

In addition to "aid for dependent children," welfare payments go to "needy" people who are elderly, blind, disabled, or otherwise unable to work and who for some reason don't qualify for payments under the Social Security program. A person must be able to prove that he really needs the money. How hard that is to do may depend a lot on which state you happen to be in, and which person at the "public assistance office" you happen to be talking to.

Everyone seems to agree that we need to reform our "public welfare" program. The system contains a lot of unfairness and inequity. There's too little money flowing in some directions, perhaps too much in others. The "proof of need" requirement sometimes results in a lot of snooping into the personal lives of the recipients. That isn't good. But what can be done about it? One suggested approach is the "guaranteed annual income"; one suggested way to achieve it is the "negative income tax."

### The Guaranteed Annual Income, And The Negative Income Tax

Maybe the government should stop providing welfare payments and simply guarantee each family a minimum income. People whose paychecks were large could still have taxes deducted, but people with paychecks too small could have "taxes" *added*! People with no job (no paycheck), could receive checks from the government. If your income is high, you pay. If it's low, you receive. See how this approach can be called the "negative income tax"?

Another approach might be to give *everyone in the country* a certain sum of money (say $1,000 a year) and call the money "taxable income." People who had no other income wouldn't have to pay any taxes on the grant but those with high incomes would have to repay most of it in taxes. Senator George McGovern suggested some such program when he was campaigning for the Presidency in 1972.

Previously, President Nixon had recommended a guaranteed annual income program, with the minimum income being $1600 for a family of four, plus $300 for each additional person. Under the proposed program, a family of four with no income would receive $1600 from the government. The family could earn up to $720 a year without any reduction at all in the subsidy payment, but as their earnings went above that, the subsidy would be reduced. When the family's earnings got up to about $4,000, the subsidy would be zero.

As a part of the President's proposed program, any able-bodied recipient would be required to register for work, and/or to participate in a job training program. Day care centers would be set up to permit mothers of dependent children to take jobs and/or training. The continuation of the family's "guaranteed annual income" would require that a person take any suitable job that became available.

What about this "guaranteed annual income" approach to getting rid of the "welfare mess"? Is it a good approach? Can it be made to work? The answer is "probably, yes" to both questions. But it isn't going to be easy to work out, not only because of the economics involved, but even more, because of the political and administrative problems. Unless it's done right it could turn into an administrative nightmare of red tape, high costs, and injustices. That would be no improvement! That's what we're trying to get rid of now.

### There Is No Easy Solution To The Poverty Problem

The poverty problem isn't going to be easy to solve. Much progress has been made, sure. The number of people classified as "living in poverty" has decreased considerably—by *almost half*—over the past twenty years. That isn't enough, but the more we reduce the number in poverty the more difficult it becomes to do more. The more we succeed, the closer we get to the "hard core" of families who have no way of becoming self-supporting—families which have no choice but to depend on the "social process" or the "political process" to provide them a distributive share of society's output.

Proposals for overcoming the poverty problem are among the most controversial public policy proposals of our time. It's obvious that the U. S. economy is productive enough to provide for everybody. There's enough so that the rich can still be rich and the middle income people can keep on living well, while the poor can all be brought up above the poverty level. The physical means to solve the problem certainly exist. But to try to design a realistic program or plan to put it all together—that's really tough. It involves hard choices—difficult trade-offs.

How do we serve the needs of the "honestly poor" without indulging the "fast buck artist"? and without destroying the incentives, and the feelings of personal worth of the recipients? How do we get the needed flexibility into our programs? And withstand selfish political pressures and bypass entangling bureaucracy? How do we balance off the insensitive snobism of the "hard-nosed" against the unworkable idealism of the "willing-bleeders"?

Can we go far enough to really take care of the problem without perpetuating it? Or without creating other problems more serious than the one we're trying to solve? Are there reflex effects to think about? Are we thinking about endangering any golden-egg-laying geese by trying to reach the best of all possible worlds?

If anyone is sure what the ultimate answer is, either he has great wisdom, or else he doesn't understand the complexity of the problem. So far, all we have been able to do is creep along, experimenting, taking various steps as they have become politically feasible. Probably that's what we will continue to do. There will be tough choices every step of the way, and it isn't likely that anybody will be completely satisfied with the results. But that seems to be the inherent nature of the income distribution problem. Remember?

## SUMMARY

This chapter hasn't given you any answers—only tough questions—questions that your generation and mine are going to have to struggle with. What is going to happen? We will direct more effort and make more progress toward these "new" microeconomic objectives: protecting the environment, bringing the population explosion under control, working out the problems of the urban areas, overcoming the problem of poverty. We *will* make progress toward these new objectives because we *must*.

The purpose of this chapter has been to sharpen your awareness of these problems—to let you see what's involved, and a few of the kinds of tough choices that are going to be required. In this chapter you have taken a quick look at some of the most pressing "*microeconomic* public policy issues" of our time. But we face some serious "*macroeconomic* public policy issues," too—chronic inflation and unemployment. That's what you'll be reading about in the next chapter. But before you go on, think about these micro-issues for awhile. The better you understand these micro-problems the more likely it is that your generation and mine will succeed in working out better solutions than we've been able to come up with, so far. We must, of course. We have no choice.

---

## REVIEW EXERCISES

### QUESTIONS (Try to write out answers, or jot down the highlights.)

1. The "desired directions of progress" in a society are always changing some. In recent years the desired directions have been changing very rapidly. Can you think of several shifts in the directions of progress, just during the last few years? Discuss.
2. "We have all been 'living high' by not paying the full costs of the things we've been using up." Explain.
3. Can you think of several ways in which you personally are helping to pay some of the cost of the environmental problem? Can you think of ways in which you are having to pay more of those costs, as time goes on? Discuss.
4. "The reason the environmental problem has grown so rapidly and become so serious is that in the 'market directed' economies of socio-capitalism, most of the things which create, and contribute to the environmental problem are free from the major control mechanism of the society—that is, are external to the market process." Discuss.
5. "The only ways the environmental problems can be handled are these: (a) force the social costs of each bit of 'environmental destruction' to be reflected in the market process, or (b) establish regulations and prohibitions to protect the environment. Both these approaches require decisive (and sometimes unpopular) actions by the political process." Discuss.
6. Explain some of the reasons why the population problem is so difficult to control.
7. Discuss some of the characteristics—the "nature"—of the urban problem, explain some of the reasons why the problem is not going to be easy to handle. Can you see examples of any of these characteristics and impediments in *your own* local area? Discuss.
8. Explain some of the important causes of, and impediments to overcoming poverty, in the "affluent societies."
9. The poverty problem is *especially* difficult to handle, because just about everything you can think of to *overcome* poverty, seems to interfere with something else—seems to generate reflex effects: "Adequate" payments to support fatherless children may result in more fatherless children; "adequate" payments to the unemployed

may result in more unemployment. When a person receives "adequate" payments to eliminate his "uncomfortable status," that automatically takes the pressure off. It relieves the urgency to work out a permanent solution to the uncomfortable situation. It tends to perpetuate the problem. This basic dilemma is inherent in all "anti-poverty" programs. It's because of this dilemma that the "degree of success" in overcoming poverty is likely to always fall short of the "desired social objective." Discuss.

# 29 UNEMPLOYMENT AND INFLATION: THE BASIC DILEMMA OF STABILIZATION

*The basic macroeconomic problem is how to keep the economy running fast enough to suit everybody, without letting wages and prices run away together.*

Suppose someone asked: "What do you think will be the world's most serious issues, the most pressing problems of modern society during the next five or ten years?" What would you say? To be sure you would mention the things you were reading about in the last chapter: the environmental crisis, the population crush, the problems of the cities, and what to do about poverty. Of course. But isn't there something, else?

## THERE'S NO ESCAPE FROM THE MACROECONOMIC DILEMMA

Do you think the basic *macro*economic issues—the problems of inflation and unemployment—are important enough to be included in the list of *major problems*? You bet they are! The basic macroeconomic problem—the dilemma of stabilization—is always with us. It's never completely solved. Since you're going to be living with it all your life, you might as well understand it. That's what this chapter is all about.

You already know a lot about this stabilization issue. In Part Five you read about inflation and unemployment, and about how monetary and fiscal policy are used to try to keep things running right. You found out that things don't always work out right. Remember?

You've read and heard a lot of news about unemployment and inflation, too. You always will. Why? Because it's a continuing, unsolved dilemma. It's always there. It plagues every modern, market-direct economy. If it's under control it isn't so bad. It's tolerable. But out of control? Disaster!

Nobody knows what the answers are. In the "pure market model," sure. There, the answers are easy. But answers that will take care of the economic stabilization problem in the unpredictable real world? No. We don't have those answers yet. That's why we're going to have to struggle with this problem for a long time. We're going to have to learn to put up with some unwanted unemployment and with some unwanted inflation. Too bad, but that's just the way it is.

### What's So Bad About Unemployment? Or Inflation?

In Chapter 16 you read about the great social cost of unemployment and inflation. Remember? Unemployment wastes resources—valuable human resources and other resources too. It frustrates and degrades people. It leads to poverty, lack of self-respect, family breakdown, crime, social and political stress.

If unemployment gets bad enough it spells collapse for the economic system. Then new ways have to be worked out to take care of production and distribution for the society and to get the economy running again. Otherwise many people will go hungry. Sooner or later the hungry people will likely revolt. That's what has happened in several places in the world just during this century. It almost happened (to some extent it did happen) in the United States during the 1930's. Read Steinbeck's *Grapes of Wrath* if you want to get a feel for what it was like.

What about inflation? It robs most people of the things they have worked for, saved for, planned for. Then it gives those things to other people (or sometimes, to the government). It forces everyone to get the "gimmies." Anyone who doesn't get them gets robbed the most.

Inflation doesn't let people be "reasonable, good neighbors" about things. It breeds clash between unions and management, between businessmen and their customers, between government officials and their constituents. It forces teachers and policemen to unionize and go on strike. It destroys the careful plans and programs of the school boards. It creates financial crisis in all the programs of the state and local governments. It turns college administration into a financial nightmare. It makes people buy more foreign-produced goods because they're cheaper than the high-priced domestic goods. It creates problems and generates stresses and disharmony in every nook and cranny of the society.

If inflation is under control, and moderate—say, if prices rise only two or three percent, or maybe even four or five percent per year—we can work out ways to live with it. But if it runs away, inflation destroys the "general acceptability" of money. What happens then? Without money, trade stops. Production stops. Employment stops. Income stops. Complete collapse of the economic system. Right? Right! That's what happened in Germany in the 1922–23 inflation. Here's an example of what it was like.

## An Example Of Runaway Inflation

Suppose you go camping up in the mountains for a few weeks. Then, when you get back to your car and back on the highway heading home, you stop at a drive-in for a double-decker burger and a chocolate shake. The counter girl says: "That'll be fifty dollars."

"FIFTY DOLLARS! Don't be ridiculous!" You leave the bag on the counter and go stomping out to your car, get in, slam the door and drive away. Soon you stop at a gas station and say: "Fill 'er up."

"You got the cash? We don't take checks from out-of-staters, and we don't take credit cards anymore. We have to have cash so we can spend the money *today*. Prices are going up too fast to wait until tomorrow."

You are confused. "Sure, I have the cash. It won't be all that much anyway. I only need about ten gallons."

"Let's see. Ten gallons, at twenty dollars a gallon, that'll be two hundred dollars. You got two hundred dollars on you?"

This nonsense is really beginning to get to you! "What do you mean, twenty dollars a gallon? What is going on around here?"

## Prices Go Up Several Times A Day

Then he explains it to you, step by step—about how prices have broken loose. How every hour, the oil company calls in a new price list. Prices are getting higher and higher, faster and faster.

While he's talking the phone rings. He answers it, then comes back shouting, "Ten gallons will cost $220 now. The price just went up again!"

You don't have $220. You ask him how much he'll give you for your spare tire. He offers you a thousand dollars. You think, "It's good that this inflation works both ways." So you sell the tire and get your tank filled.

Then you think: "By tomorrow, gas is going to cost a lot more. I'd better buy some gas cans and spend up the rest of my money right now. That way I'll be sure to have enough gas to make it home." So that's what you do. It was a wise thing to do, too. Why? Here's why.

This time next week, gas is going to cost $140 a gallon! And next month it will cost $2,400 a gallon. A couple of months later it will cost *seven million dollars* a gallon. A few weeks after that it will go to a hundred million dollars a gallon, then to a billion, then to a hundred billion, then to a trillion—but long before then, the economy will be in a state of total disruption—total collapse.

## The German Inflation Of 1922–23

The kind of runaway inflation you've been reading about is an example of exactly what happened in Germany in 1922–23. In the fall of 1923 a new administration took over in Germany. They called in all of the old, inflated marks and issued a new kind of marks in exchange. Would you like to try to guess what the exchange rate was? Would you believe a *trillion to one*? That's right.

Suppose you had lived in Germany then, and had a million dollars worth of marks in your savings account before the inflation. What would all that money be worth in the fall of 1923? Think of it this way: Suppose there were ten thousand people and each one of them was a millionaire —each one had a million dollars worth of marks before the inflation. Then, in the fall of 1923, if *all ten thousand* of those millionaires turned in their money on the same day, what would they get back? At the exchange rate of a trillion to one? They would get back *one penny's worth of marks* to split between them. Each million dollars, reduced to one ten-thousandth of a cent!

Suppose something like that happened in the United States, today. If you had enough money to buy the whole GNP of the United States before the inflation, then you'd have just about enough to buy a double decker burger and a chocolate shake, afterwards. Such is the nature of runaway inflation. Is it any wonder it disrupts the market process and brings the total collapse of the economic system?

### The Problem: Moderate But Chronic Inflation And Unemployment

It doesn't seem likely that either runaway inflation or serious, prolonged depression will occur in the United States or in any of the other advanced economies. Not that it's impossible. It's just that I think we will do enough of the right things to prevent any such catastrophe.

No, the thing we have to worry about is not total collapse of the economic system. It's the continuing, eroding effect of moderate but chronic inflation and moderate but chronic unemployment, both going along together, all the time. And what's so difficult about it is this: Whatever you do with monetary or fiscal policy to try to make one better (say, reduce unemployment) is likely to make the other worse (increase inflation). That's the basic macroeconomic problem—that's *the basic dilemma of economic stabilization.*

## THERE'S NO "FULL EMPLOYMENT AND STABLE PRICES" RATE OF SPENDING

What supports (holds up, or pushes up) total employment and output and income in the economy? Total spending? Of course. And what supports (holds up, or pushes up) prices in the economy? Total spending? Of course. Aha! See the problem? That which pushes up employment also pushes up prices. There you have it. That's the reason for this basic dilemma of economic stabilization.

In the pure market model there would be no "stabilization dilemma." A level of spending just great enough to bring full employment would not bring inflation. A level of spending just low enough to prevent inflation would not bring unemployment. In the pure market model, there's a level of spending which will bring "full employment with stable prices"—that is, no unemployment, no inflation. But in the real world I'm sad to say, it never works out quite that way.

## The Inflation-Unemployment Overlap: The Basic Dilemma

In the real world there is a wide "overlapping area," where "inflationary-level spending" and "full employment-level spending" overlap. What I mean to say is this: Suppose total spending expands enough to eliminate unemployment. Then that rate of total spending is already too great for price stability. As total spending increases, inflation gets going even before full employment is reached. Or as total spending slows down, unemployment starts increasing even before all the inflationary pressures are relieved. See the problem? If we're using monetary or fiscal policy (or both) to slow down total spending to cure inflation, long before we ever get the inflation problem under control we're likely to have more unemployment than we can tolerate! We'll get into the question of *why* all this happens. But first, let's be sure you know *what it is* that happens.

Here's another way to say it. A level of spending *too low* to support full employment, is still *high enough* to support inflation. What we wind up with is a level of spending that gives us some of both evils—unemployment and inflation both going on at the same time! So what can we do with monetary and fiscal policies? Try to reduce unemployment? We get more inflation. Try to reduce inflation? We get more unemployment. Oh, miserable dilemma! What to do?

There's no easy answer. But there's one thing that's fairly obvious. Just this: When this "unemployment-inflation" dilemma exists, *we're going to have to use something in addition to monetary and fiscal policy* if we're ever going to get it straightened out. That's for sure.

## An "Employment-Inflation" Graph Illustrates The Basic Dilemma

It's too bad that in the real world there is no "full employment and stable prices" rate of total spending. It's too bad, but that's the way it is. We are constantly between these two "evils," always suffering to some extent from both. This is the basic dilemma of economic stabilization policy. Fiscal or monetary action to cut back spending and curb inflation brings more unemployment. Fiscal or monetary action to increase spending and reduce unemployment, brings more inflation. You know that this dilemma isn't going to be easy to solve. But before we get into that, here are some graphs that will help you to see the problem more clearly.

Fig. 29-1 **TOTAL SPENDING, EMPLOYMENT, AND PRICES: THE "MODEL" CASE**

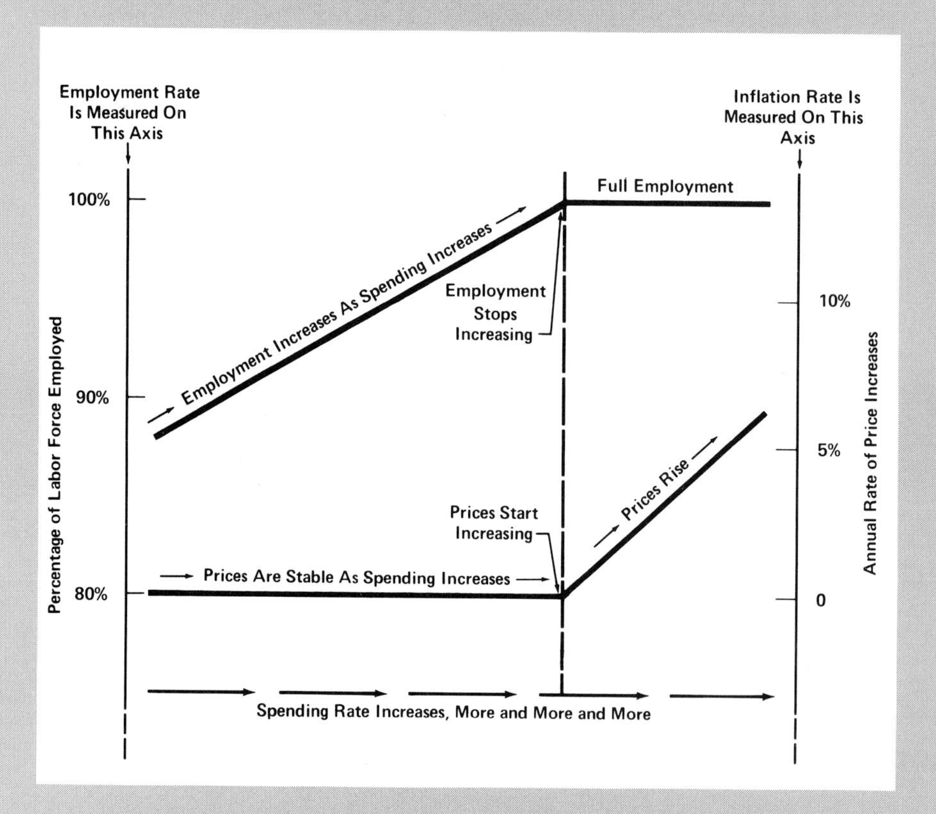


In The "Model Pure Market System" There Is No "Inflation Unemployment Overlap."

In the "model pure market system," as spending keeps increasing, employment and output will keep increasing, but prices will remain stable until full employment is reached. Prices will start to go up, bringing inflation, only if total spending keeps increasing after full employment has been reached.

If the real world worked like the model, stabilization still wouldn't be all that easy. Finding out exactly what the right level of spending would be, and then getting total spending to stabilize at that exact level, wouldn't be quite as easy as falling off a log! On the other hand, it would be comforting to know that a "full employment and stable prices" rate of spending really did exist, somewhere.

In the real world, no such rate of spending exists. The next graph shows you how it looks in the real world.

**Fig. 29-2 TOTAL SPENDING, EMPLOYMENT, AND PRICES: THE "REAL WORLD" CASE**

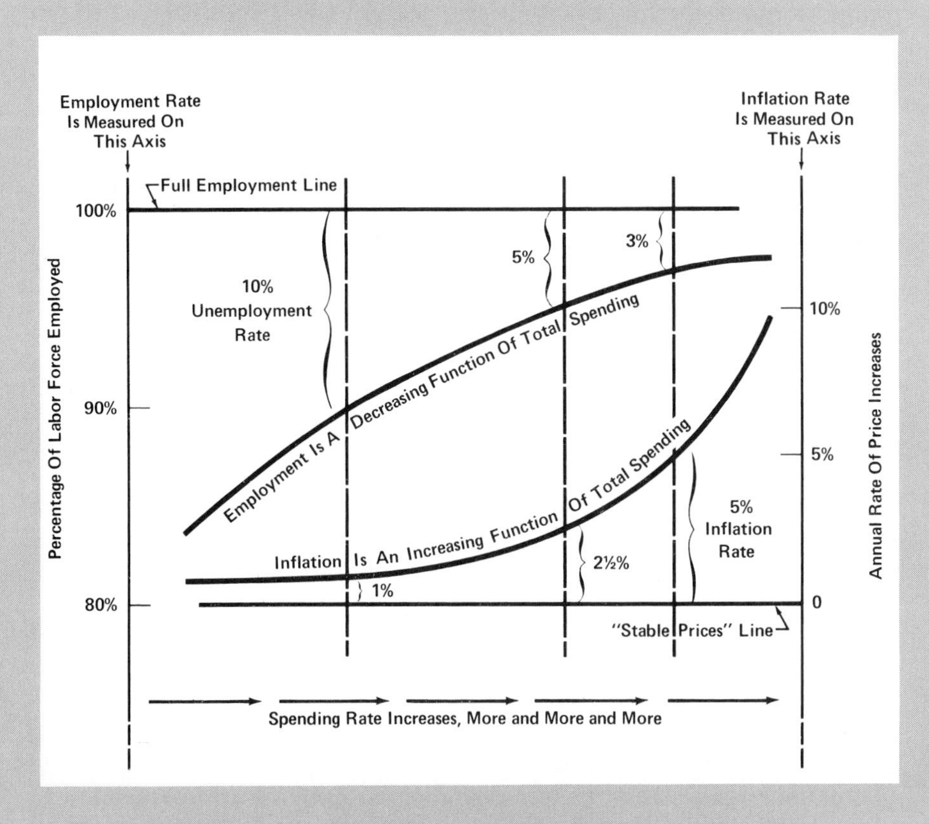


In The Real World, Inflation Speeds Up Before Unemployment Gets Slowed Down.

In the real world, there is no level of total spending which will bring full employment and stable prices. If the spending rate is already high, a further increase in spending will add only a little to employment, but a lot to inflation!

In the real world, as spending increases, employment increases, but the employment increase gets smaller and smaller as spending gets higher and higher. You could say: "Employment is a decreasing function of total spending."

What about the inflation rate? As spending increases, prices rise faster and faster. You could say: "The inflation rate is an increasing function of total spending."

If you wanted to sound more like an economist, you could talk about the "increasing marginal rate of inflation" and the "decreasing marginal rate of employment." Say it any way you want to, just remember that as spending increases more and more, each extra increase in spending is likely to add less and less extra employment, and more and more inflation.

The two figures you just studied show two very different situations. The first (pure model) figure shows a level of spending which brings "full employment and stable prices." The second (real world) figure shows some unemployment and some inflation *at any level of spending* you can choose. Why?

Why doesn't the real world look like the model? Essentially, for this reason: There are things *other than total spending* which are always pushing upward on unemployment and on prices. Pretty obvious, if you think about it, isn't it? Sure. First, let's talk about some of the things other than total spending, which push upward on prices.

**Everyone Is In The "Upward Price-Inching" Game**

Why do real-world price levels respond so differently from those in the pure model? There are several reasons, but most of these "reasons" are reflected in and work through administered prices. You remember about administered prices. You see them just about everywhere you look in the real world. Almost all sellers of all products have some market power—some power to push up the price of what they sell. This is true of consumer goods sellers, and it's true of the sellers of all the factors of production—labor, land, and capital.

Each seller wants to be sure he doesn't get left behind in the game of "upward price-inching." All of us are "price-inchers"! We give all sorts of reasons for pushing our prices up—some of our "reasons" are excuses, and some are valid explanations. Labor cites the increased cost of living and increasing output per man and other things. Businesses cite increased costs of labor, of transportation, of utilities, of interest and of other inputs. Both labor and businesses have the power to push up their prices, so that's just what they do. How much would total spending have to be cut back to make everybody stop this price-inching game? Who knows?

**Low-Wage Workers Are Pushing For Higher Wages**

These days, everyone is trying to increase his "market power," and to push up his price. Low-paid workers are getting unionized and pushing for higher wages. Whenever one group succeeds in forging ahead, everyone else tries to jump ahead a little farther. Civil service workers, teachers, firemen, policemen, sanitation workers, other government employees have been paid low wages for a long time, but now they have begun to demand wages comparable to the wages in manufacturing and construction and other high-wage industries.

People are becoming less willing to accept the idea that the person who performs one job should "by rights" receive a lot more income than the person who performs some other job. The lower income people are getting organized and are taking vigorous action to increase their income shares. Can you see how wages might be going up, even in the face of declining total spending? Sure.

There's great pressure pushing upward on wages and prices, and there's great resistance to any downward movement of either one. This is the nature of the real-world economy in which we live, and we're going to have to develop and use tools which will enable us to live with it. If monetary and fiscal tools can't do it alone, then other tools must (and will) be used. Do you suppose this helps to explain why the United States and most of the other countries of the world exercise some kind of direct controls over wages and prices from time to time (*most* of the time), these days? Of course!

We'll get into the issue of controls later on in this chapter. But first, what about the unemployment problem? It's easy to see how prices might keep on inching up by themselves, but surely, no one wants to "inch himself out of a job"! So if there's unemployment, when spending increases, why doesn't employment and production and output increase until full employment is reached? That's easy to figure out too, if you think about it.

**Some Unemployed Workers Aren't Suited For The Available Jobs**

Just as the "inflationary bias" in the real world can be explained in terms of one thing: "administered prices"—also the "unemployment bias" in the real world can be explained in terms of one thing: *all labor is not alike*. As total spending increases, more people are employed, true. But the ones most suited to work in the expanding industries are hired first. As spending increases more, only as prices go higher will people be hired at jobs they aren't well suited for. Total spending would have to go very high for all the unemployed teachers to be hired as bricklayers!

Another real-world consideration is that some businesses are always cutting back, while some others are always expanding. Some people are always changing jobs for one reason or another. While they're between jobs, they're unemployed. But this kind of temporary unemployment is not a real problem. It's only when the "between jobs" period stretches on and on—that's when some action needs to be taken to get things going again—and hopefully, without throwing gasoline on the fires of inflation.

You already know from Figure 29-2, what it looks like when spending is increased to overcome unemployment. If total spending is already high, an increase in spending is likely to bring a little more employment and a lot more inflation. Remember? It's sort of a trade-off. You give up some price stability to get some added employment—or vice versa. Do you think this "trade-off ratio" could be shown on a graph? Sure. All trade-off ratios can be shown on graphs. Of course! The "curve" we use to show "inflation-unemployment trade-off' is the "Phillips Curve."

## The Phillips Curve Shows The Inflation-Unemployment Trade-Off

The "Phillips Curve" is named for the British economist, A. W. Phillips, who developed the "curve" in the latter 1950's. What Professor Phillips did was to look back over the years and see, for different times, what the rate of inflation had been and what the rate of unemployment had been. As you would expect, he found out that when unemployment was high, the inflation rate was low. Of course! And that when the economy was booming and the unemployment rate was very low, the inflation rate was much higher. Sure. You can understand that, and you can understand why.

More spending brings more employment. Sure. But at the same time it brings more opportunities for workers and businesses to raise their wages and prices. So that's just what they do. The rate of inflation picks up. When total spending slows down that brings more unemployment. But when things are slow, there are fewer opportunities for workers and businesses to raise their wages and prices. So when the unemployment rate gets higher, the rate of inflation gets lower.

The Phillips Curve illustrates this idea: *We can have a smaller unemployment rate, only if we are willing to accept a larger inflation rate.* It's a "trade-off" situation, where we would like to have *minimum amounts* of both — minimum unemployment and minimum inflation. But to get less of one, we know we must put up with more of the other. Oh unhappy day!

Is this the "real-world dilemma of economic stabilization"? You bet. It's one of the really tough economic problems of our time. We keep trying to learn to live with it and to devise better ways of solving it or coping with it. But so far, it's still a dilemma. A very serious one.

## A Production Possibility Curve For Unwanted Products?

Now that you understand what the Phillips Curve illustrates, what do you suppose it will look like? Negative slope? Of course. As you give up some of one thing (inflation) you get back more of the other (unemployment). This curve is going to look like a "production possibility" or "opportunity cost" curve. Right? Sure. But in a way, it's very different from the production possibility curve.

The production possibility curve shows you *how much you can get of the good things you want,* and how much of one you must give up to get more of the other. You would be happy if the production possibility curve would shift upward and outward. You could then have more of either product, or a larger combination of both.

What about the Phillips Curve? It shows you *how much we must take of the bad things we don't want*—unemployment, and inflation. Would we be happy if the Phillips Curve shifted upward and outward? Certainly not! More unemployment, or more inflation, or a larger combination of both? Not if we can help it!

We would like for the Phillips Curve to shift downward and inward. In fact, we would be most happy if it would shift all the way down to zero and disappear! Then we could have full employment and stable prices and the whole dilemma would be solved. *We all wish that would happen.* Of course. But if wishes were horses, beggars would ride. Right?

Figure 29-3 shows the Phillips Curve. It's just another way of looking at the same picture you saw in the employment-inflation graph a few pages back. In fact, this Phillips Curve was derived from the two curves in Figure 29-2. Compare the two figures on the following pages and you will see that both graphs are telling exactly the same story.

## THE PHILLIPS CURVE AND THE U.S. STABILIZATION DILEMMA

The Phillips Curve *illustrates a concept.* It can't *predict* what unemployment rates will go with what inflation rates. But it does suggest that sometimes, if the curve is high up and far out (as in the early 1970's), some approach *other than* monetary and fiscal policy to influence total spending must be used.

When both the unemployment rate and the inflation rate are high, cutting back on total spending to curb inflation may bring a disastrous depression. But increasing total spending to overcome the unemployment may bring a disastrous inflation!

When the Phillips Curve gets itself shifted way up and way out, as in the early 1970's, that's when the stabilization dilemma is really serious. That's when the economic theorist's indirect tools of economic adjustment and stabilization get pushed aside by the policymakers. That's when the last resort approach of "good old American pragmatism" takes over and aims *directly* at the visible conditions—the unemployment and the inflation—and goes to work on them. Direct controls are established to slow or stop increases in wages and prices. Direct action is taken to create new jobs for the unemployed. Is that what happened in the early 1970's?

### The Wage-Price Freeze Of August, 1971

In August of 1971, after more than two years of trying to stop inflation by using "tight money" policy, and after repeatedly vowing that he would never impose direct controls on wages and prices, President Nixon imposed direct controls on wages and prices Why? Because after

Fig. 29-3 **THE UNEMPLOYMENT-INFLATION TRADE-OFF: THE PHILLIPS CURVE**

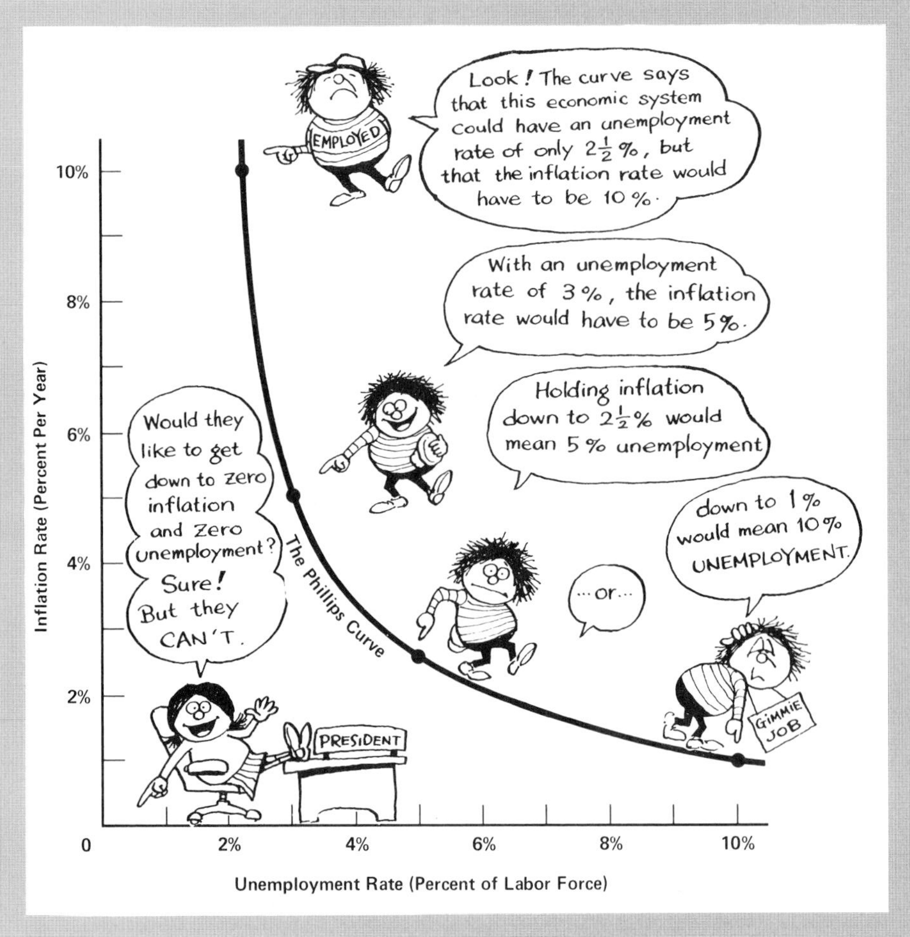


It's A Dilemma All Right. You Can't Have Your Cake And Eat It Too. But It's Bad When You Can't Have Either!

The Phillips Curve says that when the inflation rate is low the unemployment rate is high; when the unemployment rate is low, the inflation rate is high. If we use monetary or fiscal policy to get less of one, we get more of the other.

The Phillips Curve you see here was derived from the "employment function" and "inflation function" curves shown in Figure 29-2. Nobody knows just exactly where this "Phillips Curve" will be for any economic system, at any particular time. All sorts of things might cause it to shift.

The Phillips Curve illustrates the fact that something *other than* total spending must be influencing (holding up or pushing up) wages and prices. Whenever these "other than" influences get stronger, the curve shifts upward and outward; whenever they get weaker the curve shifts downward and inward.

Whenever the "other than" influences are changing, it may be more helpful to try to understand why the curve shifts, than to try to understand why it is shaped as it is. The next graph shows you a recent real-world example of "the shifting Phillips Curve."

Fig. 29-4 **SOMETIMES THE PHILLIPS CURVE SHIFTS OUTWARD**

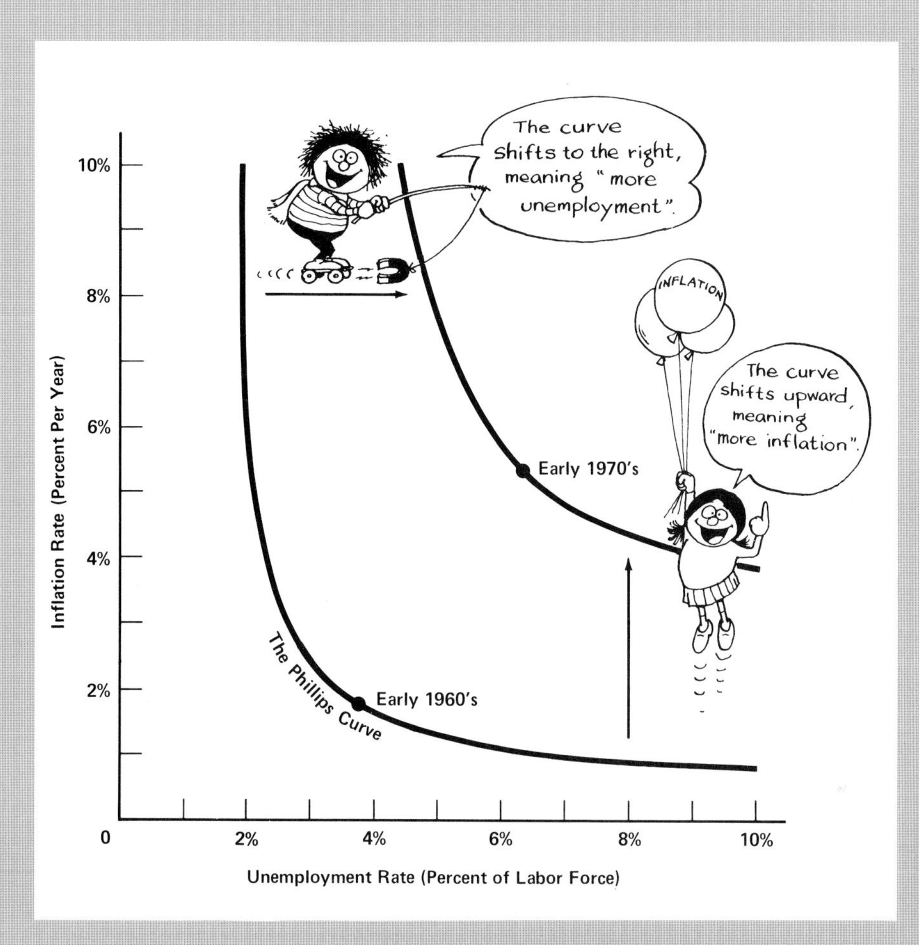


When The Curve Shifts Up And To The Right It Signals More Unemployment And More Inflation Too.

In the early 1960's, the unemployment rate in the United States was less than 4 percent and the inflation rate was less than 2 percent. Then, in the early 1970's, the unemployment rate was about 6 percent and the inflation rate was more than 5 percent. A big shift in the Phillips Curve, upward and outward!

If the Phillips Curve is going to be shifting around all over, what good is it? It illustrates a concept—the concept of the basic dilemma of economic stabilization—of the unemployment-inflation trade-off. We really don't know where the curve is going to be or exactly how it will be shaped. Still, it's a useful way to illustrate an important concept.

waiting so long for the "indirect medicine" of tight money to show that it was working, he reluctantly concluded that he had no choice but to impose direct controls. When the domestic economic problems were compounded by the international financial crisis of the U.S. dollar (which you'll be reading about in the next chapter) he really didn't have *any* choice. He had to do something.

But why? What had happened to bring such a high rate of unemployment and a high rate of inflation? What had made the Phillips Curve shift up and out so far? And how is it possible to get it shifted back down? Or, to say it differently, how can we solve this inflation-unemployment dilemma?

If we could get the Phillips Curve shifted all the way to zero, the dilemma would be solved. We could have full employment and stable prices. I'm sure nobody thinks we're ever really going to do that. Not completely. But maybe if we understand what pushes it around, maybe we can help some. So what causes the curve to shift upward and outward? What causes such high unemployment and high inflation at the same time? Let's talk about that.

## What Causes The Phillips Curve To Shift?

Anything that would get people to stop playing the "upward-price-inching" game would shift the curve down. If everybody would stop trying to play "wage-price leapfrog," always jumping to try to stay ahead of everybody else, that would really help a lot. Then we might be able to have a high level of total spending, with a low level of unemployment and a low level of inflation at the same time. It could do a lot of good if we could just get people to quit pushing all the time for higher wages and prices.

It could do a lot of good if we could make it easier for people to be employed, too. That would shift the Phillips Curve to the left, and give us less unemployment, with no more inflation. Anything that could be done to help the unemployed people become more productive, or more mobile, or more aware of available jobs, or more responsive to the changing demands for labor in the economy—anything like that would help. Even lowering the wage rate for inexperienced, untrained, unskilled workers would help. Any of these things which would reduce the "natural resistance" to full employment, would bring unemployment down without pushing up prices—that is, would shift the Phillips Curve to the left.

It isn't difficult to figure out that manpower training and job placement programs are needed to reduce the resistance to full employment and to shift the Phillips Curve to the left. But what about the inflation problem? How do we get the people and businesses to stop playing the "wage-price leapfrogging game"? How do we get rid of these "built in" inflationary pressures which push the Phillips Curve way up? Can anything be done about that?

You know that if it were not for the economic power held by all the big businesses and small businesses and labor unions and the government and just about everybody, then wages and prices would respond a lot more to the "natural forces" of supply and demand. But nearly everybody has some power to push up the prices of what he sells. So what are we going to do about it? Eliminate big businesses? and labor unions? and everyone else who has any power to influence wages or prices? I don't think so.

### President Nixon Outlawed the Wage-Price Leapfrogging Game

What can we do? I guess we can do exactly what President Nixon did. We can outlaw the wage-price leapfrogging game! We can establish controls to limit the "wage-price increasing powers" of businesses and unions and other organizations. We can establish wage-price controls. If wage and price controls can succeed in holding down wages and prices, that means the Phillips Curve shifts down close to the "zero inflation" line. Inflation is controlled, without forcing unemployment to increase.

Now you have an overview of the problem. The Phillips Curve shifts *upward,* showing more inflation, whenever everyone gets in the upward price-inching game—when wages and prices start leapfrogging each other. The curve shifts *outward* (to the right on the graph) whenever people aren't prepared for the kinds of jobs available—when a lot of young, untrained people are entering the labor force, and especially if there's a high minimum wage for new workers.

What made the curve for the U.S. economy shift so far upward and outward in the early 1970's? Why didn't the serious problems of unemployment and inflation begin to show up back in the 1950's and early 1960's? Why did the crisis wait so long to happen? To understand the answer to this question, you'll need to know a little bit about what was going on in the U.S. economy in the 1960's.

### Some Causes Of The Recent Inflationary Push

Why did the Phillips Curve for the U.S. economy shift upward in the late 1960's and early 1970's, bringing more inflation? Several reasons. During the mid and latter 1960's, government spending increased rapidly. Tax increases were delayed and more and more money was created through the banking system. Total spending in the economy kept increasing. The economy was running along just about as "fully employed" as it could get, so what happened as spending kept increasing? Inflationary pressures began pushing prices up. Of course.

Was this a shift in the Phillips curve? No. Not at all. It was just a move along the curve—a move which brought a little more employment, and a lot more inflation. Once the inflation was triggered off, it began to get worse and worse. Each time wages in one industry would move up, the workers in other industries would push for higher wages, each trying to outdo the others. Each time one business would raise prices, others would do the same. As wages increased, prices increased; as prices increased, wages increased. Everybody kept on fighting to stay ahead, while blaming everybody else for causing inflation.

In some industries, demand and productivity were increasing, so prices and wages increased. But then, everybody (whether they happened to be in high productivity, high-demand industries or not) started pushing to get just as much increase as the other guy. When the auto workers get a wage increase, the school teachers want one, too—and increased productivity be damned! The auto worker's wage can go up, yet the cost of each car may stay the same. How? Because of increased productivity per man. But if the school teacher's wages go up, unless he teaches more students, the cost of education for each student must go up. That's inflation. And that's what was happening throughout the economy. Once inflation gets going, everybody gets on the band wagon. The high-demand, high-productivity industries may be the pace-setters, but soon, throughout the economy wages and prices start leap-frogging each other across the landscape until both threaten to go out of sight over the horizon!

See what's been happening? Increased spending sets off the inflation. Total demand is too great for the total supply, so shortages develop and prices start going up. Then soon everybody starts trying to get ahead of everybody else. The inflation speeds up more and more. Inflation begets inflation. Once it gets going it tries to run away.

## Kinds Of Inflation: Demand Pull, Cost Push, Structural

Once inflation gets going, there are three kinds of inflation that all start working together, reinforcing each other. Demand-pull inflation results from too much total spending. Cost-push inflation occurs when businesses raise their prices because their *costs* are going up—usually because the wage rate goes up. Structural inflation can occur when the "structure" (the industrial make-up) of the economy is changing (as it always is, of course). "Structural inflation" can result from shifts in the buying patterns of the consumers. As demand increases in one segment of the economy (say, the automobile tire industry) this pushes up prices and wages in that industry. But in the industries where demand is declining (say, the wagon wheel industry), wages and prices don't go down to offset the increases in the expanding industry. Of course not! What really happens is that the people in the declining demand industry fight to get their wages and prices to go up, too! Cost-push inflation, triggered by a structural shift in the economy? Sure.

### President Nixon And The Monetarists

When President Nixon took office in January, 1969, he brought in a new group of economic advisers—economists with a different idea about what ought to be done, and how. Several were "monetarists," members of "the Chicago School" of economic theory, anti-Keynesians, allies of Milton Friedman, believers in limiting the size of the money supply, but except for that, a *hands-off policy toward the economy on issues of both micro and macro economics.* "Let the money supply increase a little each year, and leave everything else alone. Never fear. If you'll do that, the natural economic forces will make everything work out all right."

President Nixon's economic advisers, alarmed by the rapid rate of increase in the money supply and the high rate of inflation, called for tight money. "Stop the money supply from expanding. That will hold down total spending. Soon prices will stop rising." The Fed tightened money so tight that interest rates went up higher than they had been in the United States for more than 100 years! The stock market went into its worse crash since the depression days of the 1930's. For more and more businesses, demand fell off. Profits turned to losses. Bankruptcies increased. The economy went into recession. The unemployment rate increased from less than four to more than six percent of the labor force. And what about prices?

Prices and wages just kept right on going up. Month after month the President and his advisers continued to assure the nation that the "Nixon game plan" was working and that the anti-inflationary tight money policy was just about to take hold. How long did this go on? Until August 15, 1971. For two and one-half years! It continued until the eve of the dramatic Sunday night television address in which President Nixon announced a complete reversal of his previous economic policies, and ordered a freeze on all wages and prices.

What had happened? What had gone wrong? The tight money policy had succeeded in holding down total spending, sure. But instead of the inflation rate slowing down, employment slowed down. Prices just kept right on going up. Why? Once the wage-price leapfrogging game started going it wasn't easy to stop. Or you could say it another way: the Phillips Curve shifted.

### The Phillips Curve Shifts

What made the Phillips Curve shift? Once the inflation got moving, more and more people were left farther and farther behind. So they tried harder and harder to catch up. Each time one more group would get its wages or prices jumped out ahead, that would leave that many more, that much farther behind and that much more determined to forge ahead. And as all this "cost push" inflationary pressure built stronger and stronger, the Phillips Curve was moving higher and higher, showing more and more inflation for any level of unemployment you might choose.

Cost-push inflation and structural inflation don't have much respect for tight money, or for increased unemployment either, for that matter. And inflation begets inflation, remember? The longer the inflation continued, the harder it was to stop.

What could be done? Just what was done. The leapfroggers could be frozen in their tracks for awhile (for 90 days), during which time someone could try to figure out who was behind and who was ahead when the freeze came. Then after the 90 days were over, in "Phase Two," some adjustments could be made to let the hindmost move up with the rest of the pack. Then after that, no more leapfrogging. No more giant steps. Only little baby steps. (And don't forget to say "May I?" to the wage-price control board!)

What happens next? What's the long range outlook? Nobody knows for sure. Of course not. But here's a guess. Most of the time, we won't have to say "May I?" but there probably will be some guidelines, some "rules of proper wage-price conduct" or some such. The government controllers won't be out on center stage, but they'll be watching from the wings. If the leapfrogging game looks like it's about to get started again, I don't think it will take very long, next time, before the wing-watchers will pounce. It's so much easier to stop it when it's just getting started than after it gets going good.

## Tight Money: Necessary, But Not Always Sufficient

Tight money really can't always do the job of preventing inflation—not in the real world. That doesn't mean we can ignore the need to limit the size of the money supply. Of course not! Remember what happened in Germany in 1923? What could wage-price controls have done then? Nothing! Government incomes policies, or wage-price guidelines designed to hold down the wage-price leapfrogging game couldn't possibly work if the money supply was permitted to expand unchecked.

If total spending is permitted to expand more and more, the pressure for "demand-pull" inflation gets greater and greater. The longer the "price control lid" is held on, the harder it is to keep the prices from breaking loose. People start to offer more money "under the table" for the things they want. When this "black marketing" gets widespread it becomes impossible to stop. It becomes the name of the game. Everybody does it. All the goods disappear from the controlled markets. The natural market forces take over. When this happens, the sooner the government recognizes it and abolishes the price controls, the better it will be.

No, controls can't replace the need for responsible monetary and fiscal policies. On the other hand, neither can responsible monetary and fiscal policies replace the firm hand of government controls in dealing with the wage-price leapfrogging game. The classic argument against wage-price controls has been this: "If we want to keep the pot from boiling over we must turn down the fire under the pot (tighten money), instead of trying to clamp the lid on (control prices)!" That's a good analogy. It makes very good sense whenever we're dealing with demand-pull inflation. But when we're dealing with the upward price-inching, leapfrogging game, the analogy doesn't fit very well.

Perhaps it would make more sense to think about a pot of live fiddler crabs, each one trying to crawl over the backs of all the others and get out of the pot. It might help to put the pot in the refrigerator to cool them down, but to be sure it would be more effective to clamp the lid on! Whenever many big businesses and labor unions and professional associations and other organizations have enough economic power to set their own wages and prices—and I'm sure you know that many do—then tight money, acting alone, is an insufficient tool for inflation control. Some way to "clamp the lid on" is also going to be needed from time to time. When those times come, the sooner the incomes policies are instituted and enforced, the better it will be for everyone.

It seems to be a safe bet that we will be living under some sort of actual or potential wage-price restraints most of the time, from now on. Suppose it turns out that way. Is that good, or bad? And good or bad *compared to what*?

## DIRECT CONTROLS ON WAGES AND PRICES

It's good to hold down inflation. So unless there is something bad about direct controls on wages and prices, why not have direct controls all the time? You already know the answer. There's something bad about wage-price controls. Wage-price controls block the operation of the market economy!

### Controls Prevent The Market Process From Working

Suppose the government would set up rigid incomes policies—saying that each business should get a certain amount of profit and no more. Each worker should get a certain wage and no more. Every product price should be no higher than it was last June. What would happen?

If everything in the economy is in a sort of "general equilibrium" when the controls are established, then the "set" prices and wages and profits will all be in line at that moment. There won't be any immediate problem. But the longer the controls stay on, the more things are likely to get out of line. Demands change. Production costs change. Technology changes. As these changes occur, prices will need to adjust to the new conditions. If the adjustment can't be made, then shortages will begin to show up in some markets and surpluses will pile up in others. The longer the prices stay frozen, the worse the distortions will get. Shortages will get worse and worse. Surpluses will get bigger and bigger.

### Unrealistic Prices Generate Black Markets

Suppose prices were frozen when there was an oversupply of corn, and the corn price is fixed at 10¢ a dozen. Everybody quits producing corn. Soon the shortage of corn is very great. But some people like corn very much—enough to be willing to pay $2.50 a dozen! Suppose I'm growing some corn for myself in my backyard plot and someone sees it growing there, ready to harvest. One night he comes by and offers to leave a $5 bill under the doormat if I will look the other way while he "steals" two dozen ears. What temptation! I just might look the other way.

As time goes on, the word gets around that you can "steal" corn from my field without anybody catching you if you will leave $5 under the doormat for every two dozen ears you "steal." Soon I'm making lots of money, and people who really like corn are getting some. But what about the price controls? They aren't working anymore. Not for corn, anyway. And not for lots of other things, too.

When the "black markets" (free markets) begin to develop, they catch on like wildfire. If the controlled prices are very far out of line with the "real-world supply and demand conditions," then black markets are almost certain to arise. So if controlled prices are going to work for very long, there must be some kind of system for staying in touch with supply and demand and cost conditions in all the markets for all the products in the economy. The controlled prices will have to be adjusted to reflect the changing market conditions. But isn't that hard to do? Yes, it surely is.

### Political Influence May Distort The System

There's another problem, too. Every seller wants his price to go up a little bit. Every worker wants his wage to go up a little bit. Every business wants its profit to go up a little bit. Everybody is putting pressures on his Congressman, his Senator, his local political party representatives, writing letters to the President, doing everything he can think of to try to get some "special consideration" for his particular "hardship case." Everybody can think of some very special reason why he needs an increase. How does a system which responds to political pressures, withstand this sort of thing? It doesn't, of course. As time goes by, *politics* is likely to decide more price adjustments than *economics*!

Wage and price controls are a bad thing—a real no-no for the "free enterprise system." They're hard to design, hard to administer, hard to police. So what are we going to do when the wage-price leapfrogging game begins? We'll use wage and price controls. We have no choice.

During 1969 and the early 1970's, we suffered through a very unhappy experience. Tight money was used to try to control the inflation, but serious unemployment developed and inflation just roared on. After that unhappy experience, it is likely that from now on the American economy will always live in the shadow of some kind of incomes policies—some kind of limits or controls (actual or potential) over wages and prices.

### Most "Free Market" Systems Use Some Direct Controls

Before the latter 1960's and early 1970's, had inflation ever been a serious problem before, in the United States? And what about in other countries? You know the answer. In every healthy, prosperous, growing economy, inflation is *always* a threat. Over the past few decades, most of the time the United States has experienced less inflation than most other countries. But I'm sure you know that in the U.S. economy, prices for most things have been going up all along—sometimes faster, sometimes slower, but always trending upward.

Over the past three decades, in most countries, most of the time, there has been some attempt to use *direct restraints* to hold down wages and prices. Practical, real-world politicians have known for a long time that the unlimited creation of money will force up prices, and that the powerful thrust of a massive surge in total spending can't be held in check by wage-price controls. But they also have known that inflation cannot satisfactorily be held in check simply by tightening money to hold down total spending. There are just too many "fiddler crab" prices, crawling over each other, trying to get out of the pot!

Of course if money can be made tight enough, total spending can be cut back so much that the economy can be forced into a depression. That would cool down the autonomous enthusiasm of the fiddler-crab prices, all right! But is that really an acceptable solution? Of course not.

### When Has The U.S. Used Direct Controls?

In the United States, as you know, a system of "total wage-price controls" was established in August, 1971. That was the first time that had happened since the early 1950's, during the Korean War. It had happened during World War II also, and during other wars, of course. But what about the period from the mid 50's, to August of 1971? Were wages and prices completely free to seek their own levels? Sometimes, yes. Sometimes, no.

In the early 1960's, President Kennedy set up wage-price guideposts spelling out the "appropriate conditions" for wages or prices to be increased, and by how much. The guideposts were not "legal requirements." They were only "urged upon" the industries and the labor organizations. But both President Kennedy and President Johnson used the power of the presidency to "twist some arms," to convince some businesses and labor leaders to go along with the guideposts. This "arm-twisting"—this unofficial, but sometimes quite powerful technique for holding down wage-price increases—is known as jawboning.

The "jawboning" policies of the Kennedy-Johnson era were continued until the beginning of the Nixon administration in 1969. Then, responding to the urging of his monetary theorist-advisers, President Nixon instituted a "tight money" policy and openly denounced all other approaches to the inflation-control problem. Repeatedly he announced that he would never use the power of his high office to try to impose "wage-price guideposts" on American businesses and labor.

You know what happened in 1969. Interest rates rose rapidly and the economy headed into a recession. But the rate of inflation didn't slow down. Instead, it speeded up. The upward price-inching, wage-price leapfrogging game was taking over? Of course. The "fiddler-crab prices" were crawling out of the pot!

The President was cooling down the pot, all right! The economy was slowing down. Unemployment was increasing. But the fiddler-crab prices kept right on crawling out. For how long? Until the President decided to ignore his monetary theorist advisors and put the lid on. That's what he did on August 15, 1971.

Unless I miss my guess, the August 15, 1971, wage-price freeze brought to an end the American economy's last noble experiment in completely free wages and prices. It seems unlikely that big businesses and big labor ever again will be permitted unlimited power to set their own prices and wages without any kind of guidelines or surveillance by the government.

What do wage-price controls, or wage-price guidelines, or "incomes policies" accomplish? They don't accomplish anything unless they work. But if they're flexible and sensitive, and supported by appropriate monetary and fiscal policies, they can be made to work. If they work, they shift the Phillips Curve downward. They can give us a reduced inflation rate (a rate we can live with), corresponding to an unemployment rate we can live with. That's a goal worth seeking.

## THE PROBLEM OF UNRESPONSIVE UNEMPLOYMENT

We've been talking a lot about the problem of inflation—about how, once it gets started, it's hard to stop. It tries to run away. With everybody pushing for higher wages and prices, and with structural changes always going on in the economy and all, it's easy to see why an inflation rate of zero would be pretty difficult to maintain. That's why the Phillips Curve never gets down to the horizontal axis—down to where the inflation rate would be zero. But what about unemployment? Why doesn't unemployment ever go all the way to zero? Why doesn't the curve go all the way over and touch the vertical axis? What can be done to reduce unemployment and make the curve shift to the left?

### There Are Always Some People Changing Jobs

Just as some kinds of *inflation* don't respond very well to *reductions* in total spending, some kinds of *unemployment* don't respond very well to *increases* in total spending. One kind of unemployment which total spending could never bring down to zero is frictional unemployment. People who are changing jobs for one reason or another, don't "flow smoothly and instantaneously" from the old jobs to the new ones. Sometimes the person may be unemployed only a day or two. Other times he may be unemployed for a week or two or maybe longer.

Sometimes the "frictional unemployment" results from structural changes in the economy. Remember about structural inflation? What about structural unemployment? What do the old line, highly skilled wagon-wheel workers do when the demand shifts to automobile tires? They get unemployed. If total spending in the economy picks up, will they get new jobs? Most of the younger ones will. Some of the older ones may. Others, no. Their "day in the sun" has passed. That's the way it is with structural unemployment.

There's another kind of unemployment you hear a lot about these days: technological unemployment. That's what happens when people are replaced by machines—"automated out of a job," so to speak. That's what's happening to the migrant farm workers. Everywhere you go you see more "mechanical pickers" instead of "hand pickers" harvesting crops. You heard about this problem a few chapters ago, remember? In every modern economy, jobs are constantly being eliminated by improved technology. As wages get higher, more capital is introduced. "Technological displacement" speeds up.

The more the economy is dynamic, growing, introducing new technology, the more serious structural and technological unemployment will be. What can be done about it? Manpower training and development programs? Job training subsidies? Helping people to find new jobs? Yes. All these things, and some other things too. But one thing's sure. An increase in total spending can't solve the problem all by itself.

### New People Are Always Entering The Labor Force

Another kind of unemployment which can't be eliminated is the unemployment of the young and inexperienced people who are constantly joining the labor force. In the early 1970's, when the average unemployment rate for the nation was a little less than 6 percent, the unemployment rate for teenagers was more than 16 percent. In several places in the country it was much higher than that. But the unemployment rate for married men was only about 3 percent.

If all the members of the labor force were identical, how simple it would be to solve the unemployment problem. Just get total spending to increase, that's all! Soon everyone would have a job. But the labor force is very diverse—young and old, male and female, skilled and unskilled, black and white, educated and uneducated, brilliant and stupid, industrious and lazy, and on and on and on. So it's obvious that just increasing total spending would create a lot of inflation before it would ever bring unemployment down near zero.

### Unresponsive Unemployment Keeps The Phillips Curve Out

What happens to the Phillips Curve as more young, unskilled people join the labor force, and as more people get trapped by structural and technological and other "unresponsive" kinds of unemployment? The curve shifts to the right. After the shift, the curve shows that at any "inflation rate" you choose, the "unemployment rate" is larger than before.

## IS DIRECT ACTION NEEDED TO HOLD THE PHILLIPS CURVE DOWN?

What does the Phillips Curve say? Implicitly, it tells us that even though total spending keeps on increasing, some unemployment persists. That means that something *other than* total spending must be keeping unemployment up. Also the Phillips Curve implicitly tells us that even though total spending is held down, some inflation persists. That means that something *other than* total spending must be pushing prices up. The Phillips Curve shows us a graphic picture of the unemployment-inflation dilemma.

Anything that is going to succeed in solving the inflation-unemployment dilemma is going to have to succeed in shifting the Phillips Curve downward and to the left. In the real world, what does that mean? It means taking *direct action* to prevent some kinds of wage and price increases—increases which are not very responsive to adjustments in total spending. It means taking *direct action* to overcome some kinds of unemployment—the kinds which are not responsive to increases in total spending. That's a big order. It won't be easy to do.

Direct programs to overcome unemployment are widely accepted these days. Manpower development and placement programs have been expanded rapidly in recent years, and with a good bit of success. But direct controls to limit big increases in wages and prices? Is that a widely accepted idea these days? Certainly not!

### People Disagree About The Need For Direct Controls

To bring the long arm of government bureaucracy into the very heart of the "free market place" by setting up wage-price guidelines? or incomes policies? or some such? to expose the price mechanism to the continuous threat of all sorts of political manipulations and shenanigans? and to do this on a *permanent* basis? Who wants that? Nobody. Of course not. That is, nobody wants it if there's some realistic and feasible alternative.

Several outstanding economists disagree about the need for continuing wage-price guidelines or controls of some sort. It is my opinion that we don't have any realistic alternative. The problems of unemployment and inflation (just as the problems of pollution, population, and poverty) are going to have to be approached in the real world. If it turns out that some system of continuing controls is essential, we will have to accept that fact, and turn to the task of devising a system that will work.

### The Micro And Macro Problems Must Be Dealt With Somehow

In this chapter and the previous one you have seen a glimpse of the most important basic weaknesses of "modern capitalism." You have seen the problems which will be responsible for most of the continuing change which will occur in the United States and in the other "modern economies of mixed socio-capitalism." These problems are going to have to be dealt with, and lived with. Dealt with, because they're intolerable; lived with, because complete, totally satisfactory solutions to these problems are not likely ever to be found—certainly not in my lifetime, or yours. But progress can and will be made. Why? Because it must.

In the two final chapters of this book you'll rise above and peer beyond national boundaries. In Chapter 30 you'll take a look at how the U.S. economy ties in with the whole world. Then Chapter 31 talks about the continuing evolution of the world's economic systems. But before you go on, be sure you understand the domestic micro and macro problems of Chapters 28 and 29.

---

## REVIEW EXERCISES

**MAJOR CONCEPTS, PRINCIPLES, TERMS (Try to write a paragraph explaining each of these.)**

the basic macroeconomic dilemma
runaway inflation

the inflation-unemployment overlap
the inflation-unemployment trade-off
the Phillips Curve

**OTHER CONCEPTS AND TERMS (Try to write a sentence or phrase explaining what each one means.)**

demand-pull inflation
cost-push inflation
structural inflation
the monetarists
the "Chicago School"
Milton Friedman
incomes policies
wage-price guideposts
jawboning
frictional unemployment
structural unemployment
technological unemployment

**GRAPHS AND CURVES (Try to draw, label, and explain each.)**

Total Spending, Employment, and Prices: The "Model" Case
Total Spending, Employment, and Prices: The "Real World" Case
The Unemployment-Inflation Trade-Off: The Phillips Curve
When the Phillips Curve Shifts Outward, That's Bad

**QUESTIONS (Try to write out answers, or jot down the highlights.)**

1. Can you think of any specific examples (examples you are personally familiar with) of wage and/or price increases which are *not* likely to be stopped by a slowdown in spending? Discuss.
2. Can you think of any specific examples of unemployment which aren't likely to be overcome, even if total spending increases? Discuss.
3. Explain how the Phillips Curve is similar to, and how it is different from, a production possibility curve.
4. Mention and explain as many things as you can think of that would be likely to make the Phillips Curve shift up or down, or to the right or left.
5. What is the basic position of Milton Friedman and the monetarists regarding stabilization policy?
6. There are many reasons why wage-price controls are undesirable in general, and why it's very hard to make them work. But, under the right circumstances they *can* be made to work. Try to explain all this, in as much detail as you can.

# 30 THE PAYMENTS DEFICIT AND THE GOLD DRAIN: THE COLLAPSE AND RECONSTRUCTION OF THE INTERNATIONAL EXCHANGE SYSTEM

*The U.S. dollar is no longer strong enough to serve as the backbone of the world's exchange system, so a new system is being worked out.*

When President Nixon announced the wage-price freeze in August, 1971, he also made some announcements that affected the international economic scene. Announcements which shattered the trade and finance arrangements which had existed for the past quarter-century! Why did he do it? What had happened to force an American president to take such drastic action? What was it supposed to accomplish? And how?

The problem was an acute turn for the worse in the chronic deficit in the U.S. international balance of payments. Something drastic had to be done. This chapter will explain it all to you. Once you understand it, you'll know we aren't out of the woods on this "international balance of payments" thing yet. First you need to understand how international finance works. After that, all the other pieces will fit neatly into place.

## HOW INTERNATIONAL FINANCE WORKS

International finance works almost exactly like any other kind of finance. People buy things and pay with money. The only difference is that if you're buying things from a foreign seller, you have to exchange your kind of money (say, dollars) for some of his kind of money (say, francs), before you can pay him. Here's how it all works.

### Banks Hold Deposits In Other Domestic Banks And Foreign Banks

Every bank has "deposit accounts" in other banks. You already know about bank deposit accounts in the Federal Reserve Banks. When a bank makes a loan, the money soon comes out of that bank's deposit at the Fed and is transferred to another bank's account. But did you know that banks also have deposit accounts in other banks in this country? and also deposit accounts in banks *in foreign countries*? Sure they do!

American banks have deposits of francs in French banks and deposits of pounds in British banks and lire in Italian banks and pesos in Mexican banks and dong in Vietnamese banks and marks in German banks and yen in Japanese banks and other local currency deposits in local banks in countries all over the world! (*Your* bank probably doesn't have deposits in banks in all of these countries, but your bank is in touch with other U.S. banks that do, so your bank can arrange to get a check drawn in any kind of "money" it wants!)

### A Bank Sells A Certified Check — It "Sells Dollars"

Suppose your father needs a "certified check" for $400 to send to your college, for tuition. He goes to the bank and "buys" the certified check. What he does is to write a check to the bank for $400 (plus a small service charge) to pay for the certified check. Then he mails the certified check to the college. (The certified check is made out by the bank, to be paid to the college, and is guaranteed by the bank. A postal money order or a bank money order would work exactly the same way, and the service charge would be about the same.)

When your college deposits the certified check in its own bank account, the check then goes to the Federal Reserve bank, where the $400 is added to your college's bank's Federal Reserve account. Then the $400 is subtracted from your father's bank's Federal Reserve account. So after it's all over, what's the result? Your father's bank has "sold" some of its "Federal Reserve account dollars" to your father, and your father has paid those dollars to your college. Get it? Okay, now here's the same situation, only different.

Suppose you happen to be going to a college in France. What then? It could work exactly the same way. Your father could send the certified check, drawn in dollars, to your college in Paris. The college would then deposit the check to its account in its Paris bank. Only it would want *francs* deposited, not dollars. So the Paris bank would deposit francs in the college's account. Then what would the Paris bank do with the certified check? Put it in a "deposit by mail" envelope and send it for deposit to its own account in a U.S. bank. Of course! Now the college has the $400 worth of francs it wanted, and the French bank has another $400 (in dollars) in it's account in a New York bank.

In this case, your father sent U.S. dollars, so the French college had to get the French bank to trade francs for the dollars. So the French bank winds up with fewer francs, and with more dollars in its New York bank account. We would say "U.S. dollars have flowed overseas to France to pay for college tuition." Of course, the dollars really aren't "overseas" at all. They're on deposit in U.S. banks. That's where the dollar deposits are, but the deposits are owned by the French bank.

### Foreigners Own Dollar Accounts In U.S. Banks

A foreigner can open an account in an American bank and deposit dollars in it, just like you can. If the French college is doing a lot of business with Americans, it might decide to open an account in a New York bank. Then your father's tuition payment could be deposited right in the college's New York bank account. (But we would still say that dollars were "flowing overseas.") Then the French college could write checks in dollars to pay Sperry Rand for their Univac computer system.

As long as the French college is planning to spend up its dollars buying things from the United States, there's no need for it to go to its bank and exchange dollars for francs. It takes its dollars, as dollars, deposits them in its own New York bank account, then writes checks to pay for American goods, just as you or I would. But suppose the French college doesn't plan to buy any U.S. goods. What then? It will have to get francs so it can buy the things it needs in France, and so it can pay the people who work for the college.

### A Bank Sells A Certified Check In Francs—It "Sells Francs"

Suppose the French college tells your father that all tuition payments must be made *in francs*. What does your father do then? He doesn't have any francs! No, but his bank does. It has francs on deposit in a French bank, just like it has dollars on deposit in the Federal Reserve Bank. It can write a certified check *in francs*, just as easily as it can write one in dollars. See how simple it is? All your father has to do is tell the banker he wants the certified check written for *francs* (instead of for dollars) and that's all there is to it!

When your father buys the certified check drawn in francs, the banker will use some different names for things (just to add some mystery to it all). The certified check likely will be called a "bill of exchange." And the banker will tell your father he's "buying foreign exchange." Be sure to warn your father not to get snowed by all this talk. The whole thing is very simple. People are "buying and selling bills of foreign exchange" (certified checks drawn in foreign money) all the time, everyday.

### When The Bank Sells All Its Francs, What Then?

As Americans buy French champagne and pay for French college tuition and buy French perfume and buy tickets on Air France, the franc deposit balances of the American banks get smaller and smaller. Also, the "dollar deposit balances" of the French banks get larger and larger. How long can Americans keep on buying all these French things? What happens when the American banks have used up all of their franc deposits in the French banks, and when the French banks have accumulated so much in dollar deposits that they don't want any more dollars? What brings things back into balance again?

You know the answer. At the same time the Americans are buying French champagne and perfumes and all, the French are buying American computers and chemicals and all. The Americans buy French goods and the French buy American goods. We're using up francs, and giving them dollars all the time. They're using up dollars, and giving us francs all the time. The same thing is going on with Japan and Germany and Canada and Argentina and Greece and Libya and Britain and Brazil and everywhere else. It's sort of amazing that it all balances out. But it really does—except for those times when it doesn't—and that's when the problems arise. That's what we need to talk about now.

## HOW THE DOLLAR CRISIS EVOLVED

When the "international balance of payments" gets out of balance, that's bad. If we're running a surplus, that means the foreign buyers are trying to use up more dollars than they can get. They're trying to buy more from us than we are buying from them. If we're running a deficit, that means we're pushing dollars into their hands faster than they are using them up. Either way, it's a bad scene.

### International Payments Get Out Of Balance: "The Dollar Gap"

In the years following World War II, the United States was running an international payments surplus. Other countries were trying to spend more American dollars than they were getting. To try to "close the gap" and help foreigners to get the dollars they needed to buy American goods, billions of dollars in grants and loans were made by the U.S. government and businesses to foreign governments and banks and businesses. The "U.S. dollar" was in short supply. Everybody wanted dollars so they could buy U.S. goods. Most of the industrial products and other things people wanted were available only from the United States. The "dollar gap" was severe for many years. If a foreign bank or business had some extra money, they were happy to hold it in the form of "dollar deposits."

Dollars were "as good as gold." They really were! The foreign banks could actually turn in their dollars and get gold if they wanted to, one ounce of gold for each $35. But nobody wanted gold. Why get gold? The dollar was as good as gold, and you could put the dollars in savings accounts or bonds and notes and other securities, and draw interest!

Yes, in the years following World War II, the dollar was really in its heyday. But not now. Why not? What happened? Something happened to eliminate the surplus in the U.S. international balance of payments? Something happened to turn the tide the other way? Exactly.

## More Imports Push More Dollars Into Foreigners' Bank Accounts

What's it like when the tide turns the other way? What's it like to run an international payments *deficit*? If we're running a deficit, that means we're pushing more dollars into the bank accounts of foreigners than they want to use up to buy our goods. Their dollar balances in their U.S. bank accounts keep getting bigger and bigger. Then what happens? It can't just keep on going that way forever. That's for sure.

After the foreign banks have several billion dollars, if the deficit continues, sooner or later the foreign banks will have to refuse to accept any more dollars. That would be disastrous. We can't ever let it come to that. (And we won't.) But here's what really happened. (It *almost* came to that!)

During the years of "international dollar shortage" following World War II, the United States government initiated and expanded its programs for pumping dollars into the bank accounts of foreigners. Most of the dollars flowed into the bank accounts of foreigners to support overseas military activities, and foreign-aid programs. As time passed, these programs were increased. But much more important, as time passed the European and Japanese economies rebuilt themselves. They embodied the latest technology, their wages stayed relatively low, and soon they were selling industrial and other goods in world markets, competing with (and often underselling) U.S. goods.

## Foreigners Get More And More Dollars: "The Dollar Glut"

American buyers began buying more and more foreign goods, pushing more and more dollars into the bank accounts of foreigners. Foreign buyers began buying more and more foreign goods, not using up all those dollars which were flooding into their bank accounts. See the problem? It's a problem all right! Then as wages and prices started leapfrogging each other in the latter 1960's and early 1970's, American goods got more and more overpriced. Both American buyers and foreign buyers switched more and more from U.S. goods to foreign goods. More dollars were pouring into the bank accounts of foreigners, but the foreigners weren't interested in using all those dollars to buy American goods.

See the problem? It's really simple when you stop and think about it. The pieces all fit together. They all contribute: the foreign-aid program, the overseas military commitments of the Vietnam War and elsewhere, the wage-price spiral at home, and the recovery and rapid growth of the economies of other nations—especially West Germany and Japan. That's the problem, all right. But what happened? And what's going to have to happen next to keep things in balance?

By the latter 1960's it was clear that the international position of the dollar was in jeopardy. As good as gold? Not anymore. Foreign banks were holding billions of dollars and they wanted to get rid of them—to trade them for marks or yen, because their customers wanted to buy more things from Germany and Japan and less from the United States. But who wanted to trade marks or yen, for dollars? Nobody! But the foreign banks had a way out. They could buy gold from the United States government for $35 an ounce, then sell the gold to Germany or Japan for marks or yen. Of course! So the dollar really was as good as gold after all! (Except that the United States didn't have anywhere near enough gold to "pay off" all those foreign-owned dollars.)

### Gold Drains Out And Convertibility Is Suspended

At an increasing rate during the 1960's, several foreign banks, especially the Bank of France, began using their dollar deposits to buy gold. That is, they were "converting their bank-account dollars into gold." The U.S. gold supply at Fort Knox began to dwindle. After World War II, the United States had about 25 billion dollars worth of "monetary gold"—more than half of the world's supply. By the early 1970's, more than half of the gold was gone—shipped to foreigners in exchange for dollars. But foreigners still held about *50 billion dollars*! Suppose all of them demanded gold? The dollar was as good as gold? Not anymore.

On August 15, 1971, when President Nixon announced the wage-price freeze, what did he announce about international trade and finance? One thing was a special *import tax* to discourage Americans from buying foreign goods. That would keep Americans from pushing so many dollars into the U.S. bank accounts of foreigners. Another announcement was the immediate "suspension of convertibility" of dollars—no more exchanging dollars for gold. That meant that if foreigners with dollar deposits wanted to get rid of their dollars they could spend them for American goods, or for stocks and bonds, or for anything else for sale in the United States, but they couldn't trade the dollars for gold. Another thing the President announced was that the dollar was going to be "devalued." What does that mean?

### Devalue The Dollar? What Does That Mean?

When your father wanted to pay your tuition to a college in France, he went to the bank and bought a "certified check" (bill of exchange) drawn in francs. Remember? He paid 400 U.S. dollars for the "check." How many francs did he get? That all depends on the "rate of exchange" between dollars and francs. Suppose the rate of exchange is one dollar for five francs, or 20 cents a franc. Then for 400 U.S. dollars he could buy 2,000 French francs (5 francs = 1 dollar; 5 × 400 = 2,000, so 400 dollars equals 2,000 francs). But what about the devaluation?

Devaluation means changing the rate of exchange so that *you get less foreign money for each dollar*. That's what devaluation means, and that's all it means. In our example, suppose the dollar is devalued to where one dollar can buy only four (instead of five) francs. Then one franc would be worth 25¢ (instead of only 20¢). After the devaluation, your father's $400 will buy only 1600 francs (4 francs = 1 dollar; 4 × 400 = 1600, so 400 dollars equals 1600 francs). But the tuition at your college in Paris is not 1600 francs. It's 2000 francs. After the devaluation, your father is going to have to pay $500 to buy the check for 2,000 francs to pay your tuition!

Devaluation always makes the cost of foreign things go up. A bottle of French champagne that costs 30 francs, would cost $6 before the devaluation and $7.50 afterwards ($6 × 5 = 30 francs; $7.50 × 4 = 30 francs). A Peugeot that costs 12,500 francs, would cost $2,500 before the devaluation and $3,125 afterwards ($2,500 × 5 = 12,500 francs; $3,125 × 4 = 12,500 francs). After the devaluation, maybe Americans won't buy so much French champagne or so many Peugeots. Maybe your father will decide you can come back home and go to school at Buies Creek, N.C., just like he did! And maybe you'll fly home on Pan Am, or United, or TWA—certainly not on Air France! Those tickets cost more, now.

### Devaluation Makes Foreign Goods More Expensive

See what devaluation of the dollar does? It causes foreign things to be more expensive. It *offsets the effects of the American inflation*. American prices are high, but devaluation makes the foreign goods high priced, too. Devaluation convinces American buyers to "buy American." But that's only half the story. What about the French buyers?

The Frenchman's franc, before the devaluation, would buy only 20¢ worth of American goods. But now? It buys 25¢ worth! A $2,000 Xerox copy machine that once cost 10,000 francs, now costs only 8,000 francs. The college in Paris may decide that now, Xerox is a better buy than the French copy machine. Suddenly you may begin to see more Pintos and Gremlins and Vegas and Dusters and Cougars and Cutlasses and 'Cudas and Mavericks showing up on the Champs-Élysées—and fewer Peugeots and Citroens and Renaults on the New York Thruway!

Wait. There's more. When the dollar was devalued in 1971 (first time since 1934), it wasn't just devalued against the franc. It was devalued against the money of *all other countries*—Japan and Germany and Britain and Chile and Taiwan and all the others. That's what happens when our currency is devalued. All foreign goods become more expensive to us; our goods become cheaper to all foreigners.

The 1971 devaluation of the U.S. dollar really wasn't as large as in the illustration I've been using. The illustration uses a 20 percent devaluation, just to make the point clear. The 1971 devaluation actually was only about 10 percent and it didn't really solve the problem. We aren't out of the woods yet. Not by a long shot! Now that you understand how international finance works and how the dollar crisis evolved, you can look back and understand a lot more about what's been going on. The next section gives some of the highlights.

## THE RISE AND FALL OF THE BRETTON WOODS SYSTEM

Following the depression of the 1930's and then the run-away European and Japanese inflations of World War II, by 1945, international finance and currency exchange rates were in a total mess. During that year (1945), representatives of the major trading nations met at Bretton Woods, New Hampshire, and formed a new system of exchange rates.

### The International Monetary Fund And The World Bank

The International monetary fund (IMF) was set up as a sort of "international Federal Reserve bank" to help countries to exchange their currencies for other currencies, and to decide when exchange rates needed to be adjusted—that is, to decide when a country's money needed to be devalued—to offset domestic inflation, or for whatever reason. The Bretton Woods agreement also set up the World Bank (International Bank for Reconstruction and Development—IBRD) to make long term loans of foreign currencies as needed for reconstruction of war damage, and for development projects in the underdeveloped countries.

### Each Nation's Monetary Unit Is "Defined In Gold"

What was the role of *gold* supposed to be, in the reconstructed international financial plan? Just this: The "monetary unit" of each nation was defined as being worth a specified amount of gold. For the U.S. dollar, the "gold price" was to stay at $35 an ounce, just as it had been since 1934. That means that "one U.S. dollar" is worth "1/35th of an ounce of gold." Then a "gold price" was set for the currency of every other country. Once that was done, the exchange rates between currencies were automatically established. Here's an example.

Suppose the "gold price" in British pounds was set at 9 pounds for one ounce of gold. That means the British pound is worth 1/9th of an ounce of gold. The pound "buys" a little more than four times as much gold as the dollar "buys," so the pound is worth a little more than four dollars. Suppose the "gold price" in French francs is set at 175 francs for an ounce. Then the franc is worth 1/175th of an ounce of gold. It takes 5 francs to "buy" as much gold as one dollar will "buy," so one dollar equals 5 francs. Also, you can see that one British pound would be worth a little more than 20 francs. See what a neat system? Once you know the "gold price" in each country's money, then you can figure out the "official exchange rates" between any two kinds of money.

This system of exchange rates has existed for centuries. (In the beginning the payments were actually made in gold, of course. It was only later that a system of credits—of "bank account balances"—evolved.) Once a system of official exchange rates is set up, then when a nation devalues its money, what it does is to announce an increase in the "official gold price." That's how the dollar was "officially devalued" in 1971—the "gold price" was increased from $35 to $38 an ounce. Before, the dollar was worth 1/35th of an ounce of gold. Afterwards the dollar was worth only 1/38th of an ounce of gold. After the devaluation it would take $38 to buy imported goods which only cost $35, before. But wait. What happened, following the Bretton Woods agreement in 1945?

The world managed to live under the Bretton Woods agreement (with some exchange rate adjustments from time to time), from the mid 1940's until the early 1970's. Then the agreement was shattered by President Nixon's August 15 "bombshell." What happened to cause President Nixon to bring to an end the Bretton Woods system?

### The Dollar Becomes The Reserve Currency—"Super-Money"

During the early years following World War II, the dollar became the "reserve currency"—sort of "the medium of exchange for buying other currencies"—for all the trading nations of the world. Any nation, bank, or individual was always willing to hold U.S. dollar deposits, because dollar deposits were always acceptable in exchange for any other kind of money. Everybody wanted dollars. See how the dollar became a sort of "super-money"?

The U.S. dollar became "the medium of exchange" for buying and selling other kinds of money. Suppose you wanted to trade pounds for marks. If you could trade the pounds for dollars, then you could always use the dollars to buy marks! Everybody wanted dollars. People wanted more dollars than were available. Everybody liked to hold their "extra cash" in dollars. The dollar was the queen of the international money markets. Then things started changing.

As the European and Asian economies recovered, as Americans began buying textiles and radios and tape players and cars and things from abroad, as U.S. overseas military and foreign aid commitments expanded, as American businesses began making more investments abroad, more and more dollars began to accumulate in the bank accounts of foreigners. See what was happening? American consumers, businesses, and the government were pushing dollars into the foreigners' bank accounts faster than the foreigners wanted to spend them up for American goods. What could they do with all those dollars? Buy gold, of course.

The first serious "dollar crisis" developed in 1968. Foreigners were using their dollar deposits to buy more and more gold. The officials of the major trading nations got together and agreed that the United States would not sell any more gold except to the central banks (like the Federal Reserve) of other countries. That helped for awhile. But it didn't solve the problem.

### The Chronic U.S. Payments Deficit Gets Worse

American consumers, businesses, and the government kept right on pushing more and more dollars into the bank accounts of the foreigners. The "chronic balance of payments deficit" continued. In the mid-1950's the deficit had been averaging around $2 billion a year. In the latter 1950's it increased to around $3 billion a year. It fluctuated around that level until the latter 1960's. Then in 1969 the balance of payments deficit jumped to more than $7 billion!

In 1970, the deficit was reduced some, but by early 1971 it was running much higher than ever before. The U.S. gold stock had dwindled from $25 billion in 1948 to less than $11 billion. Foreigners were holding some 50 billion U.S. dollars. Then the dam broke. The U.S. payments deficit, just in the first half of 1971, was $12 billion! International confidence in the dollar was badly shaken. Something had to be done, *immediately*.

### President Nixon Shatters The System

President Nixon took decisive action on August 15, 1971—just in the nick of time. He "suspended convertability" of the dollar (stopped selling gold to *anyone*) and announced that the "official exchange rate" between dollars and other currencies, was being abandoned. "Let the dollar float. Let dollars exchange for whatever the foreign bankers will give in exchange for them." Later, the "gold price" was set at $38 an ounce, meaning the dollar would be "officially" worth 1/38th of an ounce of gold. (But nobody could get any gold for their dollars, of course.)

What did the people in foreign countries think about what President Nixon did?—breaking the Bretton Woods agreement, stopping gold payments, putting a 10 percent tax on imports? They were shocked at first. (Just about everybody was!) The foreigners didn't want to lose their U.S. customers. Of course not. And they didn't like the value of all those billions of dollars they were holding, to decrease, either. But they realized that the United States had been backed all the way to the wall. Something had to be done, so everybody more or less went along.

## NEEDED: A NEW SYSTEM OF INTERNATIONAL PAYMENTS

In December of 1971, a new "Bretton Woods-type" meeting was held at the Smithsonian Institute in Washington. There, representatives of the "big ten" trading nations began to try to put some of the pieces back together. The agreement was made to officially devalue the dollar by raising the dollar price of gold from $35 to $38 an ounce. The Japanese yen and the German mark were revalued upward, and some other changes were worked out.

The new "Smithsonian" arrangement seemed to work all right for several months. The U.S. economy was under "Phase Two wage-price controls," and the "inflationary erosion" of the value of the dollar seemed to be finally under control. But thirteen months later (in January, 1973) when President Nixon announced "Phase Three," relaxing the controls over wages and prices, confidence in the long-run value of the dollar collapsed. Foreigners began getting rid of their dollars—trading them for other currencies—as fast as they could.

In February, 1973 another official devaluation of the dollar was announced. (The gold price was raised to about $42 an ounce, so the official value of the dollar was reduced to 1/42nd of an ounce of gold.) But the devaluation wasn't enough. Even at the cheaper price nobody wanted to buy and hold dollars. So it was decided to let the value of the dollar "float" again—to find its own "equilibrium price" in the world's money markets.

By June of 1973 the total decrease in the international value of the "floating dollars"—that is, the *actual* amount of devaluation since the Spring of 1971—amounted to about 25 percent. For example, in the Spring of 1971 the dollar would buy 3.6 West German marks; in June of 1973 it would buy only 2.7. The number of Swiss francs the dollar would buy dropped from 4.3 to 3.1.

The kind of international payments system which will be developed over the next few years is still very much in question. Only some temporary steps have been taken so far. As to the future, of this much we can be sure: A better arrangement for international exchange—one which does not depend on the dollar as the reserve currency—must be worked out. It will take time—probably several years—to work it out.

### The Changing Role Of Gold

What is gold supposed to do, in helping to finance international trade? Simply this: when somebody gets too many francs and he wants marks instead, he can trade the francs for gold, then trade the gold for marks. Gold is supposed to be used as "reserve money," and to play the role of "super-money"—the medium of exchange, for exchanging one country's money for another's. Prior to World War I, gold performed this role well. But since that time it never has.

Ever since World War II, and up until the mid-1960's, the U.S. dollar was used as "reserve money"—as the world's "super-money." It was used as the medium of exchange for trading one country's money for another's. But you know what happened to that arrangement! So what's going to happen next? What can the countries of the world use as "reserve currency"? as "super-money"? What can they hold that will always be dependable, always acceptable in exchange for any other kind of currency? Would you believe SDR's? Maybe so!

## What About Paper Gold? The SDR's?

Why not let the International Monetary Fund (IMF) do something like the Federal Reserve does. Let the IMF print up some "super-money" (called SDR's) which everyone will agree to accept. Then let each country "buy" some "super-money" from the IMF. That way, when it sells the SDR's the IMF gets lots of all kinds of money. The IMF can have bank accounts in all countries, in all kinds of money (in case anybody needs to borrow some). And what do all the countries get? Bank accounts in the IMF, of course! Bank accounts in SDR's—in "super-money." Then if the French banks run low on marks, they can write a "super-money" (SDR) check to Germany, to buy marks. Or they can write a "super-money" (SDR) check to the IMF and buy some marks from them.

What would we call this "super-money"? Special Drawing Rights (SDR's) of course! The SDR's give the holder the right to go to the IMF and "draw out" any kind of money he wants. Since the SDR's are always acceptable at the IMF, they're always acceptable *everywhere.* They're good things to have. As good as gold, and not nearly as much trouble.

The IMF "super-money" (SDR) system was suggested as early as 1945, at Bretton Woods, by none other than John Maynard Keynes. This new idea wasn't actually tried, though, until after the first "international dollar crisis" in 1968. Now it's being used in a limited way, experimentally. It's too early to know what form it will ultimately take, but it seems certain that some form of "SDR, super-money, paper gold" system will play an important role in the world's future international financial system.

## The Big Shift In U.S. Trade Policy

President Nixon gave a pretty powerful speech on August 15, 1971. Right? You already know from the last chapter that his "new economic policy" (wage-price controls and all) was a dramatic shift in policy. Now you know that what he had to say about international finance was pretty dramatic, too. Dramatic, far reaching, basic. It rang the death-knell on the Bretton Woods system and plunged the international financial community into confusion.

The August 15 pronouncements brought to an end the period during which the United States played the role of "benevolent protector" of the economies of the other nations of the world. August 15 initiated the beginning of the period in which the United States is going to become a tough competitor in international markets. For many years the U.S. economy has been so strong, as compared with its competitors, that the United States has been able to get away with "breaking the rules of prudent international finance." But not any more.

Today there are several strong economies in the world. Now it is necessary for the United States to be more careful to protect its own position in the world economy. And another thing. The United States is going to have to stop ignoring the international consequences of its domestic economic policies—especially, inflation is going to have to be controlled. For some people, it will not be easy to get used to. We would all like to have our cake and eat it too, right? Of course.

### For The U.S.: A Tough Agenda For The Future

In order to solve its long-run balance of payments problem, here are some of the things the United States is going to have to do:

(a) hold down inflation so that American goods will be reasonably priced as compared with the prices of foreign goods;
(b) work with other countries to establish proper exchange rates between the dollar and the other currencies so that American goods will not be either "too cheap" or "too expensive" to foreigners and foreign goods will not be either "too cheap" or "too expensive" to Americans;
(c) work out some ways to reduce the total amounts of dollars the U.S. government is pushing into the bank accounts of foreigners to finance military and foreign-aid programs;
(d) keep investment opportunities in the United States profitable enough so that American businesses will not find it necessary to invest so much of their earnings overseas; and
(e) work to eliminate trade restrictions by foreign countries which deter the entry of American goods.

A nation in the international market operates sort of the way a family or a business operates in the domestic market. A family can spend more than it takes in for awhile, or a business can have its costs higher than its revenues for awhile. But there comes a time when "the chickens come home to roost." The payments have to balance out. A business cannot perpetually operate with costs greater than revenues; the members of a family cannot perpetually spend more than they receive.

The United States cannot perpetually be pushing more dollars into the bank accounts of foreigners than the foreigners want to use to buy American goods. If the foreigners aren't buying enough from us to use up the dollars, then we must not push as many dollars into their hands. As this book goes to press the dollar seems to be getting stronger, aided by shortages and price increases in the international oil markets.

## WHAT'S THE OUTLOOK?

Foreign trade is just like any other kind of trade. The reason people engage in trade is because of the great gains they derive from trade. Of course! The one "complication" of *foreign* trade is that as a part of the exchange transaction, one kind of money must be exchanged for another. Usually that's no "complication" at all. The only time it becomes a problem is when the payments don't balance.

If Americans are pushing too many (or too few) dollars into the bank accounts of foreigners, that's when there's a problem. That's when the complications arise. That's when something needs to be done to bring things back into balance.

### There's No Completely Satisfactory, Permanent Solution

It's too bad that foreign trade has to be complicated by the need to exchange one kind of money for another. But that's the way it is. The world is made up of many different economic systems. Each one tries to operate more or less independently of the others. Each one has its own wage rates and interest rates and standards of living and monetary and fiscal policies and national objectives and goals and all that.

As long as these semi-autonomous economic systems exist, the "money exchange complication" will be with us. For a long time. Right? All we can do is hope to have an "international exchange system" that will minimize the complications. But you can see that the problem will never be really "solved." Not completely. Not permanently.

Setting "international rates of exchange" for all the different kinds of money in the world is, in a way, like having a "wage-price freeze." If things are more or less "in equilibrium" at the time the freeze comes, then there's no problem. At least, not then. But what happens next? The basic economic conditions keep changing, creating more problems.

The basic economic conditions *always* keep changing. One country will have more inflation than the others; one country will introduce more high-technology capital than the others; one country will find some new natural resources, etc., etc., etc. As the basic conditions change more and more, the fixed exchange rates get more and more out of line with reality. Pretty soon one nation's money gets in short supply in the world's money markets (like U.S. dollars after World War II) while another country's money gets in surplus supply (like U.S. dollars in the early 1970's).

So what happens? Just as a wage-price freeze (if it's going to work for very long) must be adjusted to reflect changes in basic economic conditions, also international exchange rates must be adjusted. If the adjustments are too slow in coming, what happens? What happened on August 15, 1971? The collapse of the system? Yes.

### A Stable But Flexible System Is Required

What we need is a system of exchange rates that will be stable enough to provide for the smooth flow of international trade, but at the same time, flexible enough to change as the basic economic conditions between and among nations, change. That's a tough order to fill! Especially when each nation is likely to push for adjustments that will improve its own position. See the problem?

The international exchange problem has something in common with the "inflation and unemployment" problem and with most of the other problems you've been reading about in the last three chapters. That is: it isn't ever going to be solved completely, permanently, once and for all.

The solution to this problem, like most of the others, is going to have to be living, growing, evolving—changing all the time. There will be good times, when things go more right than wrong—and bad times, when things go more wrong than right. The early 1970's were bad times. The system collapsed. Many meetings were held to try to work out something.

Meetings are still being held. Progress is being made. Let's hope that things can be worked out so that the international exchange system can see some good times again—at least for awhile. Maybe it'll be possible to work out a system flexible enough so that it won't have to collapse again. Let's hope so.

---

**REVIEW EXERCISES**

**MAJOR CONCEPTS, PRINCIPLES, TERMS (Try to write a paragraph explaining each of these.)**

the "dollar gap"
the "dollar glut"
"devalue the dollar"
the Bretton Woods System

the dollar became "super-money"
the U.S. payments deficit and gold drain
the changing role of gold
Special Drawing Rights (SDR's)

**OTHER CONCEPTS AND TERMS (Try to write a sentence or phrase explaining what each one means.)**

certified check
bill of exchange
buying foreign exchange
international payments surplus
international payments deficit
International Monetary Fund (IMF)
official exchange rate
World Bank (IBRD)
"suspend convertability"
"let the dollar float"

**QUESTIONS (Try to write out answers, or jot down the highlights.)**

1. Anything, any transaction of any kind, that *pushes American dollars into* the hands (bank accounts) of foreign governments or banks or businesses or organizations or individuals, is a "negative item" on the U.S. international balance of payments. Mention and explain several kinds of transactions that would show up as "negative items" on the U.S. balance of payments.
2. Anything, any transaction of any kind, that *uses up the American dollars* held by (in the bank accounts of) foreign governments or banks or businesses or organizations or individuals, is a "positive item" on the U.S. international balance of payments. Mention and explain several kinds of transactions that would show up as "positive items" on the U.S. balance of payments.
3. "When you think of all the things which go as 'negative items' on the balance of payments and all the things that go as 'positive items,' it seems like it would be sort of a *miracle* if the thing ever came out in balance, anyway!" Discuss.
4. Trace the highlights of international finance, from the end of World War II to the present. Show how the dollar gap became the dollar glut, and why SDR's are being tried, to replace dollars, which replaced gold as the "super-money"—as the "payments balancing medium" in international finance.
5. Explain how "devaluation" is done and what it is supposed to accomplish and how.
6. As far as international finance is concerned, the United States is now in a "brand new ball game." Can you explain *what* the United States is going to have to do, and *why* it is going to have to do each of these things, to try to keep its international payments in balance? Try.
7. It isn't likely that the international payments problem will ever be solved completely, permanently, once and for all. Can you explain why?

# 31 THE CONTINUING EVOLUTION OF THE WORLD'S ECONOMIC SYSTEMS: CHALLENGE TO CAPITALISM?

*As the stage of development changes, the problems and objectives change — so economic systems and economic philosophies change, too.*

You were reading about the problem of international exchange, in the last chapter. That problem isn't going to be easy to solve. Right? But it's going to have to be solved — at least "reasonably well" anyway. How? Who is going to take the lead in getting a new international exchange system worked out? The governments, of course. The political process must do it. Who else?

Think about the macroeconomic problems you were reading about in Chapter 29. The problems of keeping the economy stable and running at an acceptable speed — not too much unemployment, not too much inflation — who is going to take care of that? The government? Of course. Who else?

What about the microeconomic problems you were reading about in Chapter 28? When the choices about what to produce, which resources to use in which ways, and how much each person will get — when all these choices come out with too much environmental destruction and too much poverty, and several other socially unacceptable choices, who is going to do anything about that? The government? Of course. Who else?

## MODERN WORLD PROBLEMS REQUIRE MORE "POLITICAL PROCESS" CHOICES

Just about all of the new, emerging problems of the modern world are going to require some sort of government action if they're going to be solved. Can you see now why economists talk about "the increasing role of the state" in making the economic choices for the society?

The choice-making role of government has been increasing in all the "market directed" economies of "mixed socio-capitalism." I'm sure you understand *why* this has been happening and I'm sure you also understand that this trend will continue. Why? Because it *must* continue. Society is demanding that something be done about the major problems, so the role of the state must continue to expand. There's just no other way to get at these emerging problems of modern society.

### As Objectives Change, "Progress" Must Take New Directions

There were times not very long ago, when the major economic problem was how to meet the basic needs of the people. That's still the way it is in the less developed countries. But in the advanced nations, not so. Not anymore. The economic problems have changed, so now our objectives must change. Now that the environmental problem and some other problems are getting bad, people are beginning to insist that something be done about them. So "progress" must take new directions. It must be aimed toward the new, emerging problems—toward a new set of objectives. But how can that be done? Only through the political process. Only by government choice-making.

Those of us who live in the rich nations are well fed, have good medical care, warm houses with running water and telephones and TV sets, automobiles and boats, all kinds of things. From here on, for us, progress is going to have to be less "consumption-oriented," less "market-oriented," more oriented toward longer range objectives. We are well fed and highly productive. We have the time and the inclination and the means to do something about the "rough spots" in our society. But in the societies where affluence has not arrived, people aren't worrying about such things. To the people who are poor, "progress" still means material things—more and better food, clothing, shelter, clean water, sanitation, medical care.

As economies grow, become more productive, more specialized, more interdependent, more "modern," the problems change. As the problems change, the directions of progress change. You already know that. But as all this happens, the *procedures for approaching solutions* must change, too.

### The State Must Make More Of The Choices

In the economic systems which rely on the market process—that is, in the advanced economies of mixed socio-capitalism—the "new" problems and objectives require more and more government action. The state makes more and more of the economic choices. That just seems to be "in the nature of things." There doesn't seem to be any other way.

Why do economic systems change anyway? To cope with the new conditions, the new issues, the new problems. Of course. What kinds of changes seem to be required to get at the emerging modern world problems? The role of the state needs to expand. The political process needs to play a greater role in the economic system.

It's sort of obvious that as the society has more and more people, living closer and closer together, becoming more and more interdependent, determining each other's opportunities for employment and incomes, infringing on each other's freedoms, using up each other's resources, polluting each other's water and air and land—it's sort of obvious that as all this happens, the society (through its political processes) must get more and more involved in deciding who gets to do what. Nothing difficult to understand about that! In the economies of mixed socio-capitalism, the role of the state—the political process—expands. In the communist countries the role of the state can't very well expand. The state is running just about everything, making just about all of the economic choices, already.

How about the underdeveloped countries?—those (more or less) tradition-bound, poverty-stricken societies where most of the world's people live? What about those societies? What kinds of emerging economic systems are we likely to see there? And what will be the role of the state? The next section gets into this question. As soon as you understand a little more about some of the problems of these underdeveloped countries—these societies the economists sometimes call "the LDC's" (the less developed countries)—you'll be able to understand the increasing role of the state there, too.

## THE UNDERDEVELOPED COUNTRIES AND THE ROLE OF THE STATE

The problem of the less developed countries is one of mobilizing resources for economic growth. The LDC's need, somehow, to generate investments—to build capital. The "free market system" is great at generating investments. Right? Isn't that how the United States and all of the other advanced "market directed" economics of socio-capitalism grew so fast? Sure it is.

When labor is plentiful and wages are low, and capital is very scarce and highly productive, businesses can be very profitable. Such profitable businesses can earn lots of money and invest it and the economy can grow by leaps and bounds! Sure. It did happen just that way in the United States and in the other advanced countries. That's what was going on all during the last century and the early part of this one. But it isn't going to happen that way in the underdeveloped countries. Not now. Not during the last quarter of the 1900's. Why? Because conditions in the LDC's are so completely different from the "western world" of the 19th century.

## The LDC's Aren't Ready For The Market Process

Certain conditions are necessary before the market process can "break loose" an underdeveloped society and start it on the path of growth. In the LDC's, those conditions don't exist. The people's desires and attitudes are not ready; the labor force isn't ready; the capital with which to begin isn't ready; the governmental stability necessary to make investments "safe" usually isn't ready, either.

More than all those other things combined, consider this: Do you think, these days, that people will accept lives of deprivation and hunger while a few capitalists get richer and richer? That's what happened in the United States and other countries during the 1800's and early 1900's. But today? No, I'm afraid not. Those times have passed. The market process alone won't do the economic development job for the LDC's. It can't. Unless the government gets involved and breaks the vicious circle, not much development is likely to occur. So here we see it again: the increasing role of the state. Economic choice-making by the political process. It looks like that's the name of the game for the LDC's, too.

The internal conditions of the LDC's aren't ready to let development "grind its way forward" through the market process. Nor are the people patient enough to let it happen that way. The economically advanced nations are always there, offering examples of how life might be—of how life is, for the "favored few."

The people of the "western world" struggled and suffered hardships for centuries as their development was going on. They put up with it. That was "just the way life was." They accepted it that way. People living in a stone age society never complained because they didn't have steel tools to work with. Of course not! But let a person use a steel tool *just once* and suddenly we have a brand new ball game.

## A Taste Of "Modern Things" Destroys Traditional Societies

Suppose you were born and raised in a very primitive society. You had always used a stone axe to cut down trees. Then one day a stranger came and let you use his steel axe. How that sharp blade slices through the wood! Unbelievable! After that, would you ever again be satisfied with your stone axe? I doubt it.

Suppose, in your society, when a person gets a small cut or scratch and it gets infected, he dies. Infection causes "blood poisoning" and kills people. Everyone in your society knows that. It has always been that way. One day you get a scratch and it gets infected. Everyone knows that you are going to die. But then a stranger visits your village. He rubs ointment on your infected scratch and gives you a shot. The next thing you know the infection is all gone away, just like magic. Wow! So what happens after that? Will you ever again be contented to die (or to watch someone you love, die) from an infected scratch? I doubt it.

In the advanced nations of the western world it took several centuries of denial and struggle and hunger and dying before people reached their present levels of survival and comfort. People didn't see, didn't know of, *couldn't conceive of* the kinds of conditions which we (the favored few of the world's people) now enjoy. Our ancestors could wait because they didn't know they were waiting. They didn't know what they were waiting for. But what about the people in the LDC's today? They know they're waiting, and they know exactly what they're waiting for. They don't plan to let "the good life" be two or three centuries in coming, either—not if there's any way they can think of to speed up the process!

Awareness of better things makes people want better things; modern medicine keeps people alive and lets the population expand. So what happens? There are more people, all wanting more. *Many many more people, all wanting much much more.* Where does the "more" come from? That's the problem. Today, perhaps we can ignore it because its only the problem of the less developed countries. Today, for most of us, the LDC's seem far, far away. Tomorrow the problem of the LDC's may be a very real problem for all of us.

About four out of every five people in the world live in these less developed countries—these countries which have hardly experienced any of the great economic advances which the "modern nations" have achieved. The modern nations have brought to the LDC's just enough things to break the traditional life styles of the past: some better tools and other things, and modern medicine to trigger the population explosion. In a very real sense, it is economic and technological progress in the advanced nations which has brought the "problem" to the underdeveloped world.

### The Need For Economic Development Becomes More And More Urgent

As population booms, the family plots of land can't feed all the people. People leave and go to the cities. That's where you can see some of the most miserable economic deprivation you could imagine—in the urban areas of the underdeveloped countries. There aren't many jobs available. It's no wonder the people are clamoring for economic development—for *any kind* of economic development. For income opportunities. For some way to get the basic necessities of life.

Economic development and population control are essential. But both of these objectives are extremely difficult to achieve. Resources (manpower and other resources) must be organized and directed toward controlling the population, developing a productive labor force, building capital. But how? There has to be *some slack* to begin with—some capable resources to work with. Where do the capable people and resources come from *to initiate* the process?

There's surplus labor in the cities and on the farms. Those people could be mobilized and trained. But who's going to get the mobilization and training started? And how? Agriculture could be made more efficient by "land reform" (changing the ownership patterns) and by introducing new technology and capital. That could produce more output. The extra output could be sold in the international markets to get foreign money. The money could be spent to buy capital.

But who is going to initiate the first step? And don't forget: the people have a strong, almost religious respect for the traditional ways of doing things. They fear change. Add to that all the taboos and illiteracy and you will get some idea about how difficult the economic development and population control programs are going to be to plan and carry out. Who do you suppose can handle it? Who is there to do it? How can it be done?

## The Essential Role Of Government

The governments of the underdeveloped countries are going to have to play a major role in getting the growth process started and keeping it going. Why? If you think about it, it's sort of obvious, really.

Throughout history, whenever major, important development changes (or any changes) have occurred to break traditions and get things on a different track, governments, using the coercive force of command, have always had to play a big role. Things must be broken loose. The government usually must be involved in doing it. How else except by the government can land reform (forcing changes in land ownership) be done, to convert tiny family plots into efficient farms and to break up and utilize the massive landholdings of the few wealthy people and "feudal chiefs"? No other way. The government must do it.

How else can the surplus labor be shifted out of agriculture and out of the city slums, and organized, trained, fed, clothed, housed, and put to work building roads, docks and harbors, water systems, power systems, training schools, other development-supporting projects?

How else can traditions and taboos be prevented from sabotaging both the population control and economic development efforts? How else can domestic consumption be held down and savings increased, and then the savings used to support the population control and economic development programs? How else can foreign aid be brought in from the advanced nations? and then channelled into the optimum uses? The government must be involved every step of the way. Even with the very best government effort it's going to be a well nigh impossible task.

What about foreign investments by "profit seeking free enterprises" from the advanced nations? Can't this kind of private foreign investment help to get things started? Sometimes, sure. But unless the local government gets some "infrastructure development" started—roads and bridges, docks and harbors, water systems, and other things needed to support economic development—and unless the local government is strong and stable and "friendly" to foreign investors, not much foreign investment is likely to occur.

Many less-developed countries are trying to "have their cake and eat it too," where foreign investment is concerned. They want foreign enterprises to come in and build the needed infrastructure, mobilize and train the labor force, build successful enterprises employing local people and generating lots of local income—but they don't want the foreign investors to receive any profits! They call profit-taking "capitalist exploitation." Such attitudes won't generate much foreign investment.

### Needed: Outstanding Local Government (Or Else, Revolution?)

Population control and economic development will never be easy to achieve in an underdeveloped country. Without a strong and stable government exerting its best efforts to plan and administer the needed programs, nothing much is likely to happen.

Unless you've been in intimate touch with some of these countries you may find it impossible to picture the conditions there. It's hard to visualize what the development process will be like—the kind of *total change* which these societies must undergo as they strike out on the path of "modernization and development." It means a massive upheaval of the society. It means the destruction of customs and beliefs which have held the society together for a thousand years. Questions of what's right and what's wrong, what's good and what's bad, what's acceptable and what's unacceptable—suddenly all are going to have different answers. I dare you try to picture that!

### The Total Social Upheaval Of Economic Development

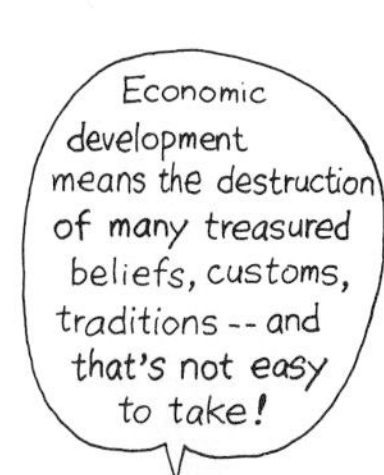

People's relationships to each other must change. Relationships among the members of the immediate family and the extended family, all changed. Traditional ideas about rights and duties and responsibilities and obligations, all turned topsy turvy—all shattered. The sources of honor and prestige and respect, suddenly all different. The whole pattern of what people are supposed to do and what they aren't supposed to do, must be cast in a different mold.

Just in your short lifetime you have seen enough social change in your society—enough differences of opinion between yourself and your parents and teachers and other "old people" about what is "acceptable behavior"—that I'm sure you have *some idea* how tough it is to work out these changing beliefs about what's good and bad, right and wrong. Can you guess how tough it must be to get an entire society to go through a complete change of mind on these things? And do it almost overnight? Impossible you say? Right!

Can you picture the kind of strong, effective leadership—the kind of power and charisma which will be required to motivate and direct and control the people as all the bonds of tradition are shattered? With the people not knowing any more what is right and what is wrong? Not knowing what they're supposed to do? The government leadership somehow must hold it all together, must keep things moving in the right direction. A tough assignment? Unbelievable!

The world really doesn't have very many of the kinds of governments or of the kinds of leaders who could pull it off. And as you might expect, such outstanding governments and leaders as do exist in the world, aren't usually found in the less-developed countries. Often the governments of the LDC's are as tradition-bound as the rest of the society. Often the governments use their powers to try to perpetuate the problems. The big landowners are likely to be the ones who run the government. Talk about land reform? Forget it! Talk about economic development which will push aside the traditional hierarchy of chiefdom and make the "ordinary people" as good as the "royalty"? Forget it!

Do you see the problem? Can you see, now, why a lot of good, honest, intelligent people will join a communist revolutionary movement (or any other revolutionary movement) which promises to get rid of the old system? In many of the underdeveloped countries the only way any improvement in the economic condition of the "common man" is likely, is for the government to change. Often it requires a revolution. That may not be a happy thought, but that's the way it is.

### Is Communist Revolution Inevitable In The LDC's?

Does all this mean that Communist revolutionaries are going to be the ones to bring economic development to the LDC's? Must the established governments be overthrown in order to break loose the "feudal bonds" of the traditional societies so that development can get started? Is a revolutionary strong-man dictator required to make it work? Of course not. That isn't the *only* way. But that is *one way* it can be (and is being) done.

Communist revolutionaries have taken over and have shattered preexisting societies, have swept aside the progress-impeding land ownership patterns, have initiated effective programs for population control, and have done many of the other things necessary for development. The cost, in terms of lost human life and lost individual freedom, has been staggering. Often the development programs have been inefficient and unsuccessful—badly planned and badly administered. But the fact remains that the "communist revolutionary approach" is one way that it can be done. For millions of people it *is* being done that way.

Isn't there some other way? Some less cruel, more humane way to achieve development? Without such tremendous cost in human lives and suffering? Of course there is—if there's enough enlightened leadership, if there's enough concerned and vocal citizenry, if there's enough planning and investment aid and support from the advanced nations, and if there's not too much corruption in the local bureaucracy. But these are big "ifs." The problem of "how to get it straightened out without a revolution" wouldn't be so tough, except that the kind of leadership needed is so hard to find—in the underdeveloped countries, or anywhere else in the world.

## The Same Problems Must Be Faced, No Matter Who's In Charge

Suppose the communist revolutionaries do take over. Then they face the tough problems. How do they get the talent needed to design effective plans, aimed toward wisely chosen goals? Then, with almost nothing to work with, how do they carry out the almost impossible task of "resource administration" for the entire nation? How do they get everything to start moving and keep moving in the right directions, toward the chosen objectives? Where do those resources and that talent come from?

It's much easier to sweep aside the old order, "eliminating" anyone who stands in the way, than it is to choose goals, design plans, and effectively administer the labor, land, and capital of a society. Fidel Castro certainly found that out in Cuba. Lenin found it out in Russia in the years following the Bolshevik Revolution. It's likely that some other underdeveloped countries will find it out too, before it's all over.

Here's an important point I don't want you to miss: In the process of economic development in the underdeveloped countries, the things which must be done, the steps that have to be taken are going to be about the same no matter who's in charge—no matter which "banner" is flying from the state house—"capitalism" or "socialism" or "communism" or "Christianity" or "Buddhism" or "nationalism" or "democracy" or whatever. But "who's in charge," and the "banner" that's flying can make a world of difference in *what procedures are followed*, in *how the changes are carried out*. In the USSR, for example, "land reform" was carried out by "eliminating" the landowners; in Taiwan, the land was "taken," but the owners were given government bonds in exchange.

It's much easier to take land by firing squad than by due process of law, of course. Quicker, too. But there may be some question about how good a society built on the "firing squad approach" is likely to be—that is, for anybody except the members of the "squad"! Maybe the slower, "due process" way is better after all.

### The Market Mechanism Can Help In The Development Process

What can we conclude about "the role of the state" (the importance of the political process) in the future of the less developed countries? The state is going to play a major role. No question about it. We can hope the leaders of the nations will know enough about basic economic concepts (or have advisors who do) so that they will be able to make effective use of the "market mechanism" to help them in achieving their objectives. As you know, whenever it can be used, the market mechanism—motivating production and conserving things by using price incentives—is likely to be the most efficient way.

But the "market mechanism" approach is good for another, perhaps even more important reason. It leaves people more free, more able to make their own personal choices. Each person can be left free to decide where he wants to live and work, which things he wants to do or make, which things he wants to use up and which to conserve, which to have and which to eat, and what to trade for what. Yet, even with all this freedom, the society's resources can be directed much more efficiently, much more "optimally" toward the chosen objectives. How's that for a good deal? More efficiency, and more freedom, too!

## ALL ECONOMIC SYSTEMS ARE ALWAYS CHANGING

All economic systems are changing all the time. The amount of government influence is changing. Customary, traditional ways of doing things are changing. The role of the market process is changing. The characteristics of all the "free mixed economies of socio-capitalism" are changing. The role of the price mechanism with its production motivating and rationing function, the institution of private property, rewards for productivity, individual freedom and responsibility—are changing constantly.

### There Is No "Ultimate Form" Of Economic System

The "controlled economies" of communism are using more of the characteristics of the market mechanism; the "free economies" of mixed socio-capitalism are embodying more goal-setting and planning, more choice-making and resource direction by the political process. What's the ultimate outcome going to be? What's going to happen to the economies of the world? That's a big question these days. Lots of people are talking about it, trying to guess at the answer. What do you think? It's hard to say. Right?

All the nations face the same problems, more or less: problems of choosing what to do with the scarce resources of the nation, of saving some output and turning it into capital, of controlling population size, of designing programs for education and training, of taking care of the incapable ones, of keeping the economy running smoothly and keeping prices stable. All nations approach these problems somewhat differently. Some do a better job than others. But all nations keep changing their approaches (their "economic systems") as time goes by.

So what's the ultimate outcome likely to be? What's the "ultimate shape" the world's economic systems will take? Where will it all settle down and stabilize? Where? Nowhere. Really, do you think there can be an "ultimate shape" of something like an economic system? Of something that's growing, changing, responding all the time to new things, to constantly changing circumstances? Of course not!

Economic systems are "living systems." Living systems never "arrive someplace"! They just keep on changing. The only questions that make any sense are questions about such things as the *directions* and *speed* and *processes* of change. Which way are we headed? how fast? and how is it all happening? Those are the things we need to be thinking about—not about some "ultimate arrival place"!

### Do Economic Systems Change As The Stage Of Development Changes?

Could it be that the "stage of economic development" in a society will influence the kind of economic system that the society will have? Does the system need to change as the "stage of development" changes? Perhaps so.

In a completely primitive society, custom and tradition seem to do the best job of making the economic choices—of looking out for the survival of the society. But if the economy is ever going to grow, the system must change. The only way much economic change is likely to come to the tradition-bound societies—that is, the only way economic development is likely to get started—is for the political process to play a larger role in the economic system.

In today's world, a growing underdeveloped country needs an economic system which contains a high degree of planning and direction and control by the political process. You already know that. But what would you call such a system? What name? I don't know. There's no name that fits, really.

Economic plans and controls certainly can help an economy on its way up the development path. But as development proceeds, conditions change. As a "planned and controlled economy" gets more developed—as it becomes more highly specialized, with high technology and high productivity—the *complexity of the planning process* begins to get overwhelming. The task of planning and directing and controlling all the people and resources and things—and doing it *efficiently*—becomes just about impossible.

In a highly complex, "mature modern economy," some "indirect way" is needed to get all the thousands—no, *millions*—of little daily choices made and carried out. The most "natural" way to do this is by using the price mechanism. Does this explain why the USSR and the Soviet satellite countries are changing their economic systems to include more reliance on the production-motivating and rationing functions of price? Of course it does.

Maybe it's time for people to start talking about the Soviet and East European economies as systems of "mixed capitalo-communism"! No. I really don't think so. Their systems probably would have to change a lot more before you'd want to call them that. But I don't suppose it really makes much difference what you call them. It's too bad, but even with all our "labels" for real-world economic systems, we really don't have any "names" that tell us what an economic system really is like. You've known that ever since Chapter 4. Remember?

### The "Changing Mix" In The Systems Of "Mixed Socio-Capitalism"

What about countries like the United States and Canada, and the West European countries and all the other countries of mixed socio-capitalism? These countries experienced their "economic growth breakthroughs" in another era, another time—another world, really. Those economies grew up basking in the philosophies and policies of nineteenth century liberalism, of laissez-faire, democracy, the Protestant ethic—of the inalienable rights (and duties) of man.

In these countries the market process and the price mechanism have been running things all along. But now we see these systems changing rapidly. The "role of the state" is increasing by leaps and bounds! Does the *stage of development* have something to do with it? So it seems.

As the "capitalist" economies have become highly industrialized and as population has expanded and become concentrated in the urban areas, these societies have become more complex, more interdependent. Most of the basic economic needs of most of the people have been taken care of. But now we have a new set of problems and needs—problems and needs which the market process can't take care of very well.

The expanding demands of modern society for better "public goods and services"—education, police and fire protection, public health and sanitation, recreation areas, highways, urban transit systems, etc., etc., etc. — require more "political process" choices. The political process is going to have to be used to tackle the "externalities" problem and the poverty problem and the whole complex of problems centered around the population crush in the urban areas. Almost any "problem of modern society" you can think of is going to require some involvement by the political process — some increase in "the role of the state."

So what's happening? The economic system is changing. The political process is making more of the choices. There's more government planning and control. The rights of private property and the freedom of the individual are being reduced. The "mix," in the economic systems of "mixed socio-capitalism," is changing. The system called "capitalism" has become (and is constantly becoming) different. That's why I've been calling it "mixed socio-capitalism." Why? Because the word "capitalism" really doesn't describe it very well at all. Not anymore.

## WHAT ABOUT THE SURVIVAL OF CAPITALISM?

Sometimes people talk about the question of "the survival of capitalism." What nonsense. The communist philosophers like to predict the end of "capitalism." What utter nonsense!

### The Capitalism Of Karl Marx No Longer Exists

Capitalism was the economic system of the mid-nineteenth century, observed and attacked (and incidentally, named) by Karl Marx. That kind of economic system doesn't exist anywhere in the world. It hasn't *for more than forty years*!

The "free societies," with (more or less) democratic governmental systems and with mixed "political process and market process" economic systems have many "bones of contention" to argue about with the communists. There's no question about that. But the argument over the *future of the system Marx called "capitalism"?* That's just ridiculous.

Nineteenth century capitalism no longer exists. Of course not. But suppose we stretch the word "capitalism" to mean "any system in which the market process plays an important role." What then? As time goes on, will the market process continue to be one of the important ways in which societies make their economic choices? For the foreseeable future, yes. The market process will continue to play an important role. No question about it. So if that's the way you want to define capitalism, then okay. Capitalism will survive.

For as long as people have personal, individual desires, and as long as incentives and rewards are effective ways of motivating people, the market process will keep on influencing social choices. But forevermore? Until the end of time? Who can say? Who cares, anyway?

Consumer demand will never completely direct the resources of any society. But consumer demand won't ever be completely ignored, either. Also, the market mechanism will be used as an "implementation tool" to carry out choices that are made through the political process. As more and more planners come to understand the *efficiency* of the market mechanism, you can expect to see it being used more and more to implement the plans.

### Will Private Property Survive?

The market process can't work without some rights of private property. Will private property survive? Of course. You know that private property rights will never be absolute and complete—but they won't ever be totally revoked, either. Just about everybody, in every society, gets to own *something*! In the United States and the other economies of mixed socio-capitalism we can expect more limitations on our private property rights as the political process makes more of the choices and takes on more problems.

What problems? Externalities must be internalized. Income must be redistributed. That means some loss of private property rights. As the society's productive efforts shift direction—from the old problems of yesterday to the new problems of tomorrow—much of the shift will be a shift from "private" objectives to "social" objectives. As the shift continues, the individual is going to lose more and more of his rights to choose. It's inevitable.

In the communist countries, private property rights are likely to be expanding. Why? Because as these countries learn more and more ways to use the market mechanism to carry out their plans, they're going to have to let people receive more personal rewards—more "private property." The income receiver must be able to keep his money, or to spend it and keep what he buys. That's his private property. Some increases in private property rights are already being allowed in some of the communist countries, just as the rights of private property are being cut back in the "free economies" of mixed socio-capitalism.

### Will Individual Freedom Survive?

What about the freedom of the individual to go his own way? to do his own thing? to choose his own path? to claim the rewards of his diligence and successes? or to suffer the hardships of his indolence and failures? What about all that? Will that sort of "individual freedom and responsibility" survive? You already know the answer.

The answer to the individual freedom question is very much like the answer to the private property question. In the "free economies," personal freedom will be reduced. In the "controlled economies," it will be increased. I'm sure you can see why. As modern society becomes more complex and more interdependent, as population expands, as the urban areas keep expanding and overflowing their boundaries, as new technological change keeps bombarding us with new problems and new opportunities—as all these things keep happening, more and more of our choices are going to come under the control of the political process.

Each individual, if left free to decide, will do his own thing—will choose his own path toward his own objectives. But when millions of people are crowded together in a little space called "an urban area," then the society can't let each individual do his own thing. You've been there. You've had lots of neighbors. You know what it's like. Right? Sure. So we have to work things out together through the political process. All of us have to go along with the restrictions decided on by the group. Of course.

## The Names Of The Systems Are "Mythical Blanket Words"

By now you can see the nonsense of talking about the "survival" of some kind of economic system. When we use labels like "capitalism" or "socialism" or "communism" we spread a blanket over and hide all the things that are going on in the economic system. When we talk about "the survival of the economic system" we're talking about something unreal—about the survival of the "mythical blanket" which we spread over and use to hide the real conditions in the economic system. Then the next thing you know the "blanket word" becomes "the name of our team." Everybody gets personally identified with the name. Everyone begins to think of it as something to believe in, to cherish, to protect, to fight for.

The Communists are really serious about this "mythical blanket" game. They really fight hard for the name "communism," and against the name "capitalism." But when you think about it, Americans are just about as serious in fighting against the name "communism" and for the name "capitalism"!

Marx invented the label "capitalism." Then he led the bitter attack on it. So what happened? All the "free society" leaders *identified* with the label "capitalism" and rose to its defense. Nobody's quite sure what either term means, or exactly what it is that's so good about one and so bad about the other, but everybody knows that communism is bad and capitalism is good (or vice versa, depending on which side you're on).

## Our "Worldly Religions": Capitalism, Socialism, Communism

These blanket words—mythical creations of two or three nineteenth century "social protest philosopher-prophets"—have become the "banners" for a kind of "worldly religion." We use these words as our justification for fighting wars "to uphold that which we know is right." Tragic.

It would be a great contribution to world peace and harmony if we could just let these "mythical blanket" words—"capitalism," "socialism," "communism"—dissolve and disappear. Then we could just look at real world economic systems and see them as they are. We could calmly study the existing conditions and the changes, the role of markets, the role of the state, the changing institutions, private property, individual freedom, individual choice. We could look at each system without prejudice. We could accept the evolution of economic systems and the changing role of the state without all this fear. Someday people might even begin to wonder what all the fuss (over "which kind of economic system") had been about.

Yes, we would all be better off if we could get rid of the labels: capitalism, socialism, communism. The labels don't really describe any of the world's economic systems anyway. If you want to see what the system really is, you need to look underneath the "blanket word" and see how the system works—how choices are made and carried out. How much social process? political process? market process? and how does each process work? If you know these answers, then you know what kind of economic system the society has. If you don't, then you don't. Whatever it is today, you can be sure that tomorrow it will be different.

## ECONOMIC PHILOSOPHIES? OR ECONOMIC SYSTEMS?

It may not surprise you, now, to find out that the words "capitalism," "socialism," and "communism" refer to economic *philosophies* much more than they refer to economic *systems*.

The philosophy of capitalism is the philosophy of the market process —the philosophy of Adam Smith's laissez-faire—of the "invisible hand," and "consumer sovereignty." It's the philosophy of individual freedom, of private property, of rewards for productivity. The idea is that people and businesses should be left free to make their own choices, to follow their own selfish interests, to do their own thing. That way, everything comes out better for everybody.

The philosophy of capitalism holds that without the freedom of markets, people really can't be free either. The government should keep the markets free, prevent monopoly, see that everyone has a fair chance, then let each person choose his own path. Each person can make his own choice about how far he wants to go in life, in which directions, what he wants to do to earn his money and how he wants to spend it. The economy adjusts its production in response to the demands of the people.

This "individual freedom" philosophy of capitalism is strong in the United States. It is also strong in the other non-communist countries, even though most of those countries call themselves "socialist." But there's no real-world economic system anywhere which even comes close to "reproducing in the real world" this "individual freedom" philosophy of capitalism.

## The "Pure Philosophy" Of Capitalism

The "pure philosophy" of capitalism is a philosophy of "rugged individualism." "Each person must stand on his own feet. If he can't stand, then let him fall, because that's what he deserves. The best thing for him and for everybody in the long run will be to let him fall."

This unmodified "law of the jungle; survival of the fittest" philosophy of capitalism was accepted and followed by most business and political leaders of the western world. This philosophy, and the harsh, 19th century system was what Marx attacked. He predicted its downfall. The "downfall" of the system didn't come the way Marx predicted, but it came just the same. Not by revolution but by evolution—by (more or less) orderly change.

Will the "individual freedom" philosophy of capitalism survive? I'm sure it will. But no real-world economic system will ever again be built on the "rugged individualist" philosophy of unmodified capitalism. Certainly not. All real-world systems embody *some* of the philosophy of capitalism. Even the communist countries use one of the most basic concepts of capitalist philosophy—incentives and rewards—more income for more productivity.

Any system based on the capitalist philosophy, today, must also embody much of the philosophy of socialism. There are still some people in the United States and other countries who don't like socialist philosophy creeping into their societies and influencing the evolution of their "capitalist" economic systems. Some cry out against it. But their voices are lost on the winds of change.

## The Philosophies Of Socialism And Communism

What is the philosophy of socialism? It's a philosophy of cooperation, of working together and sharing, of working for the good of the society, of giving up individual freedoms for the benefit of society. The society, through the government, should own most of the means of production, and the people should work, not for their own betterment, but for the betterment of everybody.

This philosophy of socialism is much more "philosophically appealing" to most people than the "self-centered, dog-eat-dog" ideology of capitalism. Everybody knows it's good to be benevolent, and to share with others. But when it gets down to the real world, the "benevolent

sharing" of socialist philosophy doesn't seem to work out. No economic system has ever been able to run on "benevolent sharing." Most people seem to be more productive, to do more to help the society toward its objectives, if they have incentives, and rewards.

What a far cry the *philosophy* of socialism is, from the *real-world economic systems* called "socialism"! Even farther, really, than the real-world capitalist systems are from the philosophy of capitalism. So what's a "socialist economic system"? It's a system in which the leaders (and maybe most of the people) proclaim their belief in and pledge their allegiance to some version of the socialist philosophy? Right!

Using the same logic, what's a "capitalist economic system"? It's a system in which the leaders (and maybe most of the people) proclaim their belief in and pledge their allegiance to some version of the capitalist philosophy? That's just about right! But from these statements of philosophy we still don't know much about what the economic systems look like. "Pronouncements of philosophy" don't tell you much about how the society makes its choices. That's why you must look down under the "mythical blanket" if you want to find out what kind of economic system a nation has.

What about communism? Is that a description of some specific kind of economic system? Or is communism more "philosophy"—more "worldly religion" than economic system, too? You have known the answer to this ever since you read Chapter 4. Yes. Communism is more philosophy than reality, too. Of course.

As an economic system, communism is "a system in which the leaders (and perhaps most of the people) proclaim their belief in and pledge their allegiance to the philosophies of communism." One "communist system" can differ widely from another. But the leaders and the people always proclaim their belief in the "mythical blanket label." So if you want to know what kind of economic system *really* exists, you must look beneath the label.

## WHERE DOES IT ALL GO FROM HERE?

Many of the tensions and conflicts in the modern world seem to be centered around these "mythical blanket labels" which don't describe anything. But each of the labels carries philosophical connotations so strong that modern nations fight about them. Each side must uphold its philosophical position—its "worldly religion." How long must this go on?

### Must We Keep Squabbling Over Our "Worldly Religions"?

Are all the communist countries going to continue to try to force their "worldly religion" on the rest of the world? Are the U.S. policymakers going to continue to take the position that the philosophy of capitalism is always right, and best? That any other approach to economic development or to anything else is bad, and wrong? That anyone who deviates from our ideas of how things ought to be done, automatically becomes our enemy? How long must all this go on?

Might we dare to hope for a little more intelligence and reasonableness? For the bitter clash over "capitalism vs. communism" to be put to rest? I don't know how the world's going to make it—with all the *very real* problems facing us—if we don't learn to stop fighting about these make-believe "worldly religions."

## I Wish Everyone Knew As Much Economics As You Do

I only wish that all of those who will be influencing the economic decisions—the leaders of all the countries all over the world—knew as much about basic economic concepts and how they work, as you do. Just the economics you've learned from this one book, if understood and applied by all the people making the decisions, could bring better lives to millions of people—could make a better world for all of us.

"But," you say, "I really don't know all that much about economics." Maybe not. Relative to what you could know, or to what you might like to know, maybe not. But if you've been really digging in and working with these concepts, and if you've really been able to see them working in your life and world, then you know a lot more useful economics than you think. Right now you know a lot more economics than most of the people in the world. You know a lot more than some of the people who will be making decisions and choosing policies that will influence the lives of thousands, maybe millions of people.

Many of the political leaders throughout the world have very little understanding of basic economic concepts. That's very unfortunate—especially so, since it seems to be necessary for the political process to get more and more involved in making the choices for society. As you might guess, some of the "economically illiterate" leaders are in the underdeveloped countries. But not all of them are. Not by a long shot! Many of them are in your own country—in the federal government, and in the government of your own state, your own city, your own county, your own township, school district—everywhere.

It's too bad that so many policymakers don't know more about basic economic concepts—about the "science of common sense." But that's the way it is. Some bad choices are bound to result. Be sorry that they don't know economics. But at the same time, be glad that you do.

This is the end of the last chapter. A short epilogue follows—no big deal—I just wanted to share some parting thoughts with you. But before you go on to that, stop and do some serious thinking about the crucial issues in this chapter, and the issues in the other three chapters of

Part Eight. These are issues you will be seeing around you all your life. The better you understand them the more insight you'll have into what's happening around you.

---

**REVIEW EXERCISES**

**QUESTIONS (Try to write out answers, or jot down the highlights.)**

1. "Capitalism — the economic system which was *named* by and *attacked* by Karl Marx — no longer exists, so Marx must have been right, after all!" Do you agree? or disagree? Discuss.
2. It seems inevitable that in the LDC's, "the state" is going to have to play an increasing role. Can you explain why this is true, and what kinds of things "the state" is going to have to do?
3. If an LDC has a strong and stable government and good economic planners, much of the development task can be approached through the use of the market mechanism — through the use of the rationing and the production motivating (incentive) functions of price. Can you describe some of the specific ways the price mechanism might be used? Try.
4. It seems inevitable that in the "free economies" of mixed socio-capitalism the role of the state is going to increase more and more. Can you explain why this seems to be true, and what kinds of additional things (which "social choices") "the state" is likely to get involved in? Discuss.
5. It seems very likely that in the rigidly-controlled economies of "Communism," the role of the market mechanism is going to increase more and more. Can you explain why this seems likely?
6. Describe capitalism, not as an economic system, but as a *philosophy*. Do the same for socialism. Then do the same for communism. (You might need to look back into Chapter 4 to review the philosophy of communism.) Now, think back to the last time you heard some people arguing over some economic disagreement. From listening to what each person said, could you tell *how much* of *which* economic philosophies each believed in? Think about it.
7. This chapter suggests that we might all be better off if we could just get rid of the "blanket labels" — the names for our "worldly religions" — capitalism, socialism, communism — and refer to economic systems by describing the highlights instead of by mythical labels. What do you think about this? I can't think of any practical way to get it done. Can you?

---

# EPILOGUE: SOME PARTING WORDS ABOUT ECONOMICS, AND YOU

Welcome to the end of the book. If you've read the whole thing, you've seen a lot of economics. If you've learned the concepts well, your eyes are now open to a lot of things you couldn't see before. You've done a lot of work. Yes. And you're well on your way. So relax for a minute. "Enjoy your achievements . . ." Smile and feel good. You have a right to be proud of yourself.

If you've really been studying hard and learning it well, by now I'm sure you have some new awareness of the world—some sharpened ability to conceptualize reality. You have a new "feel" for things that are going on around you. You'll be carrying several of these concepts with you, using them as "a way of looking at things" for as long as you live. You have a sound framework that will help you as you further develop your own good common sense, day after day, year after year.

### The World Is Your "Econ Lab"

The world is full of illustrations and examples of economic concepts. I'm sure you remember when I said, back in the beginning of the book: "there's no real-world issue or problem that is *purely* economic." Remember? Sure. It's very true, too. But almost every real-world issue or problem you can think of has *some* important economic aspects. Yes, you can be sure that you will be running into real-world illustrations of economic concepts, all the time.

Everywhere you go (now that you have built this new "window through which to observe the world") you will keep on seeing things you never noticed before. Whenever you go into a bank you will feel different, now. You'll know that when the teller is pushing those accounting machine buttons to print a deposit receipt for you, somewhere in the bank a figure is automatically being added to your checking account. And you'll know that the figure is your money! As you leave the bank you'll see your friendly branch manager sitting there behind his desk and you'll be tempted to go over and ask him if he has created any money today. Go ahead! Ask him. I dare you.

Whenever you go into your favorite shoe store you'll see those shoe boxes all around and you'll wonder how much the store keeps in inventory. And you might wonder if the inventory is expanding, or contracting. Ask the manager. I'll bet he would enjoy telling you about it!

When you go to the post office and buy a book of 8¢ stamps you're going to remember about "the 75¢ dollar" you're using to buy stamps. Or "the 37¢ dollar," if you happen to be thinking of 1960 as your base year. (If you *really* want to be sad in the post office, think about the 17¢ dollar you use to buy what was, only a few years ago, a penny postcard!)

From now on, whenever you hear the news about employment and unemployment, about Americans spending too much money abroad, about the price of wheat or oil or steel going up, about the increasing GNP, about interest rates going up and bond prices going down, and about all such things, you'll have the good feeling of being "in the know."

Whenever you hear a political demagogue promising to do all kinds of great things for the people, you'll wonder where he's going to get all the scarce resources needed to do all those great things. Whenever laws are passed influencing wages or prices, or restricting or subsidizing the supply of something, or regulating market conditions in any way, you will have some idea about whether we will soon be seeing surpluses, or shortages. You'll know that "Bowden's Law"—the *reflex principle of real-world markets*—is going to be in there working to level things out. Right?

From now on, whenever you choose corn flakes instead of eggs for breakfast, you might think about how you're influencing the resource-use choices in the economy—about how you're supporting the "Kellogg team" and deserting the "egg team." (You might even wonder if everybody else is switching to corn flakes today, and if, pretty soon, we're all going to be up to someplace in eggs!) From now on, whenever you have a worrisome choice to make, you"ll be aware that you're weighing alternatives, considering opportunity costs—thinking marginally. That's good. Some of your choices may come out better if you think about them that way. I hope so.

You know a lot of economics. You really do! Not enough to handle all the problems economists deal with. Of course not. But as you go through life with your mind turned on you'll be learning more and more economics all the time. Someday you may understand more economics than some people who majored in it! (You can get there quicker by being an econ major—but anyone who knows anything at all about "the science of common sense," knows that a person can't major in everything!)

### It Takes Scarce Resources To Solve Most Problems

Now that you know some economics, does that make you "materialistic"? There's plenty of talk going around these days against materialism. Surely no one should focus all his attention in life toward material things. Everybody knows that big money and big houses and big cars and big steaks and long trips on big airplanes don't bring happiness. No argument on that score. But everyone who thinks about it will realize that if he wants to try to seek an objective—*any* objective—it's going to require some scarce resources—energy, thought, effort, capital, other resources. *"Material inputs" are essential in order to move toward almost any objective you can think of.* That's just the way it is.

Energy is required to move things, to do things, to make things, to change things, to achieve things. And energy is scarce. So we aren't going to be able to move and do and make and change and achieve all the things you might wish plus all the things I might wish plus all the things everyone else might wish. But it's likely that *every major problem you can think of could be at least partly solved if enough scarce resources (material inputs) could be turned in its direction.* And it's almost just as certain that *no progress toward solving any of these problems is going to be made without the use of some scarce resources (material inputs).*

If it is true that some scarce resources are needed, then we can't possibly solve all the problems we would like to solve—at least, not all right now. Not all at once. That's the way it is with scarcity. There just aren't enough resources, enough inputs to do all the things all the people want done. So what can we do? What *must* we do? We must *choose.* Of course. We must decide which problems to solve and which to tolerate for awhile.

### The Stark Reality Of Economics

When you learn economics you experience a stark confrontation with reality. You come face to face with scarcity—and *scarcity is the natural condition of reality.* Once you face it, confront it, recognize it—from then on you live with an awareness of opportunity costs, of substitution, of having to choose between things. You can't go along making believe any more. You have to be honest with yourself about things. You know you can't have your cake and eat it too.

You become painfully aware that every time labor or building materials or electric energy or any other factors shift, to aim at a new objective or to attack a new problem, someone must pay the cost. Someone must *give up* something. People *don't like* to give up things! None of us wants to choose this *or* that. It's so much nicer to have this *and* that. But we can't have this *and* that, of *everything*!

## Who Will Make The Sacrifices?

Oh, it's so easy to choose to give up things we never had! "I'll give up the luxury of driving a big car if it will bring a better school system. So everyone else ought to be willing to give up their big cars, too! Of course, I don't happen to have a big car. I only have a motor bike. I sure don't plan to give *that* up! That's not a luxury. It's a necessity!"

Everyone wants someone else to make the sacrifices. Sure. But we aren't going to solve any of society's problems by "volunteering away" the things other people have, while clinging to our own. We aren't going to solve the pollution problem by wishing it on the big businesses or the local governments or the Federal government or anyone else. The same is true of education and law enforcement and everything else. These problems are society's problems. We're *all* sharing in the benefits of high productivity and individual freedom. And *we're all going to share in the costs*. You can be sure of it.

Solutions to society's problems are not going to be coming along any faster than we ourselves are willing to pay the costs—in taxes, in product prices, in the opportunity costs of *other* goods and services we forfeit, and sometimes in giving up some of our freedoms. It's too bad we're going to have to pay the costs. But we are. You know that now. You know that these are the facts of economics—the facts of the real world—the facts of life.

## A Word Of Farewell

Now that you are turned on to some of the basic economic concepts, your thoughts never again will be able to escape the fact of scarcity. You'll always have some awareness, now, of the discomfort of opportunity cost. You'll always be aware that realistic plans and programs to improve anything, or to solve any problem, require some means—effort—some resources—some "scarce factors of production." You will never forget that these factors must come from somewhere. But from where? And then to be used to meet *which* urgent need? These are the really tough questions.

Answering these tough questions can be a very difficult and trying task. If you stop and think about it for a minute you will realize that this difficult and trying task is the thing we've been talking about all along. Don't you see? It's the economic problem! It's the basic problem of economics — the problem of choosing. It's the problem of scarcity — of not being able to do all the things we'd like to do — the problem of having to decide, to choose which things to do when we can't do everything — the problem of choosing which objectives to pursue and which to forego, which things to have and which to give up. Sure. That's what economics is all about. Surely you'll never forget that.

I hope you will go further in economics. This one book couldn't take you but just so far. But no matter where you go, no matter which objectives you seek, no matter which paths you choose, I hope your choices turn out to be *truly* the right ones, for you. The right choices are the ones which, many years from now, will let you look back and say:

> "How lucky I am to have been wise enough then, to make the right choices, for me."

I wish you wisdom in all your choices. Today, and always.

# INDEX

D



E

**N**